UNDERSTANDING MICROELECTRONICS

UNDERSTANDING MICROELECTRONICS

A Top-Down Approach

Franco Maloberti

University of Pavia, Italy

A John Wiley & Sons, Ltd, Publication

Library of Congress Cataloging-in-Publication Data

Maloberti, F. (Franco)
    Understanding microelectronics : a top-down approach / Franco Maloberti.
        p. cm.
    Includes index.
    ISBN 978-0-470-74555-7 (cloth)
1. Microelectronics. I. Title.   TK7874.M253 2012
    621.381–dc23

                                                                2011024081

A catalogue record for this book is available from the British Library.

Print ISBN: 9780470745557
ePDF ISBN: 9781119976486
oBook ISBN: 9781119976493
ePub ISBN: 9781119978343
Mobi ISBN: 9781119978350

Set in 10/12 Computer Modern Roman by Sunrise Setting Ltd, Torquay, UK.

Printed in the UK

To Pina, Amélie,
Matteo and Luca

And in memory of my
father, Alberto

CONTENTS

Preface

Electronics is a young discipline. It was initiated in 1904 when, after some related inventions, J. A. Fleming conceived the first electronic device: the vacuum tube diode. This is a two-terminal component made by a hot filament (cathode) able to emit electrons in the vacuum. A second electrode, the plate (or anode), collects electrons, causing a flow that depends on the sign and the value of the voltage applied across the terminals. Such a device can conduct current only in one direction (the rectifying effect), but actually cannot fully realize "electronic" functions. Two years later L. Deforest added a third terminal, the grid, and invented the vacuum tube triode. This innovation made possible the development of "electronic" functions, the most important of which is the ability to augment the amplitude of very small electrical signals (amplification). For decades after that, electronic circuits were based on those bulky, power-hungry vacuum tubes, operating with high voltage. These were able to evolve into more sophisticated components by the addition of extra grids to allow better control of the flow of electrons from cathode to anode.

At that time the focus of electronic designers was on being able to connect a few active devices (the vacuum tubes) with a large number of passive components (resistors, capacitors and inductors) to build up a circuit. It was necessary to understand the physical mechanisms governing the devices and to know the theoretical basis of network analysis. In short, the approach was from the physics that provides background knowledge to the design theories that enable circuit design.

The situation was almost unchanged even when William Shockley, John Bardeen and Walter Brattain invented the transistor in 1947. Moreover, the focus still remained on devices and circuits for a couple of decades after the introduction of the Integrated Circuit (IC, an electronic device with more than one transistor on a single silicon die). Then, with time and at an increasing pace, the complexity of electronic systems became greater and greater, with the number of transistors greatly exceeding that of passive components. Nowadays many ICs are made only of transistors, with a total count that approximately doubles every two years. Some digital circuits contain billions of elementary components, each of them extremely small.

The result is that the technology evolution has shifted focus from simple circuits to complex systems, with most attention given to high-level descriptions of the implemented functions rather than looking at specific details. Obviously the details are still important, but they are considered after a global analysis of the architecture and not before. In other words, the design methods moved from a bottom-up to a top-down approach.

There is another relevant change caused by electronic advance: the increasing availability of apparatus, gadgets, communication devices and tools for accurate prediction of events and for implementing virtual realities. The social impact of this multitude of electronic aids is that people, especially new generations, expect to see results immediately without waiting for the traditional phases of preparation, description of phenomena by formal procedures and patient scientific observation. We can say that the practice of studying the correlation between cause and effect is increasingly fading. Fewer and fewer people want to ask "What happened?" They are just interested in immediate outcomes; the link between results and the reasons behind them puzzles people less and less. This obviously can prevent the search for new solutions and the origination of new design methodologies.

This unavoidable cultural shift is not negative in itself, but it reduces the effectiveness of traditional teaching styles. The impatience of students who expect immediate results (and fun) contrasts with the customary methods that start from fundamentals and build specialized knowledge on top of them. This is a natural and positive modern attitude that must be properly exploited in order to favor the professional growth of younger generations. In short, if a bottom-up presentation is not well received, it is necessary to move to a top-down teaching method, and that is what this book tries to do.

The top-down approach is based on a hierarchical view of electronic systems. They are seen as a composition of sub-systems defined generically at the first hierarchy level. Each sub-system, initially considered as a "black box" that just communicates with the external world via electrical terminals, is then detailed step by step, by going inside the "black boxes." That is the method that inspires this book and its organization. In fact, Chapter 1 starts from the top, presenting an overview of the microelectronics discipline and defining goals and strategies for both instructor and student. It is suggested that this short chapter be carefully read, to get the right "feel" and attitude needed for an effective learning process. Chapter 2 deals with signals, the key ingredients of electronic processors. They are represented by time-varying electrical quantities, possibly analyzed in other domains. Emphasis is therefore on the signal representation in time, frequency and z-domain. That chapter is probably one of the most difficult, but having a solid knowledge of the topic is essential, and I do hope that the required efforts will be understood by the reader.

Chapter 3 is on electronic systems. The goal pursued is to describe different applications for making the reader aware of the block diagram and hierarchical processing used in the top-down implementation of electronic systems. Important issues such as system partitioning and testing are introduced. Chapter 4 discusses signal processing. It studies linear and non-linear operation and the method used to represent the results. Signal processing operations are, obviously, realized with electronic circuits, but the focus at this level is just on methods and not on the implementations, circuit features and limits affecting real examples.

Electronic functions realizing signal processing are presented in Chapter 5. The analysis is initially at the "black box" level, because the first focus is on interconnections. The chapter also studies how to satisfy various needs by using analog or digital techniques and ideal elementary blocks. Chapter 6 goes further "down" by describing the use of analog key structures for giving rise to elementary functions. These are the operational amplifier (op-amp) and the comparator. The chapter also discusses the specifications of blocks that are supposed to be a discrete part assembled on printed circuit boards, or cells used in integrated systems.

Transformation from analog to digital (and vice versa) marks the boundary between analog and digital processing. Chapter 7 describes the electronic circuits needed for that: the A/D and the D/A converter. The chapter deals with specifications first, and then studies the most frequently used conversion algorithms and architectures. Because of the introductory nature of this book, the analysis does not go into great detail. However, study of it will give the student the knowledge of features and limits that enables understanding and definition of high-level mixed-signal architectures.

Chapter 8 deals with digital processing circuits. As is well known, digital design is mainly performed with microprocessors, digital signal processors, programmable logic devices and memories. These are complex circuits with a huge number of transistors, fabricated with state-of-the-art technology. The majority of electrical engineers do not design such circuits but just use them. Thus the task is mainly one of interconnecting macro functions and programming software of components that are known at the functional level. In the light of this, the chapter describes general features and does not go into the details of complicated architectures. The study is thus limited to introductory notions as needed by users. More specific courses will "go inside." Memories and their organization are also discussed.

Study of the first eight chapters does not require any expertise at the electronic device level. Now, to understand microelectronics further it is necessary to be aware, at least at functional levels, of the operating principles of electronic devices. This is done in Chapter 9, which analyzes diodes, bipolar transistors and CMOS transistors. This chapter is not about the detail of physics or technology. That is certainly needed for fabricating devices and integrated circuits, but not for using them. Therefore, the description given here is only sufficient for the understanding of limits and features that is required by the majority of professional electronic engineers. The elements given, however, are a good introduction to the specialized proficiency needed for IC design and fabrication.

The next two chapters use basic devices to study analog and digital schemes at the transistor level. The goal, again, is not to provide detailed design expertise, because integrated circuits implement functions at a high level. What is necessary is to be familiar with basic concepts (such as small signal analysis) and to know how to handle simple circuits. It is supposed that more detailed study, if necessary, will be done in advanced and specific courses. Chapters 10 and 11 reach the lowest level of abstraction studied in this book. It does not go further down, to a discussion of layout and fabrication issues. Those are the topics studied in courses for integrated circuit designers.

Feedback is introduced in Chapter 12. This topic is important for many branches of engineering. The chapter does not consider specialized aspects but just gives the first elements and discusses basic circuit design implications.

In Chapter 13 the basics of power conversion and power management are presented. This seemingly specialized topic was chosen for study because a good part of the activity of electrical engineers concerns power and its management. Supply voltages must always be of suitably good quality and must ensure high efficiency in power conversion. Power is also very important in portable electronics, which is now increasingly widespread. The topic, possibly studied in more detail elsewhere, analyzes rectifiers, linear regulators and DC–DC converters. At the end the chapter also describes power harvesting, a necessity of autonomous systems operating with micro-power consumption.

The last chapter describes signal generation and signal measurements. This is important for the proper characterization of circuits whose performance must be verified and checked so as to validate design or fabrication. Since sine wave signals are principally used for testing or for supporting the operation of systems, methods for generating sine waves are presented. Features and operating principles of key instruments used in modern laboratories are also discussed.

That is, concisely, the outline of the book. However, we must be aware that an important aid to the learning process is carrying out experiments. This is outlined by the saying: "If we hear, we forget; if we see, we remember; if we do, we understand." Unfortunately, often, offering an adequate experimental activity is problematic because of the limited resources normally available in universities and high schools. In order to overcome that difficulty this book proposes a number of virtual experiments for practical activity. The tool, named Elvis-Lab (ELectronic VIrtual Student Lab), makes available a virtual laboratory with instruments and predefined experimental boards. Descriptions of experiments, measurement set-ups and requirements are given throughout the book. A demo version of this tool is freely available on the Web with experiments at www.wiley.com/go/maloberti_electronics. ElvisLab provides an environment where the student can modify parameters controlling simple circuits or the settings of signal generators. That operation mimics what is done with a prefabricated board in the laboratory. The tool is intended as a good introduction to such experimental activity, which could also be performed in real sessions, provided that a laboratory and the necessary instruments are available.

The combination of this text and the virtual laboratory experiments is suitable for basic courses on electronics and microelectronics. The goal is to provide a good background to microelectronic systems and to establish by a top-down path the basis for further studies. This is a textbook for students but can also be used as a reference for practicing engineering. For class use there are problems given in each chapter, but, more importantly, the recommended virtual experiments should enable the student to understand better.

Acknowledgements

The author would like to acknowledge many vital contributions to devising and preparing the manuscript and artworks. The support of Aldo Peña Perez, Gisela Gaona Martinez (for the outstanding artwork), Marcello Beccaria, and Edoardo Bonizzoni are highly appreciated. The author also acknowledges the support of the Electronic Department of the University of Pavia and the contribution of master students to developing the ElvisLab software. Last, but not least, I thank John Coggan for the excellent English revision of the manuscript.

F. MALOBERTI
Pavia
May 2011

List of Abbreviations

µP Microprocessor
ΣΔ Sigma–Delta
AC Alternating Current
A/D Analog-to-digital
ADC Analog-to-digital converter
ALU Arithmetic Logic Unit
ASIC Application-Specific Integrated Circuit
ATE Automatic Test Equipment
Auto-ID Automatic Identification Procedure
A/V Audio/video
BB Base-Band
BER Bit-Error-Rate
BJT Bipolar Junction Transistor
BWA Broadband Wireless Access
CAD Computer-Aided Design
CAS Column Access Strobe
CCCS Current-Controlled Current Source
CCVS Current-Controlled Voltage Source
CMRR Common-Mode Rejection Ratio
CMOS Complementary MOS
CPLD Complex Programmable Logic Device
CPU Central Processing Unit

D/A Digital-to-analog
DAC Digital-to-analog converter
DC Direct Current
DDS Direct Digital Synthesis
DEMUX Demultiplexer
DFT Discrete Fourier Transform
DLP Digital Light Processing
DMD Digital Micromirror Device
DNL Differential Non-Linearity
DR Dynamic Range
DRAM Dynamic Random-Access Memory
DSP Digital Signal Processor
DVD Digital Video Disc
EDA Electronic Design Automation
EPROM Erasable Programmable Read-Only
Memory
EEPROM Electrically Erasable Programmable
Read-Only Memory
ESD Electrostatic Discharge
ESR Equivalent Series Resistance
FF Flip-flop
FFT Fast Fourier Transform

FIR Finite Impulse Response

FM Frequency Modulation

FPGA Field Programmable Gate Array

GAL Generic Array Logic

GBW Gain Bandwidth Product

GE Gate Equivalent

GSI Giga-scale Integration

HD2 Second Harmonic Distortion

HD3 Third Harmonic Distortion

HDD Hard Disk Drive

HDL Hardware Description Language

HTOL High Temperature Operating Life

IC Integrated Circuit

IEEE Institute of Electrical and Electronics Engineering

IF Intermediate Frequency

INL Integral Non-Linearity

IP Intellectual Property

I/O Input/Output

ISO International Organization for Standardization

I–V Current–Voltage

JFET Junction Field-Effect Transistor

JPEG Joint Photographic Expert Group

LCD Liquid Crystal Display

LDO Low Drop-Out

LED Light-Emitting Diode

LNA Low Noise Amplifier

LSB Least Significant Bit

LSI Large-Scale Integration

LUT Look-Up Table

Mbps MegaBit Per Second

MEMS Micro Electro-Mechanical Systems

MIM Metal–Insulator–Metal

MIPS Mega Instructions Per Second

MMCC Metal–Metal Comb Capacitor

MOS Metal–Oxide–Semiconductor

MPGA Metal-Programmable Gate Array

MRAM Magneto-resistive RAM, or Magnetic RAM

MSI Medium-Scale Integration

MS/s Mega-Sample per Second

MUX Multiplexer

NMH Noise Margin High

NMH Noise Margin Low

NMR Nuclear Magnetic Resonance

NRE Non-Recurrent Engineering

OLED Organic Light-Emitting Diode

op-amp Operational amplifier

OSR Oversampling Ratio

OTA Operational Transconductance Amplifier

PA Power Amplifier

PAL Programmable Array Logic

PCB Printed Circuit Board

PDA Personal Digital Assistant

PDIL Plastic Dual In-Line

PDP Plasma Display Panel

PFD Phase-Frequency Detector

PLD Programmable Logic Device

PLL Phase-Locked Loop

PMP Portable Media Player

POS Product-of-Sums

ppm Parts per Million

PROM Programmable Read-Only Memory

PSRR Power Supply Rejection Ratio

PSTN Public Switched Telephone Network

R/C Remote-Controlled (toys etc)

RAM Random-Access Memory

RAS Row Address Strobe

RC Resistor-Capacitor

RF Radio Frequency

RFID Radio Frequency IDentification

RMS Root-Mean-Square

ROM Read-Only Memory

RPM Revolutions Per Minute

R/W Read/Write

Rx Reception

S&H Sample-and-Hold

SAR Synthetic Aperture Radar

SAR Successive Approximation Register (Chapter 7)

SC Switched Capacitor

SDRAM Synchronous Dynamic Random-Access Memory

SFDR Spurious Free Dynamic Range

SiP System-in-Package

SLIC Subscriber Line Interface Circuit

SNDR Signal-to-Noise plus Distortion Ratio

SNR Signal-to-Noise Ratio

SoC System-on-Chip

SoP Sum-of-Products
SPAD Single Photon Avalanche Diode
SRAM Static Random-Access Memory
SSI Small-Scale Integration
T&H Track-and-Hold
THD Total Harmonic Distortion
Tx Transmission
USB Universal Serial Bus
USI Ultra Large-Scale Integration
UV Ultraviolet
VCCS Voltage-Controlled Current Source

VCIS Voltage-Controlled Current Source
VCVS Voltage-Controlled Voltage Source
VCO Voltage-Controlled Oscillator
VCVS Voltage-Controlled Voltage Source
VLSI Very Large-Scale Integration
VMOS Vertical Metal–Oxide–Silicon
WiMAX Worldwide Interoperability for Microwave Access
WLAN Wireless Local Area Network
X-DSL Digital Subscriber Line

CHAPTER 1

OVERVIEW, GOALS AND STRATEGY

Bodily exercise, when compulsory, does no harm to the body; but knowledge that is acquired under compulsion obtains no hold on the mind.

—Plato

1.1 GOOD MORNING

I don't know whether now, the first time you open this book, it is morning, afternoon, or, perhaps, night, but for sure it is the morning of a long day, or, better, it is the beginning of an adventure. After a preparation phase, this journey will enable you to meet electronic systems, will let you get inside intriguing architectures, will help you in identifying basic functions, will show you how electronic blocks realize them, and will give you the capability to examine these blocks made by transistors and interconnections. You will also learn how to design and not just understand circuits, by using transistors and other elements to obtain electronic processing. Further, you will know about memories used for storing data and you will become familiar with other auxiliary functions such as the generation of supply voltages or the control of accurate clock signals. This adventure trip will be challenging, with difficult passages and, probably, here and there with too much math, but at the end you will, hopefully, gain a solid knowledge of electronics, the science that more than many others has favored progress in recent decades and is pervading every moment of our lives.

Understanding Microelectronics: A Top-Down Approach, First Edition. Franco Maloberti.
© 2012 John Wiley & Sons, Ltd. Published 2012 by John Wiley & Sons, Ltd.

If you are young, but even if you are not as old as I am ... (well, don't exaggerate: I have white hair, I know, but I am still young, I suppose, since I look in good shape). If you are young, I was saying, you have surely encountered electronics since the first minute of your life. Electronic apparatus was probably used when you were born, and even before that, when somebody was monitoring your prenatal health. Then you enjoyed electronics-based toys, and you have used various electronic devices and gadgets, growing in complexity with you, many times a day, either for pleasure or for professional needs, ever since. Certainly you use electronics massively and continuously, unless you are shipwrecked on a faraway island with just a mechanical clock and no satellite phone, with the batteries of your MP3, Personal Digital Assistant (PDA), tablet or portable computer gone, and no sophisticated radio or GPS.

Well, I suppose you have already realized that electronics pervades the life of everybody and aids every daily action, and also, I suppose, you assume that using electronics is not difficult; electronic devices are (and must be) user friendly. Indeed, instruction manuals are often useless, because everybody desires to use a new device just by employing common sense. People don't have the patience to read a few pages of a small multilingual booklet. Moreover, many presume that it is useless to know what is inside the device, what the theoretical basis governing the electronic system is and what its basic blocks and primary components are, and, below this, to know about the materials and their physical and chemical properties. In some sense, an ideal electronic apparatus is, from the customer's point of view, a black box: just a nicely designed object, intuitive to operate and capable of satisfying demanding requests and expectations.

> **What do you expect from a microelectronic system?**
>
> I suppose, like everybody else, you expect to be able to use the system by intuition without reading boring instruction manuals, to have an answer to your request for high performance, and to pay as little as possible.

Indeed, it is true that modern electronic equipment is user friendly, but, obviously, to design it, to understand its functions in detail, and, also, to comprehend the key features, it is necessary to have special expertise. This is the asset of many professionals in the electronics business: people who acquire knowledge up to a level that gives the degree of confidence they need so as to perform at their best in designing, marketing, promoting, or selling electronic circuits and systems.

Therefore, we (you and I) are facing the difficult task of transforming a user of friendly electronics or microelectronics into an expert in microelectronics. For that, it is necessary that you, future electronics professional, open (and this is the first obstacle), read, and understand a bulky book (albeit with figures) printed on old-fashioned paper. This is not easy, because anyone who uses a computer and the Web is accustomed to doing and knowing without feeling the need to read even a small instruction manual.

I have to admit that the method followed for decades in teaching scientific and technical topics is perceived as out of date by most modern people. I am sure you think that starting from fundamentals to construct the building of knowledge, step by step, is really boring! There are quicker methods, I assume you think. Indeed, following the traditional approach requires one to be very patient and not to expect immediate results as with modern electronic aids. Nevertheless, it is essential to be aware that fundamentals are important (or, better, vital). It is well known that a solid foundation is better than sand: a castle built on sand, without foundations, will certainly collapse. That is what old people usually say, but, again, studying basic concepts is tedious. So what can I do to persuade you that fundamentals are necessary?

Perhaps by narrating a tale that I spontaneously invented many years ago during a debate at a panel discussion. That tale is given here.

The man who owned 100 cars

A rich man was so rich that he owned 100 cars, one for every moment of his life, with three drivers per car available 24 hours a day. The drivers' job included unrolling a red carpet on the small paths from one car to the next and having every car available every moment of the day and night. One marvelous day the wife of the rich man gave birth to a beautiful child. This brought great happiness to the man, his wife and the 300 drivers of the 100 cars.

Two years later, as the second birthday of the lovely boy approached, it was time to decide on the birthday present and the rich man already had thought of a small car with golden wheels. He asked his wife: *"What do you think?"*. The lady promptly replied: *"I would prefer a pair of shoes."* *"What?"* cried the man, *"I have 100 cars and miles of red carpet! My son does not need to walk! Shoes are for the poor people that have to walk."*

After the panel, when the discussion was over, a colleague of mine approached me, saying: "Excellent! You exactly got the point. Fundamentals are essential. You are right; having cars does not justify bare feet." He fully agreed with me, and certainly liked the way I described the need to know fundamentals even if powerful tools are available for helping designers.

The risk is that computer tools, embedding overwhelming design methods, favor the habit of trying and retrying until acceptable results are obtained. Therefore computer support often gives rise to results that appear very good without requiring the hard intellectual work that is supported and favored by a solid technical background.

Indeed, fundamentals are essential, but knowing everything is negative: it is necessary to settle at the right level. Saturating the mind by a flood of notions creates too many mindsets and, consequently, limits creativity. A discussion on creativity would take pages and pages, and I don't think this is the proper place to have it. However, remember that a bit of creativity (but not too much) is the basis for any successful technical job. Blending basic knowledge, creativity, quality, and execution must be the goal. This makes the difference between a respected (and well paid) electronic engineer and a pusher of keys.

Remember that anybody is able to push buttons, so becoming a key pusher does not add much to professional capability. Even a monkey can do that! So, the key point is: *where is the added value? What makes the difference?* Obviously, for a successful future, it is necessary to acquire more than the capability of pushing buttons. For this, computer-aided tools should not be used for avoiding thought but for improving the effectiveness of the learning process. This is very important, and, actually, the goal of this book is to provide, with a mix of fundamentals and computer-aided support, the basis of that added value that distinguishes an expert.

Now, I think that is enough introduction, and after this long discussion (it may be a bit boring) I suppose that you, my dear reader, are anxious to see the next step. So, ... let's organize the day. And, again, good morning.

1.2 PLANNING THE TRIP

When planning an adventurous trip, for safety and to ensure your future enjoyment it is recommended that you check a number of points. First, you have to define the trip in terms of a wish list; for example, you need to define whether you want to camp out at night, bunk in a rustic hut or stay in a five-star hotel. Also, you need to state whether you plan to stop in a small cafe and chat with local people or whether you desire to visit a museum. For this special adventurous trip, I suppose your wish list includes:

- the desire to become an expert on electronic systems, to know their basic properties, to be able to assess them and to recognize their limits;

- the wish to know more about the signals used and processed in electronic systems so as to understand whether a parameter value is good or bad and to learn how to generate test signals and use them for performance verification;

- the ability to read circuit diagrams so as to see, possibly at a glance, where the critical points are and to estimate expected performance;

- the desire to know about the basic blocks used in a system, to optimize the key performances by using computer simulation tools and to know how to interconnect those blocks so as to obtain given processing functions;

- the willingness to know in detail how transistors work and to learn the modern integrated technologies used to realize transistors and integrated circuits;

- curiosity about modeling transistors and the physical and chemical basic principles underlying their fabrication.

Well, I am not sure that all the above points are your goals, but, frankly, even a subset of them is a bit ambitious and will surely require significant efforts to achieve. But don't be discouraged. After the initial steps the path will be more and more smooth, and with the help of this book you will (hopefully) obtain good results.

After deciding on the type of trip (device oriented, integrated circuit oriented, system oriented, or another type), it is necessary to verify that you are in the proper shape to enjoy the experience. For this, there are a number of requisites that are essential. The most relevant are:

- a reasonable mathematical background with the ability to solve first- and second-order differential equations;

- knowledge of Kirchhoff's Laws, some knowledge of Laplace and Fourier transforms, and familiarity with writing mesh and nodal equations and solving such equations;

- good knowledge of the use of a computer, how to install programs and how to use the Web.

As a side note on the last point, after emphasizing that, obviously, you have to become familiar with simulation tools, I have a recommendation: do not blindly rely on numerical results. The description of a system is based on models that are always an approximation, and the numerical results are sometimes not accurate, or even credible. Therefore, use your brain first, and believe only results that conform to your personal intuition. However, your "computer brain" is not infallible, and computer simulations can help with understanding when mental reasoning possibly fails.

To do

Refine the wish list on the basis of your future activity (state what is the professional profile you would like to pursue). In preparation of this electronic trip, check and expand the list of prerequisites. Assess your shape and make sure you are ready.

Finally, after setting up the wish list and specifying the prerequisites, it is necessary to check that the preconditions are properly satisfied, not just formally but by answering the question: am I in good enough shape? What is required is not very much but is essential for achieving profitable results. As on an adventure trip, where you must be able to walk kilometers over varying landscapes and jump over some obstacles, in this electronic trip you must be able to solve a system of equations, to write nodal equations without panic, and to guess reasonable approximations that do not end up with a million volts or a hundred amps. Therefore, before starting the trip, assess yourself. If you find some weakness, quickly repair it with extra effort and exercises, and ensure that you are ready soon for this exciting electronic adventure.

1.3 ELECTRONIC SYSTEMS

The building that is the knowledge of electronics consists of many floors, with electronic systems on the top. Each floor may be connected by bridges to other knowledge buildings: those of mechanics, chemistry, biology, and also the humanities. Just below the top floor of the electronics building there are the functional blocks used to compose systems. These functional blocks are typically described, at a high level of abstraction, by language, and represented as an element of the block diagram of the entire system – a drawing depicting the sub-systems and their interactions. The flow of signals from one block to another block of the system describes signal exchange. Blocks transform the input signals to generate different outputs.

Figure 1.1 shows a possible block diagram of a system consisting of four sub-systems. There are one or more inputs that can be analog or digital (we shall study this distinction shortly) and are represented by a simple line or by arrowed channels that correspond to one or more wires carrying signals. The inputs are used in one or more blocks (in the diagram we have four blocks, A, B, C, and D); the output(s) of each sub-block are the inputs of other blocks and possibly also the input of the same block, for feeding back information to the input (see block B in the figure). The system outputs are the inputs of another system, control an actuator (we shall also see what an actuator is), or are stored in a memory for future use.

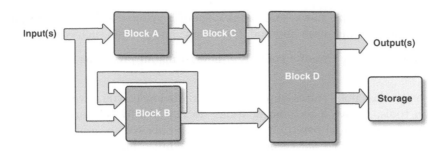

Figure 1.1   Typical block diagram of a system.

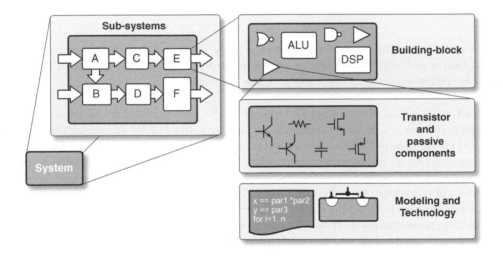

Figure 1.2   Hierarchical description of an electronic system.

The hierarchical description expands the sub-system into electrical functions that are used to produce a given transformation of the input(s) into output(s). They are graphically represented by symbols like the one shown in Figure 1.2. Below this level we have the circuits that realize functions with passive and active (transistor) components. Then the passive and active components are modeled by a set of equations, often represented by a symbol. The components are fabricated using a given technological process. They can be discrete elements that perform simple basic functions or integrated circuits that realize higher-level functions by the cooperative action of many components that are fabricated together on a single chip.

The way the components are assembled is part of the job. We can have a Printed Circuit Board (PCB) or use more modern and compact ways to assemble the system. The PCB has metal traces for interconnections and houses components that may be assembled using the surface mounting technique. Another possibility is to house many chips on the same package to obtain a System-in-Package (SiP). Figure 1.3 shows various examples of systems assembly and wire-bonding. Observe that it is also possible to stack chips one on top of another so as to exploit the third dimension. The choice between different solutions depends on a trade-off between cost, volume, and system reliability.

Figure 1.3   (a) Radio frequency module with direct wire bonding for possible SiP; (b) different types of package frames; (c) assembly on a PCB of a flip chip ball grid array circuit; (d) details of very dense wire-bonding. Reproduced by permission of © ST Microelectronics.

Self training

The assembly of a system can involve different techniques. The simplest one uses a PCB. Find on the Web more information about the following.

- What kind of material is used for PCB fabrication?

- What is a multi-layer PCB?

- What is the minimum number of PCB layers required to obtain any possible interconnection, and why, in your opinion, do designers use multiple layers?

- What can we do with those extra layers?

In addition, search the Web and find out what a System-on-Chip (SoC) is, and what a System-in-Package (SiP) is. What are the fabrication techniques used?

Write a short note on the results of your search.

1.3.1   Meeting a System

Very often during the day, even if you are not fully aware of it, you encounter electronic systems. The first time such a system touched your life was probably right after you were born, perhaps when a nurse entered a waiting room crying out to an anxious man (your father): "It's a girl!" or "It's a boy!" That man, a bit confused, might have smiled and looked up at the digital clock on the wall, with a large seven-segment display showing, maybe, "9:38".

The digital clock is an electronic system based on a precise time reference: the quartz oscillator. Probably you know that quartz is the crystalline form of silicon dioxide (SiO_2) used to show a time reference because of its anisotropy (dependence of properties on direction). What happens is that anisotropy also causes piezoelectricity. The name piezo comes from the Greek and means pressure; therefore piezoelectricity refers to electrical effects caused by pressure and, conversely, pressure that determines electrical consequences. When a piece of crystal is subjected to a voltage a stress is produced, and under certain conditions the crystal begins vibrating mechanically and electrically in a steady manner. The good thing is that the temperature dependence of the oscillations is very low: the variation at around 25°C is only 5 ppm/°C (ppm means parts per million). Therefore, a quartz crystal experiences an error of 25 ppm with 5°C change. Remember that in one day we have 86 400 seconds; therefore, one second is 11.57 ppm of a day. Accordingly, the error produced by a quartz crystal kept at a temperature 5°C different from the nominal value is about 2 seconds per day.

Because of its accuracy the quartz oscillator is used as the basis of precise clocks. The frequency of oscillation depends on the cut, size and shape. For example a disk of crystal with 1.2 cm diameter and 1.06 mm thickness oscillates at 10 MHz (fifth overtone). For watches the frequency normally used is lower, 32.768 kHz, which corresponds to a period of 30.52 μs. That frequency is chosen because 2^{15} periods make a second.

The above elements are sufficient for drafting the scheme of a clock that uses seven segments and two blinking dots to display the memorable time of 9:38. Figure 1.4 shows a block diagram with some details. The key, as already mentioned, is the clock oscillator that generates pulses spaced by 30.52 μs. The next block counts those pulses $32\,768 = 2^{15}$ times and after that generates a pulse at the output. The rate of those pulses is one per second, which is used for the blinking dots. A counter by 60 determines the minutes and another counter by 60 the hours. The content of the counter gives the signals that control the two right-hand digits. Moreover, the pulse of hours is obtained by a modulo 12 counter for determining the two left-hand digits of the clock. As an alternative, the last counter may count by 24 to show the hours of the entire day.

Four seven-segment displays, suitably lit, represent minutes and hours. For this it is necessary to specify special blocks, called seven-segment drivers, that receive the signals from the counters and transform them into segment control. Obviously the signals generated by the drivers must be strong enough (in voltage and current) to power the segments properly; they must be bright and visible even in daylight.

The block diagram is not complete, because it does not include setting the clock and possibly dimming the segments. Moreover, it may be that an advanced implementation includes automatic segment illumination control, using a sensor that measures the illumination of the environment and regulates the power sent to segments.

A second example of a system that some readers will have encountered for a while after being born is the baby incubator. It is used to care for babies in a suitable controlled temperature. I suppose you can easily imagine what its basic functions are: to measure the temperature,

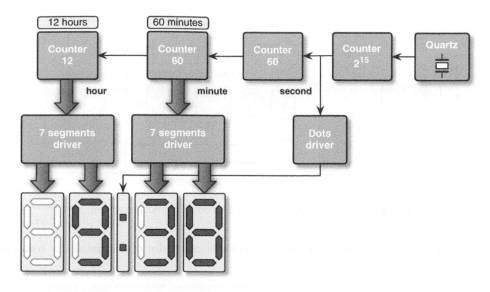

Figure 1.4   Block diagram of a digital clock.

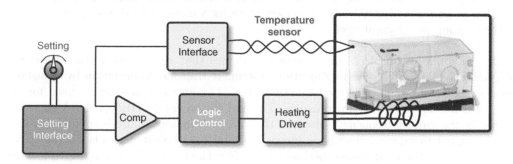

Figure 1.5   Block diagram of an incubator with controlled temperature.

compare the temperature with the desired value and increase or reduce the heating. The block diagram representing those functions is shown in Figure 1.5. The temperature is measured by a sensor that transforms a physical quantity, the temperature, into an electrical quantity, a voltage. The signal generated by the sensor is often not appropriate for use, and, for this, the *sensor interface* must change the output of the sensor into a more convenient signal (higher amplitude, lower output impedance, digital format, ...). The desired temperature enters the system by a setting control defined, for instance, by a knob or a rotary switch. A setting interface possibly transforms the setting into a form compatible with the signal given by the temperature sensor. The block called *comp*, a triangle in the figure, compares the two inputs and produces a logic signal informing the *logic control* whether the measured temperature is higher or lower than the setting. The logic control switches the heating of the incubator on or off by the *heating driver*.

An important feature is that the system uses feedback, i.e., the system sends the output signal back to better control the operation. The feedback is given by the measured temperature. Moreover, the feedback loop is hybrid: it is made up of electrical quantities and physical

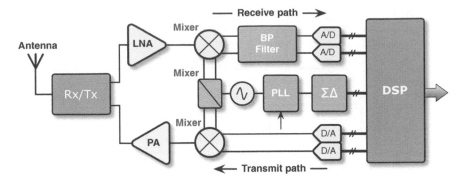

Figure 1.6   Block diagram of a wireless communication system.

quantities (heat and temperature). Electronic systems normally obtain feedback using only electrical signals.

> **Use hierarchy**
>
> The best way to describe a complex system is to split it into basic functions, without many details, and to go down hierarchically inside each block for a more detailed description until you reach the bottom, the physical level.

Another example of an electronic system is a device that we use many times a day: a portable telephone or wireless communication system. It transfers information by translating electrical signals from and to electromagnetic waves. The antenna, the device used for that purpose, operates for both the transmission (Tx) and the reception (Rx) path. A significant parameter is the carrier frequency, often in the range of many hundreds of MHz or some GHz. The signal occupies a frequency interval (signal band) that depends on the carried information and on the modulation method used before the mixer that gives rise to a replica of the input signal at frequencies around that of the carrier. A possible block diagram (albeit approximated) is shown in Figure 1.6. Next to the antenna two triangular blocks indicate the LNA (Low Noise Amplifier) and the PA (Power Amplifier). Then we have mixers, used to translate the signal at higher or lower frequencies. The blocks called A/D and D/A are the analog-to-digital and digital-to-analog converters, used to change the signal format from analog to digital and vice versa. The block PLL (Phase Locked Loop), controlled by another block, $\Sigma\Delta$, generates the carrier of the mixer. A big part of the operation is carried out by the DSP (Digital Signal Processor), possibly made up of many hundred of millions or billions of transistors. The signals from the DSP and the data converters can be multiple, and, rather than representing them by an arrowed symbol, the figure uses two oblique lines crossing the wires.

Notice that Figure 1.6 uses several blocks whose symbols and functions are difficult to understand fully at this stage of your studies. Don't worry. You will learn about those functions shortly. What is needed now is just an awareness of the hierarchical description of a system. This is similar to what is done with maps. When depicting a country the map just indicates the biggest cities and the most important highways, mountains, big rivers and lakes. Then, going down to the regional level, the maps provide a more detailed view with medium-sized cities, hills, small rivers, and so forth. Below this, the city map level gives details about streets and maybe single buildings.

Self training

Use the Web or other tools to learn about the functions of electronic keys. They can be contactless (using a short-range wireless communication link), with contact, or a mixture of the two. Account for the following options.

■ The key needs power but does not have a battery on board.

■ The key is used for a car.

■ The key operates as a remote control.

Describe the features of the system, indicate the possible options, and draft the block diagram with the main flow of signals indicated.

1.4   TRANSDUCERS

The inputs or the outputs of an electronic system are often electrical portraits of real-world quantities: physical, chemical, or biological. For example, time is a physical quantity represented by a sequence of electrical pulses at a constant pace; temperature is depicted by a voltage that increases as the temperature becomes higher; pressure is measured by the value of the capacitance or the resistance of special materials sensitive to stress. The concentration of a given gas can be detected by the conductance change of thin porous layers that adsorb that gas. Moreover, the output of a system can be a movement of mechanical parts, a variation in pressure, the generation of modulated light, or the activation of a process. The devices that interface real-world quantities with an electronic system are called *transducers*. To be more specific, if the transducer generates an electrical quantity it is called a *sensor*; when it produces an action or, more generally, gives rise to a real-world quantity it is called an *actuator*.

Sensors and actuators

A sensor senses a real-world quantity and generates an electrical signal with given sensitivity. An actuator generates a real-world quantity under the control of an electrical signal.

In the above situations the electronics is just part of a wider system, as shown in Figure 1.7: a chain of blocks with, on one side, a sensor that senses a real-world quantity and produces an electrical signal. This signal is the input of an electronic system that, after some processing, gives rise to a suitable control for driving an actuator, whose output is a physical or maybe a chemical or biological quantity.

1.4.1   Sensors

The real world produces signals in various forms; some are interesting or beneficial, other unwanted or risky. Very important for our daily activity are the acoustic and visual signals; for those we are well equipped with sophisticated senses that perceive and transform the information and carry it to the brain. Other relevant signals are the concentrations of chemical

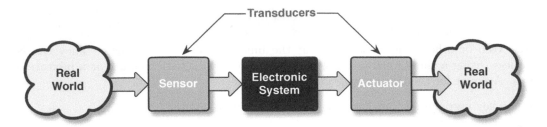

Figure 1.7 Entire system involving real-world signals.

agents that can be pleasant or dangerous. Some chemicals are detected by the nose and the tongue, which are sensitive to gases or solutions of pure elements or a mixture of elements, even in very small concentrations (as for some perfumes and odors). For other chemicals, such as carbon monoxide or water, the nose and the sense of taste have a negligible or null sensitivity; we say that those chemicals are odorless or tasteless. Obviously, for dangerous substances it is important to extend the senses' capability, to enable us to detect their presence and to give warning or take action promptly. For this purpose, often an electronic system enhances or replaces the human one by processing signals and performing actions with the help of sensors and actuators at the two ends.

Let us look at some examples. Temperature is an important physical quantity that, fortunately, is quite easy to measure. A simple temperature sensor is the thermocouple, which exploits the effect discovered by the Estonian physician Thomas Seebeck in 1822. Probably you already know that effect: a temperature difference established between the junctions of two metals determines a voltage across the terminals. Thus, a Nickel–Chromium thermocouple generates 12.2 mV with 300°C at one junction and room temperature (27°C) at the other. Another way to measure the temperature is by using the p–n junction of a semiconductor material (we shall learn later what a p–n junction is). The current across the junction with a fixed bias voltage increases exponentially with temperature; then a logarithmic (the opposite of exponential) circuit enables us to represent the temperature on a linear scale.

Even sensing light (especially in the visible range) is not difficult. There are many simple devices, among them photodiodes, which, again, are based on p–n junctions (or rather more complicated structures). When a photon with sufficient energy strikes the p–n junction in a special electrically activated region, called the *depletion region*, a pair of carriers is freed and produces a photoelectrical current. Figure 1.8 shows various packaged single photodiodes and linear and two-dimensional arrays. The cross section of a simple photodiode in Figure 1.8, whose thickness is in the range of microns (10^{-6} m), shows that the depletion region extends across two different types of doped material and almost reaches the surface hit by light. This feature is important, because the light penetrates just a tiny layer of material. Obviously the package that protects the device from dust and aggressive agents must have a window transparent to the wavelength that the photodiode wants to detect.

Many photodiodes arranged in a rectangular array make an image sensor (as used in digital cameras). Each photodiode detects one pixel of an image that is decomposed into discrete small areas. The number of pixels is, as I am sure you know, the number of dots into which the image is divided (in rows and columns). Therefore, the actuated image is an approximation of the real one through its decomposition into dots. If the image aspect ratio is 4×6 (the postcard format) with 6 M pixels (M means million), the detected or displayed image consists of an

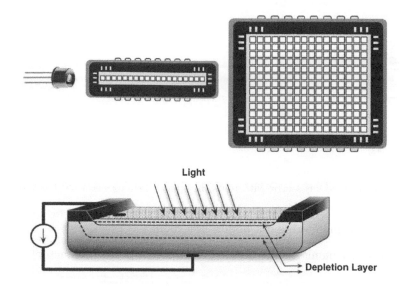

Figure 1.8 Different types of photodiodes and cross section of the device.

array of 2000×3000 dots. Such a resolution, when transferred onto 4×6-inch photographic paper, gives rise to dots separated by 50 μm (or 2 mils).

What is a p–n junction?

This, I suppose, is what you are asking yourself. We shall study this in detail later. For now it is enough to say that a p–n junction is the abrupt change of doping in a semiconducting material (such as silicon or germanium). One part has an n-dopant (such as arsenic) added, and the other a p-dopant (such as boron). Notice that "junction" here does not mean a simple joining of different materials, but corresponds to a transition of dopant within a mono-crystal. These unfamiliar words will be explained in a later chapter.

Remaining in the area of optical applications, there is an interesting sensor, the Single Photon Avalanche Diode (SPAD), which is capable of detecting the hit of a single photon. The sensor is again a p–n junction, whose biasing is close to the so-called breakdown voltage. A single electron generated by a photon triggers an avalanche of electrons. After a while the avalanche extinguishes. Therefore a pulse of current denotes the occurrence of a single photon.

Self training

Make a search on the Web for different types of photodiodes. Look at the different sensitivities and light wavelengths. Find the right solutions (possibly in terms of cost/benefit ratio) for the following applications:

- crepuscular switch for lighting the pathway in your garden;

- simple barcode reader (assume the use of red illumination in the system);

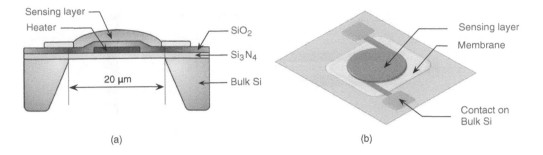

Figure 1.9   Micromachined gas sensor with micro hot-plate: (a) cross section; (b) top view.

- infrared sensor for monitoring hot bodies in a room;
- sunlight ultraviolet monitor (for choosing a sun lotion protection factor).

The sensing of chemical quantities is very important for monitoring the environment and for safety. For example, it is important to detect hydrogen, hydrocarbons, nitrogen oxides, carbon monoxide, oxygen, and carbon dioxide in a variety of ambient gas conditions and temperatures. There are many types of sensor used for chemical sensing; they can be resistive-based or capacitive-based structures. Since the same sensitive structure is often influenced by different chemicals, in order to increase the sensitivity an array of gas sensors with different responses to different compounds (an electronic nose) can be used. The output is obtained by means of complicated calculations involving the single sensor responses. In some cases the sensor is microfabricated or micromachined using Micro Electro-Mechanical Systems (MEMS) technology and/or based on nanomaterials to improve sensitivity and stability.

Figure 1.9 shows an example of a micromachined gas sensor with a micro hot-plate. The structure exploits the change in resistivity of porous materials (such as tin oxide) when they absorb a gas. First the micro hot-plate preheats the sensing layer to expel all gas. Then the measurement can take place after the gas has been absorbed. The layer resistance is measured with high sensitivity, and this is followed by translation of the result into the actual gas concentration. All of these steps require electronic support and some computation, and perhaps the details of how to do that are a bit puzzling. However, what is necessary at this point is not to give answers but to be aware of the possible complexity of electronic systems that use sensors.

1.4.2   Actuators

As already mentioned, an actuator generates a real-world quantity. The more common kinds of actuators are for audio or video outputs. For audio we have, for example, the loudspeaker and noise canceling headsets. For video signals we have many types of display that light two-dimensional arrays of colored dots. Examples are the Plasma Display Panel (PDP) or the thinner and lighter Organic Light-Emitting Diode (OLED) display. There is also the DLP (Digital Light Processing, by Texas Instruments). This is a video system using a device, the DMD (Digital Micromirror Device), made up of a huge number of micromirrors whose

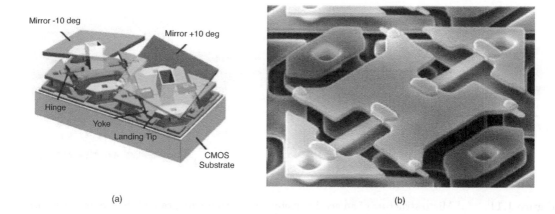

Figure 1.10   (a) Micromirror structure (DMD™, Digital Micromirror Device) used in DLP. (b) Microphotograph of a DMD. Reproduced by permission of © Texas Instruments.

sizes are as small as 10–20 μm. Figure 1.10 shows the three-dimensional view of a DMD, microfabricated by a CMOS-like process over a CMOS memory (CMOS is a term that will be explained fully in a later chapter). Each light-switch has an aluminum mirror that can reflect light in one of two directions, depending on the state of the underlying memory cell. Electrostatic attraction produced by voltage differences developed between the mirror and the underlying memory cell determines rotation by about 10 degrees. By combining millions of DMDs (one per pixel) with a suitable light source and projection optics, it is possible to project images by a beam-steering technique. Reducing the beam steering to a fraction of the pixel projection time produces grayscale. Colors, by the use of filters, are also possible.

Another actuator widely used for video is the LCD (Liquid Crystal Display) panel, used in projectors, laptop computers and displays. LCDs are used because they are thin and light and draw little power. The contradictory term "liquid crystal" indicates that the material takes little energy to change its state from solid to liquid. Its sensitivity to temperature is one of its features, as verified by the funny behavior of your computer display if you use it during hot days on the beach. The main feature of the LCD is that it reacts to an electrical signal in such a way as to control the passage of light. This property is used in the transmission, reflective or backlit mode, to obtain grayscale images. The use of filters enables colors.

The abovementioned DLP is a sophisticated example of MEMS, which integrates mechanical elements, sensors, actuators, and electronics on a common silicon substrate by microfabrication technology. Electronic circuits are made by an Integrated Circuit (IC) process. Micromechanical components are fabricated using the same processes, where they are compatible with micromachining: a selective etching away of parts or the addition of new layers to form the mechanical and electromechanical devices. We have many MEMS used as sensors or actuators. A further well-known example of MEMS, actually a sensor, is the accelerometer used in crash air-bags for automobiles.

In addition to fully integrated solutions we can have hybrid micro-solutions like the one shown in Figure 1.11. In this case, rather than realizing sensor and circuit on the same silicon substrate, the components are separated and are possibly fabricated with different technologies: the sensor is a MEMS with little electronics on board, and most of the processing is done by a

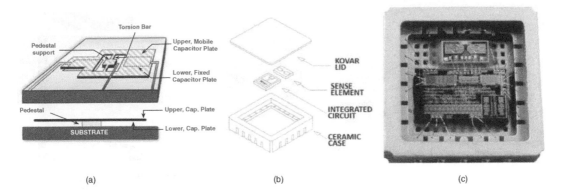

Figure 1.11   (a) Micro structure of an accelerometer; (b) assembly diagram of the System-in-Package. (c) Microphotograph. Reproduced by permission of © Silicon Designs, Inc.

conventional integrated circuit. The parts are micro-assembled on a suitable substrate (such as ceramic) and sealed on the same package to obtain a so-called System-in-Package (SiP).

Obviously, in addition to the actuators fabricated on silicon, we have many other types of actuators fabricated by conventional methods, some simple and affordable, others complicated and expensive.

> **Testing**
>
> Before delivering electronic circuits or systems it is necessary to verify their functions by testing. If the input is an electrical quantity it is not difficult to generate something similar for this purpose. The operation is much more complicated if the process involves non-electrical quantities, in which case it requires the use of a controlled non-electrical signal at the input.

Important aspects of the production of electronic systems are the packaging (sealing of the system so as to ensure protection and mechanical stability) and the testing, i.e., the verification of system performances that must correspond to those expected (the specifications). In the case of systems with sensors or actuators the packaging can be problematic because it is necessary at the same time to protect the system from undesired aggressive agents and to allow the desired quantities to interact with it. Testing is also a problem because it does not involve just electrical signals. For such systems, packaging and testing should be accounted for at an early stage in the design (and this is a good recommendation for any system, even one without transducers).

Although this book does not specifically deal with sensors and actuators, it is necessary to be aware that transducers can be essential parts, and that the design and specifications of the entire system depend on the features (such as sensitivity, accuracy, and conditions of operation) of the sensors and actuators used.

1.5   WHAT IS THE ROLE OF THE COMPUTER?

I suppose that at this point a natural question concerns the role and use of the computer. Indeed, the complexity of modern electronic systems cannot be handled with paper and pencil,

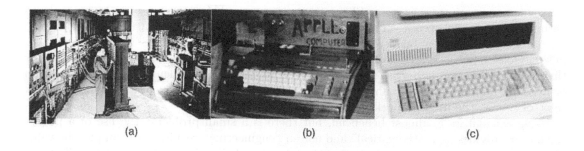

(a)   (b)   (c)

Figure 1.12   Evolution of the computer: (a) ENIAC – 1946; (b) Apple I – 1976; (c) IBM 5151 – 1981.

and probably just reading this book on paper instead of looking at a bright monitor seems funny. The answer to the question is: yes, of course, computers and simulation programs for circuits and systems are important tools for electronic design, and, aware of this, the educational method followed by this book expects their massive use.

However, before talking about computer programs it is necessary to linger a little while on the role of these tools in both designer activity and the learning process. Definitely, computers are amazing machines that make the modern age what it is. Without them we would not have access to the knowledge and comforts that we now take for granted. Computers process information for us and they do it fast – much faster than we can; they solve problems and provide solutions. The progress of computers over the years is such that problems and calculations that used to take many months or years can be solved in fractions of a second. That is the good news, but let's think a little bit on this point.

Many years ago (but not so very many), the first computers were a set of big boxes like armchairs surrounded by air conditioning and controlled humidity. The instruction set was entered by punched cards, and a suite of programs took hours, or even a day. The user had to deposit his or her deck of punched cards in a tray, and somebody in a white coat periodically came out of the mythical computer room to take the decks inside one by one. Just a small mistake required a new run and hours of waiting. The speed of the computer (or actually of all the steps: punching cards, walking to the computer center, printing results on paper, delivery of results) was very slow, and often the speed in realizing a mistake was so fast that the user understood the run was useless before receiving the answer. "That's very bad!" I suppose you will cry, but, indeed, the so frequent frustrations caused by mistakes taught people to be very careful before launching a run. The computer evolved rapidly, much more ahead than what is shown in Figure 1.12, but the first generations of computers left a good amount of time for thinking about results before receiving them. And that is what was good: having time to use and exercise the brain.

Friendly suggestion

Do not allow the computer to be faster than your brain. After receiving any simulation outcome, stop and wonder. You will find something useful hidden in the results. Certainly – and surprisingly – you will find something that helps your understanding.

Now the speed of machines is much greater than the speed of the brain, and the answer is received before the brain is able to formulate a logical prediction. Consequently, it seems convenient to many to just try and see what happens. This attitude is more general: to perform an addition, even a very simple one, very often a pocket calculator is switched on so as to discover with a few finger strokes that $13 + 26$ equals 39. Isn't it so? That's funny and, in some senses, not a serious problem. Everybody, I suppose, assumes that using the brain for that simple operation is useless. This is truer for more complex calculations. Multiplication or division of numbers with many digits would require at least a piece of paper and a pencil; thus, of course, the use of a pocket calculator is the best solution.

Using calculators is almost beneficial, but there is a small problem: the use of a pocket calculator provides a "mathematical" and not an "engineering" result. For example, the value of a current can turn out to be 21.47321832 μA but is it believable that the last two digits, or even more than two, make sense? The minimum measurable current can be in the nano- or the picoampere range but not in that of the attoampere (10^{-15} A). Therefore, using a calculator possibly conceals the meaninglessness of results.

The situation was different, or even the opposite, in the past, when division and multiplication were calculated by using the slide rule: a funny instrument that engineers used, made of two parts, one sliding inside the other. The basis of the slide rule is the logarithm, or, better, the property that $\log(a \cdot b) = \log(a) + \log(b)$ or $\log(a/b) = \log(a) - \log(b)$. The two parts of the slide rule were used to perform the addition or subtraction of logarithmic segments, and this involved possible errors in aligning segments and in reading results. The accuracy was limited and evidently could not exceed three or four digits – much less than what can be got from the computer.

I won't go further down in this archaeological description of ancient tools used before the computer, but I should make a remark. The results obtained with the slide rule were clearly approximate and definitely had lower accuracies than those obtained with a normal computer. "That was not good!" I suppose you think. But you are partially mistaken. As already outlined in the case of current, having too-precise results is nonsense. Precision is not the real world. In engineering disciplines, precision that is much higher than the roughness of real things or the fast fluctuation in time of real quantities is just a waste of resources.

There is another point to mention. We have to give the right credit to microelectronics. Indeed, computers are not just software but are primarily hardware or, better, electronic circuits. Progress in the computer area is largely due to the improvements in electronics that increased their effectiveness in terms of larger numbers of transistors, smaller areas and volumes, lower consumed power and diminished cost per transistor. Maybe you know about the prediction of the so-called "Moore's Law" made by Gordon Moore, the Intel co-founder, who in 1965 stated the following:[1]

> *The complexity for minimum component costs has increased at a rate of roughly a factor of two per year ... Certainly over the short term this rate can be expected to continue, if not to increase. Over the longer term, the rate of increase is a bit more uncertain, although there is no reason to believe it will not remain nearly constant for at least 10 years. That means by 1975, the number of components per integrated circuit for minimum cost will be 65 000. I believe that such a large circuit can be built on a single wafer.*

The prediction is normally summarized by saying that progress is almost exponential with a doubling of performances every two years (or, using a more aggressive forecast, every 18 months), and, indeed, the rule was almost confirmed by the market and advances in

---

[1] From "Cramming more components onto Integrated Circuits", *Electronics Magazine*, 1965.

technology for more than four decades. The result is that for some categories of integrated circuits the complexity now exceeds several billion transistors and speeds have reached several hundreds of GHz.

Obviously this growth cannot continue forever. Also, the improvement is not the same for all of the parameters. For example, transistor technology growth is not accompanied by an equal improvement in processing speed or power reduction. To improve performance it is necessary to use more and more transistors, and the consumed power can become a significant limit.

1.6   GOAL AND LEARNING STRATEGIES

What I have said above is not an attack on the computer and modern simulation tools, but is just given in order to define goals and learning strategies better. For this purpose, it is worth remembering that expertise in any technical field means a proper blending of six different kinds of proficiency: background notions, unconscious expertise, specific knowledge, teamwork attitude, creativity, and ability in using tools.

- **Background notions** Electronic background knowledge means understanding systems, architectures, processing methods, languages and implementation techniques. It is not only a matter of knowing new material so as to participate in generic meetings but also of having a clear, complete spectrum of solutions so as to be able to identify the most appropriate approach.

- **Unconscious expertise** Behind the background notions there is this kind of technical expertise. It results from the slow and long-range absorption of notions studied at school, apprehended by reading books, and resulting from success stories as well as from technical failures (which are much more important than easy successes).

- **Specific knowledge** Every electrical engineer has specific expertise in one of the different facets of the profession. Indeed, in addition to the basis it is necessary to go deeper into detailed knowledge of solutions and practical implementations.

- **Teamwork attitude** The design of complex systems demands a spectrum of knowledge that is not possessed by one single person. This is why activity in electronics is often carried out by teams of engineers, and the attitude towards working with others is an important quality. Among other requirements it is important to be able to communicate, to write reports and to make technical presentations.

- **Creativity** The success of a product greatly depends on quality, reliability, and often style. For quality it is important to ensure optimum execution; for reliability it is necessary to follow strict procedures and controls; for style the exterior aspect, colors, and trends are important. But in addition it is necessary to include innovation, and this is the result of creative activity. Therefore a bit of creativity, balanced by execution, procedures, and care for fashion, is essential.

- **Ability in using tools** Any activity requires proper tools. For electronics, various Computer-Aided Design (CAD) programs obtain a reliable forecast of the operation of circuits and systems before they are fabricated. Other tools verify the correctness of the transformation of a logic behavior into a schematic and then into its physical realization. Others facilitate the PCB design.

The first two proficiencies don't need further recommendations; it is probably worth discussing the others a little bit more.

1.6.1   Teamwork Attitude

Today's electronic designers cannot work alone. There are so many demands, tasks, and varied sources of information required that the job needs the support of others. We have all learned at home and at school to compete as individuals for awards, attention, and prizes. But the reality is that teamwork obtains the optimum.

Team skills are quite different from those of competing individuals. They involve cooperation, mutual support, and accountability to the team. Therefore students need to learn together to acquire the skill of developing new ideas and new methods of working together. An individual alone has but limited perception of the range of possibilities in a situation. A team develops a variety of abilities and so widens the scope of available information, options and ideas. As a result, the quality and effectiveness of individuals is enhanced because a team can help to explore hidden factors.

One of the key benefits of teamwork is synergy. It multiplies the resources of participants through the interaction of a variety of contributors, who see a problem from diverse perspectives. However, as the size of the team becomes large (the optimum is around five people) the efficiency of it diminishes because of the negative contribution of "anchors." Moreover, it is important to have in the team complementary attitudes. There must be a "manager" who keeps the project on track and on time; for this a highly organized person is necessary. A warning generator is also needed, and a problem solver and an hard routine worker. Having a creative person is also important for exploring new avenues and for stimulating the team.

The above points will be helpful in case part of the learning process involves team activity (which I strongly encourage). The formation of the team and its working activity must be carefully considered to obtain the best effectiveness.

1.6.2   Creativity and Execution

The key word of many electronics companies is innovation, but this word implies different issues and strategies. You, as a future electrical engineer and very likely an employee of a high-tech company, must be prepared to contribute to innovation. Having innovative products is a company strategy that involves both creativity and execution. Generating new architectures, new electronic functions and new circuits is important, but poor execution of the invented ideas can be a source of failure. Therefore, it is important to value creativity but not to rely heavily on it. A correct balance of the two aspects is the optimum, but in the market we can find successful companies that focus more on creativity and others excelling in execution.

Innovation, creativity and execution are topics studied in many business schools, and this is not the right place to go into further details. However, while having the issues in mind you should analyze your attitude and, possibly, try to reinforce your proficiency in areas where you are not so strong. To improve execution it is necessary to do exercises that refine the ways in which methods and techniques are implemented. For creativity, unfortunately, there is no one recipe; there is just the recommendation to avoid saturating your mind with notions and to keep vital your technical and scientific curiosity.

Another point to remember is that both creativity and ability in execution are part of your added value: the assets of your future professional career.

1.6.3   Use of Simulation Tools

Proficiency in the use of simulation tools is not just a matter of knowing how to use the tools and how to generate results but, much more, it is about how to analyze results and to be able to understand limits and opportunities. When experts design a circuit or a system they already have a given function and a given response in mind. The simulation serves to verify the idea and to find out the possible limits of the approximate reasoning. If the result is not what is expected, the expert uses rules of thumb to modify the design in the proper direction. Moreover, simulation results are used to refine the rules of thumb and account for second-order effects. Therefore simulation tools aid the designer, and the programs used for it are called Computer-Aided Design (CAD) tools. These tools support the design at various levels of abstraction. The study of architecture is normally done by using languages or behavioral descriptions; the study of circuits uses electrical networks and models of the electronic components.

When the design activity becomes routine, without a concrete contribution from the designer's skill, the computer helps in performing repetitive steps automatically, and, in this case, we talk about Electronic Design Automation (EDA). For example, when there are design rules for avoiding fabrication errors (possibly caused by inaccuracies in the process), those rules are verified with a design automation program. The design of a PCB requires some skill in component placement, but the physical design of the layout (the routing and dimensioning of traces, the choice of the layer where they run) can be done automatically.

Keep in mind ...

the key difference between CAD and EDA tools. CAD programs aid designers in their technical activity, providing elements useful for assessing the operation of the circuit and helpful guidelines for the design flow. Design automation tools are used to replace the designer in routine tasks.

Design automation programs must use algorithms capable of meeting any possible situation and level of complexity. Therefore, for simple cases a manual solution can be more effective than an automatic one, but for very complex architectures automation is the only possible solution. Since analog circuits normally require a small number of components while digital circuits count hundreds of thousands or many millions of transistors, automation is much more suitable, and more often used, for digital architectures than it is for analog schemes.

1.7   SELF TRAINING, EXAMPLES AND SIMULATIONS

This chapter has already used a special inset called "Self training" to suggest Web searches for private study on specific topics. There are also boxes with clarifications or warnings. These insets and boxes are, in some sense, small breaks needed when the complexity of new concepts causes stress. They also help in the complex mechanism of learning a new subject.

Indeed, learning is the result of a proper blending of five features:

- learning as memorizing, storing simple information that can be reproduced without significant changes;

- learning as a quantitative increase in knowledge, or just collecting information – in this case the information is stored as rough notions;

- learning as acquiring skills and methods that can be retained as background knowledge and used when necessary;

- learning as making sense or abstracting meaning – this kind of learning involves relating parts of the subject matter to each other and to the real world;

- learning as interpreting and understanding facts in a different way, which involves comprehending the world by reinterpreting knowledge.

There is a substantial difference in quality (and difficulty) between these five features, which we can summarize by "knowing that" or "knowing how." Obviously, the aim of this book is toward the more difficult "knowing how," but a bit of "knowing that" is necessary because it provides new facts and practical details and also helps in contextualizing study and avoiding unnecessary theoretical abstractions.

In addition to "Self training" the book has "Examples" with solutions, and, much more, "Computer Experiments" for simulations of systems or circuits in an environment that looks like an electronics laboratory. Moreover, as in all textbooks, there are problems given at the end of each chapter.

1.7.1 Role of Examples and Computer Simulations

It is well known that examples and training sessions facilitate the learning process. This is also what is implied by the saying: "If we hear, we forget; if we see, we remember; if we do, we understand." Well, if you read a book or are taught any technical subject without seeing examples, your performance, merely after reading, is not good. Acquired theoretical knowledge is often accompanied by perplexity, insecurity, uncertainty, and doubt.

If the above is true for any technical subject, it is surely so for electronics. There are two extra elements here causing uncertainty: the complexity of systems and inaccuracy in fabrication. The complexity obliges us to use approximations that, often, are difficult for others to comprehend. The inaccuracy in fabrication makes the predicted results not exact, as in mathematics, but only within a given range, part of which can be unsuitable for use. Since electronic designers must become accustomed to uncertainty and inaccuracy, examples and computer simulations are extremely beneficial for the learning process. The expected benefits gained are to:

- **Overcome incredulity** Very often the results of a theoretical description made with doubtful approximated assumptions are unconvincing. The outcome seems too simple, not general enough, or maybe simply unbelievable. Numerical examples together with plausible computer programs help to change those possible mindsets.

- **Understand the limits of approximation** The simple models used for hand calculations are often too approximate and can generate misleading results. Nevertheless, simple equations and simplified rules are keys for directing the designer's activity. The use of more precise computer aids verifies the rules of thumb and facilitates understanding when they fail.

■ **Reinforce knowledge** This is a general beneficial effect of using examples and computer simulations. For electronics, since often knowledge is not codified by reliable equations or reasonable rules, using examples and computer verifications is particularly helpful.

■ **Learn rules of thumb** The use of rules of thumb is very common in expert designers' activity. They know, for example, what is necessary to increase the gain of an amplifier or improve its speed. They know what the limits of a circuit or the technology used are and they are aware that it does not make sense to continue trying to improve performances above a given limit. This kind of expertise is the result of many observations, mistakes, and achievements. Therefore, with examples and simulations it is possible to accumulate knowledge that is codified in a set of rules of thumb.

1.8 BUSINESS ISSUES, COMPLEXITY AND CAD TOOLS

Before talking about CAD tools it is worth remembering that the activity of any electronics professional requires a little bit of knowledge about financial implications. Of course high-tech products must be sold at a profitable price, and this critically depends on the proper blending of technical and non-technical issues. Designers tend to focus on performance-related themes, while the concern of managers and marketing people is mainly with business. An important factor for success is ensuring low time-to-market. In a highly dynamic business like microelectronics a delay of a few months in releasing new products reduces the return on investment, perhaps to zero. Therefore, having quick design cycles is extremely important.

Another relevant element is, obviously, the cost. Financial studies distinguish between two cost components: fixed and variable costs. Fixed costs are made up of all the expenses not related to the number of circuits or systems sold. The variable costs directly depend on the manufactured parts, and are proportional to the volume of production.

What makes the microelectronic market problematic are the large fixed costs. In addition to traditional ones there are the high costs of research and development, and, much more, of the equipment needed for fabrication. The various fabrication steps that lead to an integrated circuit use a set of masks that define patterns on the silicon surface. This set of masks, especially for nanometer technologies, costs a lot of money. Thus the design must be perfectly functional the first time it is tried, so as to avoid the costs of a new set of masks. The fabrication plant used to process integrated circuits is extremely expensive with a relatively quick obsolescence. Consequently, the use of the plant must be optimal, with a large number of circuits fabricated together (batch fabrication). Similar requirements hold, though to a lesser extent, for the design and fabrication of microsystems that assemble discrete components.

1.8.1 CAD Tools

The requirement for a short time-to-market and cost-effective developments impose the need for a quick and reliable design cycle without the necessity for redesign. For this it is necessary to perform extensive simulations that verify the expected operation in all the possible conditions.

The designer uses computer programs to verify the correctness of system architectures and to estimate logic and electrical behavior. These programs also consider the effect of fabrication inaccuracy and estimate the probability of malfunctioning. They estimate parasitic terms and provide the information for re-simulating the circuit with the parasitic limits included. The tools also help the designer in critical parts of the job; for example, CAD tools optimize the physical placement of cells and make sure that performances are as expected.

1.8.2   Analog Simulator

The simulation tools used to study the electrical response of circuits derive from a program named *Spice* (Simulation Program with Integrated Circuit Emphasis) developed many years ago at the University of California, Berkeley. It uses a description of the electrical network, the netlist, that describes components and interconnections. The netlist generates a set of nodal equations based on linear or non-linear models of the components. The models use the electrical variables of the devices and, also, employ additional variables and nodes internal to the devices, introduced to specify them better. The resulting large system of integer-differential equations is solved by using the methods of numerical analysis.

The solution methods use the so-called LU factorization system, which factorizes the system of equations into two matrices: L, a lower triangular matrix and U, an upper triangular one. When it is required to solve non-linear equations the simulator employs methods like the Newton–Raphson algorithm, which proceeds in a iterative manner until it obtains the required accuracy. Since studies in the time domain need to estimate time integrals, the tool performs the numerical estimations with trapezoidal elements and appropriate time steps. The time step is critical for obtaining the overall accuracy. However, if it is too small the simulation time becomes unacceptable. The circuit simulator speeds up the process with an automatic control that adapts the time step to the signal's rate of change.

All the steps of this simulation are automatic. Moreover, graphical interfaces facilitate the description of circuits and the analysis of results. Suitable post-simulation tools plot waveforms of voltages and currents. For linear or linearized systems, the tool simulator operates in the frequency domain. It enables, among others, the following types of analysis.

- **DC analysis** This is a Direct Current (DC) study that determines the operational point of the circuit. It is done before applying time-variant signals. It is automatically performed in order to determine initial conditions.

- **AC analysis** AC stands for Alternating Current. The analysis refers to a linear circuit or assumes linearized behaviors. This is the so-called small signals analysis (it is defined in a later chapter).

- **Transient analysis** This is the study of the response to inputs that change with time. For example, it determines the response to a step on top of the initial conditions previously determined by the DC study.

- **Noise analysis** This is made to determine the contribution of unwanted signals, called noise. We shall learn that noise affects any electronic component, being caused by fundamental limits. Knowing the level of noise that affects a signal is essential.

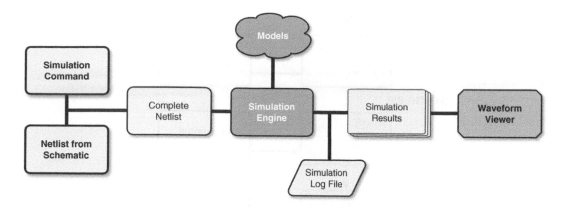

Figure 1.13   Block diagram of *Spice* architecture and functions.

- **Monte Carlo analysis** This is a statistical study of performance. Values and performances of real components differ from the expected ones because of errors in fabrication. This analysis statistically changes design parameters to reproduce the effect of fabrication errors over a set of circuit samples.

Figure 1.13 shows a simplified flow diagram of the *Spice* program. The schematic generates a component netlist that is combined with the simulation commands to generate the complete netlist. The simulation engine, the core of the tool, builds the system of equations and calculates the solutions. The results depend on the equations of the components' model. The output is given in two files, a log and a file of simulation results. A graphic section uses the latter to show, on the computer monitor, waveforms of the results obtained.

1.8.3   Device and Macro-block Models

The accuracy of the models of circuit elements determines the closeness of simulation results to experimental measures. The models portray, at various levels of detailed description, the electrical behavior of a single device – such as a resistor, a capacitor, or a transistor – or can represent the task of a macro-block made up of many components.

Models can describe with high precision static, dynamic, and temperature behaviors. They refer to specific technologies and follow the evolution and the improvements. For example, the scaling of the technology of transistors requires the use of very complicated models that also include three-dimensional effects. Since the accuracy of the model influences the precision of results, manufacturing companies and customers fix models and model parameters to guarantee the expected performance. The transistor models of advanced processes are more and more complex, with a very high number of parameters. Often, the circuit designer does not understand the role and relevance of various parameters, and assesses only the global description.

The use of complex device models obtains accurate results but, for a large system, makes the simulation time extremely long (a week or more, even with powerful computers). For this reason a detailed study is normally postponed to the last step of the design flow. The architecture feasibility can be verified with less accurate results, by using, for example, a system-level simulator or simplified macro-block models.

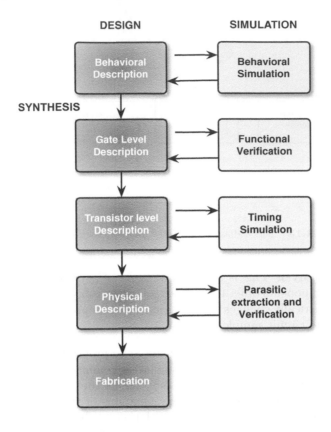

Figure 1.14   CAD tools helping various phases of a digital design.

1.8.4   Digital Simulation

Since digital integrated circuits can include many millions of transistors, studying such large circuits cannot go into the details of a single transistor's operation. What is normally thought sufficient is to ensure logic and timing functionality. Luckily, representing a logic one or a logic zero does not need much detail about amplitudes. It is just necessary to have the expected function at the right time with all possible input conditions.

Digital design and digital simulation follow a top-down approach. At a high level we describe functions with a behavioral language. Then automatic tools transform (or, better, synthesize) the behavioral description into a gate-level netlist, followed by the transistor-level representation and the physical definition of the digital circuit. Simulation tools assist all those phases as shown in Figure 1.14. Notice that the focus at the transistor level is mainly on timing verification.

Simulation tools are used not just for integrated circuits but also for systems made up of prefabricated parts. For the realization of digital functions it is possible to use special devices such as gate arrays, microprocessors and DSPs. Specific CAD tools help with the use and settings of these complicated devices (we shall learn what they are and how they work shortly).

The verification of performance uses signals at the device or system inputs. These must enable control of the correct operation of the entire circuit. Suitable stimuli are used for both analog and digital circuits. However, the complexity of digital systems means that special attention must be paid to the choice of input signals and where to apply them. It may happen that the simulation or the experimental verification is not able to check correct operation of the complete circuit. If some sections remain inactive, there is no test of the function of those sections. To exercise all the individual components of the circuit it can be necessary to use complicated stimulus patterns and to specify signal injection in internal nodes, used just for testing purposes. However, the use of complicated test patterns produces over-long simulation and testing times.

1.9 ELECTRONIC VIRTUAL STUDENT LAB (ElvisLab)

The saying "if we do, we understand" is so true that I put quite a lot of effort into allowing you to do that. CAD tools and circuit simulators are very good aids. *Spice*-like programs that run on a personal computer are available for free, and thus you can use them to study simple circuits. For transistors or other special components you can use models provided by silicon foundries or those available on the Web. The book facilitates simulation sessions by proposing several examples and problems. However, real doing is doing experiments. With an electronic circuit this means changing inputs and seeing what happens at the output; using strange operating conditions and seeing how the output deteriorates. This is what you can do in an electronics laboratory by playing with source generators and instruments connected to a circuit. You can build the circuit yourself, or the circuit might be pre-built, with instructions that guide the experiment.

This book uses an alternative option: doing experiments in a virtual electronic laboratory with plenty of virtual instruments and pre-built circuits or systems. With these you can change the inputs, see what happens and try to understand why. There are two advantages: the circuit never blows up, and instruments do not break down. You can stay in the laboratory and do experiments for as long as time you need or like to do so, without waiting for it to be available.

The name of our tool is *ElvisLab*, an acronym standing for ELectronic VIrtual Student Lab. It is accessible at http://ims.unipv.it/~ElvisLab/ and it will be made available on request to the professor or the instructor who uses this book to teach you microelectronics. The use of *ElvisLab* is straightforward, and, as you certainly expect, does not require the reading of an instruction manual. The home page of the program gives the catalog of experiments, divided into chapters. Some experiments are proposed in this book, but you can also find other, extra ones.

Almost all of the exercises divide the screen into five sectors. The top left area is the console, with buttons or numerical steppers that change the input parameters. The other sectors present the scheme of the experiment, and display waveforms or diagrams. Three buttons on the top right allow you to return to the home page, to get help and hints, and to obtain some general information on the use of the program. If you click on the bar of a small window that window replaces the large ones, so as to show the details in it better.

Figure 1.15 is a screen view. The console shows numeric steppers for changing the example parameters. At the top of each of the four buttons is the name of the variable and the actual value that can change in a linear manner between the limits indicated by the bar. The main window contains the schematic of the circuit, using pictures of instruments that look like what

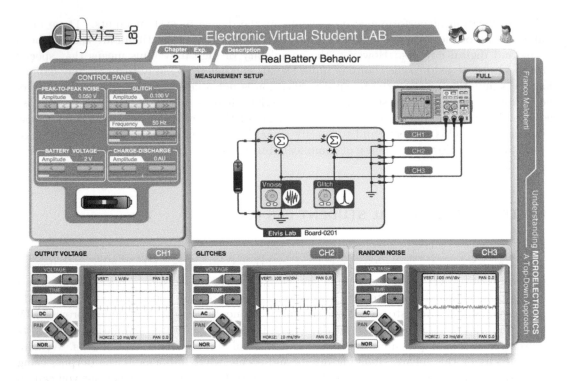

Figure 1.15   Screen snapshot of the first *ElvisLab* computer example. Reproduced by permission of © Microelectronic Department, University of Pavia, Italy.

can be seen in a real laboratory. What is seen in the figure is an oscilloscope, an instrument that displays how repetitive waveforms change with time, beginning from a defined point in the waveform period. The result is an almost stable plot representing the waveform in a defined time interval, as established by the oscilloscope scale controls.

The oscilloscope in the figure has three channels, whose waveforms are displayed in the three small windows. They use time as the horizontal scale and voltage as the vertical scale. The program uses default axis intervals, but when the vertical amplitude becomes too small or too large you can adjust the waveform view by clicking the buttons. When the waveform is in the main window, moving the marker displays the coordinates of that point of the curve.

The help button pops up the example description also reported in this book. The document requires you to do something specific, but, of course, you are free to change and look at whatever you like. I think that the description here is more than enough. You now just have to go to the *ElvisLab* address and check out its features using one of the numerous computer experiments.

Obviously the virtual laboratory is an option that does not exclude real experiments in a real laboratory. If you have that possibility you will have a double chance "to do in order to understand."

PROBLEMS

1.1 Write a list of results that you expect to obtain from the study of this book. Identify the possible weakness of your present knowledge and define a plan for making yourself ready.

1.2 Describe the operation of the electronic key of a car. What is the function of the buttons? What is the energy stored in the battery (find this data from a search on the Web)? Estimate the average energy necessary for a single opening.

1.3 List the sensors that you have at home, and find out which type of electrical quantity they detect. Distinguish between normal sensors and ones realized with silicon technology.

1.4 Modern cars use several actuators controlled by electronic circuits. List four of them and find out the type of electrical quantities necessary for driving them. What is, in your opinion, the actuator that needs most power?

1.5 Draft the architecture of a system for the automatic control of a fresh-water aquarium. It is required to feed the fish every second day (the food is in the form of small capsules) and top up the tank's evaporated water every week. Water must be filtered for 10 minutes in every hour. Use a 32.768 kHz quartz oscillator.

1.6 Sketch the architecture of a system for monitoring the dose of ultraviolet light. Include a beeper that indicates when the maximum dose is exceeded. Suppose that sunlight can change every minute. Use an ultraviolet sensor to generate a signal approximately proportional to the square of light intensity.

1.7 What are the functions performed by an air-bag in a car? What kind of sensors and actuators are needed? Draft a block diagram of the complete system with the key functions outlined.

1.8 What is, in your opinion, a very small current and a very high voltage for a modern integrated circuit? How often is it necessary to replace a size 10 micro-battery used to continuously power a system that drains $2\ \mu A$ from 1.5 V? Do a search on the Web to find possible answers.

1.9 Find on the Web the sensitivity in volts or amperes and the percentage accuracy of any kind of the following commercial sensors: pressure sensor, temperature sensor, flow sensor. Estimate the volume and the power requirements of each of them.

1.10 How many transistors are in one of the electronic devices in your pocket or purse? Estimate the volume that was required 20 years ago to obtain same function.

1.11 Find on the Web the number of transistors in the first microprocessor and the number of transistors in the latest generation of microprocessors. Estimate the growth rate per year.

1.12 Suppose you are to create a design team consisting of four people. What is your relevant proficiency? What kind of expertise would be a good match for you? Make a job description for each of the four people in the team.

1.13   Write a short note on what, in your opinion, is relevant to the learning process. Focus on the role of computer and simulation tools. What is, in your opinion, the difference between attending a lecture and reading a book?

1.14   Do a search on the Web and find two different CAD tools for analog design. Try to understand their features and advantages. What, in your opinion, makes them suitable (or not) for a learning process?

1.15   Use a flow diagram that mimics the sequence of actions of a CAD tool to derive the top-down steps that you suppose are necessary for designing the electronics circuitry for a washing machine.

1.16   Transistors are modeled with many parameters. Do a search on the Web and find a simple model used for bipolar transistors. Try to understand the meaning of the set of parameters and learn the role of the most relevant.

1.17   Do a search on the Web to find a description of the design flow of a digital system. Analyze the hierarchical organization of the flow. Be aware of simulation tools and their function in the design flow.

1.18   What is the role of packaging and assembly in microelectronic systems? Do a search on the Web to find the amount of power dissipated by a microprocessor and the type of package used for it.

1.19   Do the *ElvisLab* tour and understand the use and function of virtual instruments.

CHAPTER 2

SIGNALS

Signals are the basic ingredients of electronic circuits. Electrical variables, often in the form of voltage or current, represent signals in a way that mimics what we perceive in nature; sometimes signals are also invented by humans without any link to the real world. We shall learn an important distinction between signals: they can be analog or digital. Moreover, they can have a continuous-time or a sampled-data representation. The study of signals is carried out in the time domain or in other suitable domains: those to which the Laplace, Fourier or z-transformations convey us. Finally, we shall learn about good and bad signals (typically, bad signals are called distortion or noise), and we shall participate in the strenuous battle between the good and bad signals, with the aim of improving the quality of electronic products.

2.1 INTRODUCTION

Signals represent *information*; that is, the state and the time evolution, qualitative but also often quantitative, of relevant attributes. For an initial discussion on signals and the associated information let us consider an ordinary example: the temperature of an oven. This is a physical quantity, whose value is measured with a suitable scale and dimension, but it is also a signal that carries information on the availability and suitability of given cooking functions. When you bake a cake the temperature of the oven is set to the optimum value (say $175°C$ or $350°F$) for a given period of time. The expected result is that the mixture

Understanding Microelectronics: A Top-Down Approach, First Edition. Franco Maloberti.
© 2012 John Wiley & Sons, Ltd. Published 2012 by John Wiley & Sons, Ltd.

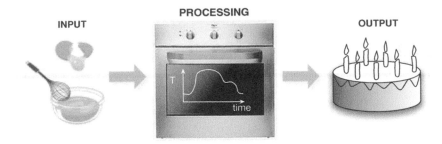

Figure 2.1   Processing of ingredients with an oven. "Temperature" is one of the signals.

of flour, shortening, eggs, sugar, and, perhaps, chocolate becomes a cake baked to perfection according to a process illustrated in Figure 2.1. If the temperature is higher than is needed or its action lasts longer, the cake comes out brown or even burnt. If the temperature is lower or it lasts for a shorter time, the cake is almost raw. Therefore the information about the final result depends on the value and duration of the signal produced by the input setting, "temperature of the oven." Notice that in the example we have more than one signal involved. The temperature of the oven produces other signals: the temperature and humidity of the mix of ingredients. Their values and time variations are the inputs of a process that gives rise to the final product and maybe to another signal, the quality of the cake: good, bad or anything in between. This signal can possibly be the input of another process – say, a contest between family members to assign the "best cook" award. Therefore, a signal carries information and its processing can produce new signals that carry another type of information, whose format can be different from that of the generating signal.

> **Remember**
>
> An electronic circuit represents information by using constant or variable signals. These signals are electrical quantities, often the value of a voltage or a current. In some cases they can be the value of a charge, a resistance, or a capacitor.

We define *nominal* as the case for which the output is what we expect when we apply a specific input. Therefore, a signal processor performing exactly its predicted function obtains the nominal system response. A defective component, an imperfect fabrication process, aging, or the interference of external agents can possibly alter the signals obtained. In our example, the oven heating system could generate higher or lower power, the door could leak more heat than expected, or a refrigerator placed close to the oven could alter the microclimate. Therefore, in addition to the nominal case we have to account for other cases, especially the worst ones. Normally we have *worst case high* and *worst case low*, which give rise to higher or lower signals than the *nominal case*. Moreover, the system operation is intended for certain conditions that are also called *nominal*. For example, we have the nominal supply voltage or nominal consumed power.

In the nominal case we have the nominal response, which, for the cooking process, is a diagram depicting the change in time of the oven temperature. It uses the horizontal axis for time and the vertical axis for temperature, as shown in Figure 2.2. The dimension used on the x axis is seconds (or minutes, or whatever the time is measured in) and on the y axis it is degrees (°C or °F). In addition to the nominal thermal profile the figure shows the responses

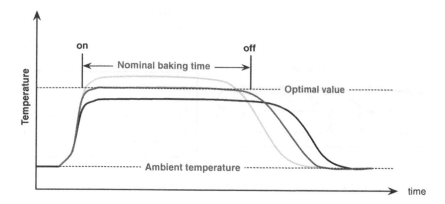

Figure 2.2   Signal representing oven temperature versus time.

in the worst case low and the worst case high. The higher or lower temperatures can possibly be counterbalanced to produce acceptable results: for example, with a higher temperature a shorter baking period, and with a lower temperature a longer one. However, to correct errors, the status of the baking process needs to be estimated, and a decision consequently needs to be taken. When baking a cake the glass window of the oven enables the output to be estimated, and this can determine action. This operation is called *feedback*, which means changing the processing flow on the fly, to meet system *specifications* and to compensate for errors caused by non-nominal or unexpected conditions.

Observe that . . .

an electronic signal and a mathematical signal are quite different. A mathematical signal is fully predictable. An electronic signal is predictable within limits established by the precision of the fabrication process and by erratic factors.

Using the cake example again, we notice that the temperature inside the oven is not uniform, being higher near the heater and lower near the door. This non-uniformity can eventually cause non-uniform baking of the cake, and surely the customer may be disappointed. From the signal point of view, we can state that oven temperature is not just a function of time but is also a function of the spatial coordinates inside the oven. For a more accurate study it would be necessary to use a four-dimensional diagram that represents the temperature as a function of space and time. Obviously, having a varying temperature inside the oven is undesirable. The design specifications typically require the oven to reach an almost constant temperature within a given range of accuracy. However, accurate study of the process can require the use of more detailed signals: a signal that provides distributed rather than concentrated information (treated without regard for particulars).

Having signals that depend on the spatial coordinates is a complication that is difficult to address. Fortunately, because of the features of electronic components we can describe them (to a good approximation) as lumped elements. This assumption is certainly valid for low speeds (or low frequencies); perhaps, for very high frequencies, in the tens of GHz range, it may be necessary to use distributed element descriptions. Consider the speed in a vacuum of electromagnetic waves: $3 \cdot 10^8$ m/s, i.e., 30 mm of material is traversed in 10^{-10} seconds. Therefore, since the dimensions of electronic systems are a few tens of mm or even smaller,

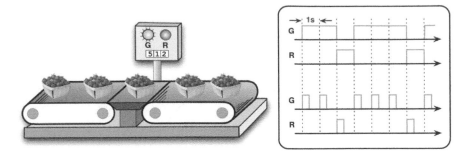

Figure 2.3 Automatic weighing station monitor.

when the rate of change is less than 10^{-10} s (which corresponds to frequencies lower than 10 GHz) lumped values are acceptable. Only when the frequencies become many tens of GHz, as happens in some communication systems, is it necessary to move to distributed element descriptions.

Until now we have focused on signals with continuous values within a given amplitude interval, the *dynamic range*. However, there are other types of signals, such as, for instance, those that have only discrete amplitudes, out of a set of finite values spanning a given dynamic range. In addition, we have signals that, for example, answer questions like: "Is this stick longer than one meter?" (or one yard). Since the answer has two values, yes and no, the information the signal carries is *binary*. That type of signal is substantially different from temperature or other physical or electrical quantities. The information content, having two or perhaps multiple values, is logic; the signal carrying it is called a *logic signal*.

Continuous-amplitude and quantized-data signals

A signal can change its amplitude continuously, or it can have amplitude that carries information only at a given time or within a time interval. In the former case we have a continuous-amplitude signal; in the latter a quantized-data signal.

Let us consider another possible example: a weight control station that measures the weight of baskets of strawberries. They must be 0.5 kg (17.65 ounces) with a maximum error of ± 20 g. The system uses a rubber belt that carries a basket onto an automatic weighing machine every second. The machine produces a green signal, and this turns red whenever the result is not within the specifications. This output signal is binary and occurs at the rate of one per second. As shown in Figure 2.3, the logic outputs becomes high in order to illuminate the green and red lights. However, the high level can remain for the entire second or can be a pulse that returns to zero after a fraction of a second. The green or the red light might blink, but the information is given anyway. In fact, in this example it is necessary to have pulses of a suitable duration, because the flash of light would not otherwise be visible. However, conceptually the pulse can be very short, at the limit – a delta. Moreover, the output does not refer to a particular moment of the one second interval but signifies the entire period. In this example the data identifies a particular sample from the set of strawberry baskets. For that reason, this type of signal is called a *discrete-time* signal, since they occur at a discrete constant rate. Systems that use discrete-time signals are called *sampled-data* systems.

2.2   TYPES OF SIGNALS

The simplest type of signal is the constant one, i.e., a signal that does not change over time. For example, the supply voltage of an integrated circuit is a *constant signal*; the reference current used to bias the operation of an integrated circuit and the power consumption of a system can also be viewed as constant signals. Nevertheless, a constant signal is just a conceptual definition. It is not possible to have signals than never change. Consider, for instance, the height of a mountain: it is almost constant, but the snow on the top changes the level, as the height depends on the snow that falls or melts. Therefore, a constant signal is often non-constant, being made up of a constant nominal value plus variations that depend on events, often unpredictable ones, which change slowly or quickly with time.

Even assuming that there are no spurs, a constant signal is just a concept. The supply voltage of an electronic circuit is zero when it is off, jumps to the supply value during operation and returns back to zero when the circuit is switched off. Therefore, the constant supply voltage is in reality almost constant only during the operational period. Moreover, the value is not exactly what is expected but differs from the nominal level by the inaccuracy of the voltage regulator. Any supply generator has a nominal voltage, but, depending on quality and cost, the real voltage can be inaccurate by $\pm 5\%$ or even $\pm 10\%$. Therefore, it is necessary to admit a range of supply voltages that produces acceptable performances.

> **Please note!**
>
> The supply voltage and references of any electronic circuit are not as expected but are always affected by errors or static shifts. These errors must be kept below a given limit, since they affect circuit operation.

In addition to the inaccuracy of the nominal value, the voltage of a real supply generator fluctuates for various unwanted reasons. There are slow changes caused by the discharge of the battery, and drifts due to change of temperature or variations in environmental conditions. There are random fluctuations, called noise, produced by unpredictable mechanisms and there are quick jumps, called glitches, typically due to the interference of clock signals. Therefore, the "constant" supply voltage can look like the plot in Figure 2.4. The nominal value is 1.8 V, but a tilt, a noise of about 2 mV, and glitches as large as 20 mV affect the ideal signal. Notice that the amplitudes of glitches are not the same because the activity of the interfering network varies. Since non-ideal situations are difficult to avoid, the designer must be able to provide circuit solutions that tolerate supply voltage variations within given limits. The same goes for other constant signals that may be used in the circuit. We shall see that electronic circuits often use voltage or current biases for their operation. The accuracy and cleanness of those voltage or current references are as important as those of the supply voltages. Thus, on one hand, reference generators must be properly designed, and, on the other, the circuit must admit some variation of references with respect to their nominal values.

The noise mentioned above is a special type of signal that does not carry any kind of information, but, on the contrary, corrupts the information. Voltages or currents in electronic circuits are produced by the cooperative action of a huge number of electrical carriers. The effect we can observe is the average of many elementary results. That average randomly changes in time because of partially conflicting actions that are not predictable. This difference from the expected deterministic value is the noise. Accordingly, a noisy signal $V_s'(t)$ is the

COMPUTER EXPERIMENT 2.1

Real Battery Behavior

This computer experiment serves to demonstrate that a constant signal does not exist in reality. A battery (or a voltage source) does not generate a constant voltage but produces a signal like the one that you see at output of the oscilloscope below. The computer experiment accounts for the contributions of three possible limitations:

- a drop time caused by the charge or discharge of the battery or by temperature variations. The drift is normally very slow. However, the result differs from an expected signal "constant" in time;
- a random fluctuation, called noise. This is an unpredictable effect associated with fundamental physical causes. Noise normally increases at high temperature;
- a sequence of undesired quick changes, named glitches, of the generated voltage. Glitches are caused by the interference of digital sections that operate under the control of a clock.

You can set the nominal voltage of the battery pack (a multiple of 2 V) and its drift. Sliders set the standard deviation of noise, amplitude and frequency of glitches.

MEASUREMENT SET-UP

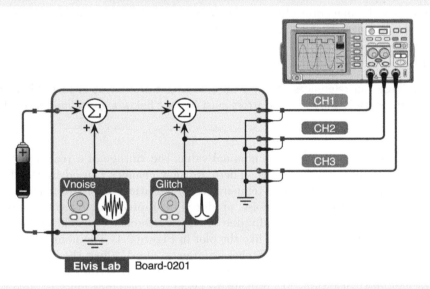

TO DO

- Set the discharge rate, noise, and glitch to zero and observe the "ideal" output of the battery.
- Set a charge or a discharge rate to observe the long-term change of the battery voltage.
- Add noise to the battery output and change its amplitude. Estimate the ratio between the power of the constant signal and the power of the noise.
- Inject glitches at different amplitudes. Observe the addition of glitches and noise.

After spending a few minutes on changing battery value, the amplitude of the tilt, noise, and glitches, imagine the difficulties in obtaining good results with an electronic circuit biased by a signal similar to the one that you have simulated. What, in your opinion, is the most critical limitation for three applications that you know?

Go to http://ims.unipv.it/~ElvisLab/ for all the Computer Experiments in this book.

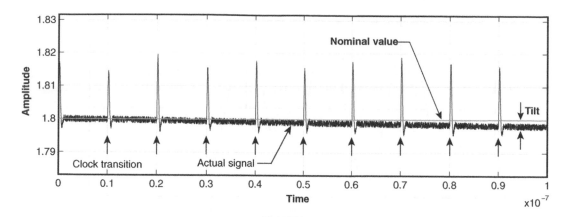

Figure 2.4   A constant ideal and real signal. In addition to a tilt, the real signal is corrupted by noise and glitches.

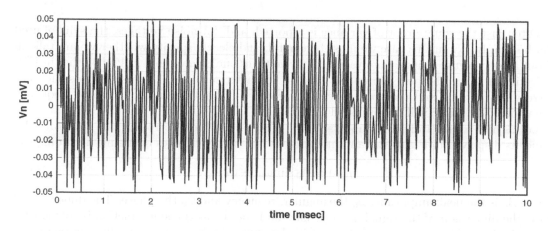

Figure 2.5   Typical plot of voltage noise as a function of time.

predicted value $V_s(t)$ plus an unpredicted amplitude: the noise.

$$V_s'(t) = V_s(t) + V_n(t). \tag{2.1}$$

Figure 2.5 shows a typical noise waveform. The scales used are almost arbitrary. However, the voltage amplitude can be the expected value for a good signal. Noise is evidently unpredictable, as the amplitude at any given time is not related to its extent at previous or successive times. Moreover, if the value of the signal is much larger than the noise, the signal waveform is easily distinguishable. On the other hand, if the signal is smaller than the noise, it will be "submerged" by spurs, making it impossible to recognize.

Noise is primarily characterized by its square variance, σ_n^2, which is the time average of the square of the noise signal. For voltages the equation is

$$\sigma_{V_n}^2 = \langle V_n^2 \rangle = \frac{1}{\tau} \lim_{\tau \to \infty} \int_0^\tau V_n^2(t)\, dt. \tag{2.2}$$

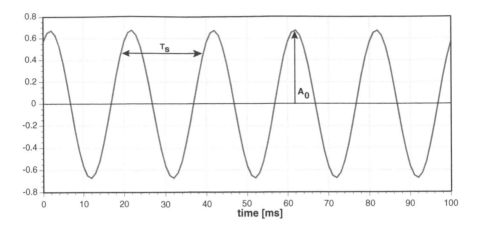

Figure 2.6   Sine wave signal.

Often just the square root of variance is reported, and, for noise voltages, the symbol used can be simply V_n. In this case we must be careful, and clearly specify whether we are talking about a noisy signal or its square variance (for voltages one is measured in V the other in V^2).

Variance is not the only parameter used to characterize noise. Others, such as the auto correlation function and the noise spectrum, provide information on the time and frequency behaviors. The latter will be discussed shortly.

Let us now consider another important type of signal used in electronics: the *sine wave*. Its value, as displayed in Figure 2.6 and indicated by the name, changes over time according to

$$y_s(t) = A_0 \sin(\omega_0 t + \varphi_0), \tag{2.3}$$

where A_0 is the peak amplitude, ω_0 the angular frequency and φ_0 the phase. The dimension of A_0 is the dimension of the signal $y_s(t)$ (it may be V or A), as the sine function is (obviously) dimensionless. The argument of the sine is also dimensionless (or, if you like, measured in radians); therefore, ω_0 is given in rad/s. Moreover,

$$\omega_0 = 2\pi \cdot f_0, \tag{2.4}$$

where f_0 is the frequency of the sine wave measured in Hz (periods per second).

The sine wave is extensively used for the study of electronic systems. What is more, it is often used for testing and characterization purposes. A sine wave at the input of some circuits gives rise to a sine wave at the output with a different amplitude and phase. For other circuits the output is not just a pure sine wave but can be viewed as the addition of many sine waves with frequencies related to that of the input. The relationship between amplitudes and phases is then used to study the performance of the electronic circuit.

Often real signals supposed to be a sine wave differ from the expected waveform because of limitations. It may happen that near the maximum the amplitude is less than expected or that additive noise deteriorates the waveform. Another possibility is that the phase is corrupted by noise. The noise limits are expressed by

$$y_s(t) = A_0 \sin[\omega_0 t + \varphi_0 + \varphi_n \operatorname{rand}(t)] + y_n \operatorname{rand}(t) \tag{2.5}$$

$$A_0 = \bar{A}_0 + A_n \operatorname{rand}(t), \tag{2.6}$$

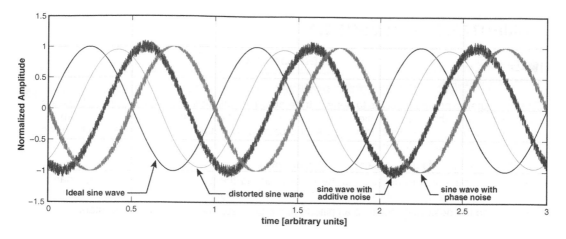

Figure 2.7   Ideal sine wave and effects of various limitations.

where rand(t) is a mathematical function representing a random signal. The possibility that amplitude is affected by noise is rare and not frequently considered.

Figure 2.7 shows various analog cases discussed above. There are four signals. The first is an ideal sine wave with unity amplitude. The second denotes a distortion that reduces the peak extent to 0.95. The third is a sine wave with additive noise, while a phase noise corrupts the last one. It can be noticed that the additive noise, like the amplitude noise, affects the vertical amplitude whereas the phase noise, which changes the horizontal accuracy, does not modify the peak amplitude.

> **Keep in mind**
>
> A sine wave signal used in electronics is always an approximation of the sine mathematical function. Its limitations are described by distorting alterations and noise terms.

Obviously the above equations are for analog signals. For signals represented in a digital manner, limitations are also possible. They can be caused by inaccurate digitization of amplitude or errors in timing. Moreover, there is an intrinsic limitation associated with the digital representation. The value of the digital sine wave at a given time is not translated into the exact value but is a little different, depending on the number of bits used.

A single sine wave is a good testing signal; however, some features of an electronic circuit are better tested by using the superimposition of two sine waves at different frequencies (*two tones*):

$$y_s(t) = A_1 \sin(\omega_1 t + \varphi_1) + A_2 \sin(\omega_2 t + \varphi_2). \tag{2.7}$$

The frequencies of the two tones are often close to each other because tones in close proximity better reveal a possible malfunctioning of circuits. The amplitudes depend on the type of test; many of them use equal or nearly equal amplitudes. Figure 2.8 shows two sine waves and the addition that makes the two-tone signal. Notice that the amplitude of the peak equals the addition of two amplitudes. The peak occurs when the tone phases, ($\omega_1 t + \varphi_1$) and ($\omega_2 t + \varphi_2$), are both $\pi/2$ or $3\pi/4$. Moreover, the mean amplitudes of both sine waves and two-tones is zero, but, obviously, it is possible to add a constant term to shift the signal up or down. Therefore,

COMPUTER EXPERIMENT 2.2

Real Sine Wave Signal

This computer experiment serves to demonstrate possible limitation on real sine waves. It uses a signal generator that gives at the output a sine wave whose amplitude and phase can be corrupted by noise. In addition there are spurs, caused by coupling with a digital clock, in the form of periodic glitches. You can:

- change amplitude and frequency of the sine wave;
- change the standard deviation of the noise affecting the sine wave amplitude and change the standard deviation of the phase noise (two different types of limit);
- change the positive or negative amplitude of the glitches. Their frequency is constant.

MEASUREMENT SET–UP

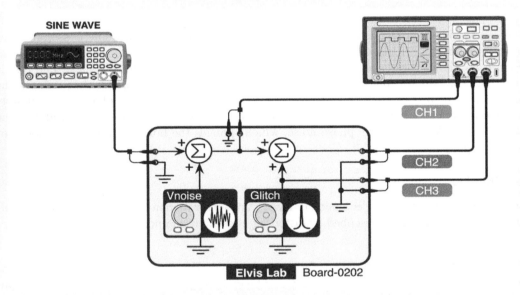

TO DO

- Set the non-ideal parameters to zero and observe the "ideal" sine wave on Channel 1.
- Set a value of amplitude and phase noise and reduce the sine wave amplitude until you are not able to distinguish between noise and sine wave. Change the frequency of the sine wave to see if there are some changes.
- Set the amplitude of the noise or the phase noise to zero. Observe the difference between amplitude and phase noise. Use both noise sources and try to recognize the two contributions for different amplitudes of the two terms.
- Inject glitches at different amplitudes. Observe the addition of glitches with noise.

After spending the necessary time on this experiment, imagine possible consequences on the use of signals that you have observed for the testing of an electronic circuit.

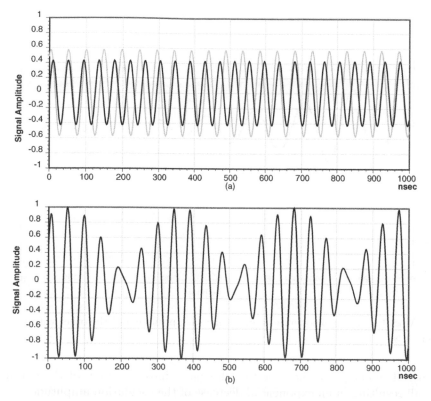

Figure 2.8   (a) The two sinusoidal components of the two tones; (b) two-tone waveform with maximum amplitude 1.

we can have

$$y'_s(t) = y_s(t) + y_{dc}, \qquad (2.8)$$

where y_{dc} is the shift. Often this is called an *offset*.

Another type of signal that is useful to know is the complex exponential, a generalization of the sine wave. It is defined by

$$y_{ex}(t) = A_0|e^{st}|, \qquad (2.9)$$

where $s = \sigma + j\omega$ is a complex variable called *complex angular frequency*. Since the complex argument makes the exponential complex and a signal is expected to be real, equation (2.9) uses the modulus. As will be known from basic math courses, if $s = j\omega$ equation (2.9) becomes the exponential representation of the sinusoidal signal $A_0 \sin(\omega t)$. Therefore, $\sigma = 0$ makes the amplitude constant, while for $\sigma > 0$ the exponential amplitude grows in the course of time. On the other hand, for $\sigma < 0$ the signal falls to zero. Figure 2.9 shows the plot of two exponential signals, one with positive and the other with negative σ. Notice that the plots use different values of A_0. Obviously, if $\sigma = 0$ we have a regular sine wave.

In the real world we have many examples of exponential signals. One is the oscillations of the Tacoma bridge, which collapsed in 1940 because of a 67 kilometer per hour (42 mph) wind. The wind caused resonance at the natural structural frequency, and the oscillations increased until the bridge collapsed. Another example is the vibration of a guitar string that receives energy from a pluck and starts oscillating in the manner of a sine wave (or the sum

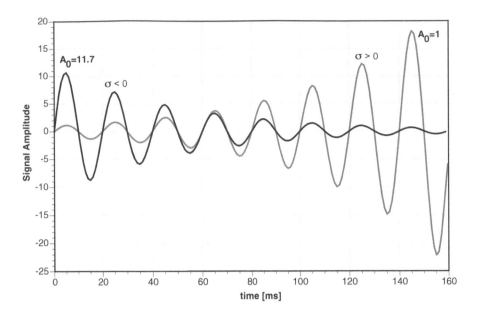

Figure 2.9   Modulus of two exponential signals, one with positive and the other with negative value of the real part of s.

of sine waves). The energy of the vibration dissipates into the body of the guitar and the atmosphere, all resulting in an exponential decrease of the oscillation amplitude.

An obvious feature of sinusoidal waveforms is that they change over time with a defined finite slope; the maximum rate of a sine wave occurs at the zero crossing, where the time derivative

$$\left.\frac{dy_s}{dt}\right|_{Max} = A_0\omega_0 \tag{2.10}$$

equals the product of amplitude and angular frequency.

For studying electronic circuits we also use the *step signal*. This, as shown in Figure 2.10(a), changes its amplitude sharply over time. The mathematical function $U(t)$, or the step function,

$$U(t) = \begin{cases} 0 & \text{for } t < 0, \\ 1 & \text{for } t \geq 0, \end{cases} \tag{2.11}$$

is normally used to represent the signal shown in Figure 2.10(a), expressed as

$$y_{step}(t) = B_0 U(t - t_0), \tag{2.12}$$

which is zero for $t < t_0$ and has amplitude B_0 for $t \geq t_0$. By adding to the signal a constant term, or offset, one obtains a pulse whose sharp transition is from a level that is not ground. Moreover, if B_0 is negative, we have a downward step.

Figure 2.10(b) shows a *pulse signal* with amplitude B_0 starting at time t_0 and lasting for interval T_0. Therefore, before $t = t_0$ and after $t = (t_0 + T_0)$ the pulse amplitude is zero, and it is B_0 in between. This pulse, too, can be described by the mathematical function $U(t)$. It is the combination of two steps

$$y_{pulse}(t) = B_0[U(t - t_0) - U(t - t_0 - T_0)]. \tag{2.13}$$

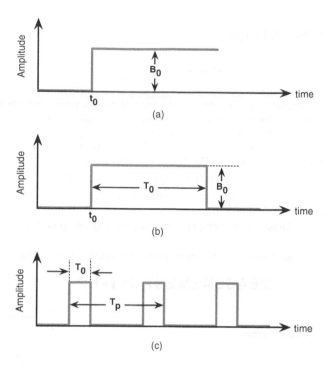

Figure 2.10   (a) Step signal; (b) pulse signal; (c) periodic pulse.

The first pulse determines a rise and the second one, with equal amplitude, cancels the first, causing a fall and a return to zero.

Often, a pulse is not a just single shot but occurs periodically with constant rate, repeating itself again and again. The distance between two successive pulses is the period of the periodic signal, T_p. Figure 2.10(c) shows the periodic version of Figure 2.10(b) (with a different time scale). The ratio between the duration of the pulse and the period is called the *duty cycle*

$$\Delta = \frac{T_0}{T_p}. \tag{2.14}$$

When the duty cycle is $1/2$, the sequence of pulses is called a *square wave*. The name is used because the period of the pulse with the low value equals that with the high value, so they are like the sides of a square. For duty cycles other than $1/2$, the sequence is called a *rectangular wave*. The inverse of the period T_p is the pulse frequency $f_p = 1/T_p$, measured in Hz. In electronics the frequencies of sine waves or rectangular waves can be very high; instead of using Hz, multiples are preferred. They are denoted by the prefixes k, M, G and T, whose values are given in Table 2.1. The table also provides the corresponding submultiples.

Observe that each of the signals described so far is defined by a few parameters; for example, the pulse (single shot) is described by the amplitude B_0 in Figure 2.10, the starting time, t_0, and the pulse duration T_0. Therefore, to define a signal it is necessary to provide the relevant parameters – the three given here for a pulse.

COMPUTER EXPERIMENT 2.3

Generation of Real Pulses

A printed circuit board generates a repetitive sequence of pulses under the control of a square wave generator. The rising edges of the square wave start the rising part of the pulse, then there is a defined steady part followed by the falling edge. Possible ringing or dumping gives a rising and falling edge. The board provides two versions of the pulse: one is the repetitive sequence, the other is an expanded in time picture for better observation of features. You can:

- set the frequency of the square wave to change the repetition rate;
- change rise time and fall time of the pulses and the duty cycle: the fraction of the duration of the pulse and the repetition period;
- determine an initial delay and the time uncertainty (or jitter) that possibly affects the rise and fall transition times;
- define the smoothness (ringing or dumping) of the rise and fall transitions.

MEASUREMENT SET–UP

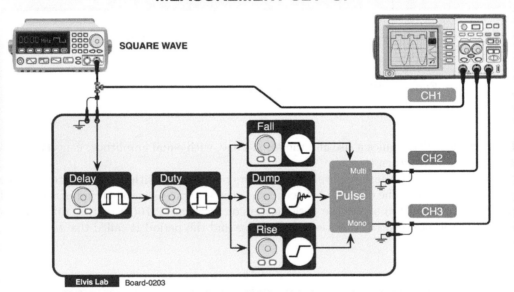

TO DO

- Set the frequency of the square wave generator, set to zero jitter and send to zero rise and fall times, and observe the pulse with various duty cycles.
- Change the rise and fall time and observe the single and repetitive pulse.
- Set the jitter control to half scale and observe the pulse at the lower and higher frequency. Change jitter and delay, to underline the difference.
- Set rise and fall times to non-zero value and observe the waveform with various dumping factors. Remember that those situations are encountered in real circuits.
- Set jitter and rise time, and suppose that an event occurs at the mid-crossing of the amplitude. Estimate the crossing time error for different settings.

Table 2.1   Scaling factors used in electronics

Prefix	Factor	Prefix	Factor
k (kilo)	10^3	m (milli)	10^{-3}
M (Mega)	10^6	µ (micro)	10^{-6}
G (Giga)	10^9	n (nano)	10^{-9}
T (Tera)	10^{12}	p (pico)	10^{-12}
		f (femto)	10^{-15}
		a (atto)	10^{-18}

Remember

An ideal step or an ideal pulse would require a sudden transition of the signal amplitude. This is impossible to obtain with electronic circuits because the speed is never infinite. The delay and the rise and fall time must be accounted for every time a pulse is used.

In addition, we observe that real circuits do not produce ideal signals but only a rough representation of them. Moreover, some signals can be generated more easily than others. For example, a sine wave is not difficult to obtain. However, rough circuits that produce sine waves can be defective and cause distortion, which is responsible for harmonic components. Fortunately, as we shall see shortly, a bandpass filter removes the harmonics and brings the signal close to the ideal version. At this stage we do not need to know what a bandpass filter is. It is enough to be aware that good sine waves are not difficult to generate.

On the other hand, steps or pulses are not easily obtained because a sudden change of amplitude would require an infinite speed of operation. Therefore, a pulse generator produces waveforms that approximately look like the one shown in Figure 2.11. The real signal has a delay, τ_d, with respect to the ideal rise; this is the reaction time of the circuit after receiving the trigger signal. Also, the pulse goes up in a finite period of time, τ_r (the rise time); then the duration of the pulse (measured as shown in the figure) can be affected by an error, ΔT_0, and the fall takes some time, τ_f, to take the pulse down to zero.

The view is still an approximation because the plots of Figure 2.11 are discontinuous in the derivative, and this is not realistic. However, for many cases the approximate plot of Figure 2.11 is adequate for studying circuits. Nevertheless, it is important to keep in mind that, even after accounting for a finite rise and fall time and recording the signal propagation delay, what is used is still an approximation of the reality.

2.3   TIME AND FREQUENCY DOMAINS

The previous section analyzed signals whose amplitude changes with time. The relevant features have been studied in the so-called *time domain*. Another way to describe signals

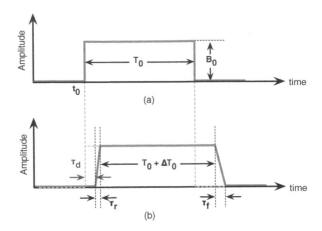

Figure 2.11   (a) Ideal pulse; (b) representation of the real limitations, delay and finite rise and fall times.

and study them implies the use of the *frequency domain*. The transformation that allows us to move from the time domain to the frequency domain is the mathematical tool called the *Laplace transform* or the equivalent *Fourier transform*.

The time domain uses the variable time, t; the Laplace transformation domain employs the complex variable $s = \sigma + j\omega$, called "generalized frequency," already used in exponential signals. The difference between the Laplace and Fourier transformations is that the Fourier domain is a particular part of the domain of the Laplace domain: Fourier uses the angular frequency ω, which is the imaginary part of s.

The main reason for moving from time to the frequency domain is a key feature of the derivative or the integral of exponential signals $y_{ex} = A_0 e^{st}$. The derivative,

$$\frac{d(A_0 e^{st})}{dt} = sA_0 e^{st} = sy_{ex}(t), \tag{2.15}$$

is just the multiplication of the exponential signal itself by s, which in this case works like a complex operator.

The integral of an exponential signal is

$$\int y_{ex}(t)\, dt = \int A_0 e^{st}\, dt = \frac{A_0 e^{st}}{s} = \frac{y_{ex}(t)}{s}, \tag{2.16}$$

which, again, uses a simple operator to transform the exponential signal into its integral, the complex variable $1/s$ that multiplies the signal itself.

From the above it is clear that using exponential signals simplifies the integration or differentiation task. All that is required is to multiply the signal by s or $1/s$ respectively. Therefore, in the s domain the derivative will become $sY_{ex}(s)$ and the integral is $Y_{ex}(s)/s$. Accordingly, exponential signals in the s domain overcome the difficulty of handling derivatives or integrals, because equations with integrals and derivatives become linear algebraic equations in the variable s (or, when using the Fourier transform, ω).

This feature is attractive, but it operates only with linear equations. The Laplace and Fourier transformations are, in fact, linear operators, and their use is not effective with non-linear systems. Actually, a relevant property is that the Laplace transform of a linear combination of time-variant signals,

$$x(t) = a_1 x_1(t) + a_2 x_2(t) + a_3 x_3(t) + \cdots, \tag{2.17}$$

with constant coefficients $a_1, a_2, a_3, \ldots$, is simply

$$\mathcal{L}[x(t)] = a_1 \mathcal{L}[x_1(t)] + a_2 \mathcal{L}[x_2(t)] + a_3 \mathcal{L}[x_3(t)] + \cdots. \tag{2.18}$$

Moreover, if signal processing is done by linear equations (i.e., we use a linear electronic circuit), the linear combination of many inputs gives at the output the linear combination of outputs produced by using the inputs one by one. Therefore, if

$$y(t) = f_{Lin}[x(t)] \tag{2.19}$$

$$y_1(t) = f_{Lin}[x_1(t)], \quad y_2(t) = f_{Lin}[x_2(t)], \quad y_3(t) = f_{Lin}[x_3(t)], \ldots, \tag{2.20}$$

then

$$\mathcal{L}[y(t)] = a_1 \mathcal{L}[y_1(t)] + a_2 \mathcal{L}[y_2(t)] + a_3 \mathcal{L}[y_3(t)] + \cdots. \tag{2.21}$$

Key point

The study of a circuit in the s (or ω) domain can be done only if the circuit is linear. For linear circuits we can use the superposition principle, which enables us to determine the responses of a system to elementary inputs and to superpose their effects to obtain a response as the superposition of inputs.

You may know that the Laplace transform is a mathematical operation that decomposes a signal into a finite or infinite superposition of exponential terms. It is defined by

$$\mathcal{L}[y(t)] = Y(s) = \lim_{T \to \infty} \int_{-T}^{T} y(t) e^{-st} \, dt. \tag{2.22}$$

The above transformation is said to be bilateral. If the lower integral limit is set to 0, the transformation is called unilateral.

Notice that the dimension of s is the inverse of time, and the dimensions of $Y(s)$ are those of $y(t)$ multiplied by time (or divided by the dimension of s). Therefore, it is necessary to integrate $Y(s)$ over any complex frequency interval in order to obtain the dimensions of $y(t)$.

The Fourier transformation produces a decomposition similar to that of the Laplace but uses $j\omega$ rather than s. It is defined by

$$\mathcal{F}[y(t)] = Y(j\omega) = \lim_{T \to \infty} \int_{-T}^{T} y(t) e^{-j\omega t} \, dt. \tag{2.23}$$

Therefore, the Fourier transform decomposes $y(t)$ into sine waves whose amplitude is given by the modulus of the complex number $Y(j\omega)$ and whose phase equals the phase of $Y(j\omega)$. The function $Y(j\omega)$ (or simply $Y(\omega)$) is called the *frequency spectrum*.

The frequency spectrum can give the signal back in the time domain. This is done by the inverse of the Fourier transform,

$$y(t) = \mathcal{F}^{-1}[Y(\omega t)] = \lim_{\Omega \to \infty} \frac{1}{2\pi} \int_{-\Omega}^{\Omega} Y(\omega)e^{j\omega t}\,d\omega, \tag{2.24}$$

which shows that a signal in the time domain is the integral of its shaped Fourier transform over an infinite angular frequency interval.

It can be useful to know how signal power spreads over the frequency. The power of a time-variant signal is the integral over time of its square. We estimate the power in the frequency domain using Parseval's theorem, which states that

$$\int_{-\infty}^{\infty} y(t)^2\,dt = \frac{1}{2\pi} \int_{-\infty}^{\infty} |Y(\omega)|^2\,d\omega. \tag{2.25}$$

If the signal is a pure sine wave $A_s \sin(\omega_s t + \varphi_s)$, its spectrum is a delta at ω_s with signal power concentrated at that frequency; the modulus of the spectrum is $A_s \delta(\omega - \omega_s)$ and the power is that of a sine wave with amplitude A_s, $A_s^2/2$.

For a two-tone signal the spectrum is made up of two deltas at the frequencies of the tones, with amplitudes equal to the amplitude of the tones. Multi-tone signals produce many lines with proper amplitudes. If a signal is periodic with a given periodicity t_s,

$$y(t + kt_s) = y(t), \quad k = 0, \pm 1, \pm 2, \ldots, \tag{2.26}$$

its spectrum is made of delta lines at $f_s = 1/t_s$ and their multiples. Depending on the signal waveform, we can have an infinite number of discrete components. In all the above cases the spectrum is made of delta lines with power concentrated at well-defined frequencies. Obviously, if there is an infinite number of components, since the total power is finite the power of components will become less and less, until at a given point it is infinitesimal.

For non-periodic signals the spectrum is a continuous one, spread over the entire frequency axis. A typical non-periodic signal is the noise, and, actually, its spectrum is a continuous one with random amplitude. Therefore, for non-periodic signals the spectrum is made up of infinitesimal amplitudes of a continuum of sine waves. In the infinitesimal angular frequency range $(\omega, \omega + d\omega)$ the spectral amplitude is $Y(\omega)\,d\omega$. The result is that the addition of sinusoidal components and a "spread over frequency power" has a spectrum, $Y'(\omega)$, made up of the addition of spectra of the two terms: a sequence of deltas and a continuous part, $Y(\omega)$,

$$Y'(\omega) = \sum_i A_{\omega_i} \delta(\omega - \omega_i)e^{j\phi_i} + Y(\omega), \tag{2.27}$$

where ϕ_i is the phase of the sine wave component at ω_i.

The plot of that kind of spectrum is problematic because of the mix of deltas and function. One solution is not to use deltas but to use pulses with finite extent, equal to the sine wave amplitude at that frequency. This is practical but can cause misunderstandings. A better method is to divide the frequency interval into small intervals (called *channels*) whose extent is, say, $\Delta\omega$. Then the power in each channel, $|Y_i|^2$, is estimated. The result is a series of amplitudes that give the power spectrum in a discrete manner. The plot is made up of the amplitudes $|Y_i|^2$ but the square root of the power spectrum, $\sqrt{|Y_i|^2}$, can also build it.

Since we use the power of $y(t)$, the value in the infinitesimal frequency range $(\omega, \omega + d\omega)$ is $|Y(\omega)|^2\,d\omega$ and the power within a generic ith channel is given by

$$|Y_i|^2 = \int_{\omega_i - \Delta\omega/2}^{\omega_i + \Delta\omega/2} |Y(\omega)|^2\,d\omega, \tag{2.28}$$

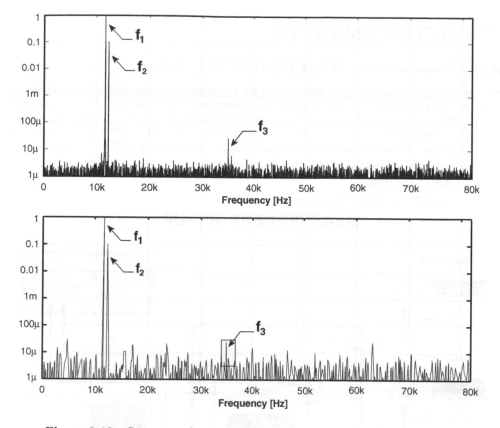

Figure 2.12 Spectrum of noisy signal taken with two different bandwidths.

where ω_i is the center of the ith channel. When the channel is tone free and $\Delta\omega$ is small enough, the amplitude $|Y_i|^2$ can be approximated by

$$|Y_i|^2 = \Delta\omega|Y(\omega_i)|^2. \tag{2.29}$$

If a channel includes a sinusoidal tone whose amplitude is A_{ω_i} and the channel extent, $\Delta\omega$, is small, the power in the channel is dominated by the tone

$$|Y_i|^2 = \frac{A_{\omega_i}^2}{2}. \tag{2.30}$$

Therefore, reducing $\Delta\omega$ diminishes the power amplitudes inside the channels that do not accommodate tones (equation (2.29)) but leaves the power of the channels that incorporate one or more tones almost unchanged.

The method discussed above is what is done by a spectrum analyzer, an instrument used to display input signal spectra. It uses a bandpass filter, whose center frequency sweeps around the frequency of interest. The bandpass filter width corresponds to $\Delta\omega$. It is important to be able to distinguish between a tone and the floor caused by noise (like the case shown in Figure 2.12). In the case of the top figure the band of the filter is low enough to make the power of the tone at f_3 dominant. If the condition is not verified it is necessary to diminish the filter bandwidth. The figure shows the spectra with two different filter bandwidths. The top

COMPUTER EXPERIMENT 2.4

Superposition of Two Sine Waves

With this computer experiment you can analyze the superposition of two sine waves provided at the inputs of a test board. The circuit makes the addition but adds noise with a white spectrum. You can observe the result both in the time domain and in the frequency domain. The oscilloscope makes the observation in the time domain. The instrument that performs the measurement in the frequency domain is named spectrum analyzer. You can:

- change amplitude and frequency of the two sine waves;
- change the level of noise;
- set the filter band of the spectrum analyzer.

MEASUREMENT SET–UP

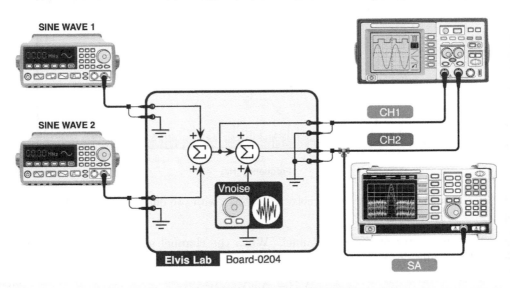

SPECTRUM ANALYZER

This experiment uses a spectrum analyzer for the first time. The instrument displays the signal spectrum, as the name suggests, by using a bandpass filter with given bandwidth whose central frequency sweeps along the observed frequency range. The display represents the power of the signal that falls inside the band-pass interval. If the filter band increases, the power of sine wave terms does not change but the power of the noise increases.

TO DO

- Set the non-ideal parameters to zero and observe "ideal" signals on the scope.
- Choose some value for the noise and reduce the sine wave amplitudes until you are not able to distinguish between noise and signal. Change the frequency of the sine waves to see if you are able to note some changes.
- Set the noise amplitude to zero and observe the effect in the time and the frequency domain. Change also the frequency of the signals.
- Change the spectrum analyzer filter bandwidth. Observe what happens to noise floor and tones representing the input sine waves.

diagram, which uses $\Delta f = 1$ Hz, shows the two tones at around 12 kHz and also the small tone around 36 kHz. That spur is well visible, being above the peaks of the noise floor. When Δf is reduced to 100 Hz (bottom diagram) the peaks of the noise (less detailed than those of the other spectrum because of the lower bandwidth) become comparable with the small tone in the 36 kHz region.

Suggestion

The filter bandwidth used to measure a signal spectrum (or to calculate it) must be such as to bring the noise floor level at least 10 times lower than the amplitude of the smaller tone you want to detect.

2.4 CONTINUOUS-TIME AND DISCRETE-TIME SIGNALS

When we think of a signal we often suppose that it is a continuous-time waveform; i.e., we assume that the signal holds continuously, and is always available to us (or the system) at will. That is an obvious way to regard signals, but it is not the only possible case, because we have a class of processing systems that instead of using a continuous representation of information utilize a discrete depiction made up of *samples*.

We have many examples of systems that work with samples in everyday life: one is measuring patients' temperature in hospitals. For a large number of patients their temperature is taken periodically, say every eight hours, so as to collect a limited number of samples per day. The reason for this procedure is that the information conveyed by a few samples per day is considered enough by the physician, who knows that body temperature (except in critical cases) does not change sharply, and that in only a few cases is it necessary to have more frequent monitoring. Obviously, in the intensive care room it may be necessary to use continuous-time monitoring.

Now, if you think of the reason why the patient's temperature is taken by using samples, you realize that is a matter of change rate. If the patient's temperature is almost constant and eventually drifts slowly, it does not make sense to measure it every minute; it would be a waste of resources (and also it would annoy the patient). If the physician expects a slow drift, then the sampling at a suitable rate is sufficient for predicting the value at any time between two successive samples. For example, the doctor can estimate the temperature value between two samples using, let us say, linear interpolation. If the temperatures at times t_1 and t_2 are T_1 and T_2 respectively (Figure 2.13), then the temperature at time t ($t_1 < t < t_2$) is approximated by

$$T(t) = T_1 + \frac{T_2 - T_1}{t_2 - t_1}(t - t_1), \quad t_1 < t < t_2. \tag{2.31}$$

The availability of three successive samples provides better information on the trend and possibly allows a quadratic interpolation (see Figure 2.13(b)). More samples are more informative, but at a given level the improvement in signal estimation becomes small. Therefore, only when you want to make a very precise estimation of a signal that is expected to change sharply does it make sense to use many samples and perform complex (and expensive) signal processing. Moreover, we have to notice that, as a side advantage of using many samples, if single samples are not very precise because they are affected by noise, processing can improve the accuracy (the simplest way is averaging many samples of a slowly changing but

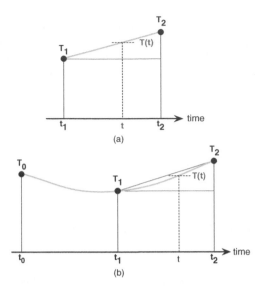

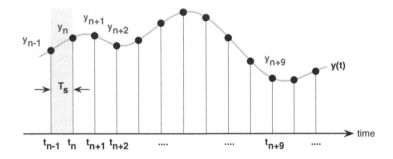

Figure 2.13   (a) Linear interpolation between two samples; (b) quadratic interpolation with three samples.

Figure 2.14   Continuous-time signal, $y(t)$, and its sampled-data counterpart, y_n.

noisy signal). The same can be obtained in a continuous-time system, but the processing is more complicated.

Figure 2.14 shows a continuous-time signal and its discrete-time equivalent. There is a sequence of discrete values whose sampling rate, f_s, is the inverse of the time between two successive samples, T_s. The rate f_s is normally constant, and this is good because handling a non-uniform series of samples is not easy. Moreover, mathematical tools for studying and processing sampled-data sets are typically available for uniform samples.

As mentioned, it is sometimes worth using high sampling rates to improve accuracy, but normally it is convenient to use a sampling rate that makes sense for the rate of change of the signal. Figure 2.15 shows a sampled-data signal represented by an excessive set of samples. It can be verified that, without limiting the effectiveness, the series can be reduced to one out of k samples (one kept, the other $k - 1$ discarded). This sampling rate reduction is called *decimation* and the parameter k is called the *decimation factor*.

An advantage of the discrete-time representation is that signal processing can be simplified. Try to estimate the average value of a waveform over the time interval $t_1 \cdots t_2$. For a

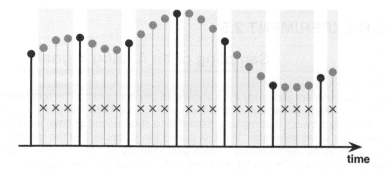

Figure 2.15   Decimation by 4: dark samples are kept and light ones are discarded.

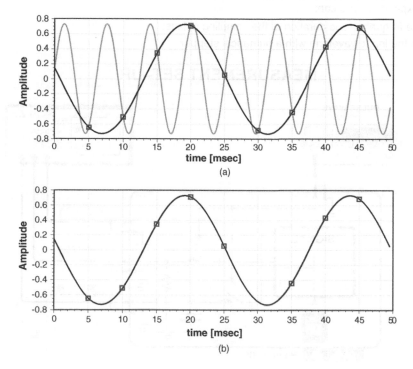

Figure 2.16   (a) Continuous-time sine wave and its sampled version; (b) same samples of another continuous-time sine wave at lower frequency.

continuous-time signal it is necessary to calculate

$$\bar{y} = \frac{1}{t_2 - t_1} \int_{t_1}^{t_2} y(t)\, dt. \tag{2.32}$$

On the other hand, with M samples in the time interval $t_1 \cdots t_2$, (the first sample is at t_1 and the last is at t_2), the average signal is simply

$$\bar{y} = \frac{\sum_0^{M-1} y(t_1 + kt_s)}{M}. \tag{2.33}$$

COMPUTER EXPERIMENT 2.5

Sampling of an Analog Signal

With this computer experiment you study the sampling of analog signals. The test board contains a generic waveform voltage generator and two circuits that perform sampling in two different ways. They perform like a switch that remains closed for a short period of time, as defined by the duty cycle of the pulse wave generator. The first output (on Channel 2 of the scope) brings the signal to zero as soon as the switch opens. That is performed by using a resistance connected to ground. The second output (on Channel 3) remains at the level of the sampling time because the capacitor operates as storing element. Therefore, the output tracks the input during a defined period and, in one case, holds the signal.

With this experiment you can:

- set the frequency and duration of the sampling pulses;
- select the input waveform with a numerical stepper.

MEASUREMENT SET–UP

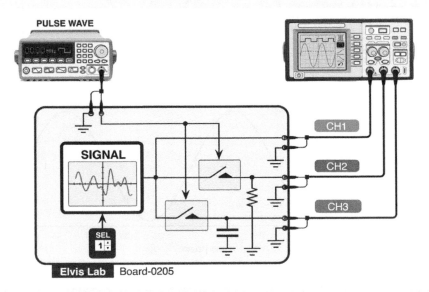

Elvis Lab   Board-0205

TO DO

- Set the frequency of the pulse generator to 10 MHz with minimum pulse duration to observe a sampled signal and its sample-and-hold counterpart.
- Change the pulse duration until it reaches 50% of the period of the pulses and notice the difference between "tracking" and "sampling".
- Set the duration of the pulse back to the minimum and reduce the sampling frequency. Observe that below a given sampling frequency you lose significant details of the input signal. Change the input waveform and notice the difference in results with smooth and busy signals.
- Set the pulse duration to 90% and change the sampling frequency to observe the same effect on tracked signals.

Comparing equations (2.32) and (2.33) shows some benefits in the sampled-data approach. It uses additions instead of integrals. Also, division by the integer M is easier than measuring $(t_2 - t_1)$ and using it in an electronic circuit that obtains division.

Until now we have assumed that the right sampling period is enough for following the rate of change of the signal. That is reasonable but it is only a qualitative proposition. Indeed, the sampling rate is important, and its value must be chosen properly.

To understand that the issue is relevant, look at the diagram in Figure 2.16(a). The thick line is a sine wave with period 6.25 ms sampled every 5 ms. It is evident that the samples are not sufficient because between two successive samples the signal makes large swings not outlined by the sequence. Obviously, that series, which does not show important features, is misleading. As a matter of fact, the samples of Figure 2.16(a) are the same as the sine wave of Figure 2.16(b). Therefore there is no way to distinguish between the signals of Figure 2.16(a) and Figure 2.16(b) (the lighter line in Figure 2.16(a)) with 5 ms rate. On the other hand, a rate of 2.5 ms would have displayed the difference. The effect just observed is called *aliasing*; it will be discussed shortly.

2.4.1 The Sampling Theorem

In order to preserve information the sampling rate must be a fraction of the period of the fastest sine wave composing a signal. Beyond this approximate rule, there is a formal study of the problem that leads to the sampling theorem, a fundamental tenet in the field of information theory. The theorem was first formulated by Harry Nyquist in 1928, but was formally proved by Claude E. Shannon in 1949. The sampling theorem it also called the Nyquist theorem or the Nyquist–Shannon theorem. It states that if the sampling frequency of a continuous-time signal is higher than twice the highest frequency of the sampled signal, then the sampled-data version completely preserves the features of the continuous-time signal.

Mathematically, the theorem is a statement about the Fourier transform of the continuous-time signal, $y(t)$. It states:

If a signal $y(x)$ has a Fourier transform $\mathcal{F}[y(x)] = Y(\omega) = 0$ for $\omega > \omega_B$, then it is completely determined by giving the value of the function at a series of points spaced $T_s = 1/(2\omega_B)$ apart. The values $y_n = y(nT_s)$ are called samples of $y(x)$.

The minimum sample frequency that allows reconstruction of the original signal, $2\omega_B$ samples per unit distance, is known as the Nyquist frequency (or Nyquist rate). The maximum time between samples, T_s, is called the Nyquist period.

The Nyquist theorem does more than just establishing the rule of having at least two samples per sine wave period. It also implicitly says that the maximum frequency that makes sense is the Nyquist frequency, $f_N = 1/(2T_s)$. In fact, the samples of an input sine wave at a frequency higher than f_N by some amount, say $\Delta f < f_N$, are indistinguishable from the samples of a sine wave at $f_N - \Delta f$ (as is the case for Figure 2.16). Therefore, the input frequency can be supposed to be lower than f_N, inside the base-band – also called the first Nyquist zone (0 to f_N). But, also, the input can be supposed to be the series of samples of a sine wave in the frequency interval (f_N to $2f_N$), the second Nyquist zone, which in turn is indistinguishable from the samples of a sine wave in the third zone ($2f_N$ to $3f_N$), and so forth.

The above can be viewed as folding the spectrum in continuous time into parts equal to the Nyquist interval. The second band is folded once. The third band folds twice (or does not fold), and so forth. To summarize, the even bands fold into the first Nyquist zone and the odd

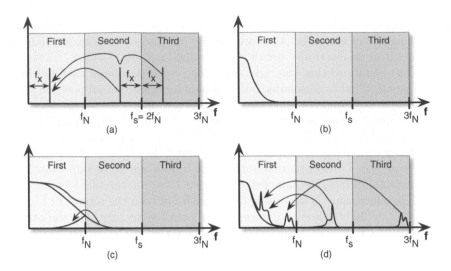

Figure 2.17   Spectra depicting various aliasing effects.

bands shift into the first Nyquist zone. Figure 2.17(a) depicts the effect for an input sine wave. If its frequency is in the second Nyquist zone at f_x distance from the sampling frequency, f_s, the spectral line goes at f_x, being folded once. If the frequency is $f_s + f_x$ the double folding gives rise to a line at the same frequency f_x in the first Nyquist zone.

Figure 2.17(b) shows a signal with spectrum entirely contained in the first Nyquist interval. The sampling leaves the spectrum unchanged because the Nyquist condition is verified. On the other hand, the case of Figure 2.17(c) has part of the spectrum in the second Nyquist zone. Its tail is mirrored in the first Nyquist interval, and, unfortunately, the superposition of frequency components alters the original spectrum. The situation depicted in Figure 2.17(d) shows a signal with two spur spectra, one in the second and the other in the third zone. The interferences folded into the base-band change the initial spectrum significantly. The spur in the second zone shows up folded, whereas the one in the third zone is just shifted (folded twice).

To avoid the aliasing that, as shown, corrupts signal information it is necessary to ensure

- that the continuous-time signal, before being sampled with a sampling frequency $f_s = 2f_N$, is band-limited with the highest spectral component less than f_N;

- that there are no spurs or interfering signals above f_N – these spurs are not part of the signal that carries good information but can possibly be added to it before sampling.

To make sure that the two conditions are verified, systems use an analog low-pass filter before the sampler. This filter, called an *anti-aliasing filter*, rejects all the frequency terms above the Nyquist limit. Since the anti-aliasing filter ideally removes all the spectral components, it may happen that, in addition to spurs, it eventually rejects part of the signal. This can be problematic, but certainly it is better to partially alter the signal than to suffer folding of the tail of the input spectrum in the base-band.

A final remark is that the sampling of a sine wave at the limit of the Nyquist range produces just two samples per period. These two samples, according to the Nyquist theorem, should be able to preserve the information. However, they are equal and opposite, to an extent that

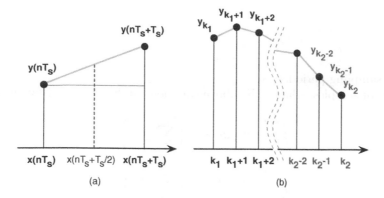

Figure 2.18   (a) Estimation of derivative with sampled-data signal; (b) estimation of integration with sampled-data signal.

depends on both amplitude and phase. Obviously, only one piece of data, the positive or negative amplitude of the samples, is not enough to extract the two variables, amplitude and phase; therefore the exact theoretical limit is not usable in practice.

2.5 USING SAMPLED-DATA SIGNALS

The derivative and the integral are, as is well known, operations that require us to perform a limit or to use infinitesimal quantities. However, the limit is not obtainable with sampled-data signals because samples are at a finite time distance. Infinitesimal quantities are not available, because, again, the time distance between samples is finite. Linear operations such as addition, subtraction, or multiplication by a constant can be done exactly with sampled-data signals. On the other hand, derivatives and integrals cannot be obtained precisely. It is only possible to approximate them. For the derivative, we use incremental ratios with samples at the lowest time separation to mimic an incremental ratio. There are three possibilities:

$$\left.\frac{dy(nT_s)}{dt}\right|_{nT_s} \simeq \frac{y(nT_s + T_s) - y(nT_s)}{T_s} \tag{2.34}$$

$$\left.\frac{dy(nT_s)}{dt}\right|_{nT_s + T_s} \simeq \frac{y(nT_s + T_s) - y(nT_s)}{T_s} \tag{2.35}$$

$$\left.\frac{dy(nT_s)}{dt}\right|_{nT_s + T_s/2} \simeq \frac{y(nT_s + T_s) - y(nT_s)}{T_s}. \tag{2.36}$$

They use the incremental ratio to assign the approximate value of the derivative to the left, to the right or at the median time of the interval $[nT_s, \, nT_s + T_s]$ (Figure 2.18(a)).

The approximation of the integral is obtained by analogy with the geometrical depiction of integration: that is, the area of the zone below the $x(t)$ plot. With samples at finite times the best way is to sum the trapezoids identified by successive samples of the signal (Figure 2.18(b)).

The result is

$$\int_{k_1 T_s}^{k_2 T_s} y(t)\, dt \simeq T_s \left\{ \sum_{i=k_1+1}^{k_2-1} y(i) + \frac{y(k_1) + y(k_2)}{2} \right\}, \tag{2.37}$$

where T_s is the sampling period.

If the number of samples is large, the above equation can be simplified by one of

$$\int_{k_1 T_s}^{k_2 T_s} y(t)\, dt \simeq T_s \left\{ \sum_{i=k_1}^{k_2-1} y(i) \right\}, \tag{2.38}$$

$$\int_{k_1 T_s}^{k_2 T_s} y(t)\, dt \simeq T_s \left\{ \sum_{i=k_1}^{k_2} y(i) \right\}, \tag{2.39}$$

which either include or exclude the last sample.

Obviously the use of approximate equations leads to approximate results. Thus, if signal processing needs to use derivatives and integrals, the corresponding discrete-time relationships produce approximate outcomes. This happens when a sample-data system uses a continuous-time function as its model and is required to obtain the same results. However, there are situations that do not use a continuous-time system as reference, because the processing is directly based on addition and multiplication by constant factors of delayed samples. These systems perform their processing by applying concepts that come directly from a discrete-time view of the operation.

2.5.1   The z-transform

The Laplace and Fourier transforms move a signal from the time domain to the s or the ω domain. For sampled-data signals the transformation that offers similar benefits is the *z-transform*. It is mathematically defined as

$$\mathcal{Z}[x(n)] = X(z) = \sum_{n=-\infty}^{\infty} x(n) z^{-n}, \tag{2.40}$$

which, using the complex variable z, defines the so-called bilateral z-transform. We can also define a unilateral z-transform with an equation equal to (2.40) but with the sum starting from 0 instead of $-\infty$.

The complex variable z plays a role similar to s in the Laplace domain. Actually, the definitions of the two transformations are similar. Remembering that

$$\mathcal{L}[x(t)] = X(s) = \int_{-\infty}^{\infty} x(t) e^{-st}\, dt, \tag{2.41}$$

we obtain the equivalence

$$\int_{-\infty}^{\infty} x(t) e^{-st}\, dt \;\rightarrow\; \sum_{n=-\infty}^{\infty} x(n) z^{-n}. \tag{2.42}$$

The right-hand side, which refers to sampled-data signals, uses a sum rather than an integral. The multiplying factors are e^{-st} and z^{-nT} respectively. Therefore, using discrete

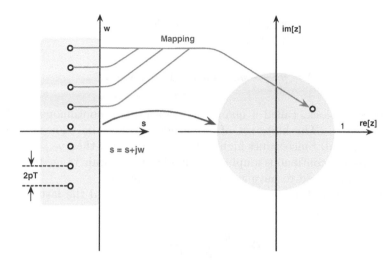

Figure 2.19   Mapping between the s-plane and the z-plane.

sampling time in the exponential ($t = nT$), the equations in (2.42) are the same, if

$$z = e^{sT}, \tag{2.43}$$

which provides a relevant link between continuous-time and sampled-data systems.

I suppose that the study of sampled-data systems will be the topic of more advanced studies. Thus you will have the chance to better understand these mathematical complications. Here we have just to outline some relevant features and acquire a background knowledge so as to be aware of mathematical tools that help in the analysis of sampled-data processors. In particular, it is useful to know about possible links between continuous-time and sampled-data systems.

Let us go back to the relationship (2.43). It is called *mapping* because it defines a link between points in the complex planes s and z. Mapping is not just a mathematical curiosity but a useful tool used to study, for example, frequency features of a sampled-data system and to control and ensure that a system is stable (stability and the required conditions are studied in a later chapter dealing with feedback).

Figure 2.19 shows relevant features of mapping. The imaginary axis $s = j\omega$ (the frequency of sine wave components) becomes the unity circle in the z-plane. Moreover, the entire left s-plane moves into the unity circle. Also, the transformation between s and z is not univocal. The infinite number of points of the s-plane with equal real parts and imaginary terms (differing by $j2\pi T$) maps just a single point of the z-plane.

The mapping equation provides an operational meaning of the operator z. Since the Laplace transform of a signal delayed by T is e^{-sT}, z^{-1} is a delay by T. Its inverse, z, is the opposite of a delay, the anticipation of a signal by the sampling period T.

2.6   DISCRETE-AMPLITUDE SIGNALS

In addition to discrete time, signals can also have a discrete amplitude. In other words, the signal amplitude is not just any value but is one of a set, $y_1, y_2, \ldots, y_N$. Typically the permitted amplitudes are equally spaced within a defined interval, called the dynamic range.

The difference between two consecutive discrete levels is called the quantization interval or quantization step, and is normally referred to as Δ:

$$\Delta = y_{i+1} - y_i, \quad \text{for } i = 1, \ldots, N-1. \tag{2.44}$$

A special processing block, called a quantizer, transforms a continuous-amplitude into a discrete-amplitude signal. The finite set of amplitudes defines the allowed dynamic range, $(y_{max} - y_{min})$, of the signal. Since values higher than y_{max} or lower than y_{min} are not expected, the *quantization* clips any continuous amplitude outside the dynamic range to the lower or the higher level. Such input is said to have an out-of-scale amplitude.

Assume that we have N discrete levels, the first $y_1 = y_{min}$ and the last $y_N = y_{max}$. The quantization step is therefore

$$\Delta = \frac{y_{max} - y_{min}}{N-1}. \tag{2.45}$$

Moreover, the quantized levels are

$$y_k = y_{min} + \Delta(k-1), \quad k = 1, \ldots, N. \tag{2.46}$$

The use of a signal with quantized amplitude seems to be a limitation because the representation of a quantity loses detail. The benefit is the ability to perform processing in a digital manner and to simplify the handling of information. However, the cost is not great because excessive accuracy is often not really necessary. For example, the age of an adult is measured in a discrete manner using the year as the quantization interval. Typically, a more exact measure of age is irrelevant. Even the use of quantization intervals such as months or weeks is an exaggeration and sounds funny; people never say, for instance, "I am 23 years, four months and two weeks old." When boiling an egg it is necessary to use the minute, and probably tens of seconds, as the quantization interval. However, finer time definition than that does not make sense, because boiling the egg for one second more or one second less does not change the final result in a perceptible manner.

Long jumps in athletics are measured in centimeters. Lower resolutions do not make sense because it is difficult to obtain better accuracy. The signal is the distance between a wooden board (possibly marked with soft modeling clay) and the mark made by competitors in a pit filled with finely ground gravel or sand. The measurement can be affected by two possible inaccuracies: detection of whether the clay was touched by the jumper and precision marking of where the jumper landed. If the sum of these two errors is of the order of a centimeter or slightly less, then the centimeter quantization is suitable and smaller Δs are irrelevant.

The above considerations indicate that the accuracy of information carried by signals depends on their specific use and possible limitations. Therefore for any application there is a correct level of accuracy and, consequently, an optimum amplitude of quantization interval.

Another element to account for is the range of variation of signals; it should be wide enough to accommodate any possible situation but not suitable for use with extremely wide ranges. For instance, using 1000 years as the full scale for the age of people is nonsense. Everybody hopes to live for a very long time, but the maximum assumed by optimists is surely lower than 256 (8 bits). Therefore, for any signal there is a suitable dynamic range, which together with the quantization interval produces a resulting number of quantization levels and thus of bits.

Amplitude	Level	Decimal	Binary
1.023		1024	1111111111
1.022		1023	1111111110
		········	········
0.326		327	0101000110
0.325		326	0101000101
0.324		325	0101000100
0.323		324	0101000011
		········	········
0		1	0000000000

Figure 2.20   Different ways to distinguish between discrete levels.

Observe that ...

discrete-amplitude (and discrete-time) signals are ideal formats for digital processing that needs clocked control and logic data. Continuous-amplitude and continuous-time signals are suitable for analog processors.

Obviously, for a fixed dynamic range, increasing the accuracy diminishes the quantization interval and augments the number of quantized levels, N, which, as is probably known, is often the power of two of a given number of bits, n:

$$N = 2^n. \tag{2.47}$$

The quantized signals are typically voltages or currents. With 10 bits, 1024 levels, and a voltage signal ranging from 0 V to 1.024 V, the quantization interval is 1 mV. Thus, for example, a continuous amplitude signal of 0.3263 mV is in the interval #327 if the 0–1 mV interval is numbered as the #1 interval. Since powers of 2 facilitate the numbering using n-bit codes, for the interval #327 the code is (0101000110), as outlined in Figure 2.20.

Figure 2.21 illustrates the transformation of a continuous-amplitude signal into a discrete-amplitude one, as obtained by a quantizer. All the magnitudes of a quantization interval are concentrated into a single level, the quantized amplitude. In Figure 2.21(a) this is the middle of the quantization interval. However, amplitudes can be concentrated towards another value, for example the top or the bottom of the quantization interval (Figure 2.21(b), (c)). As a result, the discrete-amplitude dynamic range is from the first to the last quantized amplitude but the continuous-amplitude range is a bit wider, as depicted in Figure 2.21.

With the transformation of Figure 2.21(a) (discrete level in the middle of the quantization interval) and dynamic range equal to $(0 \div 1.024)$ mV and $\Delta = 1$ mV, the quantization of an input voltage of 735.83 mV gives 735.5 mV, which is the closer discrete level. The same

COMPUTER EXPERIMENT 2.6

Quantization

This computer experiment studies the quantization, i.e., the transformation of the continuous amplitude of a sampled signal with continuous amplitude into a sampled signal with discrete amplitude. The number of bits of the quantizer determines the number of discrete levels. This test board contains two identical samplers, but in one case the clock is "perfect", in the other case jitter affects sampling and moves the sampling time backwards and forwards. If the sampling error occurs when the signal has amplitude close to a threshold of the quantizer there is an error in the result. You can:

- set frequency and delay of the pulse generator and jitter;
- choose the signal waveform with a numerical stepper;
- choose the number of bits of the quantizers (equal for both) with a numerical stepper;
- define the lower and upper level of the quantization interval (make sure that the lower value is higher than the upper one).

MEASUREMENT SET–UP

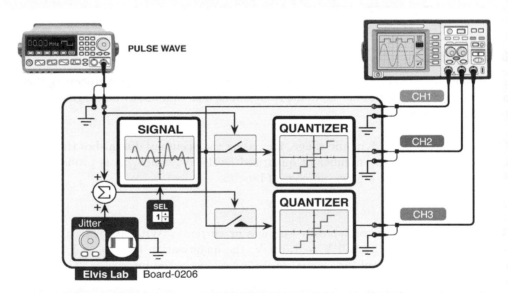

TO DO

- Set the sampling frequency at 1 MHz and zero jitter. Change the number of bits and observe the equal output signals. Slightly change the frequency to obtain different output signals.
- Change signal waveform and observe the output signal with sampling at different frequencies and delay in the sampling times.
- Set the jitter to various values and observe how that timing error alters the samples.
- Do the observations with different numbers of bits and observe the dependence of the error caused by the jitter at low and high resolution.

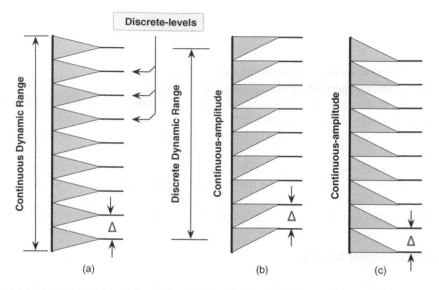

Figure 2.21   (a) Conceptual transformation from continuous-amplitude to discrete-amplitude signal – the quantization level is in the middle of quantization interval; (b) and (c) quantization level at top and bottom of quantization interval.

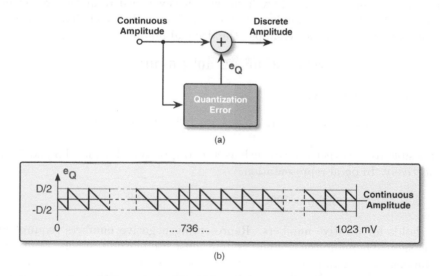

Figure 2.22   (a) Conceptual diagram of quantizer; (b) plot of quantization error versus continuous amplitude (full scale 1023 mV, quantization step $\Delta = 1$ mV).

goes for a 735.29 mV input. Therefore, to move from 735.83 mV to 735.5 mV it would be necessary to subtract 0.33 mV, while to move from 735.29 mV to 735.5 mV it is necessary to add 0.21 mV. This suggests that quantization can be viewed as the addition of a suitable quantity to the continuous-amplitude input. The added quantity is the quantization error, ε_Q.

The scheme of Figure 2.22(a) depicts a model that performs the operation. The block *Quantization Error* determines the quantized amplitude to be added to or subtracted from the sampled value of the continuous amplitude. Moreover, the quantization error amplitude is limited to the range

$$-\frac{\Delta}{2} \le \varepsilon_Q \le \frac{\Delta}{2}, \tag{2.48}$$

as the quantized level is in the middle of the quantization interval.

Figure 2.22(b) shows the plot of the quantization error as a function of input amplitude. The added error is within $\pm 0.5\Delta$ and is zero when the continuous amplitude equals the discrete amplitude. For the cases shown in Figure 2.21(b) and (c) the discrete amplitude is on one of the extremes of the quantization interval. Therefore the quantization error is only positive or negative, with range $(0, \ldots, \Delta)$ or $(-\Delta, \ldots, 0)$.

Finally, observe that the quantization error is an unavoidable error inherent in the quantization process. It becomes smaller if the quantization interval diminishes, i.e., if the number of levels increases and the dynamic range remains unchanged. Therefore more bits do reduce the quantization error, but it never gets to zero unless we use an infinite number of bits.

2.6.1 Quantized Signal Coding

We have seen that quantized signals can be represented by their sampled-data amplitude or the code, decimal or binary, numbering the quantization intervals. It is well known that decimal numbers use ten symbols and the binary ones use two symbols. The decimal and binary numbering systems are certainly well known. However, just to make sure, let me recall something basic. When writing a number the position from right to left indicates the weight, equal to the radix power. Thus, for example, the decimal number 326 means

$$326_{10} = 3 \cdot 10^2 + 2 \cdot 10^1 + 6 \cdot 10^0, \tag{2.49}$$

and the corresponding binary code is

$$326_{10} = 101000110_2 = 1 \cdot 2^8 + 1 \cdot 2^6 + 1 \cdot 2^2 + 1 \cdot 2^1. \tag{2.50}$$

It is also possible to use other radixes such as 8 (octal) or 16, which need 8 and 16 different symbols respectively. In octal representation,

$$326_{10} = 506_8 = 5 \cdot 8^2 + 0 \cdot 8^1 + 6 \cdot 8^0. \tag{2.51}$$

The above holds for positive numbers. Representing negative numbers requires an extra symbol for the sign. Therefore the decimal system must use eleven symbols: ten digits, from 0 to 9, and "minus." An extra symbol is not very expensive for decimal numbering but is not convenient for binary, because handling just two symbols is easy with digital circuits. Rather than a new symbol, an extra bit that does nothing but represent the mathematical sign can be used. Thus, for example, if the leftmost bit is "0" the sign is positive, while if it is "1" the sign is negative. For example, $0\,1001_2$ is 9_{10} and $1\,1001_2$ is -9_{10}. This method works, but it can be misleading, because 11001_2 can be confused with 25_{10} written without a sign bit. We therefore have to make clear, with the sign–magnitude system of negative binary numeration, how many bits we are using for the magnitude, and, therefore, what the largest number we will be dealing with is. For the above case, to indicate 1_{10} we have to use the notation $0\,0001_2$, and the maximum positive is $0\,1111_2 = 15_{10}$. Since the sign of zero is either positive or negative, the sign–magnitude system gives rise to two zeros, one positive and one negative, as shown, for 8 bits, in Figure 2.23. Moreover, the sign–magnitude approach is not very practical for

Decimal	Signal-Magnitude	Ones' Complement	Twos' Complement
127	0 11111111	0 11111111	0 11111111
126	0 11111110	0 11111111	0 11111111
...	....	....	....
2	0 00000010	0 00000010	0 00000010
1	0 00000001	0 00000001	0 00000001
+0	0 00000000	0 00000000	0 00000000
-0	1 00000000	1 00000000	
-1	1 00000001	1 11111111	1 00000000
-2	1 00000011	1 11111110	1 11111111
...	....	....	....
-126	1 11111110	1 00000001	1 00000010
-127	1 11111111	1 00000000	1 00000001
-128	———	———	1 00000000

Figure 2.23 Different methods for coding negative binary numbers.

arithmetic purposes, as the addition of a positive and a negative number requires a technique that is different (and more complex) from the addition of two positive numbers.

As an alternative, negative numbers can use the ones' complement system, for which negative numbers are complements of their positive counterparts. Thus, for example, the negative of $01001_2 = 9_{10}$ is $10110_2 = -9_{10}$. However, like sign–magnitude representation, the ones' complement has two different representations of 0. Even the addition of ones' complement numbers, which uses a conventional binary scheme, is not optimal because any resulting carry over must be added to the sum obtained.

There is another method for representing negative numbers, which works better with additions, the twos' complement method. For negative numbers the twos' complement uses the ones' complement of the bit pattern increased by one (Figure 2.23). In this system there is only one zero and there is also an extra digital number (-128_{10} for an 8-bit system).

2.7 SIGNALS REPRESENTATION

The coding of discrete-time, discrete-amplitude signals is useful for digital circuits that perform arithmetical operations. However, views of signals of any type do not use numerical coding, because it is difficult to perceive and is in some sense an abstruse representation. That is what happens when reading information on a car dashboard: comprehending the signals of the speedometer, tachometer, and fuel gauge from pictorial outputs is more immediate than reading the numerical values of these quantities. Therefore, we typically use plots of the values of actual amplitudes. The diagrams have a given range on the x and y axes and represent the signal with a dot at a distance from the origin that is proportional to the ratio

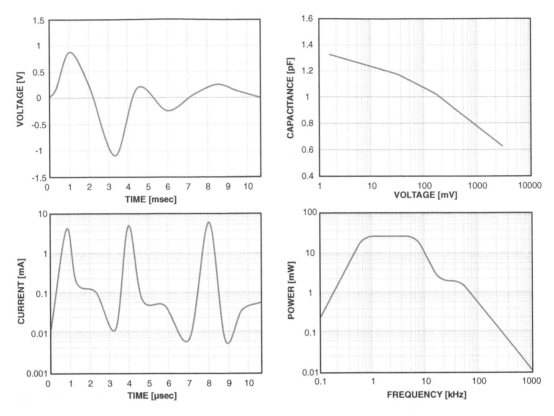

Figure 2.24   Pictorial representation of signals with linear and compressed scales on the x and the y axes. The quantity represented on the axis has (or has not) the given dimensions.

between the x and y values and the full scale. This certainly holds with linear diagrams. In many cases, however, especially when there is a big difference in signal amplitude at different operational conditions, it can be convenient to use non-linear scales. You may not know that the sinusoidal components of the sound pressure of a helicopter are 100 000 times bigger than those of a soft whisper. So if a full-scale diagram uses the sound generated by a helicopter, the softly whispered signal is scarcely distinguishable from the zero line.

In order to display very large and very small signals on the same diagram and make them both noticeable, some diagrams use compressed scales. These reduce the amplitude of large signals by following some rule, very often a logarithm or an exponential. Figure 2.24 shows four different examples of diagrams with linear and compressed scales. For example, a current with glitches that change the from 0.0035 to 4 is conveniently represented as shown in the bottom-left diagram. The others show other possible examples. In some cases the rule used does not follow a mathematical principle but is just an arbitrary compression or expansion of the scales to accommodate the waveform. This can be a good way to give the feeling of the operation but, obviously, is not good for quantitative studies since it is also misleading. Actually, this is what happens with some maps used for tourist information. Their intention is to display the center of the city and also to mark some faraway monuments. Confused by non-linear scales, people often believe that a monument is within walking distance and are disappointed to find that after hours of walking they are still not there. So using a non-linear

Table 2.2   Various symbols indicating the reference of dB

Symbol	Reference	Note
dB_V	1 V	For voltage signals
dB_{mV}	1 mV	For low voltage signals
$dB(1 V_{RMS})$	1 V	Better specifies the reference
dB_W	1 W	For electrical power
dB_{mW} (dB_m)	1 mW	For low power signals
dB_J	1 joule	For radio power or energy
dB_{FS}	Full scale	For relative comparison
dB_C	Carrier amplitude	For telecom applications
dB_{rN}	Reference noise	For signal quality measures

scale without well-defined rules is not recommended unless the diagram clearly marks how the scale distorts information.

2.7.1   The Decibel

It has been mentioned that an effective solution for representing large and small amplitudes on the same diagram is the use of a logarithmic scale. Since the logarithm is a function applied to dimensionless quantities, if the signal has dimension (such as volts [V] or watts [W]) it is necessary to transform it into a dimensionless quantity. This is done by using a normalizing constant that divides the signal. That quantity has the same dimension as the signal and a meaningful amplitude. For example, it can be 1 V, 1 mV, or 1 µV, or, for power, 1 mW or 1 µW.

Often, instead of a plain logarithm, diagrams use decibels (dB), whose definition differs for amplitude and power (i.e., the square of amplitude). The two definitions are

$$x|_{dB} = 20 \cdot \log_{10}\left[\frac{x}{x_0}\right], \quad P_x|_{dB} = 10 \cdot \log_{10}\left[\frac{P_x}{P_{x_0}}\right], \tag{2.52}$$

where the left-hand equation is for signal and the right-hand one for power; x_0 is the normalizing constant and $P_{x_0} = x_0^2$ its power. The left-hand equation (2.52) yields

$$P_x|_{dB} = x^2|_{dB} = 10 \cdot \log_{10}\left[\frac{x^2}{x_0^2}\right] = 20 \cdot \log_{10}\left[\frac{x}{x_0}\right], \tag{2.53}$$

which obtains the same dB value for the signal and its power. This is why we use two definitions of dB: to avoid confusion in measuring a signal or its power in dB, provided that we use the same normalizing constant.

As mentioned, the normalizing constant has a suitable value. Often the notation used outlines its magnitude to make clear what is represented. Table 2.2 gives the most used symbols and their specifications. Notice that communication applications, where a signal called a carrier is the reference for performance, often use dB_C: the signal measure in dB referred to the carrier.

> **Remember**
>
> A signal and its power (the square of the signal) measured in dB with the same normalizing constant have equal numerical values.

The decibel is common in acoustics, and 0 dB (the normalizing amplitude) is typically set at the human perception threshold. The reason for using dB is that the ear is capable of detecting a very wide range of sound pressures. The sensitivity of the ear is very high for extremely low signals and decreases with loud sounds. It is always surprising to learn that the softest audible signal modulates air pressure by 1 μPa (Pa refers to Pascal), while air pressure on the earth's surface is 100 000 Pa. Therefore the ear can detect a sinusoidal change of air pressure whose amplitude is 100 billion times lower than the average air pressure. It is also good to know that ear sensitivity to the amplitude of sine waves (in a given frequency range) is almost logarithmic. An exponential increase in sound pressure produces just a linear increase in feeling. But remember, for your safety, that loud rock concerts produce sound at 110 dB and the pain threshold is 120 dB. Moreover, 180 dB causes instant perforation of the eardrum.

The dB and linear scales are obviously related. It can be useful to keep in mind the relationship between the two scales. For signals 20 dB means a difference by a factor of 10, 40 dB is a 100 times difference, and so forth. For power the numerical value of dB is halved.

Table 2.3   Correspondance of dB (signal and power) and linear value

Amplification			Attenuation		
dB Signal	dB Power	Lin value	dB Signal	dB Power	Lin value
2 dB	1 dB	x 1.26	-2 dB	-1 dB	x 0.79
4 dB	2 dB	x 1.58	-4 dB	-2 dB	x 0.63
6 dB	3 dB	x 1.99	-6 dB	-3 dB	x 0.50
8 dB	4 dB	x 2.51	-8 dB	-4 dB	x 0.40
10 dB	5 dB	x 3.16	-10 dB	-5 dB	x 0.32
12 dB	6 dB	x 3.99	-12 dB	-6 dB	x 0.25
14 dB	7 dB	x 5.01	-14 dB	-7 dB	x 0.2
16 dB	8 dB	x 6.31	-16 dB	-8 dB	x 0.16
18 dB	9 dB	x 7.94	-18 dB	-9 dB	x 0.12

Table 2.3 provides amplification and attenuation coefficients with steps of 2 dB. Notice that 6 dB is almost 2; 12 dB is almost 4 and 18 dB is almost 8.

2.8 DFT AND FFT

The estimation of the frequency spectrum of analog and digital (discrete-time discrete-amplitude) signals can be problematic. We know that plain sine waves give a well-defined spectrum, a line at the sine wave frequency plus a floor caused by electronic noise or tones produced by a spur, and that with the superposition of sine waves the task is easy. The problem arises for generic waveforms that require Fourier or Laplace analysis.

For continuous-time signals we use the Fourier transform, whose definition, repeated here for the reader's convenience, is

$$\mathcal{F}[x(t)] = Y(j\omega) = \int_{-\infty}^{\infty} x(t)e^{-j\omega t}\, dt. \tag{2.54}$$

The same equation with a set of digital numbers representing a sampled-data signal gives the result. Since the set has discrete times (with a given sampling period, say T), it comes out as

$$x(nT) = \sum_{-\infty}^{\infty} \delta(t - nT)x(t). \tag{2.55}$$

Its use in (2.54) changes the continuous Fourier transformation into the discrete-time Fourier transformation

$$\mathcal{F}[x(nT)] = X^*(\omega) = \sum_{k=-\infty}^{\infty} x(nT)e^{-j\omega nT}, \tag{2.56}$$

which provides a mathematical expression for the spectrum of a sampled-data signal $x(nT)$.

Evidently, expressions (2.54) and (2.56) cannot be exactly estimated because the former equation needs the input from time $-\infty$ and the latter requires an infinite number of samples. What is done is to approximate the calculations using a reduced period of time. Moreover, for analog signals instead of performing the estimation prescribed by (2.54) it is convenient to digitize the signal and estimate the spectrum of its sampled-data counterpart.

A digital signal is a finite set of samples (which we conventionally assume as starting at $t = 0$, taken at times $nT, n = 0, \ldots, (N-1)$). Even if an infinite series is not available, it is possible to obtain a good approximation of equation (2.56) by using a series that is long enough. The result is called the Discrete Fourier Transform (DFT).

Notice that equation (2.55) requires an infinite input series. However, since only a finite set is available, we build an infinite series by duplicating available data. For this, the set extends outside the interval $0, \ldots, (N-1)T$ by supposing it is N-periodic. Thus, $x(iT)$ repeats itself again and again, becoming $x[(i + kN)T] = x(iT)$, for $0 \leq i < N$. Since the extension of the input series to being N-periodic makes the signal periodic with period NT, lines at frequency multiples of $1/(NT)$ make the spectrum. It is calculated by

$$\mathcal{DFT}[x(nT)] = X_{DFT}\left(\frac{i}{NT}\right) = \sum_{k=0}^{N-1} x(nT)e^{-j(2\pi i)}, \tag{2.57}$$

where the sum is limited to the input sequence and not to the replicas.

Observe that the first line of the spectrum ($i = 0$) corresponds to zero frequency (dc), and, since the sampling frequency of the input is $1/T$, the ($N-1$)th line is (almost) at the sampling frequency $1/T$. Therefore, if N is large enough, the lines of the DFT represent the input spectrum with good details. The result spans the base-band and the second Nyquist zone. Then, as is well known, the spectrum repeats itself in higher Nyquist zones.

All the above mathematics is certainly difficult to assimilate, and equations cannot be memorized easily. That is not really mandatory (I suppose). What is important to know is the following.

- The input signal is transformed into an N-periodic signal to estimate the DFT. This is acceptable if the length of the series is sufficient to make spectral details evident.

- A comb of lines makes the DFT spectrum. The separation of those lines is proportional to the inverse of the number of points in the series. Increasing the number of points gives a more detailed spectrum.

- Every DFT point accounts for a portion of spectrum around it, spanning $1/(2NT)$ on each side. It is like a channel of a spectrum analyzer that measures the power with given filter bandwidth. More channels reduce channel width and, therefore, diminish the power in that channel.

- If a signal, after a transient, settles to a constant level, the steady portion must be long enough before repeating the signal over and over. It may be useful to include in the series a suitably large number of constant samples to represent a settled tail well.

Equation (2.57) requires $2N$ computations per point of DFT. Therefore, for the complete spectrum $2N^2$ computations are necessary. Modern computers perform massive calculations very quickly, but for long series (many thousands of samples) computation time can be a limitation. The number of computations diminishes with special algorithms, normally referred to as Fast Fourier Transforms (FFTs). The number of operations goes down from $2N^2$ to $2N \cdot \ln(N)$, a significant advantage for large values of N. For example, with $N = 2^{14} = 16\,384$ the number of computations diminishes from more than 268 million to about 159 thousand.

About FFTs

For effective FFT calculation, use the power of two series of elements. The longer the series, the more detailed is the spectrum obtained.

Having efficient FFT algorithms is obviously desirable as they keep short the computation time with a large number of points in the series. The choice of number of points is important as, for example, the Danielson–Lanczos lemma works well with powers of two points. Other algorithms require base-4 or base-8 series of samples. There are various algorithms, among them the one used in many simulation packages that employs the *FFTW* method, which also augments its effectiveness with series made of the power of two elements.

2.9   WINDOWING

The discrete Fourier transform uses the N-periodic transformation to make a finite set of samples into an infinite periodic sequence. If the signal settles to zero or to a constant

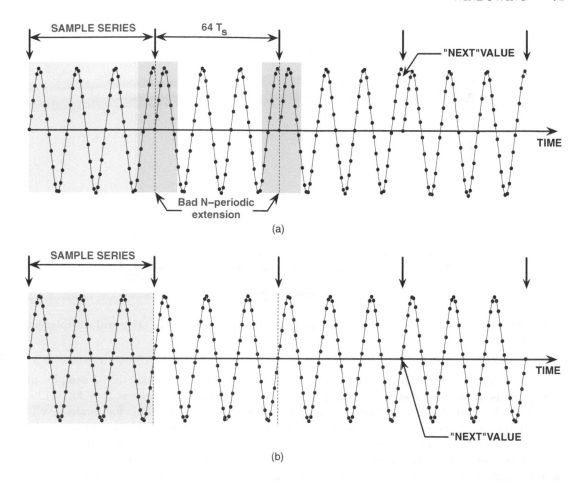

Figure 2.25 (a) Series of 64 samples of a sine wave that gives rise to a bad N-periodic extension; (b) samples of a sine wave obtained by coherent sampling.

value, many samples of the tail become equal and can seem to be useless. However, those samples (and possibly extra ones) are beneficial because they reinforce the information that the signal settled. On the other hand, when the signal is a sine wave, or more generally a periodic waveform, the series is a slot within a periodic sequence. The choice of how many points to use can cause problems because the number of elements used (normally a power of two) might give rise to bad signal representations. Consider, for example, the 64 samples of a sine wave shown in Figure 2.25(a). Since the N-periodic version of the series is a replica of the first set (sample #65 equals sample #1, sample #129 repeats #1 again, and so forth), the waveform obtained evidently differs from the original, because there are sharp transitions from an amplitude that is almost the maximum to zero. This occurs at transitions $64 \rightarrow 65$, $128 \rightarrow 129$, and at all the ends of each repeated sequence. These big jumps alter the signal features significantly; jumps also affect the spectrum, which is often poor and barely usable. What Figure 2.25(a) shows is an extreme case, because the number of points is only 64. Nevertheless, even series of 4096 points can generate significantly altered spectra.

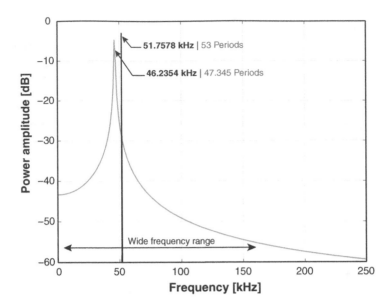

Figure 2.26   Spectra of a series of 1024 sine wave samples, one with and the other without coherent sampling.

Figure 2.25(b) shows a special case where the signal accommodates exactly 64 samples in three periods of the sine wave. The sample #65 equals #1 and the same is true for #129. Therefore, the N-periodic signal matches the original signal perfectly. This fortunate case is normally referred to as *coherent sampling* because there is a well-defined relationship between sample and sine frequency. When this happens the DFT (or FFT) spectrum is (almost) exact.

As mentioned already, coherent sampling uses an integral number, k, of sine waves equal to the total period of the series. Therefore, if the sampling frequency is $f_s = 1/T_s$ and $N = 2^n$, the sine wave frequency, $f_{sig} = 1/T_{sig}$, must satisfy the condition

$$f_{sig} = \frac{k}{NT_s}. \tag{2.58}$$

Figure 2.26 shows the FFT spectra of two series of 1024 sine wave samples. The sampling period is 1 μs ($f_s = 1$ MHz); peak amplitude of both is 0.68 and frequencies are 46.2354 kHz and 51.7578 kHz respectively. For the first frequency there are 47.345 periods in the 1024 μs time interval, while for the second frequency the number of sine waves is 53. The FFT spectrum of the latter is the expected good line typical of a pure sine wave. On the other hand, the spectrum of the first signal shows an unacceptable spread over a wide frequency range that reaches the dc and the Nyquist frequency with non-negligible amplitudes ($-60\,dB_{FS}$ and more).

Conditions (2.58) for coherent sampling constrain the usable frequencies for given sampling periods and numbers of samples. Moreover, if k is not a prime number but k_1 is its largest prime factor ($k = k_1 \cdot k_2$), the series is N/k_2-periodic, i.e., the entire series is made up of k_2 repeated sequences of N/k_2 samples. Multiplication of data by k_2 does not add information relevant for the FFT calculation, so the accuracy is affected.

Indeed, sinusoidal signals are not a true portrait of real signals because they carry negligible information (just amplitude, frequency, and phase). However, the limitation is relevant when testing circuits (for example, data converters) by the use of sine waves. Therefore it is advisable to use a prime value of k as an effective stimulus of the circuit. Moreover, the coherent sampling condition must be verified.

Follow this rule

When using the sine wave for testing spectral features of circuits use 2^N samples of a prime number of sine wave periods for the FFT.

Non-coherent sampling alters the FFT spectrum. However, coherent sampling is not always possible or even convenient. For such cases the use of windowing partially alleviates the limitations. Windowing is the preliminary multiplication of the input set by a proper function that tries to make the series N-periodic-like.

Strictly speaking the original series of samples, $x_s(nT), (n = 0 \cdots N - 1)$, is actually windowed, as it is the multiplication of an infinite series of samples $x(nT), (n = -\infty \cdots \infty)$ by a rectangle or "brick wall" window

$$x_s(nT) = x(nT) \cdot Rect(nT), \tag{2.59}$$

where the function $Rect$ is

$$Rect(nT) = \begin{cases} 1 & \text{for } 0 \leq n < N, \\ 0 & \text{otherwise.} \end{cases} \tag{2.60}$$

However, a "brick wall" window is not a good choice, because it causes sharp transitions in samples. Instead, tapering ends, as produced by other window functions, are beneficial even if the signal is altered. There are many types of window, such as the Hann, the flattop and the Blackman, that take the signal to zero at the end of the series. This ensures that the amplitudes of the first and last samples, being zero, are equal. Alternatively other windows, such as Hamming, just reduce the amplitude of the ends by a large factor (10 for Hamming).

The choice of windowing function plays an important role in determining the quality of the overall results. The use of a Hann window changes a sine wave signal as shown in Figure 2.27(a). The amplitude of the window is one in the middle of the series, and it goes to zero at the ends. The spectrum is not the expected perfect line, but, as shown in Figure 2.27(b), it is a significant improvement over the rectangular case.

Windowing operates properly when the signal used is periodic but is absolutely unsuitable for outlining spectral alteration with signals that have non-periodic features, such as transients or glitches. In particular, if a transients or a glitche occurs at the beginning or the end of a series the window significantly attenuates that feature and removes its effect in the signal spectrum. On the other hand, windowing with periodic signals can be used, even if sampling is coherent. It is beneficial when, possibly in addition to a coherent sine wave, there are spur harmonics.

Figure 2.27(a) indicates that windowing modifies signal total power. It is therefore necessary to normalize the spectrum. This is particularly necessary for quantitative estimation of amplitudes. However, what designers typically look at is not the absolute value of sine waves but the relationship between spectral components.

COMPUTER EXPERIMENT 2.7

Windowing

This computer experiment serves to demonstrate windowing. This is the tool used for avoiding bad *N*-periodic extensions and improving the spectrum of the Fast Fourier Transform (FFT). You can experience six different types of windowing functions: Rectangular, Hann, Hamming, Cosine, Blackman, Flattop. The sampling frequency is fixed at 1 MHz. The experiment allows you to:

- select the amplitude and frequency of two sine waves and the phase of the second sine wave;
- select the sampling frequency defined by the rising edge of a square wave generator;
- choose between the different windowing functions;
- choose the number of points of the input series;
- observe on the monitor the shape of the windowing functions used.

MEASUREMENT SET–UP

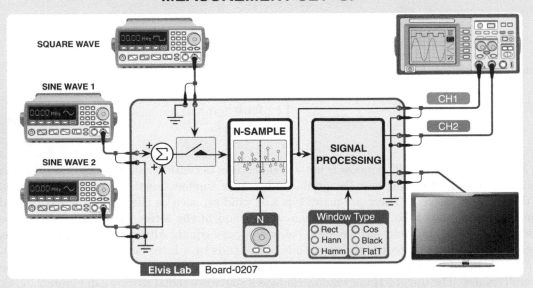

TO DO

- Set the first sine wave amplitude to zero and the number of points of the input series to $N=2^{10}$. Observe the output signal in the time domain with a rectangular window. Set the input frequency that gives coherent sampling by verifying that the first and last samples are (almost) equal also by changing the phase.
- Add the second sine wave and change amplitudes, frequencies, and phase. Observe that coherent sampling is not verified anymore.
- Use the various windowing functions and observe the waveform on the monitor. Observe the signal before and after windowing to see the effect of the operation. Estimate the difference between first and last samples with the various windowing functions.
- Change the number of points of the input series and signal to observe what changes.

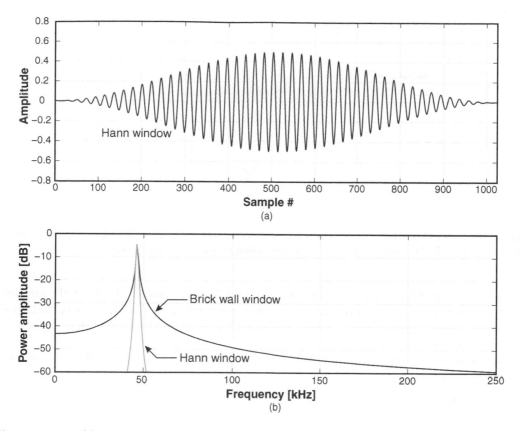

Figure 2.27   (a) Sine wave in time domain passed through a Hann window; (b) spectrum of a non-coherent series of 1024 sine wave samples with "brick wall" and Hann windowing.

2.10 GOOD AND BAD SIGNALS

When you listen to music you are pleased by the harmony of sounds or by the rhythmic sequence of tones, which your parents or maybe your grandparents would definitely consider terrible and unbearable. That music is, for you, a good or perhaps an excellent signal, able to stimulate your fantasies and create a good environment with your friends.

When an airplane receives signals from a primary radar or a transponder, following interrogation of the radar beacon to provide aircraft altitude, the pilot relies on the control system that helps in landing.

In reality, the signals that drive the headphones or those received in the cockpit are often mixed up with other signals that have nothing to do with music or the control of airplanes. These signals, called "noise," are obviously undesired, and the circuit or system goal is to make them negligible or ineffective.

Another possible situation occurs when, in addition to music reproduced by headphones, there are other sounds, such as whispering, a bell, or maybe the loud voice of a parent yelling at you to lower the volume. Those sounds carry information, but they are unwanted and interfere with the music. The whirring, the buzz, or the shouting voice are undesired components that

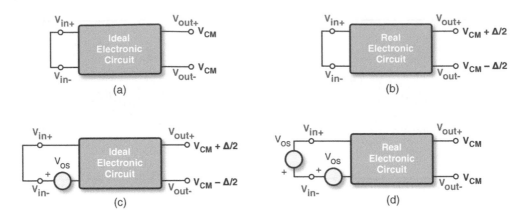

Figure 2.28   (a) Offset free circuit; (b) real electronic circuit with offset; (c) offset input generator summarizing non-ideal conditions; (d) offset compensation by external voltage source.

are generally referred to as noise, but they are different from real noise. Similarly, the airplane cockpit can receive signals from cellphones or other transmitting devices that carry undesired information. Since they carry information, although it is not the information that the system desires, they are not real noise.

In addition, other types of errors alter signals. We can classify them all by using four different categories:

- static difference (offset);

- interference (or cross-talk);

- harmonic distortion;

- noise.

2.10.1   Offset

Electronic circuits frequently use symmetrical structures to ensure or verify that signals are balanced (as in a traditional set of weighing scales). However, inaccuracies in fabrication and anomalous operating conditions (such as temperature gradients), result in errors that affect symmetrical operation. Asymmetries change effective signals by a shift, responsible for output imbalance, which arises even if inputs are symmetrical. Since the causes are mainly steady, the effect can be described as an equivalent constant (or very slowly changing) signal, called an *offset*. If the offset is at the input of the circuit it is called an input-referred offset.

Figure 2.28 explains this limitation. It is expected that the outputs of an electronic circuit are equal to V_{CM} when inputs are equal. However, static errors cause a $\Delta/2$ difference between the outputs (Figure 2.28(b)). The electrical consequence of errors is summarized by a fictitious voltage generator V_{os} in series with one of the inputs, as shown in Figure 2.28(c), of the ideal circuit. Obviously, offset can be electrically corrected by a suitable generator that compensates for it, as shown in Figure 2.28(d). The scheme uses a real voltage source with equal amplitude and opposite polarity to that of the fictitious generator representing the errors.

Techniques called *trimming* or *calibration* can also correct offset. The former is a permanent adjustment of electrical components (such as resistors or capacitors) after fabrication. Calibration is a temporary adjustment of static errors, but is also able to compensate for slow drifts. Calibration is divided into two categories: off-line and on-line. Off-line calibration is performed when a circuit is not being used, as there is a specific phase for the correction. This is like the periodic adjustment of guitar strings before playing music. On-line calibration is more complex, because the circuit continues working while calibration is in progress. The correction must occur even in critical cases that require very high accuracy but do not allow operational discontinuity. The task is difficult and often requires the use of a duplicated section of the system architecture that momentarily replaces the one under calibration. A calibration engine and the reconfiguration of systems are often provided by digital circuits that control proper calibration algorithms.

Notice that ...

offset is an electrical quantity (voltage or, rarely, current) that summarizes the effect of almost static errors. It is electrically represented by a fictitious source applied at a convenient point in the circuit, often the input terminals.

2.10.2   Interference

Interference occurs when a desired signal is mixed up with unwanted signals. The interfering part carries information not related to that of the main signal. Occasionally interference is improperly called mixing – a bad name because mixings are typically done on purpose; for example, the audio mixing console enables signals originating from separate microphones (each being used by a vocalist or a musical instrument) to be combined into a single signal. Interference is undesired because information carried by interfering sources disturbs the main signal. In electronic systems interference is often caused by limited electrical separation, which is unable to avoid coupling – such as a parasitic resistive or capacitive link – between lines that carry signals. Moreover, we have antenna radiation or disturbance at input. These possibilities are illustrated in Figure 2.29. Signals travel on metal lines in integrated circuits or in the traces of a Printed Circuit Board (PCB). If long paths are close to one another, parasitic coupling transfers part of the signal from one trace to others as shown in Figure 2.29(a). Interfering signals also affect the circuit through supply connections (Figure 2.29(b)). Supply terminals are intended to provide a dc signal, but often they carry unwanted terms such as the always present 50 or 60 Hz tones. Moreover, there is interference from nearby electrical motors that disturb circuit supply when spurs are not properly rejected by the dc voltage supply generators. A possible third source of interference (Figure 2.29(c)) is parasitic antennas.

Execution is important

In the design, fabrication and assembly of systems it is extremely important to reduce the negative effects of interference. When using a PCB, use multilayers to screen sensitive parts from interfering signals. With integrated circuits use metal layers for shielding. Never run critical traces next to interfering traces.

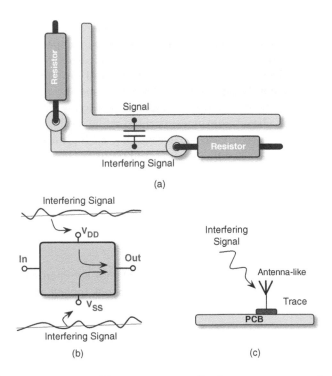

Figure 2.29   (a) Interference caused by parasitic coupling between traces on the PCB; (b) interference at input or caused by an interfering signal over the supply voltage; (c) interference due to an antenna-like effect.

The amplitude of interfering signals can be large, and can bring the operation of the circuit into bad regions. In this case interference is not just adding unwanted information but also alters the wanted information. The remedy is fourfold:

- protect the circuit from interference by taking special care in layout and fabrication;

- maintain a suitable separation of lines carrying signals on PCBs or integrated circuits;

- reject interference from the supply voltages;

- use electrical shields surrounding critical parts of the circuit, especially to avoid coupling at very high frequencies.

If interference amplitude is low, the disturbing effect is normally acceptable, because it is likely to be possible to discriminate the information needed. Moreover, when the spectra of good signals and interfering signals are at different frequencies, information can be extracted by selective processing (low-pass, bandpass or band-rejection filters).

2.10.3   Harmonic Distortion

Harmonic distortion occurs when sine waves applied at the input(s) of an electronic circuit give rise, in addition to an increased or reduced shifted replica of input sine waves, to other sine waves at frequencies related to the input frequencies. The result is undesirable in many

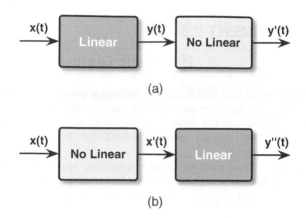

Figure 2.30   (a) Linear block before a non-linear section causing harmonic distortion; (b) pre-distortion before linear processing.

situations; for example, harmonic distortion in audio amplifiers alters the timbre of sound and causes low quality. In communications, distortion generates spurious tones that can be at frequencies near weak channels, disrupting the weak information-carrying signal.

Harmonic distortion is caused by non-linear responses that occur in addition to the desired linear processing. A system causing harmonic distortion can be divided into the cascade of two blocks, one linear and the other non-linear, as shown in Figure 2.30(a). Notice that the position of blocks is important, because the non-linear function cannot be moved before or after the linear block, as shown in Figure 2.30(b). With non-linear operations the amplitude of signals matters. Output in the two cases of Figure 2.30 turns out to be different.

As is well known, if the input of a linear system, such as the first block of Figure 2.30(a), is a sine wave, $A_0 \sin(\omega_{in}t)$, then the output is also a sine wave

$$y(t) = kA_0 \sin[\omega_{in}t + \phi(\omega)], \qquad (2.61)$$

where the phase shift $\phi(\omega)$ accounts for delay, dependent on frequency, caused by the finite speed of the system. Then we have to account for the non-linear block. We assume that speed performance is frequently determined by the linear section; therefore the non-linear input–output relationship is time independent. Suppose for the second block

$$y' = f(y) = y + a_2 y^2 + a_3 y^3 + a_4 y^4 + \cdots, \qquad (2.62)$$

which is a polynomial approximation with parameters a_2, a_3 and $a_4 \ldots$; they are called harmonic distortion coefficients.

Equation (2.62) is a linear superposition of non-linear terms; since addition is linear, the output will be the superposition of effects and each term will be estimated separately. Obviously the first one produces a replica of the input. On the other hand, a single input sine wave, $y = A_0 \sin(\omega_0 t)$, processed by the second term gives

$$a_2 y^2 = a_2 \frac{A_0^2}{2}[1 - \cos(2\omega_0 t)], \qquad (2.63)$$

which is a constant signal, $a_2 A_0^2 / 2$, and a sine wave at doubled frequency (the second harmonic tone) with amplitude $a_2 A_0^2 / 2$. Observe that the ratio between harmonic and main amplitude,

COMPUTER EXPERIMENT 2.8

Harmonic Distortion

With this Computer Experiment you can study the distortion effect of a non-linear function. It gives rise to second and third order terms with variable coefficients. The sine wave generator has a suitable interface, called GPIB, used to define the parameters of the signal generator under the control of a portable computer; namely you can sweep the sine wave amplitude. In this experiment you can:

- set the values of second and third harmonic coefficients. Notice that the coefficients are controlled in dB. Moreover the amplitude of the fundamental is 1 (0 dB);
- set, in dB, the sweep of input amplitude between two limits with 1V as reference. The sweep lasts 2 seconds;
- observe on the first window the output in the time domain (CH1 of the scope) or the spectrum;
- observe on the monitor the amplitude of fundamental and second (or third) harmonic term versus the input amplitude.

MEASUREMENT SETUP

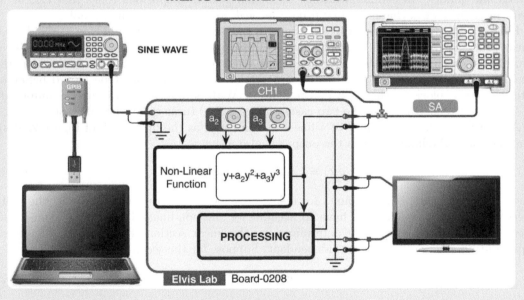

TO DO

- Set to zero the distortion terms and observe the "ideal" case in the time and frequency domain for different amplitude sweeps.
- Set a pair of distortion values with zero amplitude sweep. Observe the harmonic tones at different input amplitudes.
- Set a suitable input sweep that allows you to observe significant values of harmonic distortion.
- Determine the IP2 and IP3 for three different values of a_2 and a_3. Is there some relationship between the values of distortion coefficients with the results you get?

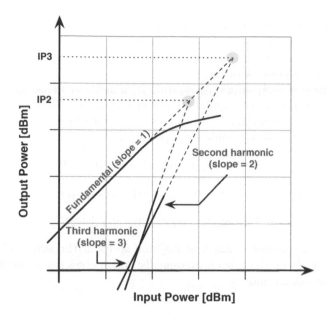

Figure 2.31 Amplitude of fundamental, second and third harmonic versus input amplitude. Note that the scales are in dBm.

$a_2 A_0/2$, increases with input amplitude. Signal and tone become equal when

$$A_0 = \frac{2}{a_2}. \tag{2.64}$$

The third harmonic distortion term gives rise to

$$a_3 y^3 = \frac{a_3 A_0^3}{4} [3 \sin(\omega_0 t) - \sin(3\omega_0 t)], \tag{2.65}$$

made up of a sine wave at the input frequency and a third harmonic contribution. The first term modifies the amplitude of the linear component; the second is, obviously, a source of limitations. Since harmonic amplitude is proportional to the cube of A_0, its effect grows significantly with amplitude A_0. Tone becomes equal to the generating sine wave when

$$A_0 = \sqrt{\frac{4}{a_3}}. \tag{2.66}$$

Very likely, amplitudes at which the conditions (2.64) and (2.66) are achieved are at very high values of A_0, which practical cases certainly never reach. However, it can be useful to plot the relationship between the input sine wave and the amplitudes of the second and third harmonic tones to find, by extrapolation, the points given by equations (2.64) and (2.66). In communications those points, measured on the input axis or, equivalently, on output, are called *IP2* and *IP3*. Figure 2.31 shows the diagram with the amplitudes expressed in dBm, as in communications. Observe that, since the diagram is logarithmic, the slope of the second harmonic is 2 and that of the third harmonic is 3, because amplitudes increase as the square and the cube respectively of A_0.

The above study can continue with higher orders. The equations become more and more complex. However, the result is that a kth order distortion gives rise to harmonics at $k\,\omega_0$ and other terms at lower harmonic frequencies. Moreover, spur amplitude at angular frequency $k\,\omega_0$ is proportional to the power k of A_0.

With two-tone signals, $y = A_1 \sin(\omega_1 t) + A_2 \sin(\omega_2 t)$ non-linear blocks give rise to harmonic tones at multiples of the sine wave angular frequencies ω_1 and ω_2 but also produce tones at combinations of originating frequencies: $(k_1 \omega_{o1} + k_2 \omega_{o2})$, with k_1 and k_2 positive or negative integers. For example, a second harmonic term produces

$$
\begin{aligned}
a_2 y^2 &= A_1^2 \sin^2(\omega_1 t) + 2a_1 A_2 \sin(\omega_1 t) \sin(\omega_2 t) + A_2^2 \sin^2(\omega_2 t) \\
&= a_2 \frac{A_1^2}{2}[1 - \cos(2\omega_1 t)] + a_2 \frac{A_2^2}{2}[1 - \cos(2\omega_2 t)] \\
&\quad + a_2 A_1 A_2 \{\cos[(\omega_1 - \omega_2)t] - \cos[(\omega_1 + \omega_2)t]\},
\end{aligned} \tag{2.67}
$$

which includes second harmonics at $2\omega_1$ and $2\omega_2$ and spurs at the sum and the difference of ω_1 and ω_2. A similar study on the third harmonic distortion gives rise to multiple components at various combinations of ω_1 and ω_2.

2.10.4 Noise

When bad signals do not carry useful information, they are called *noise*. In a room with many chatting people it is difficult to comprehend what a single person is saying because the superposition of many voices destroys information from a single one. Therefore, instead of generating many distinct components carrying informative content, the result is *cocktail party* noise. That is just one example of noise. We also have other relevant types of noise, the sources that affect any electronic circuit, produced by unavoidable fluctuations of voltage or current. Electronic noise is caused by unpredictable motion of electrical carriers that give rise to three different types (or, better, three different noise classifications). They are:

- thermal noise;

- burst noise;

- pink or $1/f$ noise.

To understand noise a good background in solid-state physics would be necessary. However, for our purposes it is enough to be aware of the key features and, possibly, to know something about the causes of noise. Here I shall try to provide the required awareness by using some humourous tales. Let me start with thermal noise.

For intuitive comprehension of conduction many teachers depict electronic devices as a box that contains carriers. If no voltage is applied, the box is horizontal. A voltage across two terminals bends the box. This bending makes the pressure between two inclined walls different. Therefore, voltage causes some flow of carriers that, in turn, corresponds to current. Now, let me replace the carriers with children in a kindergarten. Sometimes kids sleep, and, obviously, the environment is quiet (this corresponds to carriers at a very low temperature, 0 absolute degrees). When the children wake up (which corresponds to augmenting the carriers' temperature) they start jumping here and there, kicking walls at random. This creates pressure on opposite sides of the box (i.e., a voltage). Of course, the box-kindergarten is very large because it must contain a huge number of carrier-kids. The overall result is that the pressure

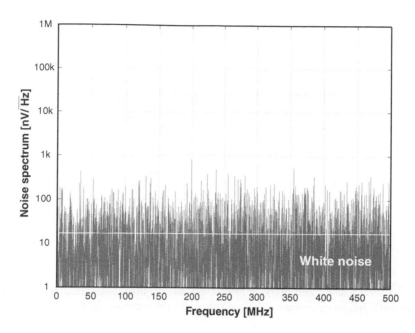

Figure 2.32 Spectrum of white noise voltage. The amplitude fluctuates in an unpredictable way, but the interpolating curve is frequency independent.

exercised by many kicks on the walls is random, and, accordingly, the voltage across the device is like that – a random variable: i.e., noise. If the temperature increases, the kids are more active, and the augmented kicking pressure on the walls increases the noise proportionally. Because of this dependence on temperature, the name for this effect is *thermal noise*. It occurs even with zero current.

An important feature of random variables is the power spectrum. For thermal noise the spectrum is white; i.e., its amplitude fluctuates randomly but the contributions at different frequencies are equal. Figure 2.32 shows a typical white spectrum.

Since the spectrum gives rise to a noise voltage, its dimensions are $[\text{V}/\sqrt{\text{Hz}}]$. The figure uses a submultiple, $[\text{nV}/\sqrt{\text{Hz}}]$, the square root of $[10^{-18}\ \text{V}^2/\text{Hz}]$. The vertical axis is logarithmic, and the average noise is about $\bar{v}_n = 16\ \text{nV}/\sqrt{\text{Hz}}$.

The power spectrum, $v_n^2(f)$, obtains noise power ($\bar{V}_n^{\,2}$), in the frequency interval $f_1, \ldots, f_2$, given by

$$\bar{V}_n^2 = \int_{f_1}^{f_2} v_n(f)^2\, df = \bar{v}_n^2 (f_2 - f_1), \tag{2.68}$$

where the dimensions are $[\text{V}^2]$ and the value increases proportionally with the frequency interval. Therefore the corresponding noise voltage $\sqrt{\bar{V}_n^2}$ increases as the square root of the bandwidth.

Above it was stated that the thermal noise spectrum is white, but no explanations were given. To understand this point, let's go back to our humourous parallel description of the effect and observe the following.

- There are no reasons to have periodic kicks at the walls, because the kids don't follow any rhythmic command. This excludes tones in the spectrum.

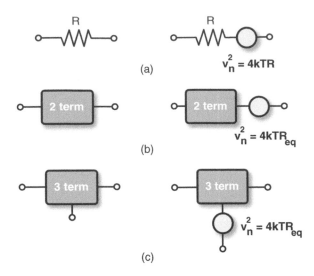

Figure 2.33   (a) Ideal resistor and noisy equivalent circuit; (b) ideal two-terminal device and its noisy equivalent circuit with R_{eq} as equivalent noise resistance; (c) the same as (b) for a three-terminal device.

■ The distribution of the kids' population is Gaussian: a few of them are very quiet, most are active, and a few others are lively. The quiet children kick the wall at a low average rate, while the exuberant ones kick it very frequently. The result is a uniform power distribution over frequency.

White noise

The limit of white noise increases with signal bandwidth. Since power is proportional to the bandwidth of the receiver, the noise voltage increases with the square root of the band. Therefore, the noise voltage level doubles with a four-times wider band.

Thermal noise affects any electronic device that behaves like a box with mobile carriers. The simplest case is a resistance, R, whose spectral density of thermal noise is

$$v_n^2 = 4kTR, \tag{2.69}$$

where k is the Boltzman constant and T is the absolute temperature.

For other devices thermal noise is defined as an equivalent noise resistance, R_{eq}, and its associated noise spectrum $v_n^2 = 4kTR_{eq}$. The equivalent circuits of a resistance or an electronic device, including thermal noise generators, are the ones shown in Figure 2.33.

Electronic devices such as transistors have three terminals, and noise modeling for them is more complex than in two-terminal devices. Usually employing one or more noise generators in series or parallel with the device nodes obtains the desired result. In some circumstances one noise source in series with one of the terminals, as shown in Figure 2.33(c), is enough.

As has already been mentioned, a good practice is to verify equations by checking their dimensions. Certainly equation (2.69) is correct, but, just as an exercise, let us verify its dimensions. That of the power spectrum is $[V^2 \cdot s]$; the right-hand side after multiplying and dividing by the electronic charge is

$$\left(\frac{kT}{q} \right) qR, \tag{2.70}$$

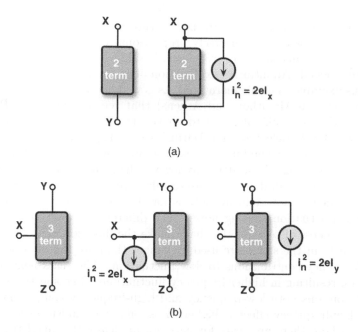

Figure 2.34   (a) Ideal two-terminal device and its burst noise equivalent circuit; (b) ideal three-terminal device and two possible burst noise equivalent circuits.

where kT/q is a voltage and the dimension of the electronic charge is current per second; therefore

$$[V \cdot A \cdot s \cdot \Omega] = [V^2 \cdot s], \tag{2.71}$$

which verifies the check.

The second type of noise is burst noise. It has a white spectrum but happens only when current is flowing (as opposed to thermal noise, which occurs even with zero current). The mechanism that causes burst noise is the jumping of a barrier. The flow of carriers is here described as if they were runners in a cross-country race. The path is uneven, with natural barriers, rambling sections, and puddles. The result is that the flow of runners is not uniform but fluctuates, because one or more runners may hesitate before jumping a barrier. The same happens for carriers in a device as they jump potential barriers encountered on their path. The white spectrum of burst noise, similar to thermal noise described by a random voltage generator, is normally represented by a random current generator, i_n, connected in parallel with the current flowing in the circuit. Its power spectrum i_n^2 is

$$i_n^2 = 2eI, \tag{2.72}$$

which is proportional to the current, I, in the circuit. This proportionality makes sense because, using the humourous example again, having more participants in a cross-country race causes a larger fluctuation in runner flow.

Figure 2.34 illustrates possible equivalent circuits without and with burst noise. The current generator with noisy current is across two terminals of the device. It can possibly be converted into a noisy voltage by circuit analysis. Its spectrum is proportional to one of the currents flowing through one terminal. Even in this case, for multiple terminals, the noise generator is placed in the proper position, usually where jumping a barrier affects conduction.

The last type of noise is $1/f$ noise. Its key feature is that, as indicated by the name, the shape of the power spectrum is inversely proportional to the frequency. Obviously the spectrum cannot go to infinity at a zero frequency, but $1/f$ behavior is almost followed until very low values are reached. An intuitive explanation of $1/f$ noise is provided here by the use of another facetious example: a comic marathon. As is well known, a marathon is a 42.195 km race (the distance between Marathon and Athens) that recalls the run of Pheidippides, a messenger, who, in the year 492 BC, had to convey the good news that a small force of Athenians had miraculously defeated King Darius' army. The difference here is that on the sides of the long path there are comfortable seats, some close to the street and others in the shade, with refreshments available, a little way away. The runners are certainly interested in the final result, but, when tired, don't hesitate to take a rest for a short or a long time.

The result is that flow is not uniform, but, as for the cross-country race discussed above, there is unexpected fluctuation causing noise. The difference is that the proximity to the street or the attractiveness of seats in the shade make the duration of stops widely variable. In particular, the seats in the shade are used for longer than are the ones next to the path, resulting in low-frequency contributions to flow fluctuation. The seats next to the path are used for quick rests, resulting in higher-frequency fluctuations. The spectrum depends on an overall effect that contains both low-frequency and high-frequency components. Since shade is attractive, the low-frequency effect is higher. A similar mechanism occurs for carriers in electronic devices, where there are seats for carriers, unpleasantly called traps, at different levels of energy. The distribution in that power spectrum of the random component is

$$i_{1/f}^2 = \frac{K_f I}{f^\alpha}, \tag{2.73}$$

where α is close to 1 and K_f is the $1/f$ noise coefficient. Again, as for the burst noise (also called shot noise), the power spectrum is proportional to the current in the device.

The value of K_f depends on technology and can possibly be improved by taking special care in fabricating devices. The $1/f$ noise is represented by a current generator, as for burst noise, or sometimes by an equivalent voltage generator, as for thermal noise.

In summary, thermal noise is a random fluctuation that affects any electronic components, independently of current. On the other hand, shot and $1/f$ noise affect a circuit only when current flows in the device. Various sources affect electronic devices, with limits extending over a wide frequency range (white spectrum) or restricted to low frequency ($1/f$ spectrum).

$1/f$ does not continue until zero

The $1/f$ noise spectrum does not increase without limit when frequency tends to zero. This is impossible because the power (the integral of the spectrum) would become infinite.

2.11   THD, SNR, SNDR, DYNAMIC RANGE

A signal is a mixture of good components, carrying desired information, and unwanted parts that degrade information and blur results. Parameters that quantify the limits provide a measure of signal quality. Accounting for distortion and noise, we can define three specifications: Total Harmonic Distortion (THD), Signal-to-Noise Ratio (SNR) and Signal-to-Noise plus Distortion Ratio (SNDR). Typically those ratios are given in dB.

Let us suppose we are using a sine wave at the input of an electronic block, giving an output of a set of sine waves at f_{in} and multiples of f_{in}. Amplitude of the component at f_{in} (assuming a voltage) is V_1, amplitude of the second harmonic (at $2f_{in}$) is V_2, and so forth. Suppose we also have noise with spectrum v_n^2.

Total harmonic distortion is defined as the ratio of fundamentals to the root-sum-square of harmonics:

$$\text{THD} = \frac{V_1}{\sqrt{V_2^2 + V_3^2 + V_4^2 + \cdots + V_k^2}}, \tag{2.74}$$

where the signal is V_1, and $k - 1$ harmonics are accounted for. Normally the estimation in equation (2.74) limits the value of k to 10.

If THD is measured in dB, calculating the ratio using powers rather than voltages produces the same result. This is

$$(\text{THD})_{dB} = 20 \cdot \log_{10} V_1 - 10 \cdot \log_{10}[V_2^2 + V_3^2 + V_4^2 + \cdots + V_k^2]. \tag{2.75}$$

The SNR is the ratio between signal power and noise power. For its calculation it is necessary to know the extent of interval frequency perceived by the system. Actually any system has a band of operation; for example, the human ear detects sound up to 20 KHz (or more for good ears). Therefore, since noise above the audible range is irrelevant, the only noise that disturbs people when listening to music is that in the range 60–20 000 Hz. For other users the perceived band can be different; for example, dogs sense a wide range of frequencies, 70–45 000 Hz, and bats much wider, 2–110 kHz. The same principle holds for any system that uses signals. Therefore, system bandwidth is what matters in estimating global disturbance caused by noise.

If the bandwidth limits are f_1 and f_2, that is the interval we have to account for. Therefore, the power of a noise generator whose spectrum is $v_n^2(f)$ is

$$V_n^2 = \int_{f_1}^{f_2} v_n^2(f) \, df. \tag{2.76}$$

Then the Signal-to-Noise Ratio (SNR) is defined by

$$\text{SNR} = \frac{P_{signal}}{P_{noise}}. \tag{2.77}$$

The measure of SNR of a system with dynamic range V_{DR} assumes the use of a sine wave as the input signal. Therefore, since the maximum usable amplitude is $A_{max} = V_{DR}/2$, the maximum signal power is $A_{max}^2/2$. It is also reasonable to suppose that noise is independent of signal. In this case SNR increases with signal, and its peak value is at full amplitude:

$$\text{SNR}_{max} = \frac{A_{max}^2}{2V_n^2} = \frac{V_{DR}^2}{8V_n^2}. \tag{2.78}$$

With signals smaller than the maximum, signal power diminishes and SNR becomes less. This again presumes that noise is not changed by signal. The assumption is reasonable, because mechanisms that cause noise are weakly related to signal. In real cases SNR deviates slightly from the above assumption. The graph in Figure 2.35 shows SNR versus the signal amplitude for different situations. One line, dashed, shows the ideal, with a peak SNR equal to 80 dB at 0 dB$_{FS}$. The solid line shows a slight loss at amplitudes close to the full scale.

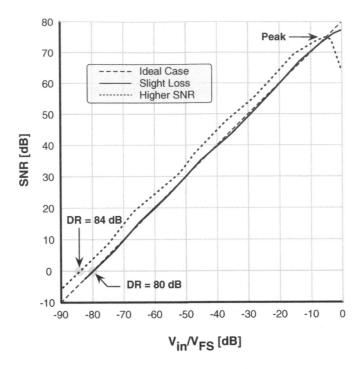

Figure 2.35   SNR for different input amplitudes.

The dotted line corresponds to a case with higher SNR at low input signals but with a peak SNR that drops for high signal levels. Instead of reaching 80 dB the peak is at 74 dB. The input amplitude that corresponds to SNR = 0 dB defines the Dynamic Range (DR). This is the ratio between maximum input and the level at which signal equals noise. As discussed, and shown in Figure 2.35, the DR can be higher than the peak SNR. Notice that the SNR in the figure is good but not excellent. Electronic circuits can obtain SNRs as high as 100 dB or more.

For many applications the combined effect of distortion and noise is important. This limit is measured by the SNDR, which is the ratio of signal power and the power of noise plus the power of harmonics:

$$\text{SNDR} = \frac{V_1^2}{2\bar{V}_n^2 + V_2^2 + V_3^2 + V_4^2 + \cdots + V_k^2}, \qquad (2.79)$$

where $V_i, i = 1 \cdots k$ are the amplitudes of sine waves at frequencies $i \cdot f_{in}$, and $\bar{V}_n^2$ is the noise power estimated by using equation (2.68). For maximum amplitude of the input sine wave, $V_1 = V_{DR}/2$.

It is evident that the SNDR is worse than the SNR, since, in addition to noise within the band of interest, it includes spur tones caused by harmonic distortion. A key goal of designers is to lose as little as possible with harmonic tones and to make the SNDR close to the SNR.

PROBLEMS

2.1 Draw the diagram of the distance versus time of a trip that you frequently make. Outline bottlenecks and fast sections. Consider different situations, including normal working days, rainy days and quiet weekends. Determine the worst case "fast" and the worst case "slow" of the journey time.

2.2 The voltage at output of a sine wave generator has amplitude of 1 V with amplitude fluctuation of ±10 mV. Estimate the error in the crossing time for amplitudes of 0.95 V and 0.5 V. The sine wave frequency is 1 kHz. What happens to the results if the frequency increases to 43 kHz?

2.3 What is the expected profile temperature in a room with a heater at one side and an open window at the opposite side? Define a simple model that accounts for window aperture and includes the effect of a big table in the middle of the room.

2.4 Draw in a squared exercise book a waveform that changes in a random manner. The horizontal axis is time and the vertical one is amplitude. Make a bar chart of the amplitude distribution. Use one sample for a square, and amplitudes equal to the average value in that square.

2.5 Draw the hypothetical diagram of a river level using four points per hour. Suppose that the level between successive measurements changes linearly with time. Remove one point out of two and redraw the diagram. Repeat the "decimation" by 2 until you think the accuracy is not satisfactory.

2.6 The passage of a person through a door interrupts a laser beam. Plot the signal received by a light detector when many people exit the room, and determine the type of information the signal provides for different flow conditions.

2.7 Plot the addition of two sine waves at various frequencies in the interval $0 \div 10$ msec. Use the following frequency–amplitude pairs: [1000 Hz; 1 V], [1.2 kHz; 0.2 V], [1 Hz; 0.2 V], [1200 Hz; 0.6 V]. Start with a diagram made up of 10 points, and double the number until you think the result is satisfactory.

2.8 Use the computer and any language you are familiar with to plot the multiplication of two sine waves at various frequencies. Use the following frequency–amplitude pairs: [100 Hz; 1 V], [1.2 kHz; 0.34 V].

2.9 A signal is the superposition of three sine waves:

$$y_1(t) = 2 \sin\left(36\,t + \pi/7\right)$$
$$y_2(t) = 0.8 \cos\left(21\,t\right)$$
$$y_3(t) = -0.2 \sin\left(72\,t\right).$$

Estimate the maximum amplitude and the first time at which the signal crosses zero.

2.10 Plot with any computer aid the following signal: $\sin\left(\left[1000 + \sin(4t)\right]t\right)$. Define a suitable time interval that allows you to make a good observation of the waveform. What you plot is a sine wave whose frequency changes sinusoidally. The operation is called frequency modulation with a sine wave as modulating signal.

2.11   The fall time of a 4 µs, 1.8 V pulse is 35 ps. Determine the delay of a pulse with 1.2 V amplitude and 100 ps rise time that crosses the fall front of the other pulse at 0.9 V.

2.12   What is the s-domain equivalent of the following relationship:

$$y(t) = 3x(t) - \int dx \int 12.2x(t)\, dx + \frac{dx(t)}{dt}?$$

What changes if the first term becomes $4x^2(t)$?

2.13   Draw a 1 MHz sine wave and its samples obtained with a sampling rate of 25 MS/s. Plot on the same diagram the sine wave whose frequency is in the third Nyquist region that gives rise to identical samples.

2.14   What is the z-domain equivalent of the following relationship:

$$y(nT + T) = y(nT) + 3x(n+1) - 2x(nT)?$$

Suppose that a second signal, $x_2(nT)$, multiplies the last term. What changes?

2.15   Find in the z-plane the mapping of $s = -2 + j16$. The sampling period is 1 s. Determine the values of s that give the same mapped value on the z-plane.

2.16   A quantizer has 3121 quantization levels spanned uniformly in the 0–1.8 V interval. Determine the value at output and the quantization error for 1.23561 V at input. Assume the discrete level is in the middle of the quantization interval.

2.17   Represent the full scale and result of the previous problem by using a binary digital signal magnitude representation. Zero output corresponds to 0.9 V.

2.18   Make the following transformations:
$372 \rightarrow dB$; $0.327\ \text{V} \rightarrow dB_V$; $0.36\ \text{V} \rightarrow dB_{mV}$; $10\ \text{W} \rightarrow dB_W$;
$0.032\ \text{V} \rightarrow dB_{FS}$ with full scale 1.3 V.

2.19   A processing block transforms a sine wave with amplitude 118 µV to obtain as output a sine wave at the same frequency shifted by $\pi/2$, and amplitude 0.932 V. What is the gain in dB? The shift changes to $\pi/4$. What is the obtained gain in dB?

2.20   The voltage across a 1.2 kΩ resistance is a sine wave with 1.2 V amplitude. What is the power in dB_m? What is the power with a triangular wave of the same amplitude?

2.21   Power transmitted by an antenna is 16 dB_m. With a supply voltage of 1.8 V, what is the average current consumed by the power amplifier, supposing that it works with 36% efficiency?

2.22   The FFT of the signal at output of an electronic circuit with a sine wave at input uses 2^{12} points. Determine at least four input frequencies that give rise to coherent sampling. The ratio between sampling rate and input frequency may range between 50 and 100.

2.23   Do a search on the Web to find the definitions of windowing functions. Plot the Hamming window defined by

$$w(n) = 0.54 - 0.46 \cos\left(\frac{2\pi n}{N-1}\right).$$

Use $N = 256$ and $N = 1024$. Write the equation that defines a triangular window.

2.24   An input signal made by two sine waves plus offset

$$V_{in} = 0.310^{-3} \sin(10^5 t) + 1.210^{-3} \sin(10^6 t) + V_{OS}$$

passes through the non-linear function

$$V_{out} = 320 \, V_{in} \in |V_{out}| < 1 \, V,$$

indicating that the output saturates at $\pm 1 \, V$ with large input signals. Estimate the maximum offset that avoids output distortion.

2.25   A non-linear block has transfer function

$$y = 6x + x^2 + 0.12x^3.$$

Plot in dB versus the amplitude the fundamental, second, and third harmonic components of the output determined by a sine wave at input. The input amplitude changes from $x_{min} = 0.01$ to $x_{max} = 10$. Determine $IP2$ and $IP3$.

2.26   What is the RMS noise voltage caused by a resistance of 100 kΩ? The bandwidth of interest spans from 1000 Hz to 1 MHz. What upper frequency halves that noise voltage? Assume the temperature is $T = 300$ K. What is the answer for $T = 400$ K?

2.27   The current generated by a photodiode is injected into a 400 MΩ resistance to give rise to an output voltage. The bandwidth used is 200 MHz. The photo-generation is caused by electrons jumping a barrier. Current is proportional to light flow F_L:

$$I_{ph} = \beta F_L, \tag{2.80}$$

where $\beta = 10^{-6}$ and F_L is in Lux. For which light intensity is the SNR 40 dB?

2.28   A device, equivalent to a resistance of 160 kΩ, has white and $1/f$ noise, the latter modeled by parallel current noise generators $i_n^2 = K_f \, I/f$, $K_f = 6 \cdot 10^{-16}$. Plot the noise spectrum on a log-log scale. Determine the frequency at which the $1/f$ noise equals the white term.

ADDITIONAL COMPUTER EXAMPLES

Computer Example - A/2.01

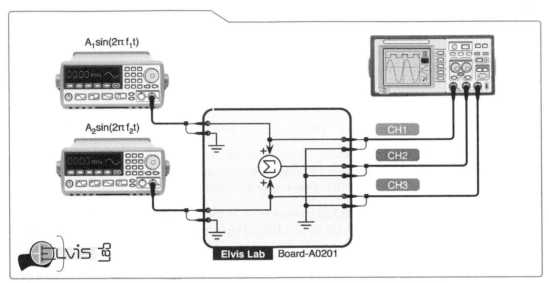

An *ElvisLab* experiment that enables you to observe the superposition of two sine waves. Change frequency and amplitudes to see addition and separate waveforms.

Computer Example - A /2.02

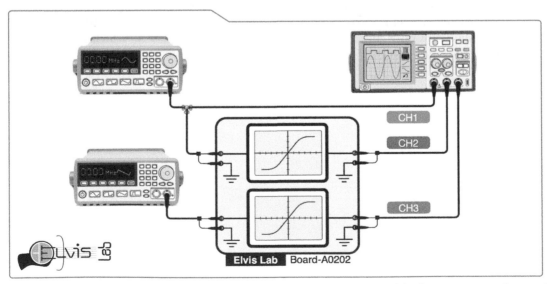

With this experiment you can investigate the effect of non-linear blocks on commonly used waveforms. Observe how distortion can alter sine waves and triangular waves.

Computer Example - A/2.03

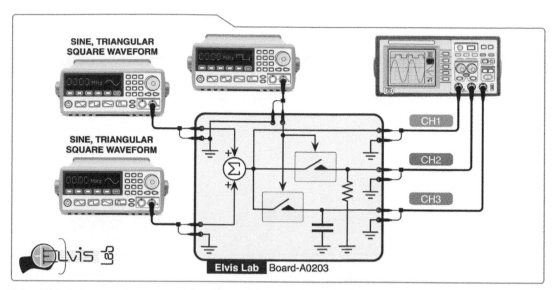

This *ElvisLab* experiment enables you to derive the sampled and sampled-and-held view of the superposition of two repetitive signals at different frequencies. Change frequencies and amplitudes to see the result.

Computer Example - A /2.04

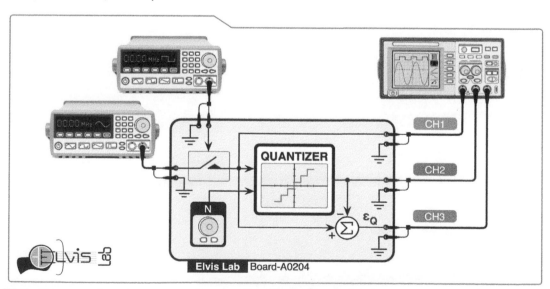

With this experiment you can observe the effect of quantization on a commonly used sine wave. The three channels of the scope show you the sampled-data signal before and after quantization and the quantization error. Change the ratio between sampling and sine wave frequency to discover some interesting features.

CHAPTER 3

ELECTRONIC SYSTEMS

This chapter describes electronic systems, block diagrams, and specifications. We shall consider systems for entertainment, communications, computation and processing, measure, safety, and control. An important topic covered by this chapter is the block diagram, which is used to outline relevant signal transformations such as amplification, filtering, and non-linear processing. It also shows generic functions for power control, clock generation, and so forth. Other useful information is given by specifications that provide, using numbers and diagrams, the requests that the system must verify in the nominal and in the so-called "worst case" conditions.

3.1 INTRODUCTION

Electronic systems, as discussed in the first chapter, enable suitable processing of signals. They have one or more inputs and produce one or more outputs. In some cases inputs are electrical quantities generated by sensors. After the proper transformation, outputs for storing data, for wired or wireless transmission, or for driving actuators are generated.

An electronic system is conveniently represented by a block diagram specifying at different levels of detail the processing performed or the auxiliary functions. The description in some cases can be incomplete, because many details are not essential at high levels of abstraction. In other situations functions are only approximated, because it is enough to mention the purpose

Understanding Microelectronics: A Top-Down Approach, First Edition. Franco Maloberti.
© 2012 John Wiley & Sons, Ltd. Published 2012 by John Wiley & Sons, Ltd.

of a block. The system description is often hierarchical, with various levels that, step by step, provide details on the functions and operation of circuits.

The hierarchy employed in systems is similar to the one used with maps, which encode and display geographical information at various level of resolution. It is well known that hierarchy helps in pathfinding or in searching for locations, because a problem is transformed into multiple levels of searching. For example, to find the hotel where somebody has to go for a business trip, it is worth starting at the regional level by searching cities without worrying about streets and squares. At city level it is possible to look for street names, their positions and relative distances, and the route from the airport to reach an address, without caring about buildings. Finally, at street level it is possible to find building details and other features, such as one-way restrictions, that are useful in order to reach the building where the hotel is located.

Similarly, an electronic system is like a city or, perhaps, a state, but with transistors or resistors instead of the bricks of the buildings. Therefore, as already mentioned, to learn how such a system works it is convenient to use hierarchy. Often, there are many levels of description with increasing details that, at bottom, reach the lumped element representation, the interconnection of two- or multiple-port components (such as resistors or transistors). The electrical operations are described by relationships between variables such as voltage, current, and charge. Basic elements are displayed in circuit schematics with connections made by wires.

3.2 ELECTRONICS FOR ENTERTAINMENT

Systems in the the entertainment category are well known to the younger generation. The best selling gifts for birthday celebrations and traditional feasts are devices with embedded electronics. Almost every boy or girl has received electronics-based toys. Also, they have used toys with an educational aim (edutainment toys) like, for example, educational games that help children to learn about animals and the living environment. Very popular are cars with remote controls that have two commands, one for gas and brake, and the other for left–right steering. Recall also the soft-bodied baby doll that responds to a child's attentions and is capable of multiple actions such as producing feeding and burping sounds, saying "mama," and making a happy babbling noise.

For adults many devices and entertainment systems are on the market. For all of them we can describe features, defining specifications and sketching block diagrams. The following section considers a limited sample of devices, mostly classified as consumer electronics, i.e., equipment intended for everyday use.

3.2.1 Electronic Toys

Let us consider a hypothetical talking baby doll with animated arms (Figure 3.1). It is a soft-bodied doll designed to call for and respond to the child's attention. It comes with some accessories (such as a feeding spoon and a baby pacifier) and uses two AA batteries. Under its velour dress there is an electronic system that performs a number of functions such as, among others, moving the arms when the doll cries, saying nice words when the spoon touches the lips, or giggling when the doll is patted on the shoulders.

The system includes a number of sensors: some switches, a pressure sensor on the shoulders, and a simple optical sensor system on the mouth that distinguishes the white color of the spoon

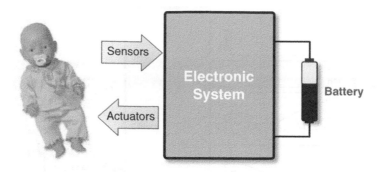

Figure 3.1   Schematic representation of an electronic doll.

from the red of the baby pacifier. Moreover, there are actuators, a micro-motor for moving the arms and a small speaker to produce sounds.

The system operation depends on commands given by sensors: an on/off switch to start/stop the operation; a "start crying" switch to give suitable commands; pressure sensors to measure possible patting and caresses; the optical sensor to detect the position of spoon or pacifier. The commands are analog or digital signals to be processed (or interpreted) so as to determine consequent actions: reproduction of pre-recorded sounds or voices, and movement of the arms.

The above general description can be viewed as the first part of the specification of an electronic system. In addition, it is necessary to define other features denoting electrical and mechanical properties and conditions. For example, the requirement to use two AA batteries indicates that the supply voltage of the electronic part is nominally 3 V. However, since the voltage of batteries drops during their life the supply voltage information should include a minimum (and maybe also a maximum) that ensures correct operation. Another electrical specification is the power consumption, which can be different when the system operates fully and when is in "stand by" mode. Power is a very important feature because the limited energy stored in batteries must be carefully managed to ensure long use before changing batteries. In addition, the volume of the speaker must be defined, and, possibly, a volume control indicated.

System specifications

The specifications (*specs*) of the electronic system used inside a baby doll with electronic controls indicate various relevant parameters. Among them we certainly have:

Supply voltage Nominal, lowest value, and maximum value.
Power consumption With normal operation, and in stand-by.
Temperature Range of normal operation, and storage temperature.

These parameters obviously describe conditions with which all electronic systems comply. There are other defined quantities that specifications also provide.

In addition to general system specifications, there are others that are relevant for particular elements that form parts of the system. For example, the speaker's features are important for

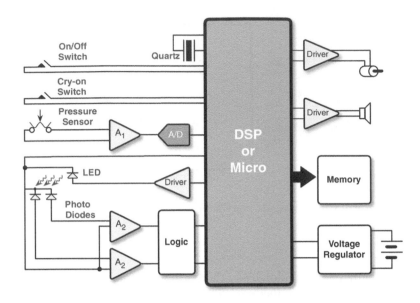

Figure 3.2   Block diagram of the electronic toy shown in Figure 3.1.

defining the specification of the speaker driver. The optical interface also needs specs on its operating conditions.

Figure 3.2 shows the block diagram of a possible implementation. The left-hand side is the input section, with various kinds of sensors. The amplitude of the signals depicting pressure and light might not be enough and may require amplification. The results are converted into digital form by analog-to-digital converters and logic functions. All signals are then processed by a DSP or a microprocessor (μP) so as to control the actuators suitably. Two drivers transform the output of the digital processing into a signal with voltage and current that are suitable for driving the motor and the speaker. In addition, there is a memory that contains the program for controlling the logic and providing pre-recorded sounds. The system also uses a voltage regulator for the necessary supply voltages, obtained from batteries. The quartz is the basis of timing control for all the functions.

The DSP or μP performs logic functions necessary to the system, described by a defined code. The program, written in the language understood by the particular microprocessor or digital signal processor that is used, contains instructions that correspond to statements such as the following.

> *Check the status of the cry-switch every 100 ms. If the switch is closed, then fetch the pre-recorded sound "cry" and play it for 10 seconds.*

To know how statements are codified, handled, and transformed into commands we have to descend to a lower level of the description hierarchy. Here it is enough to be aware of the key functions necessary for operating a simple toy rather than the details of its operation.

Other popular electronic toys are remote-controlled (R/C) toys: cars, missiles, airplanes, helicopters, snowmobiles, motorbikes, and many others. The remote link can be via radio or infrared signals. In the case of radio, the short-range communication required enables designers to use a carrier at 27 MHz, or, in the USA, to employ the frequency band 26.96–27.28 MHz

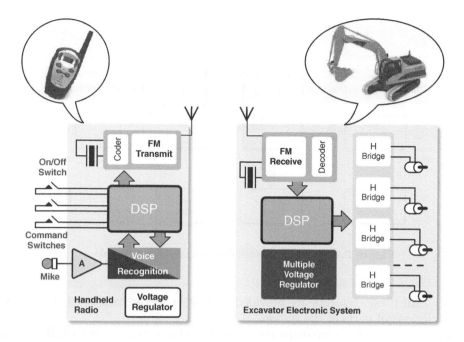

Figure 3.3 Block diagram of hypothetical voice-activated remote-controlled excavator.

allocated by the Federal Communication Commission (FCC) for transmission of control or command signals.

The inputs usually come from a small console that, in the simplest case, consists of two joysticks or a wheel and a trigger. Inputs from the console are transformed into signals suitable for wireless transmission. The receiver, which comes after the antenna, translates the radio signal into commands for driving the actuators of the electronic toy.

Let us consider, as a hypothetical example of an advanced toy, a R/C voice-activated excavator. It consists of a command unit plus the excavator toy itself, which has an articulated arm mounted on a rotating platform on top of a carriage with tracks. The possible movements concern the carriage (with constant speed), rotation of the platform, and action of the articulated arm. Figure 3.3 shows an approximated block diagram of the systems in two modules: the handheld radio and the excavator electronics. The handheld radio has some buttons used for on/off switching, for sending commands, and for the training of vocal commands. During normal operation a recognized command is processed and transformed into digital form before transmission to the excavator. The radio-frequency part uses, in this example, the Frequency Modulation (FM) method. The excavator module receives the FM signal, decodes the command word and transfers the result to the DSP that sends control signals to the drivers of the motors. Possibly the supply voltages used by the electronics and the motors are different, and, with a single battery pack, it is then necessary to provide a voltage regulator producing multiple voltages.

The block diagram of Figure 3.3 is incomplete, as it does not include some blocks that, probably, are necessary and will be used in the system. For example, there is no memory block, which would certainly be required to store algorithms for voice recognition and for retaining the digital coefficients used in signal processing. Moreover, the voltage generation block is generic and does not provide details of the voltages, maximum generated currents,

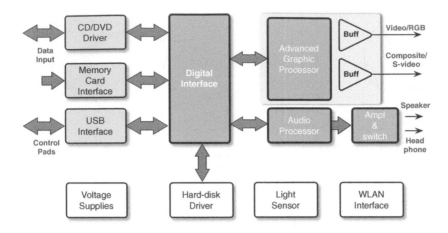

Figure 3.4   Simplified block diagram of a game console.

and efficiency. All these elements are provided at a lower hierarchical level that gives more detailed descriptions of parts of the system.

Obviously, the toy should be accompanied by an instruction manual, but reading manuals is not an activity welcomed by young users. Because of that, this modern toy uses an electronic instruction manual, supported by multilingual voice synthesized output.

For the fabrication of the system, however, it is necessary to fix features that are relevant for avoiding errors or misunderstandings. The agreement on those parameters gives rise to precise system specifications that avoid misinterpretations of the design.

System specifications

The specs of the R/C toy, in addition to generic values such as supply voltage, power, and temperature of operation, may include the following.

On-board CPU Number of MIPS (Mega Instructions Per Second), number of bits.
On-board storage Capacity of flash memory, Random-Access Memory (RAM).
Wireless operational range Indoors, line of sight.
Maximum moving speed Various moving parts, in meters per minute.
Operational time With alkaline battery, with rechargeable battery.
Mechanical specifications Dimensions, weight.

3.2.2   Video Game and Game Console

Video games and game consoles are popular electronic interactive entertainment systems used either by children or adults. Everybody knows what they are, and it is not necessary to spend a minute in their description, or in illustrating their benefits (such as the enhancement of visuomotor skill) or drawbacks. Since a large variety of models with different features are on the market, generic description of architectures is also difficult.

The evolution of the video game and the game console followed and favored the evolution of integrated circuit technology and progress in data storage. The technology evolution satisfies the increasing needs of data processing. For high video quality there is a huge amount of data to load, and for this video games benefit from the Blu-ray disc™, able to store as much as 50 GB.

The video game is a complete complex system; the game console is simpler because it uses the features and audio/video outputs of a personal computer. A possible block diagram of a game console is shown in Figure 3.4. There may be on-board actuators such as a Blu-ray disc or a light sensor for adjusting the video brightness. The input from control pads is via Universal Serial Bus (USB) ports and an associated interface. Possibly the console uses a massive internal memory, which can be solid state or a hard disk. For these it is necessary to use suitable electronic control circuitry. Moreover, the console communicates with the network by a Wireless Local Area Network (WLAN) interface. The system operates almost completely in the digital domain, but the outputs are either analog or digital as required for properly driving the audio and video actuators.

Self training

Make a search on the Web for different types of video games. Look at different features, sensor interfaces and actuators. Make a list of:

- the type of sensor and actuator used;

- basic functions implemented;

- the type of communication and data storage devices;

- the required supply voltages, power consumption and operational time.

Sketch the block diagram of a modern video game.

3.2.3   Personal Media Player

Other example of electronic systems for adult entertainment are Portable Media Players (PMPs), also called audio/video jukeboxes or portable video players. These are handheld audio/video systems that can record and play back audio/video (A/V) from television, a Digital Video Disc (DVD) player, a camera, or a media file; in addition they can be used to play games. The files are uploaded into a flash memory via the USB port or are downloaded from the Internet. Initially the PMP was a simple audio player but smaller and lighter. Then it evolved in response to consumer demand to become a device capable of handling video, audio, and data.

The block diagram of the system includes audio and video sensors and actuators. Data and controls can also be entered by the touch screen. Figure 3.5 indicates the various types of actuators used and their electrical interfaces. Observe that the monitor with touch screen option communicates in both directions with the electronic part, which is indicated only by a black box. What is inside that black box is detailed, for a hypothetical architecture, in Figure 3.6. Even for this block we identify various functions; among them we have amplification,

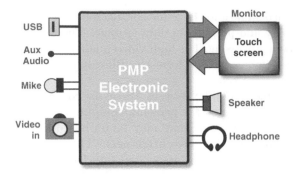

Figure 3.5   Block diagram of a PMP with just the details of the sensors and actuators used.

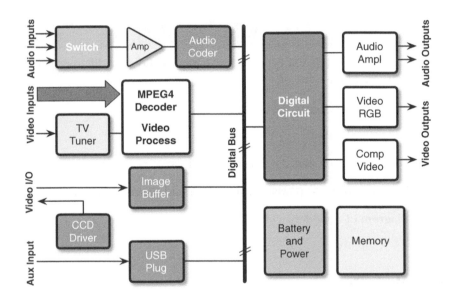

Figure 3.6   Possible simplified block diagram of a PMP system.

which, as we shall study in detail shortly, increases the amplitude of signals for better handling or use. We have the TV tuner, dedicated to the transformation of a signal received via an antenna into a format useful for controlling the display. Black boxes that operate in the digital domain indicate other functions. The outputs of those blocks are transformed appropriately for driving the actuators. The block diagram also includes blocks not connected with others. This, obviously, does not reflect reality; for example, the memory interacts with other parts of the circuit, and the battery and power supply are obviously connected to the system to power it. However, block diagrams often specify only the interconnections that are relevant for understanding the functions at that level of the hierarchy. Figure 3.6 indicates memory and power supply just so as to outline the fact that the system includes them.

Systems for personal entertainment are normally classified in the category of "Consumer Electronics." Do a search on the Web for function descriptions and block diagrams of the following electronic "consumer" systems:

- laser printer;

- Bluetooth™ headset;

- digital still camera.

Write a short note on the results of your search.

3.3   SYSTEMS FOR COMMUNICATION

Communication systems are used every day for talking with others, displaying images, sending written messages, or transfering data. People have always had the need to communicate with others, and electronics significantly facilitates it. Just a couple of centuries ago communication was quite difficult and slow, mainly done by mail that could only go as fast as the speed of a train or of a sailing ship. Today with modern systems we have revolutionized communications, so that information can reach a single place or multiple addressees in seconds.

Communication systems use different transmission media to convey data from a sender (or transmitter) to a receiver. They can be material media such as an optical fiber or a copper cable, or the vacuum (or the atmosphere) that conveys electromagnetic waves such as light and radio waves. We therefore distinguish between wired and wireless systems. Moreover, within the transmission media we have multiple communication channels that are used to separate different and unrelated flows of information. Often this separation is obtained by allocating channels in specific frequency intervals, typically defined by international agreements that regulate the use of the electromagnetic spectrum. The transmission medium does not transfer the data perfectly; it causes attenuation and adds unwanted components. Therefore, with electronics, in addition to the actions required for transmitting and receiving signals, we have to compensate for the limits of the communication channel.

All the above points, and other issues, are studied in communications courses and are not considered here. However, the electronics of communication systems requires a strong interaction between the communication engineer and the electronic designer so as to find the best trade-off between communication needs and their circuit and system implementations.

The following section discusses popular communication systems and presents suitable architectures and high-level building blocks.

3.3.1   Wired Communication Systems

Traditional telephony (PSTN – Public Switched Telephone Network) is the oldest and most popular wired communication system. The connecting line reaching the fixed telephone is a pair of wires that are twisted to limit the electromagnetic interference. In addition, in some regions there are networks of cable for high-bandwidth TV signals. Since there are millions of kilometers of twisted pairs intended for voice service in the world, it makes sense to try to extend their use to other kinds of signals, namely video or data. Therefore, although the trend is moving towards an increased use of wireless communication systems, wired systems are still as relevant as they have ever been.

The telephone is for the voice, which, as is known, has a limited frequency bandwidth. Moreover, voice telephone quality is just a fraction (3.2 kHz) of the human vocal range (about 15–20 kHz) The telephone channel capability matches the voice signal band that it was intended to transmit. For data transmission, that low channel capability is a real limitation. Therefore, the challenge is to be able to accommodate a large bandwidth (carrying data at a high data-rate) on a low-bandwidth medium. This is done by the use of conventional modems or (for broadband) X-DSL (Digital Subscriber Line) modems. Figure 3.7 shows a possible block diagram of a DSL modem. Notice the transformer at the input that obtains a galvanic insulation of the electronics from the twisted pair. This is possible because the modem has its own power supply generator, and the power provided by the telephone line is not necessary. The block of the diagram immediately after the transformer is called a hybrid. Its function is to transform the two wires on the telephone-line side into four wires, two for the transmit and two for the receive path (these pairs are indicated in the diagram by arrowed lines, each representing a pair). The hybrid is also supposed to eliminate coupling between the transmit and receive connections of the four-wire paths. Another block to notice is the SLIC (Subscriber Line Interface Circuit), the driver and the interface for the telephone set.

Become accustomed to acronyms

This chapter has already used acronyms and abbreviations. It is obviously necessary to know what they mean. However, there are so many of them that it is difficult to remember the exact meaning of each. What is necessary is to be aware of what they refer to, and to consider them as new words.

Other examples of wired communications use optical fibers or the power network as their physical support. In the latter case digital data can be conveyed via the home power wires so as to create a household network using powerline communications modems, which can connect Internet appliances without the need for rewiring the house.

3.3.2   Wireless: Voice, Video and Data

The quality of wired communications is satisfactory, but the system requires a physical medium. This is a limitation that pushes technology in the wireless direction, in urban, suburban, and rural environments. As the reader probably knows, the wireless telephone service uses a cellular partitioning of the territory, with transceivers in each cell serving mobile cell phones in that cell. When a mobile exits a cell and enters another the system tracks it, and the mobile changes the channel to ensure continuity in connection and communications.

The investment required for base stations that ensure total signal coverage is high, and its compensation requires attractive services, including voice, video, and data; it is also advisable to do a careful check of performance, and of the trade-off of performance against cost, for various alternatives. Moreover, since radio spectrum is a limited resource it is vital to obtain optimum use of the allocated spectrum and channels. The result is often a mixture of analog and digital apparatus, such as the scheme for a base station already considered in Figure 1.6.

COMPUTER EXPERIMENT 3.1

Go Inside the Cellular Phone

This computer experiment allows you to look inside a piece of electronic apparatus that you meet every day: the cellular phone. This device is considered a necessity by most people throughout the world. Cellular phones are capable of doing a lot of things, including storing data, taking pictures and recording videos, surfing the Internet, sending e-mails. It even works as radio, MP3, MP4 player as well as a mobile games player and navigator. However, what is inside the cell phone between the touch screen and the battery is almost unknown. This experiment tries to overcome that limit, permitting you to:

- explore the hierarchical structure of a typical cellular phone (it may be that when you do this experiment the cell phone has already evolved to something more modern ...);
- have an idea on the main blocks that control the typical interface devices, such as the display, USB ports, SIM card, headsets, camera, etc.;
- go down a few levels of the architecture, starting from the highest level: a typical picture of the cellular phone and its possible PCB. At the lower level you have a generic block diagram with macro-functions implemented by other diagrams one level below.

THE CELLULAR PHONE

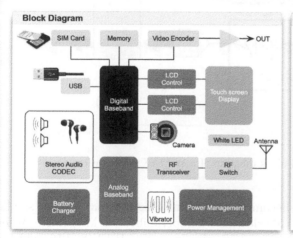

HOW TO EXPLORE

The starting diagram of this experiment has the top level in one window and the top functional diagram in a second window. Click on one of the blocks of the functional diagram and enter into it. Observe the new block diagram and notice the different functions performed. You can further descend down the hierarchy for other levels until the lowest available in this example.

The bottom of the description is an integrated circuit performing a specific function as required by the description one level above. The overall combination of functions performs what is required one level above and so forth.

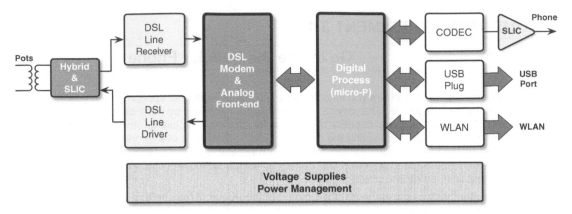

Figure 3.7 Simplified block diagram of a DSL modem.

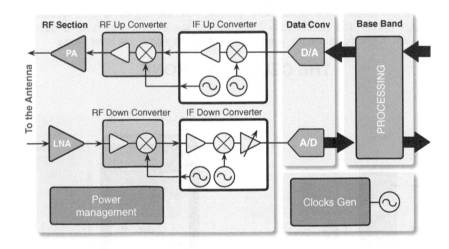

Figure 3.8 Possible simplified block diagram of a WiMAX base station.

In congested areas the density of users becomes too high, making necessary to use smaller cells or special standards. One of these is Worldwide Interoperability for Microwave Access (WiMAX), a wireless digital communications system, also known as IEEE 802.16, which is suitable for wireless metropolitan area networks. WiMAX provides Broadband Wireless Access (BWA) at a range of up to 3–10 miles (5–15 km) for mobile stations. There is also a need for small-area coverage like that of wireless local area networks, which in most cases limit the service to only 100–300 feet (30–100 m).

Figure 3.8 shows a possible simplified block diagram of a WiMAX base station. In addition to the names of blocks, diagrams specify more about what is inside. Obviously at this level many of the functions are just names and acronyms; what is important is to understand the level of complexity and the idea that different functions must be realized using analog or digital circuits. The block diagram of Figure 3.8 divides the system into four sections: RF, data converters, base band and clock generation. Notice that the RF section includes functions, called "up" and "down" conversion, obtained by a special circuit, the mixer, whose symbol is a crossed circle. It has two inputs, and the cross indicates multiplication. The

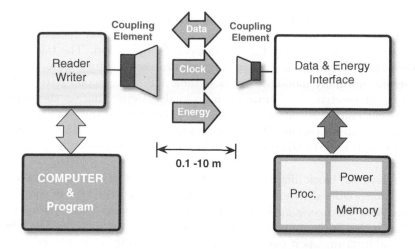

Figure 3.9 Block diagram of reader/writer and transponder of an RFID system.

displayed architecture shows up and down conversion in two steps, involving an Intermediate Frequency (IF). The diagram also outlines a section used to generate clocks in the system. These are important components for high-frequency circuits; generation and distribution of clocks is crucial for obtaining good performance. The power management section indicates the need for special attention in power handling. This is an important point for the effective operation of the Power Amplifier (PA), the block used to provide enough power to drive the transmitting antenna and reach the receiver.

Self training

There are many standards for wireless communication with associated frequency bands. Do a search on the Web and find the main features of two of them. Consider communications that:

- convey only voice over a short range (about 5 Km);

- convey voice, video, and data at high speed.

Write a short note on the results of your search.

3.3.3 RFID

RFID stands for Radio Frequency IDentification. This acronym is the name given to small systems used for storing and retrieving information about tagged items. Examples are luggage, the products of a store, and medical tools in the emergency room, but also food, animals, or people, for the purpose of identification and tracking. An Automatic Identification Procedure (Auto-ID) is necessary in many sectors, as barcode labels are often inadequate. The optimal solution is to store data in electronic format and to transfer the data without contact between data-carrier system and reader/writer. Power is often provided in a wireless manner.

RFID systems are made up of two components: the tag, or transponder, placed on the object to be identified and the writer/reader that sends the interrogation, receives and interprets data, and frequently also send energy for the transponder operation. The tag consists of two key parts. One is the electronics used for storing and processing information, for sending and receiving the signal, normally modulated at a suitable RF. The second part is the coupling element necessary for receiving and transmitting signals. Depending on the frequency used the coupling element can be a coil or an antenna. There are various standards for the frequency used. These systems can use frequency ranges as low as 135 kHz for a simple passive transponder or as high as 2.45 GHz for demanding applications, as high frequencies enable higher data rates.

Figure 3.9 displays the block diagram of an entire system. The distance between fixed station and mobile transponder depends on the frequency used and the type of coupling element. For example, in contactless access authorization systems or in smart label systems the distance ranges from a few inches (or centimeters) to a couple of yards (or meters). For labels used in monitoring surgical instruments it can be a couple of tens of yards (or meters).

Often the tag is batteryless and receives energy from the fixed station. It is therefore important to have low-power, low-voltage operation. This may critically affect the maximum distance between reader and transponder.

System specifications

Specifications of particular systems detail special parameters. They are a mixture of information about the electrical operation of the analog and digital sections and the mechanical details. The specs of an RFID tag may include the following.

Data Encryption If included, a few details on the method.
Collision avoidance Algorithm concerning the possibility of reading more than one tag in the same reader's field.
Technology Passive, semi-passive, active.
Supported Standards International Organization for Standardization (ISO), Institute of Electrical and Electronics Engineering (IEEE), ...
Power of the transponder Adjustable or fixed, power in stand-by, in mW.
Bending diameter In mm with a limit on tension of the label.

3.4   COMPUTATION AND PROCESSING

The most frequently used and popular electronic systems are for computation and processing. They are based on extremely complicated integrated circuits that contain hundreds of million or even more than tens of billion transistors. The huge number of basic components is such that the hierarchy used for representing circuits is essential, with, at the top, only global functions.

An important requirement of this type of system is the capability of storing, in some cases temporarily, digital data, and for this they use memories, devices obtained from the most advanced semiconductor technologies. Moreover, computation is done by using a combination of

COMPUTER EXPERIMENT 3.2

The History of the Microprocessor

This computer experiment illustrates a brief history of the microprocessor. The invention in 1971 by Intel Corporation of the legendary 4004 (the world's first four-bit microprocessor) was the beginning of the new era in computer applications. Gordon E. Moore, co-founder and Chairman of Intel Corporation, authored the so called Moore's Law, published in 1965. This law says that the number of transistors of an Integrated Circuit (IC) doubles approximately every two years. The law has been followed for decades and this experiment allows you to verify that. The continuous growth of complexity is the key to the enormous computation capability of many digital electronic devices. This experiment shows the evolution as a trip on an elevator: every floor you find a device of the successive generation. You can:

- go up and down through the different floors of ElvisLab's elevator;
- look at architecture, microphotographs and different packages used by the most famous (and widely used) microprocessors;
- acquire a general knowledge of the main features of the microprocessors, such as clock speed, die dimensions, number of transistors used on the chip, etc.;
- verify the long-term evolution in computing hardware as predicted by Moore's Law.

ELVISLAB'S ELEVATOR

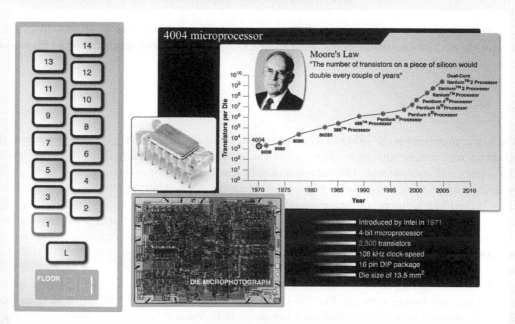

HOW TO EXPLORE

On the left side you have the elevator control. Push the up or down button and change the floor's number. Stop for a while in each floor and discover the features of the device displayed there. That is a quick trip along microprocessor history. Enjoy it.

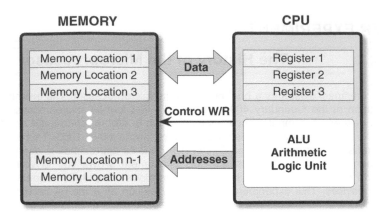

Figure 3.10   Example of simplified CPU with three registers.

circuits and procedures, or, as the reader certainly knows, by the combined action of hardware and software. The design of complex hardware cannot be done by a single person, or even by a single technical discipline. Indeed, there is not only diversity in specialization (such as expertise in either analog or digital circuits) but there are distinct professions; there are system architects, high-level hardware description language experts, specialists in CAD tools, and so forth.

The following section gives short descriptions of integrated circuits used in schemes for computation and processing systems, with some examples of block diagrams used in high-level representations.

3.4.1   Microprocessor

The microprocessor is the brain of a computer, and its evolution in terms of speed, processing capabilities and complexity have determined the present explosive growth of performance in modern personal computers. The introduction of the first microprocessor dates back to the early 1970s, when they were used for pocket calculators. Now they are very complex devices with an incredibly high computation speed.

In addition to their use in computers, the microprocessor is employed in generic applications (in the case of general-purpose processors) or in specialized control-oriented applications; in these cases it is called a microcontroller. Indeed, the two names indicate slightly different devices: the microprocessor normally needs some external components, such as memory, or devices for receiving and sending data for its use. The microprocessor is the heart of a computer or a complex processing system. On the other hand, a microcontroller is all of that in one; it does not require external components because all necessary peripherals are already built into it.

An important part of the operation of computing systems is the memory that is embedded in a microcontroller. Another key part is the Central Processing Unit (CPU) with its built-in capability to multiply, divide, add, and subtract. The overall architecture is quite complex; however, to have an idea of a possible simplified organization of blocks, let us consider the diagram in Figure 3.10. The displayed CPU uses three registers. They are memory locations used to temporarily store the input and output of arithmetic operations. The Arithmetic Logic

Unit (ALU) is the block used for arithmetic. The memory interacts with CPU via busses: one for addresses and another for data. As is known, a bus is a group of wires (normally the number of wires is a power of two). Around the blocks of Figure 3.10 many other blocks perform various functions. Examples are the input/output unit, the timer, and the watchdog, a circuit that checks for the occurrence of possible flaws in normal operation. This block, not indicated in the figure, is for validating the execution of programs at every clock cycle and, in case of errors, restarting the cycle.

Self training

Use the Web to compare parts from different manufacturers. Prepare a comparison chart of microprocessors based on different key features. In particular, consider:

- the number of transistors;

- the CPU speed;

- the bus speed.

Find the prices of three microprocessors and correlate performances and cost.

3.4.2 Digital Signal Processor

In addition to the need for higher microprocessor speed it has also become necessary to have a dedicated processor to perform repetitive, high-performance, numerically intensive tasks. The new device is called a Digital Signal Processor, or DSP, and, as its name suggests, it is used for heavy-duty processing in digital form. To be precise, the DSP is a number-crunching machine capable of performing Fast Fourier Transforms (FFTs) or complex functions such as audio and video signal compression and decompression, which were previously only possible with mainframe computers. Algorithms implemented using DSPs can be extremely complex, but their basic operations are simple. They mainly consist of the four simple arithmetic functions $(+, -, *, /)$, carried out at incredible speed.

Figure 3.11 shows a typical DSP block diagram. The functions in the figure are generic, with few details about data flow or the busses used for data transfer. There may be other block diagrams to outline special features such as, for example, power-down capability.

Because of the dissimilar needs of applications, various different DSPs are available in the market. They are classified by many factors, such as data width (the number of bits the processor handles) and the type of arithmetic performed (fixed or floating point). For example, we have DSPs with 32-bit processors, 24-bit processors, and also 16-bit processors. For portable systems, such as biomedical monitoring devices powered by tiny batteries or an energy scavenger, consumed power is very important. These applications necessitate special care over the technology to ensure very low leakage power – that is, power lost whether the circuit is active or idle.

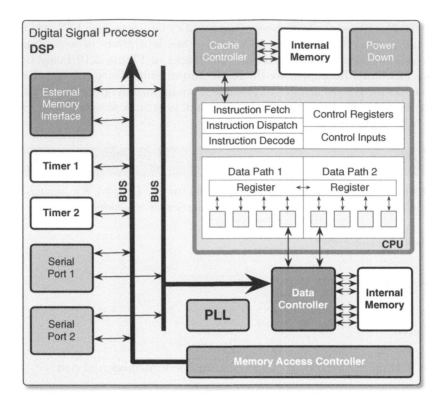

Figure 3.11   Possible block diagram of a *DSP*.

3.4.3 Data Storage

The storage of electronic information is essential for many applications. The systems used for this are called memories, and they differ from one another in the type of supporting media they use, in speed of access, and in data capacity. As discussed before, microcontrollers and DSPs have memories on board. They use Random-Access Memory (RAM), which is often volatile; i.e., it loses the data when the power goes off.

In addition to the data storage necessary for signal processing it is necessary to store data for archival purposes. Mass storage devices with huge data capacity, such as the Hard Disk Drive (HDD), give the scope for this. As you will know, the HDD stores digitally encoded data on rapidly rotating platters with magnetic surfaces. The disk rotates at several thousand Revolutions Per Minute (RPM) with a media transfer rate of several Gbit/s. The capacity of a HDD is many thousand of GB or even TB (G means billion, and T is trillion).

Integrated technologies enable miniaturized solid-state memories that, thanks to very advanced packaging techniques (like the three-dimensional stacking of chips shown in Figure 3.12) satisfy the need for mass data storage. The memory sticks thus produced are increasing in capacity, and for many applications they now replace the HDD. Basic devices and circuits for solid-state memories are studied in some detail in a later chapter.

COMPUTER EXPERIMENT 3.3

What is Inside?

This computer experiment looks for the answer to the question: "What is inside?" The ElvisLab board of this experiment incorporates two selectable black boxes (you can choose one or the other) driven by two signal generators and observed by a scope. An unknown function, which processes the two input signals, is implemented by each black box. The two generators give sine wave signals at output, connected to the Channels 1 and 3 of the scope. Your task is to change the input signal, observe the output and derive the simple processing performed by the black box. With this experiment you can:

- change amplitude, frequency, and DC level of two sine waves. One of the two phases can be also changed;
- switch between the two black boxes performing different processing functions;
- change the unknown function of each black box. The operation occurs automatically when you start the experiment or you switch between one black box to the other. The board chooses two parameters of the unknown processing function in a random manner.

MEASUREMENT SET–UP

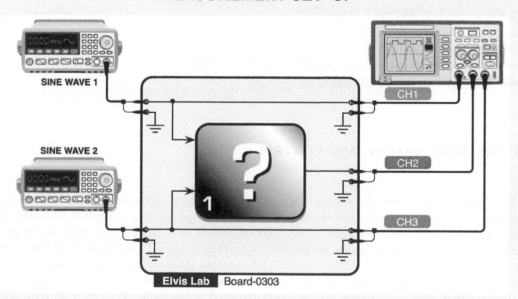

TO DO

- Choose one of the two black boxes and set the amplitude of one sine wave to zero.
- Vary amplitude and frequency of the other sine wave to look at results. Do the same for the other input signal. Apply both signals and look for possible superposition behavior.
- In the case of addition of results, extract the performed function by changing the input amplitudes and verify a possible guessed processing.
- In the case of non-linear behavior, make a reasonable processing assumption based on experimental measures and verify the assumption with new experiments. Try, for example, equal frequencies of the input sine waves or set one or both frequencies to zero, for using only a DC signal.

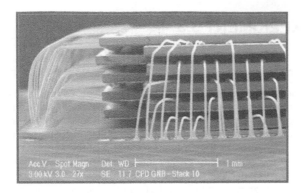

Figure 3.12 Stack of 10 chips and multiple bonding. Reproduced by permission of © ST Micro-electronics.

System specifications

The specs of a stick memory may include:

Maximum transfer rate in MB/s;
Maximum capacity in GB;
Maximun data transmission clock serial or parallel, in MHz;
Durability number of plug/unplug operations (in a room or a harsh environment).

3.5 MEASURE, SAFETY, AND CONTROL

Electronic circuits used for the signals produced by sensors are often for safety or control purposes. The core elements are sensors that transform physical or chemical variables into electrical variables, often voltages. The band of signals is normally low, but amplitude is weak, sensitive to other variables, and, often, blurred by noise. Thus, the task of electronics is to augment the amplitude and to separate information from spurs and interfering terms.

Increasing amplitude means the use of amplifiers with large gain that do not add noise or shift the signal level. Therefore, the function required is low-noise, low-offset amplification. When the signal is large enough, a data converter may change the information to digital form for suitable further processing. When signal is mixed with spur it is necessary to exploit some difference in their features. For example, a dc or low-frequency signal corrupted by noise is revealed by averaging or filtering. Averaging removes random components with zero mean. Filtering highlights frequency components featuring the quantity being measured.

Suppose a system is to detect a low-level light, but interference and noise make the measurement difficult. If the light is chopped with a square wave, as conceptually indicated in the arrangement shown in Figure 3.13, the input signal becomes the signal plus spurs during one half period, and only the spurs during the other half. The electronics can discriminate between these two situations by using repetitive measures and averages. Other techniques, some discussed below, involve open-loop or closed-loop interaction with sensors.

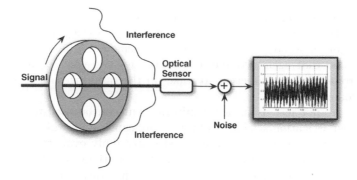

Figure 3.13   Chopping of input signal to distinguish it from noise.

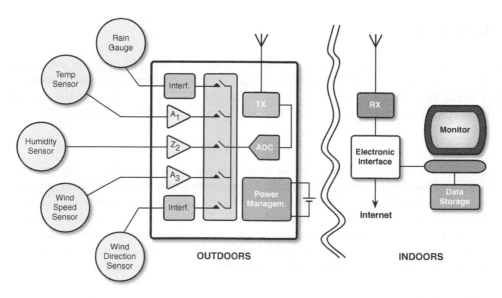

Figure 3.14   Block diagram of weather station consisting of outdoor and indoor sections with wireless communication.

3.5.1   The Weather Station

Quantities frequently measured are those featuring weather conditions. Sensors used include the anemometer for measuring wind speed, the hygrometer for humidity, the rain gauge, the solar radiation meter, the thermometer, the UV index meter, and so on. All of these quantities are often collected and displayed by a single device, the weather station. This may be interfaced to a personal computer, on which data can be displayed, stored, or transmitted via the Internet.

Figure 3.14 shows the block diagram of a possible weather station. It is made up of two sections, one outdoors, with five wired connected sensors and wireless communications with the indoor section, which is connected to a PC and thus the Internet. The signal of some sensors needs amplification or transformation, obtained by using suitable blocks. An interesting feature is that after analog preprocessing the system uses only one data converter to transform five signals into digital form. This, in addition to saving hardware, indicates that measuring weather quantities can be done sequentially at a rate determined by the measuring needs. Since the outdoor section uses batteries, it is worth measuring data just once in several hours and keeping the station in stand-by between two measurements. The communication system is one way. It may be that such a system could also use two-way communication, with a signal being sent from the indoor section to wake up the outdoor part so that it can take measurements on demand.

Weather stations are not just useful for monitoring but can be used to improve efficiency in agriculture. Proper irrigation schedules obtain significant savings of water without limiting effectiveness. For this, an electronic system estimates daily evapotranspiration by using microclimate parameters obtained from multiple weather stations, wirelessly connected, and determines the amount of water required. In this case the station is just a part of a more complex system, which also uses actuators for irrigation control.

3.5.2   Data Fusion

Often, sensors are not just sensitive to a single physical or chemical quantity but respond, with different sensitivities, to many quantities or substances. Therefore, a sensor produces an electrical signal that depends on various relevant inputs, $X_1, X_2, \ldots, X_N$. If the maximum sensitivity is to one of these, X_1, it is said that the sensor senses that quantity, but in reality the output is

$$Y = f(X_1, X_2, \ldots, X_N). \tag{3.1}$$

For example, a sensor that senses carbon monoxide can also sense alcohol, and its output depends on temperature. For medical diagnosis, information given by a single parameter, either X-ray imaging, Nuclear Magnetic Resonance (NMR), or simpler measures such as temperature or blood pressure, are not enough for reliable diagnosis. Each parameter denotes many possible diseases, and, in some cases, will not be significant. In the same way, the information that a system wants to detect is often not univocally determined by the output of just one sensor but by the combined and concurrent action of many sensors.

> **Did you know that . . .**
>
> data fusion mimics the learning process of the brain? Concurrent measures of different quantities give rise to decisions after a learning phase. The same is done with circuits or computation procedures called neural networks.

The technique used to improve reliability and robustness is called *data fusion*. This method uses signals from multiple sensors and obtains the result from a synergic combination of information. Data fusion is, indeed, what is done by the human brain in order to recognize a quantity. It processes information from sight, sound, smell, and taste in order to make decisions. In a similar way electronic systems obtain data fusion. They are a combination of sensors, electronic interfaces, processing of data, and generation of outputs.

The use of multiple sensors is the best way to detect the presence of a specific condition or object, and to classify, track, and monitor it. Therefore, to identify entities such as aircraft, weapons, and missiles, military systems use multiple sensors for object identification based on different principles, such as radar, infra-red, and electromagnetic sensors, and Synthetic Aperture Radar (SAR). The fusion of data confirms inferences, increasing confidence in the final response.

COMPUTER EXPERIMENT 3.4

General Car Control

With this experiment you can identify the main electronic blocks used in the engine control of a car. As you probably know the combination of hardware and software creates complex electronic systems. Basically there is a microprocessor, several drivers, and input/output interfaces for the control of various functions. For modern cars we have fuel injection, ignition timing, idle speed, air conditioning, engine coolant temperature, etc. There are different signals coming from engine sensors that must be processed in real time, i.e. within a time that is short enough to allow "immediate" reactions. All those functions are implemented in the engine control electronics.

Also for this experiment you can:

- go down in the hierarchical description of the architecture and explore the organization of the electronic engine control;
- perform a brief overview of the electronic blocks used for controlling various parts of the car and become aware of the functions required.

GENERAL CAR CONTROL

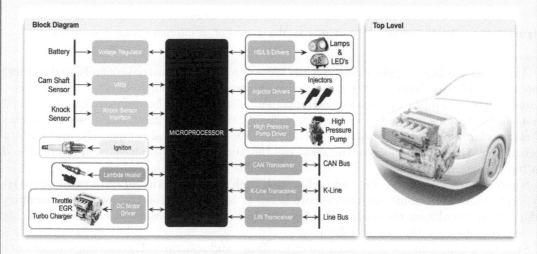

HOW TO EXPLORE

The top level of this experiment is the view of the engine and the block diagram of the entire control system. At lower levels you find various additional block diagrams and their functions.

You have just to click on one of the highlighted blocks and enter into it. Go down again one or more levels to explore the system organization. Starting from the top level you will learn about various electronic blocks that perform the electronic control of the car engine.

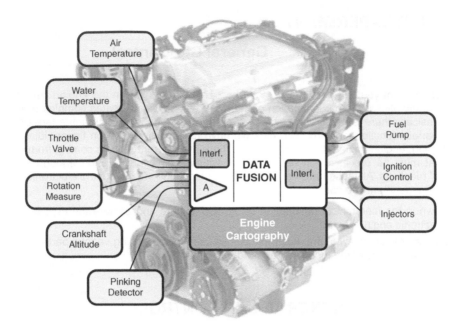

Figure 3.15   Block diagram of data fusion for the control of an internal combustion engine.

Among non-military applications, it is the areas of robotics, industrial controls, and medical applications like ophthalmology that mostly benefit from data fusion. Robots use video images, acoustic signals, and infra-red and electromagnetic signals to gather information. Industrial controls use various measures of status for optimal performance. Electronic noses use arrays of chemical sensors with diversified sensitivity to different chemical compounds, and pattern recognition. Often the sensors of the array use the same device, but mechanical filters placed on top change its sensitivity.

Sensors employed should be compatible with each other in terms of data-rates, sensitivity, and range. Electronic circuits support that. It is also essential to have complementary information to enhance the quality of inference. In some cases sensors that almost duplicate information improve the system reliability.

Systems for data fusion are a mixture of hardware and software. The hardware is for data acquisition, and the software is for data processing and for implementing algorithms. The processing can be done using a PC or, for stand-alone systems, by having on-board processing facilities.

Notice that ...

the activity of an electronic engineer increasingly requires a wide spectrum of knowledge and the consequent capacity for teamwork.

To give you an example, consider data fusion for the control of an internal combustion engine in a modern car. There are many sensors, as shown in Figure 3.15. They measure different physical quantities that enter the electronic system. After interfacing and preprocessing the data fusion block performs data analysis and, using the information from an engine map,

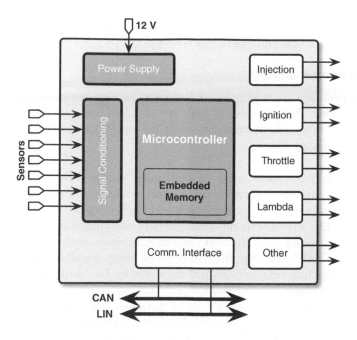

Figure 3.16 Block diagram of a control unit used in cars.

determines parameters that control the operation. Thus the system gives rise to outputs by putting together information obtained from many heterogeneous sensors to produce a single composite picture of the engine controls.

3.5.3 Systems for Automobile Control

As well as the above example using data fusion, both combustion engine and electric cars use electronics to monitor and control their functioning. Systems use various sensors and transducers to operate the engine, brakes, and transmission, or to enhance vehicle stability and safety, and passengers' comfort. Such systems, in some cases separated into an Engine Control Unit ECU), an Electronic Throttle Control (ETC), and a Transmission Control Unit (TCU), manage ignition timing, variable valve timing, and the level of boost in turbocharged cars. The throttle control determines, using commands from the ECU, the position of the throttle valve without any mechanical cables from the pedal. The transmission control unit determines the transmission shifting sequence.

In electric vehicles the electronics takes care of battery power management, since the operative life of rechargeable batteries depends critically on charging and discharging control. Voltages and currents are continuously measured to ensure best effectiveness. This is also necessary to guarantee safety because the high concentration of energy can give rise to explosions or fire.

The electronic units in cars are assembled in small boxes located in the engine or the passenger compartment. In the former case environmental conditions are extremely harsh, with low or very high temperatures. Because of this, instead of PCBs ceramic boards that withstand very high temperatures are often used. Moreover, these units must be waterproof and acid resistant. The architectures of all of these units have similar block diagrams; as shown in Figure 3.16, they include a microcontroller and embedded memory as the core, have interfaces for multiple sensors, and operate actuators.

Electronics in harsh environments

In-engine or in-transmission electronics can be exposed to high temperatures, well above the conventional 125°C limit. Temperatures as high as 150–175°C can require special technologies and packaging. In all cases, the temperature must be kept below 400°C to avoid permanent damage to silicon integrated circuits.

Since modern automobiles have many control units, it would be necessary to use many wires to provide communications between them. This would create difficulties because of the resulting vast amount of wiring, increased weight, and hence higher overall costs. This would also affect efficiency. The solution to this problem is to use single- or dual-wire serial busses, like the early CAN (Controller Area Network) and its successors, such as J1850 and the cheaper LIN (Local Interconnected Network). Instead of using many parallel wires to obtain parallel communication, a serial bus transmits digital data, bit by bit, sequentially over one or two wires connecting all devices.

The use of serial busses is beneficial for short distances and low bandwidth signals. The exchange of data uses suitable protocols and data rate. For instance, CAN can transmit at up to 1 Mbps (MegaBit Per Second), while LIN is for low communication speeds (20 kbps). The choice between types of bus depends on signal bandwidth and the required reliability (high for the control of steering and brakes). For non-critical cases, such as door control (lock, window lifting, and mirror control), lighting, climate regulation, seats, and rain sensors, the use of an inexpensive bus can be convenient. LIN is an example of this. Since it uses a self-synchronization mechanism, it does not require a quartz oscillator. It can be implemented in conventional microcontrollers with no additional hardware except for a driver device. It is always advisable to make technical choices while bearing in mind such economic constraints.

3.5.4 Noise-canceling Headphones

Other examples of measure and control are given by echo cancellation, noise suppression, and sound enhancement techniques, all made possible by electronics. Noise-canceling headphones improve the sound quality in noisy environments by analyzing the surrounding noise by using a small microphone, detecting the noise and interferences, and canceling them in an active manner. Figure 3.17 plots the signals necessary to describe their principle of operation. The signal received by the ear is a mixture of desired sound with noise plus interference. The effect of the undesired part is canceled by recognizing it and generating an inverted replica, the anti-spur. The addition of received signal and anti-spur gives the desired signal. Thus, the inverted waveform actively cancels the noise and the interfering signal.

The key function of noise-canceling devices is to recognize an unwanted signal and to reproduce its opposite effect at the speaker. By their construction, headphones block out, albeit partially, external noise, because the ear-cups absorb the external signal, especially at high frequencies. Therefore, in addition to passive attenuation it is necessary to enhance the ear-cup effect with the anti-noise signal – mainly necessary for low-frequency interfering signals.

The active cancellation is obtained by two methods: the feedback type and the feedforward type (Figure 3.18). In the former case there is a microphone near the ear. It detects the signal

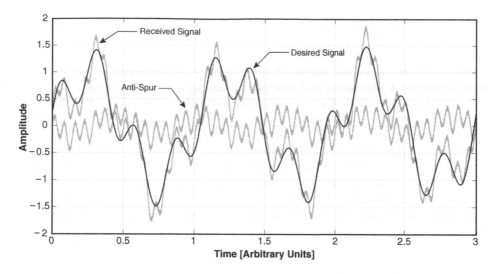

Figure 3.17 Signals involved in the noise-canceling method.

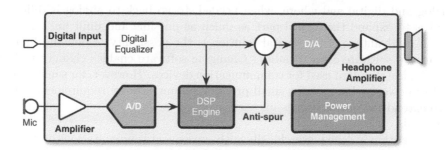

Figure 3.18 Noise-canceling headphones: feedback type and feedforward type.

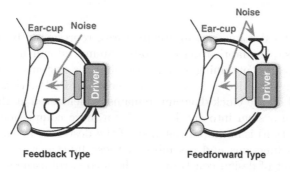

Figure 3.19 Block diagram of feedback noise-canceling headphone control.

and sends it to the driver, which changes the control until the expected sound is obtained. The latter approach uses an external microphone that measures the interference, and, after processing that predicts the attenuation of the ear-cup, mixes the opposite result with the signal. Obviously the performance of the feedback method is superior, but the approach is less suitable for miniaturization.

Figure 3.19 shows the block diagram of a possible feedback driver. The sound detected by the microphone near the ear is amplified and converted into digital form. This signal and the digital input, passed through a digital equalizer (which is an additional option on the headphones), are inputs of a DSP engine that determines the anti-spur signal. This is added to the input before D/A conversion and headphone amplification. The block diagram outlines the power-management block that provides supply voltages for the entire system.

Notice that the described architecture uses feedback. The output is fed back to the input to improve the quality. In this case the goal is to remove unwanted sounds. More generally, feedback serves to improve quality and for processing or control. An important point to notice is that correction must be quick, because if there is excessive delay the correcting signal is not a good replica of what is desired, causing a degradation of performance. Therefore, processing and generation of the anti-spur must be done in a very short time. This is not a problem when the system uses digital techniques. As you probably know, digital circuits operate at extremely high speeds.

3.6   SYSTEM PARTITIONING

Earlier sections discussed many system architectures, represented by block diagrams that outline specific features. For fabrication of those systems we can use either advanced microelectronic technologies for obtaining systems on chips or systems in packages, or we can employ surface mounting of active or passive components on PCBs. However, the distinction into blocks, as specified by the block diagram, remains evident even at the physical level.

The partitioning of a system into sub-blocks (for a more comprehensive description but also for effective fabrication) is an important task of the designer. It involves two key issues: identification of partial functions and assembly of those functions to realize systems. The result has to satisfy a set of design constraints such as cost, robustness, performance, power consumption, and so on. Cost, remember, is one of the most relevant constraints.

An easy distinction is between analog and digital, but deciding where to have the border between analog and digital and where, when needed, to go back to analog is difficult. The present trend is to expand the digital part as much as possible and limit analog processing to interfaces. An advantage of the digital solution is the possibility of defining functions by software (as in the software-defined radio). Changing software enables changes to functions like, for example, the standard used for communication devices. However, for some applications where ultra-low power or low cost for small product volume are key requirements, it may be convenient to extend the analog processing section or, at the limit, to realize all the functions by analog methods.

The system partition is done hierarchically with consequent representations of various block diagrams. The first step involves decisions at a high level that account for non-electronic issues. Indeed, such decisions are trade-offs between benefits and limitations, including, as already mentioned, cost. For example, a high performance A/D converter can relax a requirement for analog filtering, but the cost of commercially available parts possibly leads to bad cost/performance ratios. The use of a function inside a block can avoid the need for drivers that consume power.

Often a standalone part implements partial functions. We have, for example, amplifiers, data converters, filters, CPUs, and memories. They provide given performance and can be used in different systems. The availability of basic blocks provides obvious indications for

COMPUTER EXPERIMENT 3.5

What is Inside? (Again)

This computer experiment offers you another ElvisLab board that implements unknown functions represented by black boxes. The board uses a single sine wave generator at input and has two outputs. The board has two external controls that change the processing functions. Notice that the controlling parameters do not necessarily give rise to a linear operation but are constant in time. There is another unknown constant parameter that controls the processing. Its value is randomly determined at the beginning of the experiment by an internal random number generator.

This experiment allows you to

- change amplitude, frequency, and DC level of the sine wave input signal;
- switch between the two black boxes performing completely different functions;
- change the behavior of each box by the setting of two external controls.

MEASUREMENT SET–UP

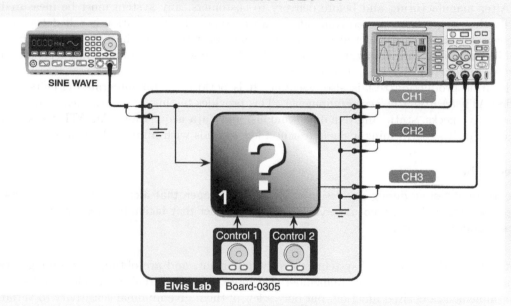

TO DO

- Choose the first black box and set the controls to zero; vary amplitude, frequency, and DC level of the sine wave input and observe the output. Make a first assumption of the two input-output responses.
- Set the controls to a given value, one by one, and refine the input-output guessed models.
- Set both parameters and observe the overall responses. Try to derive the complete input-output transfer functions. Change the sine wave signal and parameters to verify your model.
- Switch now to the second black box and repeat the above steps. What are two possible functions that you cannot identify with the input signal provided with this experiment?

system partitioning. Also, basic blocks are available as IP (Intellectual Property). This means not a real off-the-shelf component but its well-defined and verified design, available to be incorporated in integrated circuit architectures.

System partitioning is the task of system architects, especially when designing complex digital systems. What has been briefly discussed here is the functional partitioning useful to distinguish separate modules that in the past were basic components such as transistors, resistors, or capacitors, and now are ICs or IP. For medium complexity partitioning is done manually, but more and more automatic tools are used. They are based on the search for minimal defined objectives, which are called cost functions.

At this level of study it is not useful to go into further detail. However, it is important to be aware of the issue and to know that for high complexity, and in other cases too, partitioning is a critical step of the design activity.

3.7   SYSTEM TESTING

After manufacturing and before delivery to customers, any system must be measured to verify that its performance corresponds to expectations. This operation is called *testing*. At the very beginning of product lifetime it may be necessary to debug the system and to improve yield. This initial characterization helps in identifying possible production problems and may involve minor, but essential, changes of design.

Routine production test is also essential. It is performed, for integrated circuits, on so-called ATE (Automatic Test Equipment). This provides testing resources such as probes for continuity checks, static and dynamic stimulus, and data acquisition. An ATE tests a large quantity of ICs, performing a single test in a few seconds with minimal human interaction.

> **Remember**
>
> Be aware of testing during any design phase. It may happen that architectures are not testable because some points are not observable. Also, remember that facilitating testing improves the cost/performance ratio.

Quality and reliability of systems are very important, and are obtained by taking care in fabrication but also in design. Remember that more than one circuit or architectural solution can implement a desired function, but only a few of them give minimal sensitivity to variation in the parameters and to aging. Therefore, using those particular solutions ensures quality. The quality is normally defined by the defect-free fraction of the number of shipped units. If the process yield is higher than the required quality level, testing can be performed on a randomly selected subset of devices, especially at the intermediate steps of the fabrication flow.

Accelerated aging tests determine how products perform under extreme field conditions and how long performance remains unchanged. The most common of these are High Temperature Operating Life (HTOL), autoclave or pressure-cooker tests. Measures before and after each stress test look for possible shifts in parameters.

Since the cost of testing equipment is very high, long testing procedures can be too expensive. To minimize costs, the design of architectures and circuits should consider testing needs at the very beginning and throughout the entire design flow. Often, systems are so complex that identifying possible failures is not easy. For this situation, techniques that reconfigure the architecture just for testing purposes, or methods called "design for testability," are essential.

PROBLEMS

3.1 Find on the Web the block diagram of an electronic toy. Analyze the block interconnections and understand the function of each block. Make a list of electronic functions needed to realize the product.

3.2 Consider a hypothetical electronic system and briefly describe its features. The use of the Web can help you in the search for a suitable system. Prepare a table with key specifications indicating, if necessary, worst cases.

3.3 An induction cooker used to replace traditional gas cooktops is flameless, does not use red-hot heating elements, and does not suffer from incomplete combustion. Find on the Web the basic principle of this system, and draft a possible functional block diagram. Suppose that the control is by an array of touch switches, and gives information on temperatures on a Light-Emitting Diode (LED) display.

3.4 A remote keyless system activates car functions without physical contact. Consider a system operating as both a remote keyless entry system and a remote keyless ignition system. Suppose it will have a button for locking or unlocking doors, another for opening the rear tailgate, and a red panic button that activates the car alarm. The key closes any open windows and the roof, and receives a tone indicating that the operation is in progress. Draft a block diagram and list key specifications.

3.5 Find on the Web the block diagram of a flash card. What kind of data interface and protocol does it use to communicate with the external world? What is the capacity? What is the transfer speed? Figure out what the error-checking operation is.

3.6 There are many types of DSP on the market. Specifications distinguish them. Find via a search on the Web three different DSPs, and compare the bits in the data-bus, package, clock speed, and power dissipation.

3.7 Suppose you want to design a system that detects the passage of a boat near the shore. You use three sensors: one detect waves, the second is a sonar that reveals objects underground, and the last senses the reflection of a lighthouse light beam. Describe using reasonable equations the sensitivity of the sensors to the boat, underground rocks, and the wind changing direction. What is the processing needed for data fusion?

3.8 What, in your opinion, is it necessary to verify in order to ensure the correct operation of a noise-canceling headphone? Suppose that it will use a mixer and three sound generators, one driving the headphone speaker and the other external speakers. How do you quantify the result, and what can be a go/no-go choice?

3.9 Do a search on the Web and describe in a short report the different communication protocols used in car.

3.10 What is system partitioning, and what are the benefits of the method? Use one of the block diagrams described in this chapter, and make a suitable partitioning for a PCB implementation.

ADDITIONAL COMPUTER EXAMPLES

Computer Example - A/3.01

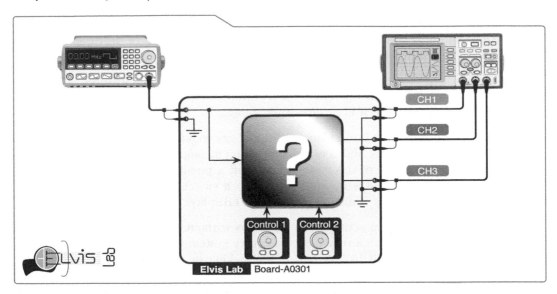

An *ElvisLab* experiment that enables you to look inside a linear circuit. You have to control the input, look at output and discover what is inside.

Computer Example - A /3.02

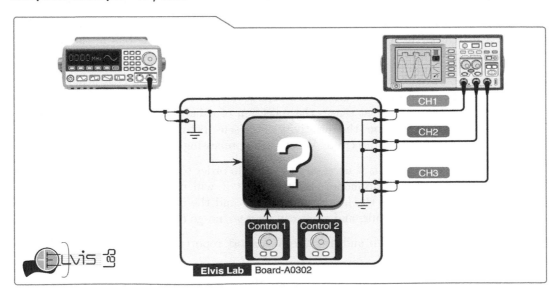

With this experiment you look inside a non-linear circuit. Remember that as it is non-linear, the superposition principle is not valid.

CHAPTER 4

SIGNAL PROCESSING

Signals generated by sensors, received by aerials, or fetched from storing media often need to
have some of their features changed for optimum use. This modification, performed by electronic
circuits, is what we call signal processing. This chapter studies different types of processing and
analyzes tools for doing it. We shall distinguish between linear and non-linear, analog and digital
processing. Moreover, we shall study how to estimate responses, we shall learn analysis methods
for linear processing, and we shall discuss filtering in the frequency and the z-domain.

4.1 WHAT IS SIGNAL PROCESSING?

Processing, in common language, is the blending of ingredients to obtain a particular end.
The results, desirably, have better features than the initial ingredients. A set of processing
operations constitutes the "recipe" of the transformation. Since the "ingredients" of electronic
circuits are signals, the input(s) and output(s) of electronic processors are signals as well. They
can be voltages, currents, or, perhaps, variable resistances or tunable capacitances. The set of
operations that change input signals into outputs are the equivalent of a recipe.

Figure 4.1(a) shows the generic block diagram of a signal processor. It has a number of
inputs (say n) and of outputs (say m). The block of Figure 4.1(a) establishes a relationship

Understanding Microelectronics: A Top-Down Approach, First Edition. Franco Maloberti.
© 2012 John Wiley & Sons, Ltd. Published 2012 by John Wiley & Sons, Ltd.

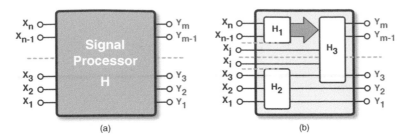

Figure 4.1   (a) Generic signal processor; (b) hierarchical representation.

between inputs and outputs given by the concise equation

$$\mathbf{Y} = \mathbf{H}\{\mathbf{X}\}, \tag{4.1}$$

where $\mathbf{H}$ is the processing "recipe." The outputs correspond to the unknown element of a mathematical problem, because, for given input variables, $\mathbf{X}$, the solution of (4.1) determines the unknown output variables, $\mathbf{Y}$.

Figure 4.1(a) depicts processing as a "black box" because operations are performed inside the block without disclosing details. However, the processing function $\mathbf{Y}$ can be broken down into a set of sub-processing functions with subsets of the inputs. The results obtained are subsets of the outputs. Some of the outputs are not outputs of the overall processor, because they are just inputs of internal sub-functions. A more specific description is given by the set of equations

$$\mathbf{Y_i} = \mathbf{H_i}\{\mathbf{X_i}\}, \quad i = 1, \dots, k, \tag{4.2}$$

where k is the number of sub-processing functions.

Figure 4.1(b) illustrates the model. Its diagram is very similar to that of one of the systems discussed in the previous chapter. The scheme, in addition to being a good way to represent a mathematical problem, uses new variables that often correspond to equivalent signals in the system. These new variables are relevant to understanding the process better and help in obtaining good operation. The study of black boxes that only have external terminals is difficult. By contrast, looking inside facilitates a step-by-step comprehension and favors optimization, because an internal malfunction reflects a system malfunction.

Don't forget

Real circuits with operational limits can still process signals. The results may or may not be acceptable, depending on the required precision. Accuracy of results depends on the accuracy of the components used.

Obtaining equivalence between mathematical equations and systems is the goal of signal processing. However, since processing is done with circuits, the fabrication and functional limits of circuits affect the accuracy of the results. One functional limit is, for example, the constrained amplitude of signals, which, typically, cannot exceed the supply voltage range. Suppose we are using a single 1.8 V supply generator; since circuits need a margin above and below the 0–1.8 V interval, analog electronics can generate good signals only with amplitudes in the range, for example, of 0.3–1.5 V. Accordingly, signal processing cannot use signals,

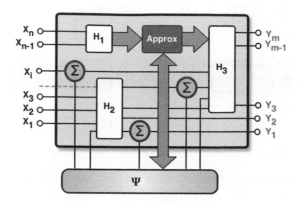

Figure 4.2 Block diagram of a processing system with unwanted spur signals.

even internally, that exceed those limits, otherwise the results produced are badly formed or clipped.

Another point is that, in addition to the signals used, there is electronic noise, which, as everybody knows, degrades the SNR. Further, because of the approximations used in digital implementations of the processing recipe, another form of noise corrupts the results. Approximations, such as rounding of data or rough coefficients, add extra errors that, under some conditions, are equivalent to noise. In these cases the SNR of digital signals is degraded.

If the various sources of error and noise make up a set of variables $\mathbf{\Psi} = \psi_1, \psi_2, \ldots, \psi_k$, equation (4.1) should be modified to

$$\mathbf{Y} = \mathbf{H}\{\mathbf{X}, \mathbf{\Psi}\}. \tag{4.3}$$

Figure 4.2 shows a system with spurs. Some of them are added to signals, and others enter processing blocks because their effect is not additive but operates in a non-linear manner. Note that, for simplicity, only spurs that really matter should be considered. The ones with negligible effect must be disregarded. This qualitative statement can be better expressed as one regarding the expected overall accuracy. If, for example, an output admits an error of 0.1%, spurs that produce errors much lower than 0.1% can be ignored. Therefore, for any particular case, only spurs whose effect exceeds a given limit must be considered. Caring for other spurs just increases cost and limits profit.

Good processors?

A good processor is able to reject spurs caused by interference and unwanted signals coming from the supply connections. The quality of processors also includes their ability to operate in harsh conditions, such as very high temperatures, radiation, and shocks.

Obviously, unwanted signals should affect output minimally. Therefore, a key designer task is to find the optimum implementation of equation (4.3). There are often disparate architectures that can implement the processing needed by that equation. Moreover, several circuits are able to realize the same architecture. Designers must therefore identify the solution that has minimal sensitivity and maximal robustness against unwanted signals.

In addition to noise, accuracy is limited by other factors, such as temperature, humidity, light, aging, electrical shocks, and interferences between analog and digital sections.

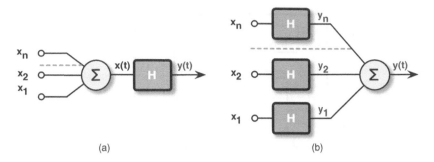

(a) (b)

Figure 4.3 (a) Linear system with superposition of signal at input; (b) equivalent processing.

Those types of error are normally dealt with by using various circuit techniques (some of them will be studied in this book) and taking special care in fabrication and use.

4.2 LINEAR AND NON-LINEAR PROCESSING

An important difference between processing blocks is their linear or non-linear operation. The key property of linear systems is the superposition principle, which states that if the input $x(t)$ is split into two (or more) parts

$$x(t) = x_1(t) + x_2(t) + \cdots \tag{4.4}$$

and $y_1(t)$ is the output with x_1 at input while $y_2(t)$ is the output with x_2 as input, then

$$y(t) = y_1(t) + y_2(t) + \cdots . \tag{4.5}$$

Therefore we can apply various components separately and by adding the results obtain the output required, as outlined in Figure 4.3. The superposition principle is important because the Fourier transform represents any signal as the superposition of sine waves. Therefore the output of a linear system is simply calculated by superposing separate sine wave responses.

Linear processors use linear operators or linear combinations of linear operators. Key linear processing functions are multiplication by a constant, derivative, and integral. Therefore, for example, the following processing of two inputs, $x_1(t)$ and $x_2(t)$, is linear:

$$y(t) = K_1 x_1(t) + K_2 \frac{dx_2}{dt} + K_3 \left[x_1(t) - \int x_2(t) \, dt \right]. \tag{4.6}$$

Linear processors in the discrete-time domain (with sampled-data signals) use the linear operator delay or a linear combination of signals and delayed signals. Therefore, for example, the following sampled-data processing function is linear:

$$y(nT) - y(nT - T) = K_1 x_1(nT) + K_2 x_2(nT - T). \tag{4.7}$$

Linear systems can be studied by means of linear transformations like the Laplace or the z-transform. The benefit is that linear transformation of linear systems gives rise to a linear system in the transformed domain. The result is that, instead of time or discrete time, the

COMPUTER EXPERIMENT 4.1

Superposition of Three Different Signals

This simple virtual experiment allows you to verify the linear superposition of three different signal waveforms. The top signal generator is a sine wave, the second one enables a square wave and the last one generates a pulse signal. The amplifier has a fixed gain equal to 2 and the supply voltage of the analog block that produces amplification is 1.8 V. You can:

- change the amplitude of the sine wave signal. The frequency of the sine wave is constant and equal to 38 MHz;
- change the amplitude and repetition frequency of the square wave. The initial delay is zero because the trigger of the oscilloscope uses that signal to synchronize the three input waveforms;
- change the amplitude and delay of the pulse. The repetition rate of the pulse is twice that of the square wave.

EXPERIMENTAL SETUP

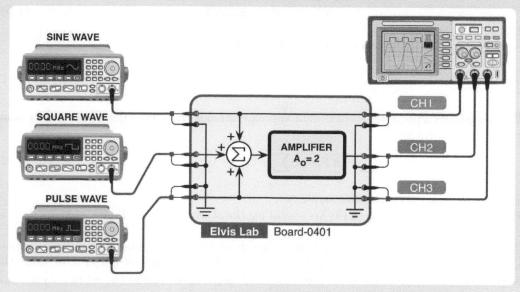

TO DO

- Set the amplitudes of square wave and pulse of both generators to zero and vary the sine wave amplitude. Observe the non-linear response of the amplifier at high amplitude and verify the gain of the amplifier.
- Reduce the sine wave amplitude and add the square wave and the repetitive pulse. Observe the resulting waveform.
- Set the frequency of the square wave to zero, to use it as a DC shift of the sine wave. Observe the output waveform with a small input sine wave but large DC shift. Notice the different level of distortion for minimum and maximum swing.
- Set the sine wave to zero and use the other input signals to obtain a sequence of delayed bipolar pulses.

transformed systems use s or z. Operating in the transformed domain permits us to define transfer functions (to be studied shortly) or to establish conditions for the stability of systems (which will also be studied shortly).

As you probably know from basic mathematics, the Laplace transform replaces derivatives and integrals by the operator s or its inverse:

$$\frac{dX(t)}{dt} \Rightarrow s \cdot x(s) \tag{4.8}$$

$$\int X(t)\,dt \Rightarrow \frac{x(s)}{s}; \tag{4.9}$$

therefore we can say that the Laplace operator changes a set of integro-differential equations into a better and more easily solvable system of linear equations.

On the other hand, the z-transform obtains equations in the z domain by using the delay as a basic operator. Indeed, as already discussed in Chapter 2, the delay is used in approximating the derivative or the integral. The z-transform of a delayed signal is

$$X(nT + T) \Rightarrow z \cdot x(z); \tag{4.10}$$

this, applied to the approximate expression of the derivative, gives

$$\frac{Y(nT_s + T_s) - Y(nT_s)}{T_s} \Rightarrow y(z)\frac{(z-1)}{T_s}. \tag{4.11}$$

For integrators a possible sampled-data approximation is the accumulation of discrete samples. Its z-transform is

$$\sum_0^\infty Y(nT_s) \Rightarrow \sum_{i=0}^\infty y(z)z^i = y(z)\frac{1}{z-1}, \tag{4.12}$$

which uses the sum of geometrical series.

Both equations (4.11) and (4.12) are useful for the study of sampled-data systems such as the s counterpart.

An advantage of linear systems is that if they have multiple inputs and multiple outputs they can be studied separately with a single input to obtain concurrent outputs, by using the superposition principle. In the time domain a single input gives rise to the set of outputs $y_{i,j}$:

$$y_{i,j}(t) = H_{i,j}[x_i(t)], \quad j = 1, \ldots, m, \tag{4.13}$$

where $H_{i,j}$ is the jth element of equation (4.1) estimated with just one input, $x_i(t)$:

$$\mathbf{H}_i = \mathbf{H}[0, 0, \ldots, x_i(t), \ldots, 0]. \tag{4.14}$$

Then the output $y_j(t)$ is the superposition of all the $y_{i,j}(t)$:

$$y_j(t) = \sum_{i=1}^n y_{i,j}(t) = \sum_{i=1}^n H_{i,j}[x_i(t)], \quad j = 1, \ldots, m. \tag{4.15}$$

Therefore we can conclude that the study of linear systems can be reduced to the study of single-input single-output processing.

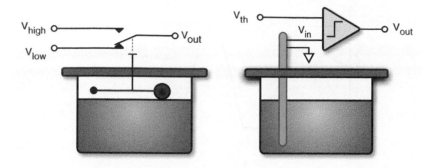

Figure 4.4   Mechanical and electronic threshold detector with mechanical and electronic sensor.

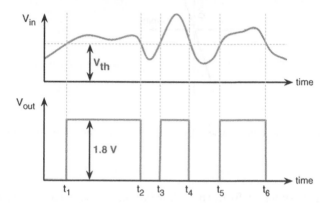

Figure 4.5   Input and output of the simple non-linear processor of Figure 4.4.

Non-linear processing is, as its name suggests, a modification of input(s) by non-linear functions. Non-linear operations include division or multiplication of signals, division or multiplication by time-varying factors, or generation of signals whose amplitude depends on special features of the input.

A simple example of the last category, widely used in electronics, is threshold detection, a problem often encountered in normal life. For example, it may be necessary to verify whether the water or fuel level in a tank exceeds some safety limit. For this, as Figure 4.4 shows, in addition to the well known electro-mechanical method, an electronic system with a suitable sensor measures the level and provides a voltage proportional to that level. Another voltage, V_{th}, is the signal denoting the threshold. The two signals are inputs of a non-linear processor that generates V_{low} when the level is below the threshold and V_{high} when it is above.

$$V_{out} = V_{low} + (V_{high} - V_{low}) \cdot step(V_{in} - V_{th}), \tag{4.16}$$

where *step* is the step function.

Waveforms of a possible input and the corresponding output are shown in Figure 4.5. The input exceeds the threshold at times t_1, t_3 and t_5, and it becomes smaller at t_2, t_4, and t_6. Therefore the output is three pulses, lasting from t_1 to t_2, t_3 to t_4, and t_5 to t_6 respectively. The low voltage is 0 V and the high voltage is 1.8 V, like the supply voltage of the circuit. In some situations, in order to avoid repetitive on and off switching, the threshold in one direction

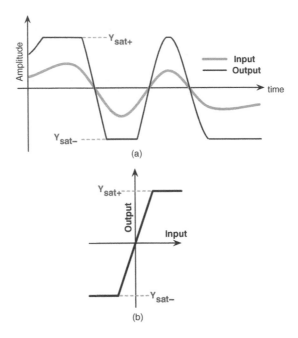

Figure 4.6   Non-linear processing caused by amplification and clipping, and its input/output characteristics.

at an intersection point is different from the following one; in this case we say that the system has a hysteresis.

Another simple example of non-linear processing is amplification with clipping. This is the multiplication of input by a constant factor until output reaches a given level. Above that level the output remains constant. This is what happens in real amplifiers that cannot generate amplitudes outside the supply limits. Amplification with clipping is shown in Figure 4.6(a), where input is correctly amplified by the gain value until the output reaches positive or negative saturation levels. The output then remains constant until the input amplitude exceeds the saturation levels divided by the gain.

The input/output characteristics shown in Figure 4.6(b) describe the operation. Output is three times the input until saturation. However, having a sharp change of slope is not a good model. Since the derivative is a delta, the solution using numerical analysis methods can be problematic. Moreover, real circuits have transitions that are less abrupt, with smooth changes of input/output slope near saturation. Therefore, models of amplification with clipping that better fit reality use non-linear functions such as $\arctan(x)$ or e^{-x}.

A third example of non-linear processing is the multiplication of two input signals $x_1(t)$ and $x_2(t)$:

$$y(t) = k \cdot x_1(t) \cdot x_2(t). \tag{4.17}$$

Since input and output have given dimensions, the need for a correct balance of dimensions obliges the coefficient k to have its own dimensions. For example, if inputs and outputs are voltages, the dimension of k must be V^{-1}.

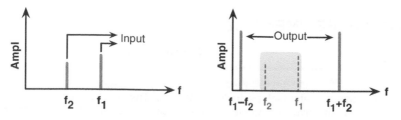

Figure 4.7   Spectrum of the input and output of a multiplier of two sine waves.

Consider the special case of multiplication of two sine waves

$$x_1(t) = A_1 \sin(\omega_1 t) \tag{4.18}$$
$$x_2(t) = A_2 \sin(\omega_2 t + \varphi_2). \tag{4.19}$$

The multiplication gives

$$y(t) = \frac{kA_1 A_2}{2}\{\cos[(\omega_1 - \omega_2)t - \varphi_2] - \cos[(\omega_1 + \omega_2)t + \varphi_2]\}, \tag{4.20}$$

which are two sine waves at frequencies equal to the addition and subtraction of the generating ones, as shown in Figure 4.7. The output amplitudes are equal, and the phases are $-\varphi_2$ and $+\varphi_2$ respectively, plus an extra $-\pi/2$ shift to account for the cosine function. The diagram that assumes that $f_1 > f_2$ has both terms in the positive frequency axis. The concept of negative frequency, not relevant at this level of study, is not discussed here. It is a topic covered in communications courses.

Often one of the two inputs is not an exact sine wave but has its spectrum spread around a given frequency. In this case, multiplication by a sine wave serves to move the signal band to higher or lower frequency ranges. Since one of the generated terms is undesired, it should be removed by filtering.

Some non-linearities are modeled with the Taylor series development of the input/output relationship $y = f(x)$:

$$y(t) = a_0 + a_1 x(t) + a_2 [x(t)]^2 + \cdots + a_n [x(t)]^n. \tag{4.21}$$

As studied in Chapter 2, with input sine waves the quadratic term causes sine waves at doubled frequency to be produced.

4.3   ANALOG AND DIGITAL PROCESSING

There is a key difference between two types of processing: analog and digital. Analog processing uses voltages and currents to obtain results; digital uses numbers, often with 2 as the basis, and the signals used are just a way to indicate numbers. Analog circuits use analog signals, i.e., continuous amplitude, continuous time, or, in some cases, sampled data. Digital circuits use logic signals, i.e., a finite number of possible amplitudes at discrete times. For 2-base numbers, digital circuits use two currents or two voltages to distinguish between two symbols. Indeed, it is not necessary to use exact levels, but it is enough to have an evident

COMPUTER EXPERIMENT 4.2

Water Tank

A small community uses a water tank as the local reservoir. The flow from the water source is not uniform. A valve, remotely driven via wireless, opens or closes the input flow with the control unit of the system. The water coming in can cause waves with various amplitudes fading in time almost exponentially. The system has three sensors: one to measure the water level and two to detect crossing of thresholds indicating minimum and maximum levels. When the minimum occurs a radio alarm is sent. The levels at which the sensors switch on and off differ by a suitable amount to avoid unwanted repetitive on-and-offing. You have to observe the signal waveforms so as to set the on and off levels of the sensors properly. In this experiment you can:

- change the water level in the tank;
- change the amplitude of waves and the dumping time constant. There is a button to simulate the generation of a wave;
- Change the MIN and MAX thresholds of the water in the tank.

SYSTEM CONFIGURATION

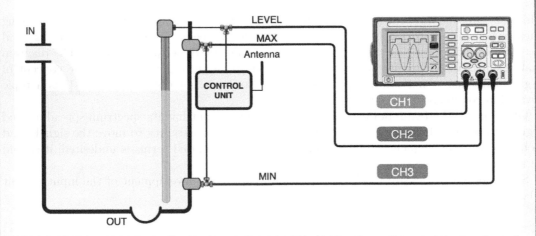

TO DO

- Review the system configuration and connections to the scope. Set the water level and the thresholds for minimum and maximum level. Observe the signals and notice that there is some noise at the output of the sensor level. The MIN and MAX signals are logic ones.
- Bring the water level to the minimum and maximum levels. Observe the output of the sensors. Change the thresholds until you get clean transition of the expected threshold signals.
- Set the wave amplitude and the dumping constant. Observe what happens when a wave occurs.
- Change the threshold values so as to have clean transitions with maximum amplitude waves.
- What is the margin on threshold values required by the wave effect? What is the fraction of the water tank that avoids on and off switching with maximum waves?

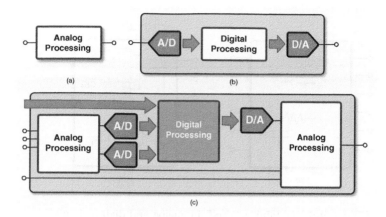

Figure 4.8   (a) Analog processing; (b) digital processing; (c) complex analog and digital processing.

difference between amplitudes to distinguish one symbol from another. Therefore the 2-base symbols 1 and 0 are just generic high and low levels. This feature provides a margin against signal fluctuation, giving robustness against errors.

When input signals are analog the use of analog processors is straightforward (Figure 4.8(a)). On the other hand, for digital processing the use of data converters (ADC) changes the analog format into a sampled-data and discrete-amplitude signal. Also, if the output is analog, a Digital-to-Analog converter (D/A) transforms digital results into analog format, as shown in Figure 4.8(b).

Analog or digital?

The trade-off between silicon area, system volume, power consumption, and cost determines whether to use analog or digital processing.

A mixture of analog and digital processing, as shown in Figure 4.8(c), is also possible. The inputs are either analog or digital. Analog signals processed in the digital domain need data converters. Obviously, for defining this kind of mixed-signal processing it is necessary to find the most convenient points for the analog/digital and digital/analog transitions. The choice should not be based on designer convenience, when the designer perhaps knows given processing techniques better, but on the search for an optimal solution. The selection may depend on the following features.

- Analog processing enables high speed; it is therefore appropriate for signals whose bandwidth is large, and is close to the technological limits.

- Digital processing ensures robustness (with a low sensitivity and degradation caused by noise). It is also suitable for complex processing, such as non-linear and adaptable operations. Therefore digital processing is for complex and robust designs.

- Low bandwidth analog processing requires low current. Even if the supply voltage used is sometimes higher than for digital signals, analog processing is optimal for micro-power, as required for portable and autonomous applications.

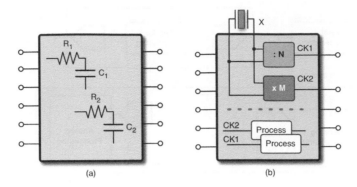

(a)   (b)

Figure 4.9   Obtaining "time" in analog and digital circuits.

■ Digital processing benefits from technology scaling, leading to integration of millions of transistors on the same chip. The combined use of hardware and software makes digital processors programmable and updatable. This means, once again, that digital processing is for complex systems and also favors software programmable devices.

4.3.1   Timing for Signal Processing

Often, processing needs a time reference (or a frequency reference). Therefore it is necessary to produce and control signals that define time. For analog circuits electrical quantities related to time are resistance and capacitance, whose product has time as a dimension. The product of resistance and capacitance is often called the *time constant*

$$\tau = RC. \tag{4.22}$$

Figure 4.9(a) schematically represents a processor that uses two time constants for its internal time or frequency requirements. It gives an answer for what is needed but the time constants obtained are not very accurate, because the accuracy of resistors and capacitors is normally low and their errors are uncorrelated. Random errors caused by fabrication limitations makes the values of resistors and capacitors

$$R_{real} = R_{nom}(1 + \epsilon_R) \tag{4.23}$$
$$C_{real} = C_{nom}(1 + \epsilon_C), \tag{4.24}$$

where the subscripts *real* and *nom* indicates the real and nominal values, and ϵ_R and ϵ_C are the relative errors in resistance and capacitance, which can be as high as 20%, depending on the fabrication process. Since the errors are uncorrelated, the inaccuracy of the time constant is a random variable. It is the quadratic combination, resulting from errors of resistance and capacitance

$$\epsilon_\tau = \sqrt{\epsilon_R^2 + \epsilon_C^2}, \quad \tau_{real} = \tau_{nom}(1 + \epsilon_\tau). \tag{4.25}$$

Therefore, since ϵ_τ can be 30%, the accuracy of the time obtained using time constants as the basis is pretty low.

COMPUTER EXPERIMENT 4.3

Polynomial Non-Linear Response

This Computer Experiment helps you in studying the non-linear processing of a sine wave signal. The experimental setup includes a signal generator for input and an oscilloscope for visualizing waveforms. A fabricated custom instrument performs the non-linear processing. It generates an analog output given by the relationship

$$V_{out} = a_0 + a_1 V_{in} + a_2 V_{in}^2 + a_3 V_{in}^3 + a_4 V_{in}^4 + a_5 V_{in}^5$$

The custom instrument allows you regulate the value of coefficients a_0 - a_5 with up and down buttons. Six digital displays show you the selected value of coefficients. The instrument generates two outputs. One is a component at frequency nf_{in} the second one is the output voltage. The value of n can be selected with a numerical stepper. You can:

- set the coefficient values of the linear and the polynomial terms. Notice that you can set the coefficient multiplier of any tone term to zero;
- change the amplitude of the input sine wave. The frequency is fixed at 10 MHz;
- choose the sinusoidal component to be displayed on Channel 2 of the scope.

EXPERIMENTAL SETUP

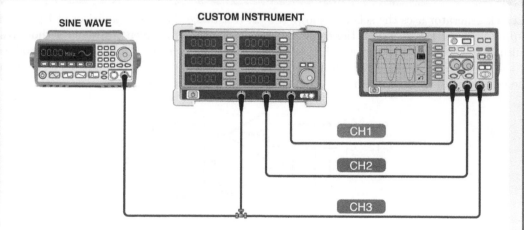

TO DO

- Set the first coefficient to 1 and the others to zero. Augment the value of the first coefficients and determine the dynamic range of the instruments.
- Set the first coefficient back to 1 and choose any value for a_2. Notice that the DC level of the output changes slightly. Understand why?
- Set the other coefficients to any value and observe the resulting effect. Measure the amplitude of each sinusoidal contribution with two sets of coefficients.
- Be careful with coefficients and input amplitude to avoid saturation of the output of the custom instrument.

Another way to obtain time is to exploit the time required by the voltage across a capacitance charged with constant current to reach a given threshold level. In this case, also, accuracy is poor, because capacitance and the current generator must be accurate.

Time is also given by quartz. Its precision is, as already discussed, good. Therefore quartz solutions are widely used, even if they are expensive because they require dedicated electronics to drive the quartz (and that is not cheap). A system normally uses one quartz as a precise time-base generator with dividers or multipliers to obtain the necessary time-controller signals. As shown in Figure 4.9(b), a main or master clock generated by the external quartz is processed by the division and multiplication blocks. Clocks may be available as output for external use or monitoring. The division or multiplication factors are normally to the power of two. However, there are frequent cases with generic integer factors or even fractions. Fractional dividers are needed for studying frequencies that are not integer multiples of the clock. The use of well-defined factors is such that the main and derived clocks are said to be "locked."

Remember

Bad shaping and inaccurate synchronization of the clocks sequencing digital circuits can disrupt processing. Therefore the generation and distribution of clocks throughout systems require special care, especially at very high frequencies.

Example 4.1

A clock generator uses the scheme below. A 2 kΩ resistance charges a capacitor of 1 pF at the supply voltage, 1.2 V. When the capacitance voltage crosses a given threshold a short reset pulse of 10 psec activates the switch that discharges the capacitor. A longer pulse is produced at the output. Determine the period of pulses for a threshold equal to 0.3, 0.6, 0.9 V. Estimate the error caused by noise with $\sigma = 10$ mV affecting threshold voltages.

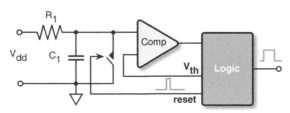

Solution

The voltage across the capacitor rises exponentially with time constant equal to the product of resistance and capacitance $\tau = R_1 C_1 = 2$ ns.

$$V_C(t) = V_{dd}(1 - e^{-t/\tau}),$$

which crosses the threshold ($V_{th} < V_{dd}$) at

$$t_d = \tau \ln \left[\frac{V_{dd}}{V_{dd} - V_{th}} \right]. \tag{4.26}$$

That determines the delay after the reset pulse. Accounting for the pulse duration, the thresholds 0.3, 0.6, 0.9 V, are crossed at times 0.58 ns, 1.40 ns, and 2.78 ns respectively.

The time derivative of the voltage across the capacitance and the threshold error give the error of the crossing time at $\bar{t}$,

$$\sigma_{t_d} = \sigma_{V_{th}} \frac{dt_d}{dV_{th}} = \sigma_{V_{th}} \frac{\tau}{V_{dd} - V_{th}}, \tag{4.27}$$

showing how error increases linearly with time constant. With $V_{th} = 0.6$ V the variance of the generated time period is 33 ps.

4.4   RESPONSE OF LINEAR SYSTEMS

The response of continuous-time linear systems, studied in either the time or the frequency domain, exploits the superposition principle. Any input signal can be seen as a superposition of constituents in the time or the frequency domain. The result is the superposition in time or frequency of the responses to single constituents.

4.4.1   Time Response of Linear Systems

A generic time-varying signal can be viewed as the addition of delta, step, or sine wave functions. In the first case a continuous-time signal $x(t)$ is split up into an infinite number of deltas, all delayed, one after the other, by infinitesimal quantities. The amplitude of the deltas equals the input when deltas occur:

$$x(t) = \int_{-\infty}^{t} x(\tau)\delta(t - \tau)\, d\tau. \tag{4.28}$$

Figure 4.10(a) shows a generic signal and a few of the delta functions that portray it. Notice that the deltas of Figure 4.10(b) are a finite subset; in reality it would be necessary to account for a continuum of deltas.

The system response, $h(t)$, to a delta, $\delta(t)$, occurring at $t = 0$ is called the *impulse response*. Moreover, the response to a delayed delta, $\delta(t-\tau)$, is a delayed version of the impulse response, $h(t - \tau)$. Therefore the superposition principle states that the response to $x(t)$ is

$$y(t) = \int_{-\infty}^{t} x(\tau)h(t - \tau)\, d\tau, \tag{4.29}$$

which superposes the delayed version of delta response weighted by the input amplitude at $t = \tau$. The integral in equation (4.29) is called the *convolution integral*.

Figure 4.11 illustrates convolution. Figure 4.11(a) is a possible delta response. It rises after the delta occurs, reaches a maximum, and after a while fades toward zero within a given period of time, because the system has forgotten the effect of the delta. Figure 4.11(b) uses a small slot of possible input that gives rise to the weighted and delayed impulse responses shown. Their superposition is the output. In reality, the convolution integral produces infinitesimal contributions, since they account for input in an infinitesimal $d\tau$ period. Superposing an infinite number of infinitesimal terms produces a finite output.

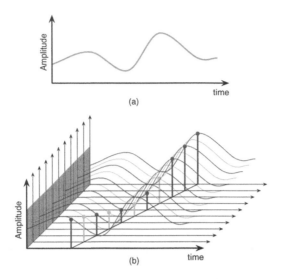

(a)

(b)

Figure 4.10   (a) Time-domain generic signal; (b) portrayal of signal with weighted delta functions.

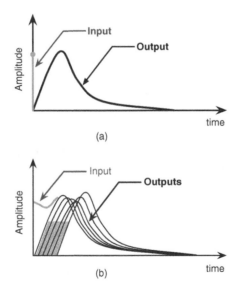

(a)

(b)

Figure 4.11   (a) Delta response of a system; (b) small slot of input and corresponding weighted delta responses to be superposed to obtain output.

In some situations it can be convenient to split the input using step functions. For example, the signal shown in Figure 4.12 is made by two pulses whose duration is t_1 and $t_3 - t_2$. Their amplitudes are A_1 and A_2 respectively. Such a signal can be conveniently broken down into four step functions. The first starts at $t = 0$; then, at $t = t_1$ there is an equal and opposite step with the same amplitude A_1. The second pair has amplitude A_2; the first starts at the

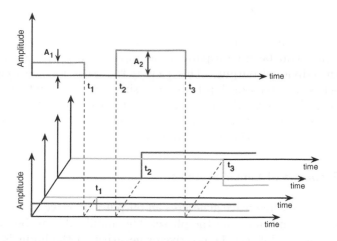

Figure 4.12   Signal made of two rectangular waves split into steps.

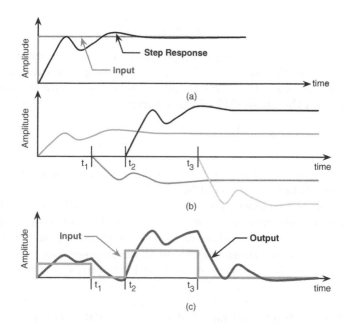

Figure 4.13   (a) Step response; (b) separate output responses; (c) superposition of separate results.

beginning and the other starts with opposite sign at the end of the second pulse, at t_2 and t_3 respectively. The pulse response builds the output signal, since the system is linear. Let us suppose that the diagram in Figure 4.13(a) is the step response. It shows a finite slope that indicates an impossible immediate reaction to the step. After some ringing the response settles to a level that, in this case, equals the step amplitude (the gain is 1). Figure 4.13(b) shows the four weighted responses, which, as expected, start at times 0, t_1, t_2, and t_3. The superposition is shown in Figure 4.13(c). Observe that the duration of the first pulse and the time interval $t_1, \ldots, t_2$ are not enough to enable the output to settle. The tail of the first step response affects the output even after the second step is applied.

Another possible way to slit up input is by sine waves. Again, thanks to the superposition property, each sine wave can be used separately as input. Then the overall output is the superposition of each individual output. This view is the basis for the frequency response analysis used to describe the relationship between a sine wave at input and a sine wave at output. Frequency analysis is normally more convenient than studying delta or step functions. Step or delta responses are informative only when the input is made up of steps or is delta-like. For more generic, periodic inputs, sine wave study is best.

4.4.2   Frequency Response of Linear Systems

In linear systems an input sine wave generates an output sine wave at the same frequency. What changes is just the amplitude and the phase. The amplitude can be bigger or smaller, as in Figure 4.14. The phase shift can be positive or negative. If the input is

$$x(t) = A_{in} \sin(2\pi f_1 t + \varphi_{in}), \tag{4.30}$$

the output is given by

$$y(t) = A_{out} \sin(2\pi f_1 t + \varphi_{out}). \tag{4.31}$$

Since the system is linear, when the input amplitude increases the output increases as well, in a manner proportional to the input variation. However, the phase shift does not change if the frequency remains the same. Therefore, to obtain the output, only the ratio between the amplitudes (the gain) and the phase shift are necessary:

$$G(f) = \frac{A_{out}}{A_{in}}, \quad \Delta\varphi(f) = \varphi_{out} - \varphi_{in}. \tag{4.32}$$

Notice that the gain and the phase shift both depend on the frequency of the sine wave. Therefore the processing is fully represented by two diagrams, one depicting the gain and the other the phase shift versus frequency. This pair of diagrams is called a *transfer function*.

A useful way to depict a sine wave is to use a vector with the same amplitude and the same phase as the sine wave itself. Since the phase of a sine wave with frequency f_1 is $(2\pi f_1 t + \varphi_{in})$, the corresponding vector rotates counterclockwise at an angular speed proportional to the frequency. The amplitude of the sine wave at any time is the projection of the vector onto the vertical axis. This is shown in Figure 4.15, where at $t = 0$ there is a vector denoted by V_{in} with a given amplitude, and phase $\varphi_{in} = 0°$. In the figure there is also another vector denoted by V_{out} with a smaller amplitude and initial phase $\varphi_{out} = 250°$. The two vectors depicting input and output rotate with same angular frequency and determine with their projections the instantaneous values of the signals. For example, at time t_1, which gives rise to the dotted-line vectors, the input phase is $130°$ and the output phase is $130° + 250°$. The projections determine the signals at that time. Moreover, the time required for a complete turn of the vectors, t_{turn}, is the period of the sine waves:

$$t_{turn} = \frac{1}{f_1} = \frac{2\pi}{\omega_1}. \tag{4.33}$$

The vectorial description of sine waves is beneficial only when the frequencies of the depicted sine waves are the same. In this case, since the vectors rotate synchronously, it is convenient

COMPUTER EXPERIMENT 4.4

Impulse Response and Convolution

This experiment permits you to understand impulse response and convolution. The board uses two equal linear circuits. The first circuit has at its input a long pulse (a rectangle), while the second one has a short pulse signal. The output of the first circuit is the convolution of the input with its pulse response. The output of the second circuit gives rise to an approximation of the delta response. The computer receives the output with short input and, after digitization, estimates the convolution. The computer displays the comparison between output of the first circuit and the estimation of the convolution.
In this Computer Experiment you can:

- change amplitude and pulse duration of the two input signals;
- choose any of the different waveforms available on the computer;
- observe with the scope the output signal of both linear circuits.

EXPERIMENTAL SETUP

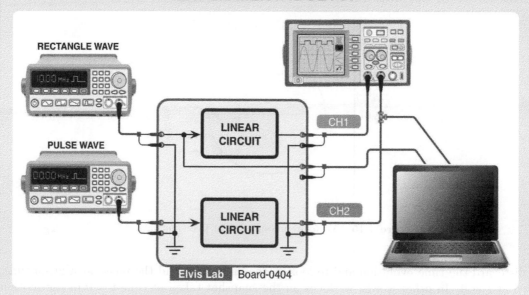

TO DO

- Set the pulse duration and amplitude of two input signals and observe the output of both linear circuits with the scope.
- Determine the maximum duration of a short pulse which gives rise to an output waveform that does not change its shape. This shape mimics the delta response of the linear circuit.
- Compare the measured and estimated convolution response on the display and determine the difference with longer duration of the short pulse.
- Increase the period of the rectangular wave pulse and observe the results.

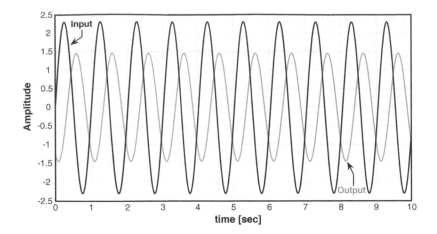

Figure 4.14   Possible input and output of a linear system. The input is a sine wave and the output is a sine wave at the same frequency.

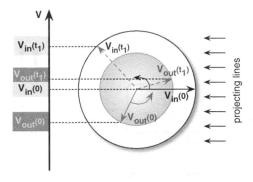

Figure 4.15   Vector representation of two sine waves.

to disregard the time evolution and to look at the amplitude and the phase at a given time, often at $t = 0$. In other words, we focus on the snapshot taken at time $t = 0$ of the motion picture of the rotating vectors that depict the sine waves.

The vector diagram provides an immediate view of the relationships between sine waves. For example, vectors starting from the origin of the complex plane and ending at 1 (phase $0°$), $0.8 + j0.8$, $j1.3 - 1.2 + j0.6$ (Figure 4.16), indicate that the first sine wave has a smaller amplitude and is delayed with respect to the others by $150°$, $90°$, and $45°$ respectively. Another relevant advantage of the vectorial view is the easy way of performing additions. The vector of the sum of two or more sine waves is the vectorial addition of the vectors depicting the addends.

Notice that a vector representation involves linear processing because a linear system changes only amplitudes and phases but not frequency. Consider, for example, the derivative or the integral. If the input is

$$x(t) = A\sin(\omega_1 t), \tag{4.34}$$

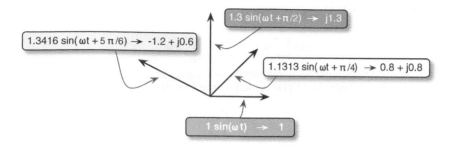

Figure 4.16 Vector and complex representation of sine waves at the same frequency.

its derivative satisfies

$$y(t) = \frac{dx(t)}{dt} = \omega_1 A \cos(\omega t) = \omega_1 A \sin(\omega_1 t + \pi/2), \tag{4.35}$$

showing a change by the factor ω_1 of amplitude and denoting a phase shift by $\pi/2$.

Since the dimensions of a signal and its time derivative differ, when we embody both with the same electrical quantity it is necessary to use a multiplying factor with a time dimension. For example, the vectorial representation of a sine wave and its derivative, constructed by two voltages V_{in} and V_{out}, follows the relationship

$$\vec{V}_{out} = j\omega\tau\,\vec{V}_{in}, \tag{4.36}$$

which features two vectors in the complex plane, one real and the other imaginary, with voltage as the dimension of both. The result, in fact, matches what foreseen by the Fourier transform. In fact, in the Fourier domain the derivative is multiplication by the operator $j\omega$:

$$\frac{d}{dt} \to j\omega. \tag{4.37}$$

A similar study accounting for the integral gives rise to an equivalent relationship between vectors depicting a signal V'_{in} and its integral V'_{out},

$$\vec{V'}_{out} = \frac{\vec{V}_{in}}{j\omega\tau}, \tag{4.38}$$

which are one on the real axis snd the other on the negative imaginary axis.

4.4.3 Transfer Function

The vectorial description of sine waves brings in the definition of complex transfer functions. These are the ratios between vectors or complex numbers depicting input and output sine waves at variable frequency. Linear operators transform the input into output by s-domain relationships whose general stucture is

$$X(s)[a_0 + a_1 s + \cdots + a_i s^i + \cdots + a_n s^n]$$
$$= Y(s)[b_0 + b_1 s + \cdots + b_j s^j + \cdots + a_m s^m], \tag{4.39}$$

which can be rewritten in the compact form

$$X(s) \sum_0^n a_i s^i = Y(s) \sum_0^m b_j s^j. \tag{4.40}$$

The above, using the definition of the transfer function in the s domain, $H(s)$,

$$H(s) = \frac{Y(s)}{X(s)}, \tag{4.41}$$

gives rise to

$$H(s) = \frac{\sum_0^n a_i s^i}{\sum_0^m b_j s^j} = \frac{P(s)}{Q(s)}, \tag{4.42}$$

which is a ratio between two polynomials of order n and m respectively.

Using the Fourier transform, instead of Laplace, provides an equivalent relationship in the frequency domain (or, better, angular frequency domain, $\omega = 2\pi f$):

$$H(j\omega) = \frac{y(j\omega)}{x(j\omega)} = \frac{P(j\omega)}{Q(j\omega)}. \tag{4.43}$$

Therefore an input sine wave with angular frequency ω, phase zero and unity amplitude gives rise to an output with amplitude and phase shown by the complex number $H(j\omega)$. If input amplitude is bigger or smaller, the output scales by the same factor. If the input phase is different from zero, the output phase has an equal shift from the phase established as $H(j\omega)$. This is because of the linearity property.

The two polynomials at the numerator and denominator of (4.42) have zeros at n and m points respectively. Therefore

$$H(s) = \frac{P(s)}{Q(s)} = K \frac{\prod_{i=1}^n (s - s_{z,i})}{\prod_{j=1}^m (s - s_{p,j})}, \tag{4.44}$$

where K is a suitable constant with proper dimensions.

Another possible form of the transfer function, often used, is

$$H(s) = \frac{P(s)}{Q(s)} = K' \frac{\prod_{i=1}^n (1 - s/s_{z,i})}{\prod_{j=1}^m (1 - s/s_{p,j})}. \tag{4.45}$$

The zeros of the numerator and the denominator, $s_{z,i}$ and $s_{p,j}$, can be real or complex. However, since the coefficients of the polynomial are real, zeros and poles possibly occur in pairs with complex conjugate values. The zeros of $P(s)$ are called *zeros* of the transfer function, because at those complex frequencies $H(s)$ is zero. Zeros of $Q(s)$ are called *poles*, because at those frequencies the transfer function amplitude is infinite. Zeros and poles are normally indicated in the complex plane s and distinguished by a circle or a cross respectively, as shown in Figure 4.17. There are six zeros and five poles: some are pairs of complex conjugates, and others are real. Observe that the zeros are in the right and left semi-s-plane, while all poles are in the left semi-s-plane. This feature is important to notice because, as we shall learn in a subsequent chapter, having poles in the right semi-s-plane can cause instability.

A sine wave or superposition of sine waves makes $s = j\omega$ in equation (4.44) or equation (4.45). Therefore, examining the transfer function on the imaginary axis enables us to

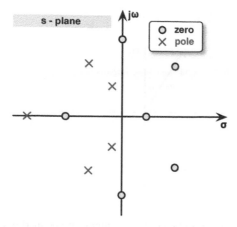

Figure 4.17   Position of zeros and poles of a possible transfer function in the s-plane.

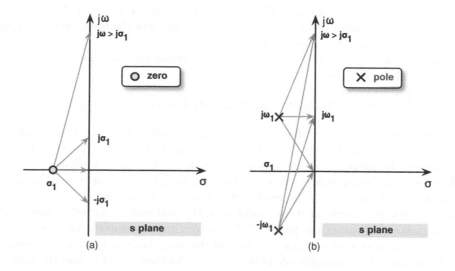

Figure 4.18   Vector representations of the effect of (a) a real zero and (b) two complex conjugate poles.

study the effect of zeros and poles with input sine waves. A real zero or pole at σ_1 produces a term like (disregarding a minus sign)

$$(j\omega - \sigma_1) \quad \text{or} \quad (j\omega\tau_1 - 1), \tag{4.46}$$

where $\tau_1 = 1/\sigma_1$ is the time constant that generates the zero or pole. Often, instead of the symbol σ with some subscript, which denotes a point on the real axis of the s-plane, the symbol ω with some subscript is used. The above becomes $(j\omega - \omega_1)$ or $(j\omega/\omega_1 - 1)$. This practice can be misleading, but it makes sense, because a frequency response deals with frequencies and the symbol ω is conventionally used to indicate angular frequencies ($\omega = 2\pi f$).

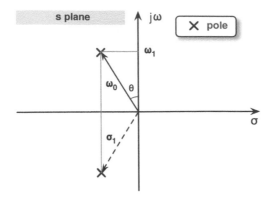

Figure 4.19   Real and complex conjugate poles in the s-plane.

Observe that $(j\omega - \sigma_1)$ is a vector connecting the point σ_1 on the real axis with $j\omega$ on the imaginary axis. The phase, as shown in Figure 4.18(a), is zero at $\omega = 0$, $\pi/4$ at $\omega = \sigma_1$ (and $-\pi/4$ at $\omega = -\sigma_1$), and tends to $\pi/2$ for $\omega \gg \sigma_1$. The modulus of the vector is at its minimum for $\omega = 0$ and increases with the angular frequency.

Consider now a pair of poles at $\sigma_1 \pm j\omega_1$. The corresponding terms in the frequency response equation (4.44) with $s = j\omega$ are

$$(j\omega - \sigma_1 - j\omega_1)(j\omega - \sigma_1 + j\omega_1). \tag{4.47}$$

These are two vectors on the s-plane joining the fixed points $\sigma_1 \pm j\omega_1$ to a variable point $j\omega$ on the imaginary axis. Figure 4.18(b) shows a graphical representation of the vectors at relevant angular frequencies. At $\omega = 0$ the vectors have the same amplitude and opposite phases. Therefore, the overall phase is zero. As the angular frequency increases, the amplitude of the vector starting from $\sigma_1 + j\omega_1$ decreases and the amplitude of the other vector increases. The product can possibly decrease if the modulus of the real part σ_1 is rather lower than ω_1. For positive angular frequencies, the phases of the vectors increase and the phase shift becomes positive. At $j\omega_1$ the first vector phase is zero and the amplitude reaches its minimum. For bigger angular frequencies the amplitude of both vectors increases, and overall the phase tends to π.

Equation (4.47) can be rewritten as

$$(j\omega)^2 - j\omega(2\sigma_1) + (\omega_1^2 + \sigma_1^2), \tag{4.48}$$

or, using the modulus of vectors, ω_0, and defining a new quantity Q, called the *quality factor*,

$$\omega_0 = \sqrt{\omega_1^2 + \sigma_1^2}, \quad Q = \frac{\omega_0}{2\sigma_1}, \tag{4.49}$$

equation (4.48) becomes

$$-\omega^2 - j\omega\frac{\omega_0}{Q} + \omega_0^2, \tag{4.50}$$

showing that at $\omega = \omega_0 (> \omega_1)$ the amplitude is ω_0^2/Q and the phase shift is $-\pi/2$. Moreover, if σ_1 tends to zero at $\omega = \omega_0$, Q tends to ∞, making the modulus almost zero.

4.5 BODE DIAGRAM

The previous section showed that frequency response is the transfer function estimated on an imaginary axis. The Bode diagram represents the modulus and phase graphically, by using convenient scales. On the imaginary axis equation (4.45) becomes

$$H(j\omega) = \frac{P(j\omega)}{Q(j\omega)} = K' \frac{\prod_{i=1}^{n}(1 - j\omega/s_{z,i})}{\prod_{j=1}^{m}(1 - j\omega/s_{p,j})}. \tag{4.51}$$

For plotting Bode diagrams it is convenient to distinguish between real and complex solutions, leading to

$$H(\omega) = K' \frac{\prod_{i_r}^{n_r}(1 - j\omega/\omega_{z,i_r}) \prod_{i_c}^{n_c}(1 - j\omega/(\omega_{z,i_c}Q) - (\omega/\omega_{z,i_c})^2}{\prod_{j_r}^{m_r}(1 - j\omega/\omega_{p,j_r}) \prod_{j_c}^{m_c}(1 - j\omega/(\omega_{p,j_c}Q) - (\omega/\omega_{p,j_c})^2}. \tag{4.52}$$

Real zeros and poles are $(\omega_{z,i_r}; \ i_r = 1 \cdots n_r)$ $(\omega_{p,i_r}; \ i_r = 1 \cdots m_r)$. Complex zeros and poles are $(\omega_{z,i_c}; \ i_c = 1 \cdots n_c)$ and $(\omega_{p,i_c}; \ i_c = 1 \cdots m_c)$. The constant K', denoting gain, has a suitable value and matching dimension.

4.5.1 Amplitude Bode Diagram

The amplitude Bode diagram uses a log scale on the frequency axis and dB on amplitude axis. With logarithms it is possible to exploit the key property stating that the logarithm of a product or division is the sum or subtraction of the logarithms of the individual terms. Therefore the products in the numerator of equation (4.51) are accounted for with a positive sign and the ones in the denominator have a negative sign:

$$|H(j\omega)|_{dB} = |K'|_{dB} + \sum_{i=1}^{n} |1 - j\omega/s_{z,i}|_{dB} - \sum_{j=1}^{m} |1 - j\omega/s_{p,j}|_{dB}. \tag{4.53}$$

The amplitude Bode diagram (often referred to as the "Bode diagram" without the specification "amplitude") becomes the superposition of Bode diagrams of single terms of equation (4.51) or, rather, equation (4.52).

The Bode diagram of real zero at $-\omega_z = -2\pi f_z$, or real pole at $-\omega_p = -2\pi f_p$,

$$H_z(\omega) = 1 + j\frac{\omega}{\omega_z} \quad \text{or} \quad H_p(\omega) = \frac{1}{1 + j\omega/\omega_p}, \tag{4.54}$$

$$H_z(f) = 1 + j\frac{f}{f_z} \quad \text{or} \quad H_p(f) = \frac{1}{1 + jf/f_p}, \tag{4.55}$$

represents moduli

$$|H_z(\omega)| = \sqrt{1 + \left(\frac{\omega}{\omega_z}\right)^2} \quad \text{or} \quad |H_p(\omega)| = \frac{1}{\sqrt{1 + (\omega/\omega_p)^2}}, \tag{4.56}$$

$$|H_z(f)| = \sqrt{1 + \left(\frac{f}{f_z}\right)^2} \quad \text{or} \quad |H_p(f)| = \frac{1}{\sqrt{1 + (f/f_p)^2}}, \tag{4.57}$$

which for $\omega \ll \omega_z$ (or $f \ll f_z$) is 1 and for $\omega \gg \omega_z$ (or $f \gg f_z$) tends to $\pm\omega/\omega_z$ or $(\pm f/f_z)$. The same goes for the value of the pole. Therefore, using the dB scale, it comes out as

$$
|H_z(\omega)|_{dB} = \begin{cases} 0 & \text{for } \omega \ll \omega_z & (4.58) \\ 20\log\dfrac{\omega}{\omega_z} & \text{for } \omega \gg \omega_z & (4.59) \end{cases}
$$

$$
|H_p(\omega)|_{dB} = \begin{cases} 0 & \text{for } \omega \ll \omega_p & (4.60) \\ -20\log\dfrac{\omega}{\omega_p} & \text{for } \omega \gg \omega_p, & (4.61) \end{cases}
$$

or

$$
|H_z(f)|_{dB} = \begin{cases} 0 & \text{for } f \ll f_z & (4.62) \\ 20\log\dfrac{f}{f_z} & \text{for } f \gg f_z & (4.63) \end{cases}
$$

$$
|H_p(f)|_{dB} = \begin{cases} 0 & \text{for } f \ll f_p & (4.64) \\ -20\log\dfrac{f}{f_p} & \text{for } f \gg f_p. & (4.65) \end{cases}
$$

Let us now assume we are using a normalizing unity angular frequency $\omega_N = 1$ rad/s (or when using frequencies, a normalizing unity frequency, $f_N = 1$ Hz) in equations (4.59) and (4.63):

$$
|H_z(\omega)|_{dB} = 20\log\frac{\omega}{\omega_N} - 20\log\frac{\omega_z}{\omega_N} \quad \text{for } \omega \gg \omega_z \tag{4.66}
$$

$$
|H_z(f)|_{dB} = 20\log\frac{f}{f_N} - 20\log\frac{f_z}{f_N} \quad \text{for } f \gg f_z. \tag{4.67}
$$

Frequently, the dimensionless normalized quantities ω/ω_N and ω_1/ω_N are denoted by just using ω and ω_1. This simplifies writing the equations. Therefore equations (4.66) and (4.67) become

$$
|H_z(\omega)|_{dB} = 20(\log\omega - \log\omega_z) \quad \text{for } \omega \gg \omega_z \tag{4.68}
$$

$$
|H_z(f)|_{dB} = 20(\log f - \log f_z) \quad \text{for } f \gg f_z, \tag{4.69}
$$

with the warning that the log does not use a frequency or an angular frequency but a number.

Keep in mind

Expressions $\log(\omega)$ or $\log(f)$ are formal mistakes, because the subject of log functions must be dimensionless. When they are used, an implicit multiplication by a constant equal to 1 s/rad or 1 s is supposed.

The above approximate equations (4.58) and (4.65), in a dB–$\log\omega$ plane, are straight lines. Four equations give horizontal lines at 0 dB and four are lines with slopes of 20 or -20. Moreover, at $\omega = \omega_z$ (or f_z) the actual value of amplitude is $\sqrt{2}$, or 3 dB. Similarly, we can verify that at pole frequency the amplitude is $1/\sqrt{2}$, or -3 dB.

The results obtained are used in the approximate Bode diagram. The said diagram uses just two straight lines with transitions from one to the other at a pole or zero location. At the

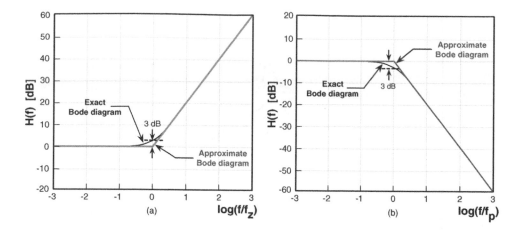

Figure 4.20 Bode amplitude diagram of (a) a real zero and (b) a real pole.

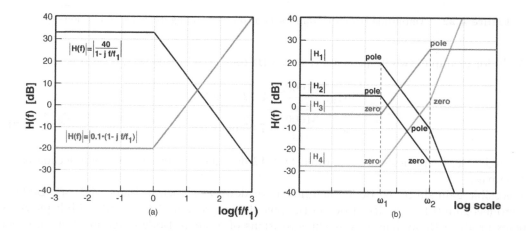

Figure 4.21 Approximate Bode amplitude diagrams: (a) single real zero or pole; (b) two real solutions.

transition point the error due to approximation is 3 dB. Approximate and exact diagrams are shown in Figure 4.20 for one zero and one pole at finite frequency. The frequency interval spans from $0.001 \cdot f_z$ to $1000 \cdot f_z$ (or from $0.001 \cdot f_p$ to $1000 \cdot f_p$). If the zero and the pole are at zero frequency the diagrams are just a line with 20 dB slopes, one positive and the other negative.

Since the frequency axis is logarithmic, the separation between two frequencies can be expressed in decades or in octaves. A decade means a factor of 10; an octave means a factor of 2. Let us remember that octave comes from a definition used in music. An octave corresponds to halving or doubling the frequency distance between musical pitches.

When $\log \omega$ increases by a unit, the angular frequency increases by 10 (a decade). Therefore, the slope of the diagram, measured in dB per decade, is 20 dB/decade. Moreover, since $\log 2 = 0.301$, doubling the frequency changes the module of $H_1(\omega)$ by $20 \log 2 = 6.02$ dB. Because of this feature we say that the slope is approximately 6 dB/octave.

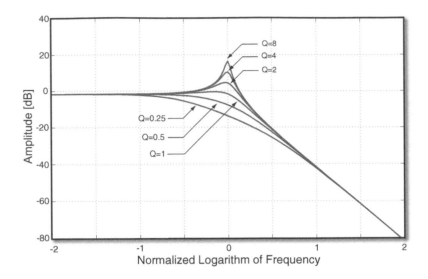

Figure 4.22   Amplitude Bode diagram for two complex conjugate zeros and various Q.

With, for example, two frequencies $f_1 = 238$ Hz and $f_2 = 31.6$ kHz, the ratio is

$$\log_{10} \cdot \frac{31600}{238} = 2.12 \text{ decades} \tag{4.70}$$

$$\log_2 \cdot \frac{31600}{238} = 7.05 \text{ octaves;} \tag{4.71}$$

accordingly, the frequencies are separated by 2.12 decades or 7.05 octaves, one decade being 3.32 octaves.

A possible gain shifts the Bode diagrams up or down as shown in Figure 4.21(a). Bode of the zero has gain 0.1 (-20 dB); Bode of the pole denotes a gain of 40 (32 dB). Multiple zeros or poles produce Bode diagrams as superpositions of a single Bode diagram. For an exact diagram the operation can be a bit complex, but drafting the approximate diagram is easy. At the occurrence of a zero, the slope increases by 20 dB; when encountering a pole, the slope diminishes by 20 dB. Therefore, with two real solutions, say at ω_1 and ω_2 ($\omega_1 < \omega_2$), and some gain, the approximate Bode diagram is like one of the four pictured in Figure 4.21(b). $H_1(f)$ depicts the diagram with two poles. At high frequency adding two drops causes a negative slope of -40 dB/dec. The diagram at low frequency is at 20 dB, denoting a 10 gain. $H_2(f)$ depicts a Bode diagram with pole before zero; $H_3(f)$ shows one the other way around: zero before pole. Accordingly, both cases have zero slopes at high and low frequencies because of the two equal and opposite changes of slope. In one case the low frequency gain is lower than at high frequency, and in other case it is the other way around. $H_4(f)$ has two zeros; consequently, the slope goes to 20 dB/dec between the two zeros and becomes 40 dB/dec at high frequency ($f \gg f_2$).

Notice that what we have discussed refers to real zeros or poles. As indicated by equation (4.52), two complex conjugate zeros at ω_0 and a given Q give rise to the term

$$H_0(\omega) = 1 - \frac{j\omega}{Q\omega_0} - \frac{\omega^2}{\omega_0^2} \tag{4.72}$$

and the modulus is

$$|H_0(\omega)| = \sqrt{\left[1 - \frac{\omega^2}{\omega_0^2}\right]^2 + \left[\frac{1}{Q\omega_0}\right]^2}.$$ (4.73)

Obviously, with complex conjugate poles the inverse of expressions (4.72) and (4.73) must be considered.

The Bode diagram is again 0 dB for $\omega \ll \omega_0$ and a straight line for $\omega \gg \omega_0$. Since the diagram accounts for two zeros or poles, the slope at high frequency is 40 dB or -40 dB. The Q determines the shape in the transition region. With $Q \ll 1$ the transition is very smooth. On the other hand, large Q causes peaking around ω_0. Figure 4.22 shows the Bode diagram with complex conjugate poles and various values of Q. The peaking starts if $Q > 1/\sqrt{2}$. Under this limiting condition, $Q = 1/\sqrt{2}$, the response is said to be maximally flat. The amplitude with maximally flat conditions is -3 dB at $\omega = \omega_0$. With $Q = 1/2$ the amplitude at ω_0 is -6 dB, while with $Q = 8$ there is a gain of about 10 (20 dB).

Notice that approximately one decade from ω_0 the Bode diagram is constant or a straight line. Therefore the exact diagram deviates from the approximate one by less than a two-decade interval (or 6.6 octaves).

4.5.2 Phase Bode Diagram

The phase Bode diagram plots the phase of $H(\omega)$ (or $H(f)$) versus the logarithm of angular frequency, ω, or frequency, f, normalized to have a dimensionless quantity.

The phase of the product or division of complex variables is the sum of the phases of the individual complex components. They are marked as positive if the term is the numerator and negative if it is the denominator. Therefore the phase of equation (4.51) is

$$Ph\{H(j\omega)\} = \sum_{i=1}^{n} Ph\{1 - j\omega/s_{z,i}\} - \sum_{j=1}^{m} Ph\{1 - j\omega/s_{p,j}\}.$$ (4.74)

Because of the additive property it is convenient to estimate the phase of a single real solution or that of complex conjugates. Considering a single zero again at $-\omega_z$ or a pole at $-\omega_p$ (the same principle with frequency instead of angular frequency is equivalent), the phase is

$$Ph\{H_z(\omega)\} = \arctan\left(\frac{\omega}{\omega_z}\right), \quad Ph\{H_p(\omega)\} = -\arctan\left(\frac{\omega}{\omega_p}\right),$$ (4.75)

which is 0 for $\omega = 0$, becomes $\pi/4$ at $\omega = -\omega_z$ (or $-\pi/4$ with the pole at $\omega = -\omega_p$), and saturates toward $\pi/2$ (or $-\pi/2$ with a pole at $\omega \to \infty$). Zeros and poles on the positive part of the real axis give the opposite phase behavior.

The phase Bode diagram uses a log scale for the frequency axis, and therefore

$$Ph\{H_z(\omega)\}_{Bode} = \log\left[\arctan\left(\frac{\omega}{\omega_1}\right)\right]$$

$$Ph\{H_p(\omega)\}_{Bode} = \log\left[-\arctan\left(\frac{\omega}{\omega_1}\right)\right].$$ (4.76)

The function is plotted in Figure 4.23. As expected, the phase at $\omega = \omega_1$ is $\pi/4$ (45°) for a real zero at $\omega = -\omega_1$ and $-\pi/4$ ($-45°$) for a real pole. Since the asymptotic value of the

COMPUTER EXPERIMENT 4.5

Discover the Bode Diagrams

This experiment uses a special board with active filter made of the cascade of six digitally programmable analog sections: one gives rise to a real zero, one a real pole and two pairs generate complex conjugate zeros and complex conjugate poles. A custom instrument, realized by the ElvisLab technicians, allows you to set the position of zeros and poles. Moreover, the instrument generates a signal for the display that shows the zeros and poles on the *s*-plane. The spectrum analyzer plots the Bode diagrams of the output and generates a sweep of the sine wave input signal in a defined frequency interval. You can:

- define the position of zeros and poles on the *s*-plane. If the real part equals -99 the filter section is a short circuit. The real part of poles is negative or zero. Zeros can be either in positive or negative half *s* plane;
- choose the frequency interval of the spectrum analyzer.

MEASUREMENT SET–UP

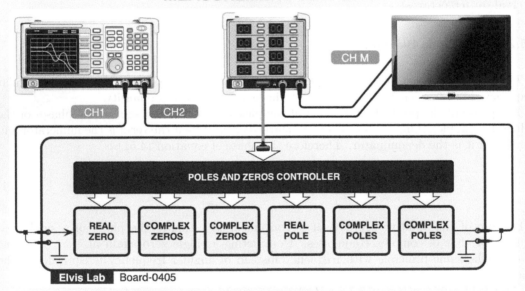

Elvis Lab   Board-0405

TO DO

- Set all the real parts of zeros and poles to -99, to bypass input to output and observe the plot of the spectrum analyzer. As expected, you get 0 dB and 0° shift at any frequency.
- Set the imaginary part of a pair of poles to 1 and observe what happens to the magnitude and phase when changing the real part. Notice the phase at the poles' frequency.
- Add a real pole and move its position near to and far away from the complex poles. Observe how the phase changes (notice the plot of the phase that ranges from 0° to 360°).
- Use multiple poles and zeros and change their relative positions. Observe what happens with zeros and poles superposed one another, and when zeros are symmetrical to poles with respect to the imaginary axis.
- Change the sign of the real part of the zeros and explain the effect that you observe on the magnitude and phase of the Bode diagrams.

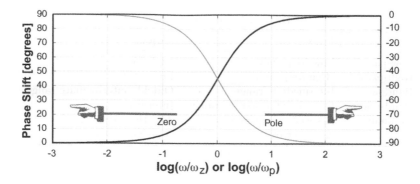

Figure 4.23   (a) Bode phase diagram of real zero at $-\omega_z$; (b) Bode phase diagram of real pole at $-\omega_p$.

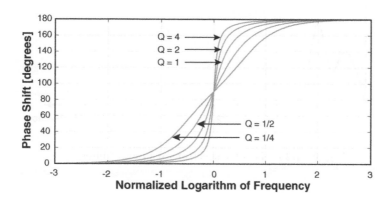

Figure 4.24   Bode phase diagram for complex conjugate zeros at different values of Q.

phase shift is $\pm\pi/2$ ($\pm 90°$), it is used to say that a zero on the negative real axis gives rise at high frequency to a $90°$ phase shift and a pole $-90°$ phase shift.

Equation (4.72) enables us to estimate the phase shift of a pair of complex conjugate zeros for a given ω_0 and Q:

$$Ph\{H_0(\omega)\} = \arctan\left[\frac{\omega/Q\omega_0}{1 - \omega^2/\omega_0^2}\right]. \tag{4.77}$$

The result starts from zero at low frequency, is $\pi/2$ at $\omega_0 - \epsilon$, and would become $-\pi/2$ at $\omega_0 + \epsilon$. This unrealistically steep change of phase is avoided by adding the constant amount π. The phase shift continues increasing and becomes π for $\omega \to \infty$. For complex poles the situation is complemented with a phase shift that ranges from zero to $-\pi$.

The value of Q controls the speed of the phase change in the region around ω_0. As shown in Figure 4.24, a higher Q increases the speed of change. An infinite value of Q (which means having zeros on the imaginary axis) causes a sudden change of phase that jumps from zero to π at ω_0.

Note that even at one decade of distance from the zeros or poles, either real or complex conjugate, the phase shift is a few degrees from the asymptotic value. It may be as large as $20°$ for complex zeros or poles and low value of Q. The residual phase shift can be a problem in some situations that we shall study in later chapters.

Table 4.1   Types of filter

Type	Pass	Reject
Low-pass	Low frequency	High frequency
High-pass	High frequency	Low frequency
Bandpass	Intermediate range	Outside intermediate range
Band-reject	Outside intermediate range	Intermediate range
Resonant	A specific frequency	All other frequencies
Notch	All other frequencies	A specific frequency

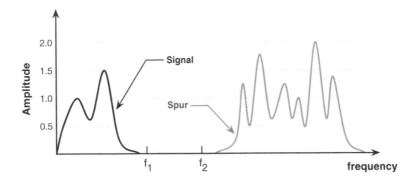

Figure 4.25   Possible spectrum at input of filter.

4.6   FILTERS

A filter, as its name implies, makes a selection from a mixture of elements by separating components with desired features from others with unwanted attributes. Therefore filters pass what is good and reject the bad. This is what is done with electronic signals that are a mixture of good and bad information.

Electronic filters select, using frequency as the discriminating parameter. If the filter passes low frequencies and rejects high frequencies it is called *low-pass*. It assumes that information is in the low-frequency range and spurs are at high frequencies. On the other hand a filter that rejects low-frequency components and passes high frequencies is called *high-pass*. In addition, there are *bandpass* filters that accept frequencies in a range; we also have *band-reject* filters, for which a given frequency interval is undesired. Finally, we have filters that focus on a single specific frequency. This type of filter is called *resonant* or *notch* if it passes or rejects a given single frequency. The various categories and features are given in Table 4.1.

Obviously a real filter is not perfect; in its *pass* region a small change of amplitude (and phase) may occur, and the *reject* region is not a brick wall with absolute cancellation of frequency components. A filter just produces attenuation, whose effectiveness depends on its closeness to the pass region and the complexity of the filter itself. Moreover, the transition between the pass and reject regions cannot be abrupt. A transition region between the pass interval and the stop interval is necessary and is used.

Consider the spectrum of Figure 4.25. It outlines a band-limited signal with frequency components below f_1. Amplitude is such that the maximum spectrum is 1.5 (in arbitrary units).

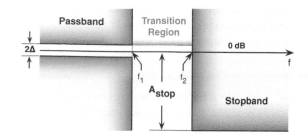

Figure 4.26   Mask of low-pass filter.

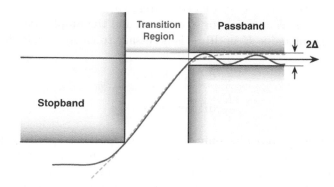

Figure 4.27   Mask of a high-pass filter and two possible filter responses.

High-frequency spurs with large amplitude significantly corrupt the signal. Suppose that an acceptable level of spur is 60 dB below the peak of the signal. Since the peak spur is $2/1.5$ (~ 2.5 dB) higher than the signal, it must be attenuated by 62.5 dB. The spurs occur at frequencies above f_2; therefore, the filter must keep the spectrum unchanged (0 dB attenuation in the 0 to f_1 range) and provide at least 62.5 dB of attenuation for $f > f_2$. The region between f_1 and f_2 is "grey" as it does not contain any frequency component. It is a transition region used as a convenient drop in filter response.

The above filter features can be better shown in the diagram called a *filter mask*. This provides in graphic form the filter requirements both for amplitude and phase response. Since for some applications phase is not relevant, in those cases only the amplitude mask is used.

Masks use a linear or logarithmic frequency axis and, for amplitude, often use the dB scale. Consider the low-pass amplitude mask of Figure 4.26. There are three frequency intervals, $f < f_1$, $f_1 \le f < f_2$ and $f \ge f_2$. The first is the passband region, the second is the transition region, and the last is the stop-band region. Dashed limits specify unwanted regions. The response in the pass region can possibly feature any amplification or attenuation but is within the $\pm\Delta$ range. The amplitudes in the stop region must be at least A_{stop}. The transition region is for the passage between $\pm\Delta$ and A_{stop}. The response can be anywhere, even if in some cases the amplitude cannot exceed given limits (like $+\Delta$, as shown in the figure).

Figure 4.27 shows the mask of a high-pass filter. It is complementary to the low-pass one of Figure 4.26: low frequencies are rejected and high frequencies are passed without significant change. The figure also shows two possible responses; both stay inside desired limits. One is smooth in the bandpass range, and the other experiences some ringing. Having a smooth response is more difficult than permitting ringing inside the passband interval. Even if the mask is the same, the difference can require more effort in designing the circuit.

Always remember that ...

filtering is a linear operation. Every time frequency or s is used we imply that the superposition principle holds. Superposition is possible only with linear, time-invariant systems.

Clearly small values of Δ, higher values of A_{stop}, or small transition regions make designing the filter difficult. Therefore it is important not to specify filter masks above what is really needed. In other words, the requirements must correspond to the right trade-off between various important quantities, such as system cost, area, volume, and power consumed.

If the transition region is X decades (or Y octaves) and the amplitude difference between the passband and the stopband is K in dB, it is necessary to ensure a roll-off of

$$\frac{K}{X} \text{ dB per decade} \tag{4.78}$$

$$\frac{K}{Y} \text{ dB per octave.} \tag{4.79}$$

Note that the filter mask provides only limits, and does not specify other constraints. Therefore many solutions can satisfy the requirements. Even if specifications impose some extra features, such as *maximally flat* output, there are many possible circuit solutions. The best choice depends on the value and sensitivity of the components used. For example, a scheme that employs resistances can require too big or too small values. An architecture that nominally fulfills the specs can deviate significantly when there is a minimal change in parameters. The dependence can be on the static value of components or the dynamic behavior of active blocks, like the speed of an operational amplifier (these will be studied in a later chapter). Moreover, linearity of components can affect harmonic distortion differently.

Example 4.2

The masks of two low-pass filters have passband upper limits at 2 kHz and 8 MHz respectively. The stopbands begin at 11 kHz and 300 kHz and require a 67 dB attenuation. Estimate the roll-off of the filters.

Solution

The two separations of pass and stop band, in decades, are

$$X_1 = \log_{10} \frac{11 \cdot 10^3}{2 \cdot 10^3} = 0.74, \quad X_2 = \log_{10} \frac{3 \cdot 10^8}{8 \cdot 10^6} = 1.57.$$

The roll-offs are respectively $67/0.74 = 90.5$ dB/dec and $67/1.57 = 42.7$ dB/dec. The first requirement is very difficult; the second is affordable.

Figure 4.28 shows another amplitude mask. It is for bandpass response. There are two transition, or gray, regions, whose extents can be different; one is below a lower limit, the other above an upper desired limit of a range of interest. The expected spectrum of the good

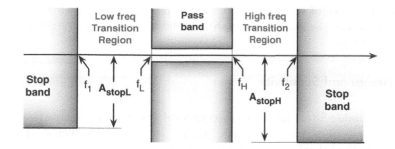

Figure 4.28 Mask specification of a bandpass filter.

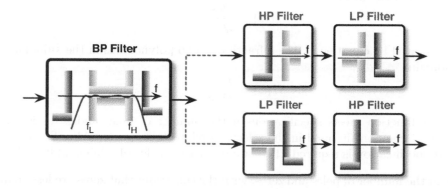

Figure 4.29 Bandpass filter and its equivalent cascade of a low-pass and a high-pass filter, or the other way around.

signal is between f_L and f_H. Required attenuation factors in stop bands depend on the system, and, as shown in the diagram, they can be different. The complexity of realizations depends on both attenuation in the two stop regions and extension of the transition regions.

Notice that a bandpass response can be thought of as the cascade of low-pass and high-pass responses. Therefore two filters next to each other produce the bandpass function. Since a filter is a linear operator, the order used in the sequence, as shown in Figure 4.29, is unimportant. However, practical issues can make one architecture better than another. Rejecting some frequency components before others can benefit specific applications.

The cascade scheme divides the filter design problem into simpler steps. It realizes the poles and zeros featuring low pass in one block, and poles and zeros producing high pass in another block. The method is not only suitable for bandpass, and can be extended to other types of filters with high complexity. The result is a simplified solution but, from a sensitivity point of view, not always the optimal one.

The value of the roll-off derived from specifications is important for predicting the complexity of the filter. We have seen that a single zero or a single pole gives rise to a 20 dB/dec roll-off. Therefore a roll-off of 30–40 dB/dec is a moderate requirement; a roll-off of 100 dB/dec is a difficult goal requiring complicated circuits.

Continuous-time (analog) or sampled-data (either analog or digital) circuits realize filter specifications. The design of these filter architectures requires study and knowledge that are not covered here. The goal of this book, and what is done in the following subsections, is to provide an introduction to the issue so as to provide the necessary understanding of problems, limitations, and methods.

4.6.1   Analog Design and Sensitivity

Continuous-time analog filters are interconnections of passive components (such as resistors, capacitors, and, in some cases, inductors) and, possibly, active blocks, such as transconductors and operational amplifiers (both to be studied shortly).

With lumped, passive elements and active devices, all basic linear operations (addition, subtraction, integration, and differentiation) are obtained. Therefore we can implement generic transfer functions of the form

$$H(s) = \frac{P(s)}{Q(s)}, \tag{4.80}$$

which is, as required by equation (4.42), a fraction of two polynomials in the s-domain or the frequency domain:

$$H(\omega) = \frac{P(\omega)}{Q(\omega)}. \tag{4.81}$$

The transfer function meets specifications if the number of poles and zeros is adequate and the values of components used in the selected architecture are suitable. Therefore, after defining what specification is required, there are further steps for obtaining a filter. They are:

- choice of the number of poles and zeros, with the constraint that zeros are less than poles (by the way, the number of poles gives the *order* of the filter);

- choice of architecture and circuit scheme;

- estimation of the components' values.

The first issue was carefully studied in the past, and excellent knowledge was generated. Methodologies based on early studies are now embedded in computer programs that enable the designer to find quick answers. Obviously, to design an optimal solution with demanding specifications one should not just rely on computer results but also undertake appropriate study and use suitable design methods.

> **Notice**
>
> The number of poles of a transfer function must exceed the number of zeros (or they must at least be equal) because it is not possible to have responses that increase indefinitely with frequency.

Architecture and circuit schemes, as indicated above, are the result of a choice because there are several possibilities, especially for high-order filters. A single pole or a single zero response is just obtained by a Resistor-Capacitor (RC) network and the possible addition of an active block to decouple input and output. For multiple zeros and poles, however, many configurations can be used. This step of the design flow has also been carefully studied in the past, giving rise to various solutions, which are well analyzed and described in specialized books.

To understand the criteria followed to select a suitable scheme it is worth remembering that all the circuits define internal variables, which are voltages of nodes or currents in wires. Thus the relationship we should study is not just equation (4.81), the link between input and output, but a set of relationships relating all variables defined by the circuit, each of them depending on one or more elements of the electrical network. For example, an internal division of a voltage V_1 by a constant factor obtained by two resistances R_1 and R_2 produces

$$V_2 = \frac{R_1}{R_1 + R_2}V_1 = \frac{V_1}{1 + R_2/R_1}. \tag{4.82}$$

The result depends on the values of two resistors, or rather on the ratio between them.

In general, for any internal variable, X_i, there are a number of parameters (resistors, capacitors, active elements) $u_{1,i}, u_{1,i+1}, \ldots, u_{1,k}$ that influence it. As a result,

$$X_i(\omega) = F_x(u_{1,i}, u_{1,i+1}, \ldots, u_{1,k}, \omega). \tag{4.83}$$

For the above, a more detailed expression, which better specifies dependence on circuit design parameters, of (4.81) is

$$H(\omega) = \frac{P(R_1, R_2, \ldots, R_p, C_1, C_2, \ldots, C_q, p_1, p_2, \ldots, p_r, \omega)}{Q(R_1, R_2, \ldots, R_p, C_1, C_2, \ldots, C_q, p_1, p_2, \ldots, p_r, \omega)}, \tag{4.84}$$

where $R_1, R_2, \ldots, R_p$ are the p resistances used, $C_1, C_2, \ldots, C_q$ the q capacitances, and $p_1, p_2, \ldots, p_r$ the r parameters featuring active blocks.

The use in equation (4.84) of the nominal values of resistors, capacitors, and parameters of active blocks obviously gives the predicted response. However, variations with respect to the nominal values, or errors in fabrication or changes of environmental conditions, modify the result that, possibly, does not meet the specifications any more.

Let us take another example: a single-pole low-pass filter made with an RC network. The position of the pole depends on the product of resistance and capacitance, i.e., the time constant. Since the accuracy of resistors or capacitors can be $\pm20\%$, $\pm10\%$ (or for expensive components $\pm5\%$), it is necessary to account for their limitations when designing the filter. In other words, the design must be an engineering one, not a mathematical one.

Note that there is no correlation between resistor and capacitor values, because their fabrication processes are different and unrelated. Therefore, inaccuracies are statistically independent, and errors must be quadratically superposed. If ϵ_R and ϵ_C are the errors in the resistor and the capacitor respectively, the time constant error is

$$\epsilon_\tau = \sqrt{\epsilon_R^2 + \epsilon_C^2}. \tag{4.85}$$

Equation (4.85) gives a possible error or, to put it more precisely, one of the possible errors whose variance can be estimated from a large collection of time constants. The use of a large set of inaccurate values gives rise to an amplitude distribution as in Figure 4.30. The distribution peak is at the average of the set of values, and σ is half the width of the curve at half peak height. The vertical axis gives the occurrence over the number of samples (2^{17} for the case depicted in the figure).

A more accurate study leads to a statistical description given by the Gaussian function

$$f(x) = \frac{1}{\sigma\sqrt{2\pi}}e^{(x-\mu)^2/2\sigma^2}; \tag{4.86}$$

COMPUTER EXPERIMENT 4.6

Sensitivity of Transfer Function

This experiment uses a board with a digitally controlled active circuit able to perform a linear transfer function. The polynomials of the numerator and the denominator are both fourth order. The transfer function is given by:

$$H(s) = \frac{a_0 + a_1 s + a_2 s^2 + a_3 s^3 + s^4}{b_0 + b_1 s + b_2 s^2 + b_3 s^3 + s^4}$$

A custom-designed instrument with an intuitive input interface sets the coefficients "*a*" and "*b*". The custom instrument contains a signal processor that estimates poles and zeros and drives a monitor to display them on the *s*-plane. The spectrum analyzer generates a swept sine wave input and shows the output Bode diagrams. You can:

- set the values of all coefficients of the transfer function;
- choose the frequency interval examined by the spectrum analyzer.

MEASUREMENT SET–UP

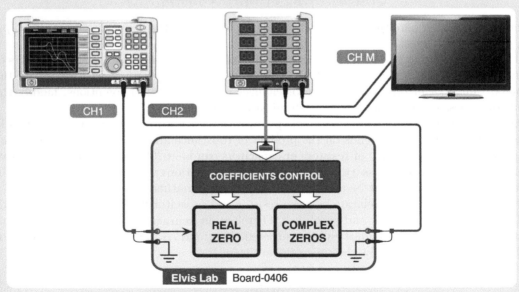

TO DO

- Set coefficients of the transfer function that produce a low-pass response.
- The other way around, choose zeros and poles and estimate the value of coefficients. Verify the Bode diagrams of the transfer function performed.
- Change the coefficients, one by one with small steps, and identify the one that gives rise to the largest modification of the amplitude response.
- Estimate the sensitivity of the design parameters for two different $H(s)$ responses. Do this by measuring the amplitude response at 10 relevant frequencies and estimate its variation for a small change of one parameter.

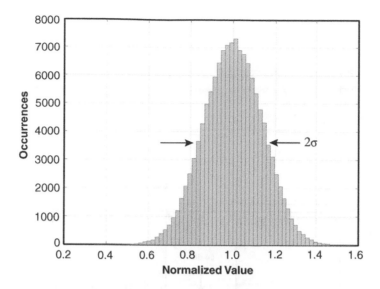

Figure 4.30 Statistical distribution of the value of an inaccurate quantity. The horizontal axis is normalized to the average of the quantity.

the plot is a bell-shaped curve; symmetrical with respect to μ, and has unity integral. The more accurate result has width at half peak of 2.35 σ instead of the 2 σ mentioned before. Even if it is not precise, assuming 2 σ as the width at half peak height is more mnemonic. Whichever is used, the statistical distribution function is important to predict the yield of integrated circuits or systems. Typically, testing requires that key performances must be lower or higher than, or within, given values. The parts that do not satisfy the conditions are rejected and worsen the yield. This topic, specific to courses in the manufacturing engineering discipline, is not studied here.

Let us go back to the above study that led to equation (4.84). Errors in the values of resistors, capacitors, or parameters of active elements affect internal electrical quantities. In turn, errors in electrical quantities modify the coefficients of the polynomial of (4.84) and, finally, alter the frequency response. Therefore, different schemes that use different internal variables generate different errors on response. This is why the accuracy of the overall response depends on its circuit implementation.

Limitations caused by component inaccuracy are quantitatively represented by the *sensitivity*, classically defined by

$$S^X_{\alpha_i} = \frac{\partial (\ln X)}{\partial \ln(\alpha_i)} = \frac{\partial X / X}{\partial \alpha_i / \alpha_i}. \tag{4.87}$$

For the transfer function described by equation (4.84), which depends on many parameters, many sensitivities are defined. They are

$$S^H_{R_i} = \frac{\partial H / H}{\partial R_i / R_i}, \quad i = 1, \ldots, p, \tag{4.88}$$

$$S^H_{C_j} = \frac{\partial H / H}{\partial C_j / C_j}, \quad j = 1, \ldots, q, \tag{4.89}$$

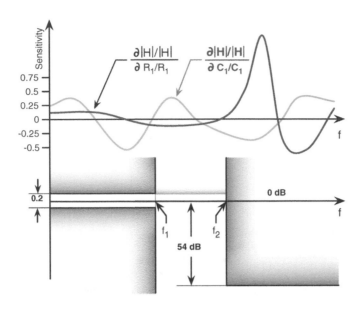

Figure 4.31   Sensitivity to amplitude response and filter specification.

$$S_{P_j}^H = \frac{\partial H/H}{\partial p_l/p_l}, \quad l = 1, \ldots, r. \tag{4.90}$$

Since $H(\omega)$ is complex, the results are complex as well. Therefore pairs of diagrams may represent modulus and phase. Figure 4.31 shows the sensitivity of a hypothetical amplitude response to variations in a resistance and a capacitance. The figure also gives specifications for the related filter. Notice that the limit is different in different frequency regions. Sensitivity to resistance R_1 is well below 0.25 in the passband and peaks to more than 1 in the stopband. Sensitivity to capacitance is always limited but its value reaches 0.4 in the passband. As outlined by the filter mask, because of requirements in the passband (± 0.1 dB in the figure) the change in $|H|$ cannot be larger than a very small amount, while the admitted change in the stopband can be very large. Accordingly, the sensitivity in the stopband to C_1 creates more problems than does sensitivity to R_1. Therefore, in addition to studying sensitivity it is necessary to understand when the limit matters.

Observation

High sensitivity in the filter passband is more critical than high sensitivity in the stopband, where attenuation is large and possible related changes become small.

If the sensitivity or inaccuracy of one element is much larger than that of others, attention focuses on that single case. On the other hand, when various errors are comparable they must be superposed. Since stochastic errors are uncorrelated, the superposition must be quadratic. However, when the critical parameter is not a single resistance, say R_1, or a capacitance, say C_1, but the time constant produced by them, it is necessary to account for the error of R_1C_1. As outlined by equation (4.85), errors in capacitance and resistance are summed quadratically.

Thus, the time constant is the limit

$$\frac{\partial |H|}{|H|} = S_{\tau_1}^H \sqrt{\left[\frac{\partial C_1}{C_1}\right]^2 + \left[\frac{\partial R_1}{R_1}\right]^2}. \tag{4.91}$$

Example 4.3

The specification of a filter requires that amplitude in the passband be (0 ± 0.1) dB. The architecture of the filter is a proper interconnection of resistors, capacitors, and operational amplifiers with nominal values capable of obtaining a response within a ± 0.05 dB interval. Sensitivity to the worst component is, in the passband, 0.34. Estimate the accuracy of that parameter required to meet the specifications. Suppose that all other design parameters have negligible sensitivity.

Solution

The margin allowed by the architecture is about ± 0.05 dB. Accordingly, the error $\Delta |H|/|H|$ caused by the sensitive parameter cannot exceed ± 0.0058. Since the sensitivity is 0.34, the maximum permitted variation of the parameter is

$$\Delta x / x = \frac{\Delta |H|/|H|}{S_x^H} = \frac{\pm 0.0058}{0.34} = \pm 0.017.$$

Therefore, the accuracy must be as low as $\pm 1.7\%$.

4.6.2   Sampled-data Analog and Digital Design

A sampled-data filter can be either analog or digital (with discrete amplitude in the digital case). In both forms, filters use delay as a basic operator rather than using a derivative or an integral. The sampled-data filter is conveniently described in the z-domain, the counterpart of continuous time. The transfer function has the form

$$H(z) = \frac{P(z)}{Q(z)}, \tag{4.92}$$

where $P(z)$ and $Q(z)$ are polynomials with given order.

The transfer function has poles and zeros and their locations are such that the frequency response

$$H(e^{j\omega T_s}) = \frac{P(e^{j\omega T_s})}{Q(e^{j\omega T_s})}, \tag{4.93}$$

which is the z transfer function estimated on the unity circle, meets specifications as depicted by the filter mask.

Remember that in the sampled-data domain the frequency interval that matters is the Nyquist interval $(0, \ldots, f_N; \ f_N = 1/2T_s)$, because the responses in other Nyquist bands are replicas, possibly mirrored, of the first zone. Therefore, the filter masks must define their requirements in the first Nyquist interval only.

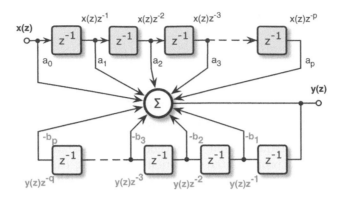

Figure 4.32   Direct implementation of equation (4.96).

The design of an analog sampled-data filter must account for component inaccuracies, as it does for continuous-time circuits. In particular, it is necessary to ensure a margin in the mask specifications to provide room for degradation caused by various inaccuracies, such as those of time constants. However, the accuracy of time constants in sampled-data circuits produces fewer problems than in continuous-time ones, because a special design method, called the "switched" capacitor, gives time constants of the form

$$\tau = T_S \frac{C_1}{C_2},\qquad(4.94)$$

which is the product of sampling period and capacitor ratio. The accuracy of sampling periods is normally high, being based on quartz clocks. Capacitor ratios are very good for integrated circuit implementations that use the same technological steps to realize capacitors. So analog sampled-data filters realized with integrated technologies offer better accuracy than their continuous-time counterparts.

Digital implementations of sampled-data filters perform processing as prescribed by the transfer function by means of addition, multiplication, and delay. Obviously, there are various architectures that can perform processing, but, just to understand one of the possible ways of solving the problem, observe that equation (4.93) has the form

$$\frac{y(z)}{x(z)} = H(z) = \frac{a_0 + a_1 z^{-1} + a_2 z^{-2} + \cdots + a_p z^{-p}}{1 + b_1 z^{-1} + b_2 z^{-2} + \cdots + b_q z^{-q}},\qquad(4.95)$$

which yields

$$y(z) = -(b_1 z^{-1} + b_2 z^{-2} + \cdots + b_q z^{-q})y(z)+$$
$$+ (a_0 + a_1 z^{-1} + \cdots + a_p z^{-p})x(z).\qquad(4.96)$$

The processing scheme of Figure 4.32 produces direct implementation of equation (4.96). Two digital delay lines, one for input with p cells and the other for output with q cells, make the circuit. Signals at various signal taps are multiplied by their coefficients, and all the terms are summed together.

Observe that the signal at input is quantified by, say, n bits; moreover, coefficients are quantified by another number of bits. If one coefficient has m bits, its product with an n-bit signal is $n + m$-bit, a resolution that can be too large. Therefore, it may be necessary to truncate the resulting word-length. Moreover, adding several terms produces results with more bits than the longest term of the addition. Again it may be necessary to truncate partial results or even the final value.

What we have said above indicates that any digital implementation is a trade-off between cost and benefit, exactly as for analog implementations. Using filter coefficients with reduced resolution implies accepting possible alterations in frequency response. Reducing the resolution of signals means adding quantization noise. Therefore the margin allowed for in specifications must be used by the designer to achieve optimal performance. In addition to fulfilling the response requirements, it is also necessary to keep signal-to-noise ratio within prescribed limits.

Another point is that operations require a suitable calculation time. Since processing cores use several clock cycles to do the job, even with processors running at many hundreds of MHz the output is not "immediate." Delay can be a problem with systems requiring quick actions or being used (we shall cover this in a subsequent chapter) in feedback.

4.7 NON-LINEAR PROCESSING

Previous sections studied various facets of linear processing (analog and digital). The focus on linear processing is because many applications aim at various functions with minimal harmonic distortion. Moreover, designers often exploit the superposition principle. However, a number of applications use non-linear techniques because of their potential benefits. So, in addition to studying linear operations, it is important to know about non-linear processing methods.

There are many systems that exploit simple non-linear functions, some of them based on threshold detection. They use single or multiple logic information for measuring or control purposes. For example, automatic door systems can use soft-start/soft-stop control and on/off information from photocells. Drivers of motors that change in a non-linear manner limit the problems of moving high-inertia parts. Therefore, the life of a driver is considerably extended by soft-start operation.

Other important applications that use non-linear processing are audio and video ones. As is well known, hearing response is almost logarithmic, and this feature is often used to optimize the transmission of vocal signals. Since large amplitudes in audio signals are not very significant, they can be attenuated without losing information. This strategy leads to logarithmic or pseudo-logarithmic analog compressors or a non-linear (or *companding*) data conversion.

Telephone transmission of voice uses a data converter with uneven quantization intervals that reduce the number of bits required. Voice signals are transmitted using eight bits, but the dynamic range corresponds to 12–13 bits. The code uses one bit for sign and seven for non-linear quantization. With seven bits we approximate the equation

$$y = \frac{\log(1 + \mu|x|)}{\log(1 + \mu)} \quad \text{for } 0 < |x| < 1, \tag{4.97}$$

which is called the μ-*law* and is used in North America and Japan (with $\mu = 255$), or the equation

$$y = \begin{cases} \dfrac{1 - \log(A|x|)}{1 + \log(A)} & \text{for } \dfrac{1}{A} \le |x| \le 1 \\[2ex] \dfrac{Ax}{1 + \log(A)} & \text{for } 0 < |x| < \dfrac{1}{A}, \end{cases} \tag{4.98}$$

called the A-*law*, which is used in Europe (with $A = 87.6$).

COMPUTER EXPERIMENT 4.7

Processing in the *z*-Domain and Saturation

With this computer experiment you can study the linear processing in the *z*-domain. The three inputs are given by discrete-time sine wave generators. The sampled-data analog circuits process exactly if their outputs are, in module, less than 1. Above that limit the output saturates to ±1.2 with smooth transition. There are two outputs: one is the superposition of the three inputs after multiplication by $H(z)$, the other is the superposition of the three inputs followed by the multiplication by $H(z)$. The test board contains four equal processors that do the analog processing at a sampling frequency of 32 MHz. The transfer function $H(z)$ is the weighted addition of three successive input samples $(1 + a_1 z^{-1} + a_2 z^{-2})$. You can:

- change the amplitude of three inputs and the frequency of the last two. The frequency of the first sine wave is 1 MHz;
- set the value of the coefficients a_1 and a_2;
- select one of the single processed outputs and display it on the scope.

MEASUREMENT SET–UP

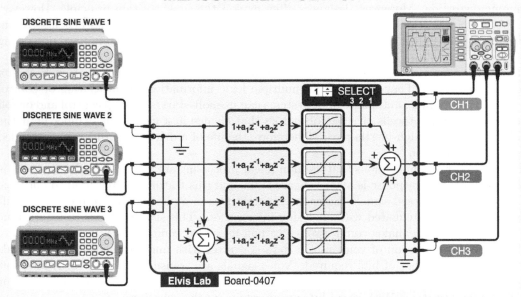

TO DO

- Set the amplitude of coefficients a_1 and a_2 to zero and observe the outputs. They are equal to the inputs of their superposition for amplitudes that do not cause distortion.
- Set the second amplitude small and zero the others, and to observe waveforms with low and high frequency with different coefficients, a_1 and a_2. Try $a_1 = \pm 2$ and $a_2 = 1$.
- Use all the three inputs with different frequencies and amplitudes. Notice differences in the responses when the analog processors start saturating.
- What is the maximum ratio between maximum output and peak input with frequencies equal to 1 MHz, 347 kHz and 2 MHz, respectively and the same input amplitudes? Does the value depend on the peak of each single-input processor?

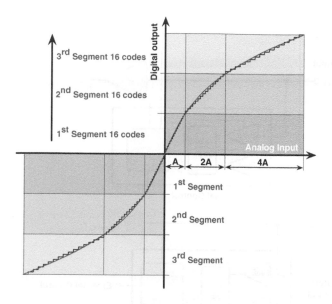

Figure 4.33   The first three positive and negative segments of the μ coding.

In reality quantizers approximate the above equations by hybrid linear and non-linear coding. The dynamic range is divided into eight coarse, uneven segments (or chords) that use equal steps inside them. Figure 4.33 shows the first part of the input/output response of the μ coder. The horizontal axis represents analog inputs, and the vertical axis plots the corresponding digital codes. Notice that, as required, the amplitude of the analog segment increases exponentially. The first segment is A, the second is $2A$, and so forth, until the last segment, which is $128A$. Each segment is divided into 16 equal steps. Therefore, three bits specify a segment and four bits are for linear coding inside the segment.

Non-linear processing is profitably used to handle noise in audio applications. Many systems on the market aim at reduction or active cancellation of noise. For noise reduction it is necessary to distinguish noise, or "hiss," from signal and to weaken noise only. One method assumes that if the signal is below a predefined magnitude it is noise and must be significantly reduced. Suitable processing dynamically estimates the noise threshold using, for example, the peak amplitude. The circuit analyzes the signal in time-slots to extract

$$x_{max}(i) = \max[x(t)], \quad (i-1) \cdot \Delta t < t < i \cdot \Delta t \tag{4.99}$$

$$x_{min}(i) = \min[x(t)], \quad (i-1) \cdot \Delta t < t < i \cdot \Delta t. \tag{4.100}$$

Processing of the sets of sampled-data signals indicates whether the time-slot contains signal or noise. Moreover, if peaks are large threshold increases, with small peaks the threshold decreases. A hypothetical stereo block diagram is shown in Figure 4.34. The peak value of the inputs is periodically measured (reset determines the beginning of each time-slot); the peak values, converted into digital form, are the input of the logic processing. The result, combined with volume control, determines the gain of the power amplifiers.

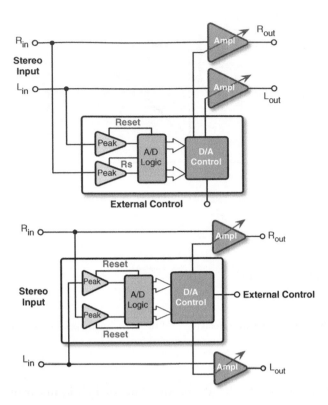

Figure 4.34   Block diagram of stereo noise-reduction audio amplifier based on signal-controlled gain.

Peak detection and variable gain amplification are obtained from suitable analog non-linear circuits. Therefore, non-linear processing can be either a mixture of analog and digital or totally digital processing, with data converters at the input and output of the system.

A second technique for noise cancellation relies on the wide-spectrum properties of noise. After selecting successive time-slots, the spectrum in each time-slot gives information on the frequency distribution of signal and noise, allowing selective filtering of input. Simple solutions just determine the upper limit of the signal spectrum and reject frequencies above that limit. Figure 4.35 shows a block diagram of a possible system that uses programmable filters whose coefficients are continuously updated.

The processing is non-linear because the operation is time variant. Coefficients of a linear circuit, the filter, are periodically updated on the basis of analysis of the input signal. The same goes for other audio-like processing, for example dynamic control of hearing aids.

Since the method described is a non-linear operation, strictly speaking it is not possible to talk of filtering, because a filter is a linear circuit. However, in practice one may consider it as filtering, even if some unexpected sounds (such as small clicks) may result from switching coefficients. Formally, we should estimate the integral convolution of input with an impulse response that changes with Δ periodicity. At the end of each time-slot, the values of state variables in the system are the starting conditions of other systems. They themselves evolve in the subsequent time-slot, with different responses to input and to the initial conditions.

COMPUTER EXPERIMENT 4.8

Non–Linear Processing in the *z*-Domain

This experiment permits you to experience non-linear processing in the *z*-domain. This operation is performed in digital communication to correct alterations of transmitted signals because of variable atmospheric conditions. The experimental setup multiplies two discrete-time signals, $A_1\sin(\omega_1 t)$ and $A_{off}+A_2\sin(\omega_2 t)$. The two frequencies are quite different.

The "log" block separates the two components, the discrete filter removes the low-frequency term before the inverse "log^{-1}" transformation. The scope displays the high-frequency sine wave, the result of the product and output of the homomorphic corrector. You can:

- change the frequency and amplitude of the two input sine waves;
- define the offset value, $A_{off,}$ of the signal at low-frequency;
- define the sampling frequency of the sampled-data system;
- select the order k of the discrete high-pass filter $(1-z^{-1})^k$.

MEASUREMENT SET–UP

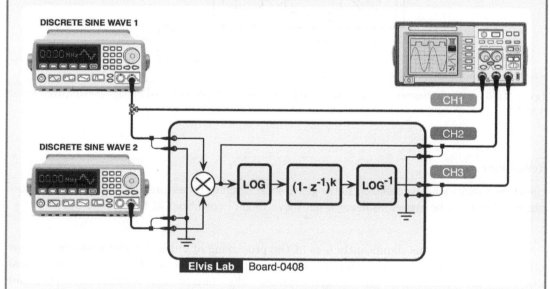

Elvis Lab Board-0408

TO DO

- Set the frequencies of the sinusoidal input signals. Choose a large ratio between two frequencies. Set the amplitude of the high frequency sine wave.
- Set the offset of the low-frequency sine wave and small amplitude to 1.
- Observe the output waveform with different orders of the high-pass filter. Notice that at high frequency the filter gives rise to amplification that is larger for high orders.
- Estimate the reduced effectiveness of increasing the lower sine wave frequency.

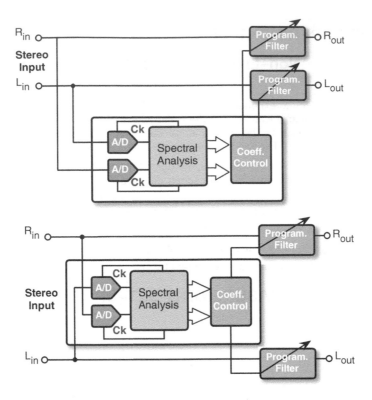

Figure 4.35   Block diagram of stereo noise-reduction audio amplifier based on signal-band-controlled filtering.

Remember

Linear processors whose coefficients change, even at discrete times, are actually non-linear processors. They can be viewed as linear when they "forget" conditions at the change times.

Non-linear operations significantly benefit the processing of video signals for storing, transmitting, or enhancing the video quality. There are many approaches that always use digital solutions to meet the need for increasing sharpness, video compression, counter extraction, and image recognition. There are various non-linear algorithms translated into software that run on computers, but in a number of situations it is necessary to use analog electronics tailored for portability, speed of operation, and low power consumption.

Self training

Many video compression algorithms are referred to by a common acronym: JPEG, which stands for Joint Photographic Expert Group, a committee that defines standards for compressing high-quality still images.

Search the Web and find more information on JPEG and video coding algorithms. Limit the search to generic information, but try to understand the level of complexity of the electronics required and how difficult real-time processing is. (By the way, "real-time" means performing processing while the events are occurring).

Write a short note on your search results.

Video compression is an important example of non-linear video processing, because the amount of data to be transmitted or stored is very large: every pixel can be digitized at a very high resolution, such as 16 bits, and images can use several million pixels. For reducing the amount of data, complex non-linear compression algorithms are used. They are able to obtain compression rates as large as 12–18 with images indistinguishable from the original, or they can go up to compression rates of 160 with recognizable images. Compression algorithms have evolved in complexity and effectiveness, and they use various techniques such as transform-based coding, sub-band coding, entropy coding, run-length coding, vector quantization, block-truncation coding, and coding by modeling a scene. All of these are listed here to give an idea of how many methods can be used. They are defined by complex equations whose computation is done with DSPs.

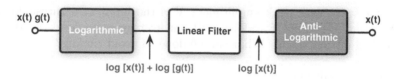

Figure 4.36   Homomorphic separation of multiplied signals.

Other examples of non-linear processing come from communications. One is correcting the alteration of transmitted signals that change because of variable atmospheric conditions, especially with high-frequency carriers. The limit causes a slow varying attention $g(t)$ of expected signal, $x(t)$. The received signal becomes

$$y(t) = x(t)g(t). \tag{4.101}$$

Non-linear processing corrects the error, assuming that the signal spectrum is at frequencies much higher than those of $g(t)$. One solution, shown in Figure 4.36, is to pass the signal (or, rather, its modulus) through a logarithmic block. The output is

$$\log|(x(t)g(t))| = \log|x(t)| + \log|g(t)|, \tag{4.102}$$

which transforms multiplication into the addition of two terms with well distinct spectra. Therefore, a simple high-pass filter rejects the low-frequency term denoting possible change of atmospheric conditions. Then the anti-logarithmic block rebuilds the signal. The method, called *homomorphic* separation, is normally carried out by using digital blocks.

PROBLEMS

4.1   Estimate the s-domain equivalence of the following processing functions:

$$y(t) = 3x(t) + \int dt \int \frac{1}{4}x(t)\,dt + \frac{d^2x(t)}{dt^2}$$

$$\int y(t)\,dt = y(t) + 3\frac{d^2y(t)}{dt^2} + 6x(t).$$

4.2   Determine which one of the following processing functions benefits from the transformation from time to the s-domain:

$$y = ax + by', \quad y = ax + \frac{b}{x}, \quad \int y\,dy = \frac{y}{a} + x'',$$

$$y - x = yx + y', \quad yy' = x + x', \quad ay + x' = b \int x\,dx.$$

The coefficients a and b are constant.

4.3   What is the z-domain equivalent of the following sampled-data relationships?

$$y(nT + 2T) = -y(nT) - 2y(n + T) + x(nT) - x(nT + T)$$

$$y(nT + T) = y(nT) + x(nT) + 3x(nT + T) + 3x(nT + 2T) + x(nT + 3T).$$

4.4   A noisy sine wave with peak amplitude 1 V is the input of a threshold detector set at 0.5 V. Draft the input waveform with an average noise level of 10 mV and the corresponding output signal. What can be done to avoid ringing?

4.5   The inputs of a multiplier are

$$x_1(t) = 3\sin(2 \cdot 10^3 t) + 7\cos(7 \cdot 10^3 t)$$
$$x_2(t) = 2.6\sin(6 \cdot 10^3 t) + 4\sin(8 \cdot 10^3 t).$$

What is the spectrum of output signals?

4.6   The output response to an input pulse is a ramp with slope equal to the input amplitude multiplied by three. Then the output remains constant until the end of the pulse. Determine the processing function.

4.7   The input signal of the linear block is a sine wave $x = 2\sin(2\pi f_{in} t)$. The response of the same block with a unity step at input is a ramp with slope 1.6 V μs that lasts for 0.7 μs. After 0.7 μs the output remains constant. Determine the expression of the waveform at the output and estimate its value at 5 μs for $f_{in} = 2$ MHz.

4.8   Plot the vectors representing the following sinusoidal signals:

$$y_1(t) = \sin(w_1 t), \quad y_2 = \cos\left(w_1 t + \frac{\pi}{4}\right);$$

$$y_3(t) = \frac{1}{2}\sin\left(w_1 t + \frac{\pi}{2}\right), \quad y_4(t) = -\frac{1}{3}\cos(w_1 t).$$

4.9   Two complex numbers $(3 + 4i)$ and $(-2 + i/2)$ represent two sine waves. Estimate the amplitude and phase of the two sine waves and the amplitudes of their addition and subtraction at $t = 0$. Determine the result by the use of the vectorial equivalents of addition and subtraction.

4.10   A linear processing block transforms the input $x(t)$ into the output $y(t)$ by two intermediate variables $y_1(t)$ and $y_2(t)$. The time-domain relationships between the variables are

$$y_1 = \int (x - y)\, dt$$

$$y_2 = \int (x - y_1)\, dt + y_1$$

$$y = \int (y_2 + y)\, dt.$$

Determine the transfer function in the s-domain.

4.11   A transfer function with three poles and two zeros has one pole at $s = -1$. The input/output relationship is

$$x(s^2 + 40s + 409) = y(s^3 + 11s^2 + 39s + 29).$$

Plot on the s-plane the position of zeros and poles.

4.12   A transfer function $H(s)$ has three zeros and four poles. They are at

$$f_{z1} = f_{z2} = 10 \text{ MHz}, \quad f_{z3} = 20 \text{ MHz}$$
$$f_{p,12} = (1 \pm 0.3i) \text{ MHz}, \quad f_{p,34} = (2 \pm 0.7i) \text{ MHz}.$$

Poles and zeros are in the left half-s-plane and $H(jw_1) = 1300$ at 500 Hz. Determine the transfer function.

4.13   The transfer function
$$H(s) = \frac{s^2 + 19s + 78}{s^4 + 23s^3 + 192s + 936}$$

denotes two zeros and four poles. Two of the poles are complex conjugate located at $s_p - 5 \pm j$. Determine zeros and poles and plot them on the s complex plane. Estimate amplitude and phase at the complex frequency $s = j3$ rad/s.

4.14   Plot the approximate Bode diagram of the transfer function of Problems 4.12 and 4.13. What changes if one of the zeros overlaps a pole?

4.15   The system in the figure below is made up of three blocks.

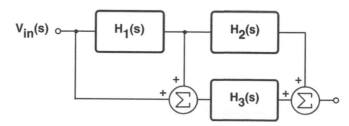

$$H_1(s) = \frac{s}{1 + 10s + 32s^2}, \quad H_2(s) = \frac{20s^2}{1 + 8s + 20s^2}, \quad H_3(s) = \frac{s}{1 + 5s}.$$

Estimate the transfer function and draft the Bode diagram.

4.16   Determine the transfer function obtained by the transformation $s' = 1/s$ applied to the linear system of Problem 4.15.

4.17   The transfer function of a signal processor has four poles and three zeros. Poles are at $f_{p1} = 100$ Hz and $f_{p2} = 10$ MHz. The remaining two are complex conjugate with modulus 80 MHz and phase 30°. The zeros are at 1 MHz, 60 MHz, and 250 MHz. Plot the approximate Bode modulus diagram and sketch the phase diagram.

4.18   Draw the mask of a low-pass filter with the following specifications: passband = 4 kHz, stop-frequency = 10 MHz, ripple in passband = ±0.5, attenuation in stopband = 23 000. The mask must be in dB with on the x axis the logarithm of frequency normalized to 1 Hz.

4.19   The ratio between the passband and stopband limits of the mask of a filter is 100. The required attenuation goes from ±0.1 dB to 56 dB. What is the order of the all pole transfer function that meets the specification?

4.20   Transform the low-pass response of Problem 4.12 into a high-pass counterpart generated by the frequency transformation $f_L/f_u \Rightarrow f_u/f_H$ where $f_u = 10^5$. (L and H indicate the variable frequency in the low-pass and the high-pass implementations.)

4.21   A transfer function of an electronic block denotes two poles and two zeros:

$$H(s) = \frac{s^2 + as + b}{s^2 + cs + d},$$

where $a = 3.3 \cdot 10^4$, $b = 2.6 \cdot 10^8$, $c = 10^4$, $d = 2.525 \cdot 10^7$. Estimate the sensitivity of H to parameters a and c. What changes in the zero and pole positions when a and c change by +5%?

4.22   Draft the block diagram that generates the following sampled-data transfer function:

$$H(z) = \frac{a_0 + a_1 z + a_2 z^2 - a_3 z^3}{1 + b_1 z + b_2 z^2 - b_3 z^3}.$$

ADDITIONAL COMPUTER EXAMPLES

Computer Example - A/4.01

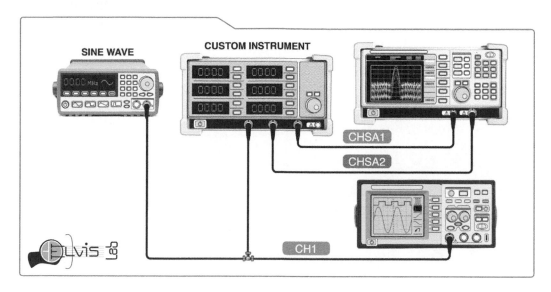

An *ElvisLab* experiment that enables design observing spectral components at the output of a non-linear circuit and the effect of a single non-linear term.

Computer Example - A /4.02

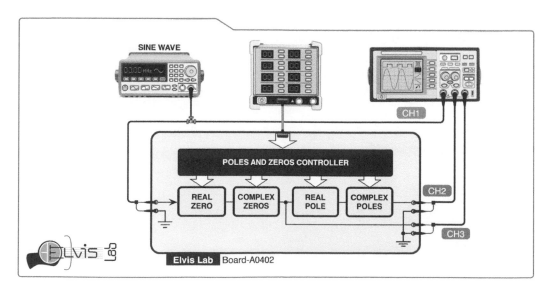

With this experiment you can observe the waveform at the output of a filter and at an intermediate point.

CHAPTER 5

CIRCUITS FOR SYSTEMS

This chapter discusses how to obtain processing by electronic circuits. We shall see that the interconnection of small electronic blocks that implement simple functions produces high-level processing. Each basic cell uses electrical variables and generates electrical variables at output, which can then be used as the input of other cells. The possible limitations caused by the real operation of electronic circuits and the interactions between basic blocks are also analyzed.

5.1 INTRODUCTION

The previous chapters studied signals and signal processing. At a high level we have seen that continuous-time or sampled-data architectures made up of black boxes perform complex functions by cascading and interconnection. The same is done by electronic circuits, but, while processing boxes use signals, electronic circuits use electrical signals produced by currents and voltages. Moreover, inside the black boxes there are other boxes, capable of performing simpler operations, handling voltages and currents as well. The interconnection of electronic blocks causes a cascade of functions, but it also gives rise to interactions between voltages and currents that can alter processing. Therefore, in addition to the simple cascade of blocks to determine signal processing, it is necessary to go further and study electronic circuits, their operation, and the types of interconnections that are needed to attain desired ends. Here we do that at the behavioral level. The details at the level of electronic components are postponed to

Understanding Microelectronics: A Top-Down Approach, First Edition. Franco Maloberti.
© 2012 John Wiley & Sons, Ltd. Published 2012 by John Wiley & Sons, Ltd.

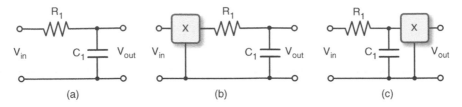

Figure 5.1   Low-pass electronic network: (a) simple version; (b) with possible input conditioning to ensure blocks fit; (c) with possible output conditioning, also to ensure blocks fit.

later chapters. Before we start we define "service" functions, and afterwards we discuss the methods used to perform simple linear and non-linear operations using electronic circuits and systems.

Describing processing with a block diagram is the premise of modular design, a method that enables multiple uses of modules already designed or fabricated. Moreover, when developing new parts, the reuse of previously implemented functions reduces the so-called time-to-market. Therefore, modular design, which corresponds to the plain interconnection of electronic blocks, is an effective way to produce electronic systems. The method is very similar to connecting the colorful interlocked plastic bricks of kids' toys. Obviously, as with such toys, it is necessary for the parts to fit properly. Therefore, one key concern of designers is to ensure the correct matching at interfaces. This is an important point, because, both with analog and with digital circuits, data transfer is often problematic. The quality of analog signals can degrade, and the transfer of digital codes can encounter difficulties because of signal timing, clock inaccuracies, susceptibility to interference (especially with low-voltage circuits), and so forth.

Modular design

Modules of an electronic system are important building parts of an integrated circuit. They are used for speeding up the circuit design, assembled on a PCB or in a more compact manner as a System-on-Chip or System-in-Package. For SoC or SiP the modules are elements of libraries providing pre-designed functions and performance.

Modular design requires a minimal signal alteration caused by block connections, and this is not a feature ensured by all circuits. This requirement holds for both digital and analog cells, but, for analog ones, even a small alteration is critical. Consider for example the passive low-pass filter shown in Figure 5.1(a). Its transfer function, assuming voltages at input and output, can be easily derived in the s domain. The result is

$$V_{out} = V_{in} \frac{1}{1 + sR_1C_1}, \tag{5.1}$$

which is 1 at low frequency and drops as $1/(j\omega R_1 C_1)$ at high frequency.

That simple network is not a good fit for many input and output terminals. A generic input source is possibly affected by the network, and a generic output connection possibly changes response, input, and output. To have no limitation at the input terminal, the circuit that brings the input should be an ideal voltage source, and this is rarely the case. To have a good fit at output the load should show infinite impedance, and this, again, is rarely the case. Therefore, a designer has two possible strategies: to accept interdependence of blocks

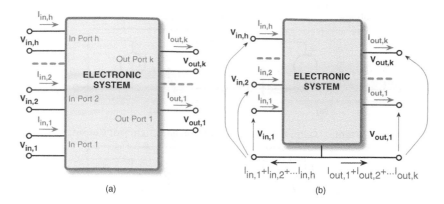

Figure 5.2   (a) Electronic system with multiple inputs and multiple outputs; (b) inputs and outputs with one terminal in common.

used in modular design and account for their effects, or to ensure that electrical variables fit, by having suitable features in the block interfaces.

In the former case, assembling a system with interconnected blocks is not very useful because it is necessary to account for the effects by altering the entire electronic network. For this, it can be necessary to solve a huge system of equations describing thousands or perhaps millions of active and passive devices and their interconnections. In contrast, if blocks do not disturb each other, the system can be studied at the electrical sub-network level, while being careful about accuracy and proper timing of the electrical variables. After that, the second-order effects of possible influences between blocks are accounted for.

Input fitting often requires a block in front of the active or passive network performing a function, like the one called X in Figure 5.1(b). The same goes for ensuring output fitting, as shown in Figure 5.1(c). Obviously, those blocks refer to well-defined input or output quantities. Therefore the circuit needs different interfaces, if the expected input or output variable is a voltage or a current.

5.2   PROCESSING WITH ELECTRONIC CIRCUITS

Generic processors have multiple inputs and multiple outputs. Therefore, the electronic circuit that performs the corresponding function should have various inputs, achieved with multiple electrical ports, and a number of outputs, consisting of other electrical ports, as shown in Figure 5.2(a). The system shown in the figure has h ports at input and k ports at output.

Electrical ports have two terminals, featuring a voltage across and a current entering one terminal and exiting from the other. However, it may happen that many or even all of the ports have one of the terminals in common, as shown in Figure 5.2(b). In some cases, for each port, it may be necessary to know the time or the frequency dependence of both voltage and current because the response can be time variant, depending on frequency in a non-linear manner. Moreover, it may happen that one electrical variable is more meaningful than the other. For example, the current can be zero while the voltage varies to carry information. Alternatively, the current carries information and the voltage is almost constant or zero. In the former case

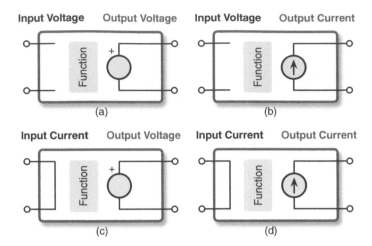

Figure 5.3   Ideal interfaces of electronic blocks with voltage or current at input and output.

the effective (and meaningful) variable is voltage; in the latter, current is the relevant quantity. This different relevance is often clarified by saying, for example, that the input of a block is voltage and the output furnishes current (or any other pairing of electrical quantities).

It is evident that having just one electrical quantity to represent information is optimal to ensure fitting. The focus must be on preserving the significant variable without devoting special attention to the other. Obviously, input and output fitting also means input and output compatibility (they should use the same electrical quantity, i.e., voltage or current, ensure the same range of variation, such as, for example 0–1.8 V or ±20 µA, and attain the same maximum rates). Moreover, fitting must be ensured, with one or more blocks connecting and loading one of the outputs. This feature is important for digital circuits that specify how many logic circuits the output can serve. It will be discussed shortly.

5.2.1   Electronic Interfaces

Voltages or currents are ideally generated by voltage or current sources. They keep voltage across terminals, or current through them, at the specified value independently of load. The amplitude and its variation in time featuring signals are not modified by any external agents. Therefore, voltage or current sources produce output that "fits all" possible loads used to quantify voltage or current respectively.

Apart from being generated, signals must be observed or measured. For voltages or currents, the measurement is ideally done with infinite or zero impedance respectively. Therefore, the ideal interfaces of electronic blocks with voltage or current at input and output are the four types shown in Figure 5.3. The blocks, which use only one input and one output, measure, ideally, voltage or current at input, perform a defined process and generate, ideally, voltage or current at output.

Unfortunately, no real block is able to do what is represented in Figure 5.3. The output voltage or the output current does not stay steady but may change when loaded. The limit is modeled by the use of an impedance in series with the voltage generator or an impedance in parallel with the current source respectively. To have a good source, the series impedance used for voltage must be low and the parallel one used for current must be high. Even real

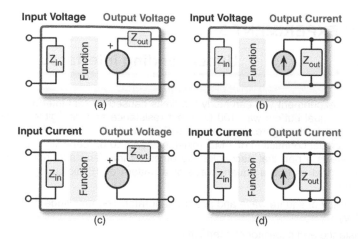

Figure 5.4   Real interfaces of electronic blocks with voltage or current at input and output. As stated, Z_{in} must be large for voltage input and small for current input. Z_{out} must be small for voltage output and large for current output.

inputs can affect the operation. Their measurement of voltage or current is normally modeled by an impedance across the input terminals. This impedance is, hopefully, large for measuring voltages and small for detecting currents. In truth, "large" and "small" are not meaningful if the relative term is not specified. Therefore, more properly, "large" is relative to the series impedance of a loaded voltage source and "small" is with respect to the impedance in parallel with the measured current generator.

> **Remember**
>
> The voltage–current characteristic of an ideal source is a line in the voltage–current plane that is parallel to the voltage axis for a current source and parallel to the current axis for a voltage source.

Interfaces of real electronic blocks are depicted in Figure 5.4. They account for voltage or current signal generation, input impedance, and output impedance. Notice that the use of different output schemes is just for convenience. A voltage generator with an impedance in series reminds us that voltage is the output quantity; the use of a current generator with an impedance in parallel indicates current at output. However, the Thévenin or Norton transformations change one scheme into the other, since they are actually equivalent.

In many electronic circuits the output does not influence the input. This is what the diagrams of Figure 5.4 represent. They use just an input impedance and nothing related to output. However, there are cases where, because of some "kick back," the input depends on the output conditions. For those situations it can be necessary to include in the input model a generator controlled by the output variable. Those input interfaces are used rarely; however, it is important to keep in mind that in some situations we may need to use them.

Let us go back to typical interfaces and assume that the input of Figure 5.4(b) is driven by the output of Figure 5.4(a). If the voltage of the generator is V_{out}, the input of the next block is

$$V_{in} = V_{out} \frac{Z_{in}}{Z_{in} + Z_{out}}. \tag{5.2}$$

COMPUTER EXPERIMENT 5.1

Understanding Interfaces

With this computer experiment you can study the limits caused by real interfaces. The ElvisLab board contains two equal buffers with 100 Ω output resistance and negligible capacitance. The interface, as outlined in the figure, is a low-pass RC circuit. The input of the oscilloscope is equivalent to the parallel connection of 1 MΩ resistance and 10 pF capacitance. Before Channel 3 of the scope there is a port that decouples the output of interface and input of scope almost ideally. The input signal can be a square wave or a pattern of 16 digital logic levels repeated cyclically. You can:

- set the frequency of square wave and pattern generator. The amplitude of the two signals is constant (1 V);
- set the resistance and capacitor of interfaces;
- define the sequence of 16 bits of the pattern generator.

MEASUREMENT SET–UP

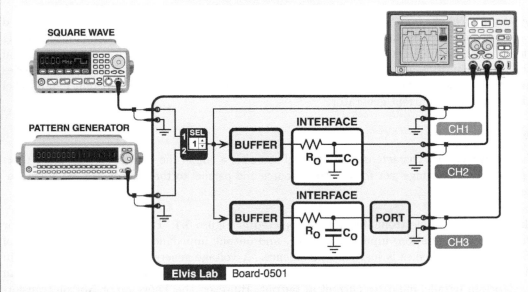

TO DO

- Set resistance and capacitance of the interfaces to zero and sweep the frequency in the 1 MHz–1 GHz range. Observe the waveforms with square wave and various input patterns.
- Set the interface resistance to 100 kΩ and increase the capacitance from zero to 5 pF. Observe waveforms with input square wave at different frequencies. Explain the results.
- Change the parameters of interface and observe the output waveforms at different input frequencies and various bit stream patterns.
- Verify that the output resistance of the buffer is 100 Ω. To facilitate the measure use very high frequency and observe the Channel 3 waveform.
- Verify that the input capacitance of the oscilloscope is, actually, 10 pF.

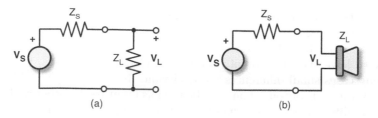

Figure 5.5   (a) Equivalent circuit for matching network; (b) example of loudspeaker matching.

With current at input, using the Norton schemes of Figure 5.4(d), the result is

$$I_{in} = I_{out} \frac{Z_{out}}{Z_{in} + Z_{out}}. \tag{5.3}$$

The limitations represented by equations (5.2) and (5.3) are unavoidable, because electronic interfaces are never ideal. However, often, having a notable but constant attenuation is not a serious drawback unless very large losses occur. What can create problems is frequency dependence inside the signal band, because the signal spectrum is modified. More importantly, in some regions of operation the interface can become non-linear. This would alter the overall response, which is expected to be linear. As is well known, non-linearity is undesired in many applications because it causes harmonic distortion and produces mixing of sinusoidal components.

For some applications, instead of voltage or current, power or energy matters. This typically occurs when the electronic circuit drives an actuator such as a loudspeaker. The amount of music played by a speaker actually depends on the power transferred to the load and not on the voltage. Hence, a typical requirement is to obtain the maximum power transfer from the driving interface to the load. For this type of situation so-called impedance matching is required. This means setting the impedance of the load equal to that of the source. The scheme used to study the condition is the one in Figure 5.5. The total power and that transferred to the load are

$$P_{tot} = \frac{V_S^2}{(Z_S + Z_L)}, \quad P_{load} = V_S^2 \frac{Z_L}{(Z_S + Z_L)^2}; \tag{5.4}$$

therefore, the maximum transfer is for $Z_L = Z_S$, which is the impedance matching condition. Since the total power is $P_{tot} = 2P_{load}$, the efficiency obtained is 50%. Thus an interface with 10 Ω output impedance reaches its optimum performance when driving a loudspeaker with the same impedance: 10 Ω. A similar condition holds when considering impedances instead of resistances.

Note that, in fact, for loudspeakers the impedance has a negligible imaginary part; i.e., it is almost a resistance. For other cases the imaginary part of the load can be significant, especially at high frequencies. Therefore, the matching conditions should be verified at any frequency of operation. Matching is an important design issue with high-frequency signals that are normally transferred through coaxial cables or transmission lines. How to obtain matching is studied in detail in specialized courses. However, it is worth knowing that a lack of matching causes the reflection of a fraction of the power, which returns to the source. Matching termination produces optimal transmission of power with no signal reflection. Matching is necessary with transmission lines, but even within microelectronic circuits, where the distances traveled are a few millimeters or, at the maximum, centimeters, it may be necessary to comply with matching, if the frequencies are several GHz.

For digital circuits, ensuring matching just means using the same type of signal and the same amplitude levels. Logic gates handle two amplitudes: "high" (one) and "low" (zero), which are normally given by full power supply and zero voltage, respectively. Therefore it is just necessary to have the same supply voltages in blocks or, perhaps, to employ voltage adapters. In many cases small differences are irrelevant. If, for example, a block uses 1.8 V and another 1.83 V the logic output of one can be used as input to the other without problems. However, when the difference is more than 0.6 V it may happen that input protection in a circuit (to be discussed shortly), in an attempt to limit the input voltage, may drain a lot of current from the higher power supply. This can cause heating of the part and permanent damage.

Do not underestimate

Interfacing digital blocks is often straightforward and much simpler than interfacing analog circuits, but this point should not be disregarded, especially when assembling different ICs on a board or microsystem, and when the clock frequency is very high.

Modern digital circuits mainly use so-called CMOS logic, but old circuit solutions or some special applications employ other logic families, such as TTL (Transistor Transfer Logic) or ECL (Emitter Coupled Logic). Since these logic families require a certain amount of input current in order to operate correctly, the driver must be compatible with the logic in use. Moreover, some digital circuits, especially at high frequencies, utilize current as an input or output variable. Therefore interfacing is not straightforward but requires proper care and in some cases needs a voltage-to-voltage or a voltage-to-current adapter. More information on logic families and interfacing requirements is provided in a later chapter.

5.2.2 Driving Capability

It is evident from what was discussed in the previous subsection that any electronic block must be able to operate the driven blocks or actuators. This characteristic is called *driving capability* (or, for digital cells, *fan-out*). Normally, output interfaces are modeled by a signal generator and an impedance that must be large when supplying current and small for providing voltage. However, it may happen that the load established by the driven stage or by the actuator is so strong that it alters the functional behavior of the electronics, because part of the circuit is in bad regions of operation. Obviously this kind of situation must be avoided or moderated.

Consider the scheme of Figure 5.6, which depicts the interface between an Integrated Circuit (IC) and another IC through a board connection that establishes an overall load capacitance, C_L. The signal generated by the driver is a logic voltage that suddenly changes from zero (the lower supply voltage) to the positive battery supply, V_B. If the block has suitable driving capability and the bonding inductance is negligible, the expected signal at the input of the next cell is exponential. For a step jumping at $t = t_1$, the result is

$$V_{in,N}(t) = V_B(1 - e^{-(t-t_1)/R_0 C_L}).$$ (5.5)

Even the current charging the capacitor C_L changes exponentially:

$$I_{C_L}(t') = \frac{V_B}{R_0} e^{-t'/R_0 C_L} = \frac{V_B}{R_0} e^{-t'/\tau_i},$$ (5.6)

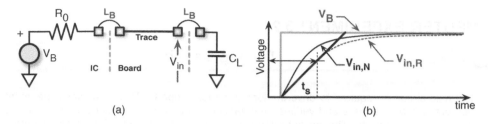

Figure 5.6   (a) Simple interface between two integrated circuits; (b) expected and obtained input response to a step signal.

where $t' = t - t_1$ and $\tau_i = R_0 C_L$ is the time constant of the interface connection.

Observe that the current peaks immediately after signal switching and is equal to V_B/R_0. Moreover, since the network of Figure 5.6(a) is linear, the response to waveforms other than a step is easily estimated using the superposition principle.

However, the system may not be able to provide peak current because the generator inside the IC delivers up to a maximum value, say I_{SR}. Thus the voltage waveform is not exponential, because, for a while, the current does not exceed I_{SR}. The voltage across the capacitor, charged with constant current, is a ramp until t_{SR}, the time at which the current drops below I_{SR}. The voltage across the resistance is $V_{SR} = R_0 I_{SR}$. After t_{SR} the waveform changes exponentially as shown in Figure 5.6(b). Therefore

$$V_{in,N}(t') = \frac{I_{SR}t'}{C_L}, \quad I_{C_L}(t') = I_{SR} \quad \text{for } t' < t_{SR} = \tau_i \left\{ \frac{V_B}{V_{SR}} - 1 \right\} \tag{5.7}$$

$$V_{in,N}(t') = V_B - V_{SR}e^{-(t'-t_{SR})/\tau_i}, \quad I_{C_L}(t') = \frac{V_{SR}}{R_0}e^{-(t'-t_{SR})/\tau_i} \quad \text{for } t' > t_{SR}. \tag{5.8}$$

The output voltage derivative is smooth with maximum value V_{SR}/τ_i (or I_{SR}/C_L). In contrast, the slope of current is discontinuous at t_{SR}. This is unrealistic, because the model simply accounts for a maximum value of I_{out} and not a smooth change near the maximum.

When the current has its largest value, the circuit is said to be slewing. In that condition the voltage slope is called the *slew-rate*. This is the maximum rate at which an electronic circuit can respond to an abrupt change of input level. Slew-rate is measured in [V/s], or in [V/μs]. Sometimes it is given in [MV/s].

The limited driving capability, described here for a step signal, also affects circuits with other types of input signal, such as sine waves or combinations of sine waves.

Typically when the interface cannot satisfy a demand for large currents two disadvantages occur. One is distortion of waveforms; the other is the longer time needed to approach the steady level. Distortion is detrimental for analog operation. Delay is negative in the digital case, especially when it affects synchronization of signals. Timing inaccuracy is a frequent source of malfunctioning in high-speed digital circuits.

Because of the above potential problems, modular design must choose ICs or functional blocks that ensure suitable driving capability. Physical parts or soft cells of various sizes are normally available, but sometimes it is necessary to design special cells that give rise to a higher slew-rate than available cells provide. Increased sizes of transistors augment I_{SR} and reduce R_0. These factors both favor improvement: the slewing time is lower, and a diminished value of R_0 makes the exponential part of the waveform faster.

COMPUTER EXPERIMENT 5.2

Driving Capability

Any electronic circuit is required to drive the load it serves properly. This computer experiment gives you practice on the issue and verifies what happens when a real circuit does not have the necessary driving capability. The driving interface of this ElvisLab board has a given output resistance and a maximum deliverable current. The input signal is a sine wave or square wave. The output of the interface drives the load made by an inductor that mimics the bonding of an integrated circuit or the inductance of a trace on a board. Then there is a capacitive load. The input capacitance of the scope is 1 pF. In this experiment you can:

- change amplitude and frequency of sine and square wave signals;
- set the value of inductance, load capacitance and maximum current that the driver can deliver, and the output resistance of the interface;
- select the type of input and the signal to be sent to Channel 1 of the oscilloscope.

MEASUREMENT SET–UP

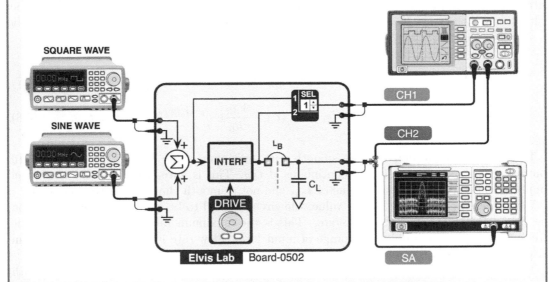

TO DO

- Set inductance, loading capacitance, and output resistance to zero. Observe the signal waveform in the time and frequency domains with the maximum driving capability.
- Leave the driving capability at maximum but set the other parameters to finite values. Observe the output voltage for different amplitudes and frequencies.
- Reduce the value of driving capability and observe the output in the time and frequency domains. Notice that the use of the spectrum analyzer identifies better when limited driving capability occurs.
- Determine the current, for a 0.6 V input sine wave at 240 MHz, which the interface must be able to deliver for giving an output with spur harmonics 76 dB below the input.

When blocks are in separate integrated circuits, the capacitive load of interconnections is significant. The load is caused by the output and the input pin plus the load of trace on the board. Depending on the package, the total parasitic capacitance ranges from a fraction of a pF to several pF. If blocks are within an integrated circuit, the load capacitance is much smaller – of the order of a few or a few tens of fF. The load depends on transistor sizes. It is minimum for digital cells and a bit higher for analog ones, because analog schemes use larger transistors. Note that low capacitance means high speed and low power consumption. Moreover, the energy needed to charge capacitors is proportional to the capacitive value.

A final remark is that, in addition to driving the capacitive load, the interface can be required to drive a series inductor, as outlined in Figure 5.6. The bonding that connects integrated circuits to the pins is the main source of parasitic inductances. Their value is about 1 nH per mm of bonding. As is known, the series of inductance and capacitance can give rise to oscillations, dumped by the series resistance, at the resonant frequency

$$f_{res} = \frac{1}{2\pi\sqrt{L_p C_p}}. \tag{5.9}$$

For $L_p = 4$ nH and $C_p = 2$ pF the resonant frequency is 1.78 GHz, a large value but a possible source of limitations at very high speeds. Since an undesired ringing modifies the circuit operation, it is necessary to account for it and, when necessary, to use design or fabrication precautions.

5.2.3 Electrostatic Discharge Protection

An important function of the input and output terminals of any integrated circuit is to protect the circuit from dangerous agents. It is known that the operation of a circuit can be impaired by aggressive corrosive elements or by the self heating that can produce too high temperatures. These circumstances are addressed by properly sealing the circuit into packages and by devices that favor the dissipation of heat. But, in addition and more critically, circuit damage can occur because of high voltages at pins. You probably already know that any circuit has maximum electrical conditions that it can withstand. Indeed, the nominal supply voltage of an integrated circuit is principally determined by the maximum voltage limit, diminished by a suitable safety margin. Exceeding the maximum limit, even for very short periods of time, damages the circuit. The failure can be gentle when errors degrade operation over a long time or catastrophic when high overvoltage may destroy the circuit.

Since modern integrated circuits use very thin layers of oxide, the usable voltage is becoming lower and lower. The dielectric strength of silicon dioxide, like that of many other insulators used in integrated circuits, is some millions of V/cm. That is a large number, but when the oxide thickness is only a few nm, just a small number of volts breaks the oxide. It is therefore necessary to prevent the voltage of the supply generator and at the input and output ports exceeding a safety limit, even for very short periods of time.

Overvoltages in circuits can be generated by capacitive or inductive coupling. Switching surges, caused by on and off switching operations, the switching of inductive and capacitive loads, and the breaking of short-circuit currents, may lead to damage in electrical equipment nearby. More frequently, the triboelectricity caused by friction when separating materials is a source of high voltage. The human body acquires a substantial charge of triboelectricity, producing a voltage that can be tens of kV in some cases, depending on the electrical properties

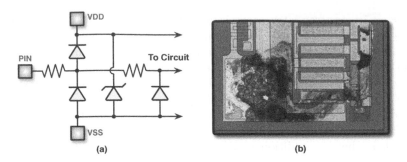

Figure 5.7   (a) Electrical scheme of possible ESD protection; (b) damage caused by ESD on an input pin.

of materials. For example, the friction of leather shoes walking on a woolen carpet may generate 1.5 kV on a humid day, and gives rise to as much as 15 kV in dry conditions.

Breaking thin oxides

Be aware of overvoltages at the inputs of integrated circuits. The oxide thickness of a MOS transistor gate goes down with the minimum line-width. It can be, for example, 4 nm. With 6 MV/cm dielectric strength only 1.5 V across the oxide destroys the layer. Even lower voltages, if they are higher than what is recommended, are problematic: they cause aging of the circuit.

A triboelectrical charge is not completely transferred to the parts touched because of the relatively high impedance of the body; but the voltage produced is often very high, much higher than the limit of damage to circuits. Therefore, handling circuits before they are assembled on the board requires special care. For example, operators and work surfaces should be grounded, and devices ought to be in properly shielded enclosures. When circuits are assembled the risk of damage is reduced, because pins are normally connected to low-impedance paths. However, the residual risk and the possibility of high voltage occurring, even in assembled circuits, makes it advisable to use input and output protection that is able to clamp high voltages with fast-acting devices.

Transient suppressors are available for PCBs, but their high cost often prevents their use unless it is really necessary. Instead, integrated circuits always incorporate a clamping function on the supply terminals and the input or output pins. Clamping is obtained by using dedicated devices or circuits. Simple solutions use zener diodes, devices that show a very high resistance until their reverse biasing exceeds a given limit. Since for direct biasing the zener diode has a low resistance, protection can require using two back-to-back zener diodes. Another useful device is a diode that has a large resistance with reverse biasing and a small resistance with direct biasing.

Figure 5.7 shows a possible protection scheme. It uses a zener diode to prevent overvoltage between supply terminals. Moreover, the scheme includes diodes that stop the voltage at the pins becoming higher than the positive supply or lower than the negative supply. The series resistances possibly absorb part of the drop caused by overvoltage and limit the current in the protection circuit. This protection scheme is for voltage signals: i.e., situations where the current entering the pin is negligible. The voltage drop across the protection resistance is zero or very small under normal conditions.

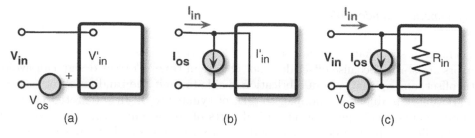

Figure 5.8   (a) Input referred voltage offset; (b) input referred current offset; (c) input referred voltage and current offset.

5.2.4   DC and AC Coupling

The interface between the output of one block and the input of another block creates the necessary coupling between electronic stages. The main purpose of the interface is to transfer information carried by the amplitude and the change over time of an electrical variable. However, the information from some signals is contained in their absolute value, or DC part; for others, what matters is the variation of the signal, or the AC part, measured against a suitable constant level. This level is normally called the *quiescent amplitude* and is set at the best voltage or current for optimal operation. Suppose, for example, that a voltage signal is properly generated in the range $V_1 - V_2$ and that the signal can be either positive or negative. In this case, its optimal quiescent value is the mean value between V_1 and V_2. On the other hand, if a signal in the same range has only positive (or negative) values, the best quiescent level is a bit above (or below) the lower limit of V_1 (or V_2). Both choices depend on the specific DC feature of the signal and are to enable the maximum swing of the signal.

When the absolute value of a voltage or a current must be preserved, the coupling used is DC, which stands for "direct current," even if voltage rather than current is the coupled variable. DC coupling means that the interface must preserve the absolute value of voltage and current and must carefully account for it as an essential part of a signal. If the interface is not able to keep the DC exact value but operates as if it had at input an amplitude changed by a fixed amount, the error is called *offset*. Offset is a constant but unpredictable shift caused by static differences between real and expected implementations. An offset is normally represented by a battery in series or, in the case of current, by a current generator in parallel with the input. Therefore, as shown in Figure 5.8, the input signal really applied, for voltage or current is

$$V'_{in} = V_{in} - V_{os}, \quad I'_{in} = I_{in} - I_{os}, \tag{5.10}$$

where offset can have positive or negative values.

For a more generic input interface the offset can affect both voltage and current, as shown in Figure 5.8(c). That scheme apparently does not make sense, because the parallel of a resistance and a current source can be transformed into a voltage generator $I_{OS}R_{in}$ in series with the resistance R_{in}, so that the combination of the two offset generators leads to a scheme equal to Figure 5.8(a). However, what is illustrated in the figure has some practical value because it can outline two different sources of error: an imbalance that causes voltage offset and a possible current leakage that produces current offset.

5.2.4.1 Systematic or random offset

Offset denotes a systematic error or a random error that shifts the DC value of a signal. All the errors are static, because offset does not change in time. Correct design can avoid the systematic offset, and special care in fabrication is able to limit the random offset.

Systematic offset is due, for example, to lack of symmetry in an integrated circuit that would require symmetrical operation in critical parts of the scheme. Ensuring symmetry is expensive or difficult, and the designer may decide to accept a small amount of asymmetry or, by mistake, disregard the rule. The resulting error, being by design, systematically occurs for all the ICs produced using that scheme or fabrication sequence. Systematic errors can also affect printed circuit boards because of unsuitable component placement or unmatched trace routing.

The second type of error accounts for random inaccuracies that occur during the PCB production or the IC fabrication. Design and fabrication "recipes" would lead to nominal outcomes, but because of unexpected and unavoidable fluctuation of parameters, the result obtained is not exactly what was desired. For example, the fabrication process of integrated circuits requires temperature cycles, and, if the furnace is not set up well, global or local variations of temperature take place. A local change of fabrication parameters, normally a gradient, alters the values of components produced on the same IC. Suppose two equal capacitors are made together, with parallel plates and silicon dioxide as insulator. If the oxide thickness is not constant along the chip but changes with distance between capacitors, Δx: $t_{ox}(x) = t_{ox,0}[1 - \delta_{ox}(x - x_0)]$, the capacitors, nominally equal, are as a result mismatched. If the centroid of C_2 is Δx apart from C_1, the capacitor values are

$$C_1 = \epsilon_0 \epsilon_r \frac{A}{t_{ox,0}}, \quad C_2 \simeq C_1(1 + \delta_{ox}\Delta x), \tag{5.11}$$

denoting a difference that increases with the distance between the centroids of the capacitors. The IC designer often reduces those mismatches by using minimal distances and symmetric or common centroid arrangements.

In addition to random errors caused by parameter drift, there are entirely random variations for both ICs and PCBs that depend on inaccuracies in fabrication. The difference between components used in PCBs is always random, because they normally come from independent sources. To reduce the random offset, it is therefore necessary to use high-accuracy and therefore expensive components or to select components and choose pairs that match.

Remember

Offset can be a serious limit with small signals that carry information at very low frequencies, almost at DC. Offset is reduced by careful design (systematic offset) and accurate fabrication of components (random offset).

After that brief discussion of offset, its possible causes and its remedies, let us go back to DC coupling and notice how important it is to ensure that interface operations match. If the maximum amplitude on one side is, for example, equal to 1.3 V and the other side permits 2.1 V, the DC system must use the worse value, 1.3 V, for maximal signal swing.

It is worth knowing that DC coupling in not always used because the constant part of the signal is relevant; it is also chosen because it is simpler and more effective. A DC interface

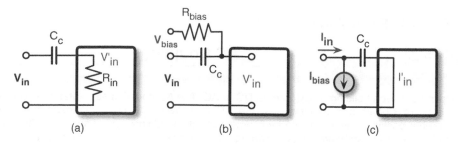

Figure 5.9 (a) AC coupling; (b) AC coupling with upper terminal biasing; (c) AC current coupling with bias current.

just requires two wires, or a single one, and no passive or active components. Moreover, the quiescent part of a signal can be used for biasing or supply purposes.

The second possible type of coupling is AC (Alternating Current), which discards the constant amplitude or changes its value to a more convenient level. The simplest method of AC coupling consists of using a capacitor that separates the DC component from the varying part of the signal and obtains a result around the reference, as shown in Figure 5.9(a). The diagram uses the input resistance, which is very high for voltage signals, to transfer the reference level. In fact, since the DC current is zero, the DC voltages of the two terminals are equal.

Some applications require the use of another reference level, different from the main reference, to bias the electronics to an optimum level. The scheme of Figure 5.9(b) fulfills that function. The desired bias voltage, V_{bias}, connects the input of the next block by a high resistance. The AC part of the signal is transferred by capacitive coupling. Therefore this scheme does not use the common reference but that shown in the diagram of Figure 5.9(a), with a resistance connected to V_{bias}. Figure 5.9(c) shows AC coupling in the case when the input generator needs a current bias.

Suppose we are to use at input a voltage generator, V_s, with low series resistance, R_s (low means that R_s is much lower than R_{in}). The response in the frequency domain is

$$V_{in}(s) = V_s(s)\frac{sR_{in}C_c}{1 + s[(R_s + R_{in})C_c]}, \qquad (5.12)$$

which replicates $V_s(s)$ if $R_s \ll R_{in}$ and $\omega \gg 1/[2\pi(R_s + R_{in})C_c]$. Therefore, in reality, the DC coupling is a network with a zero–pole transfer function. The zero is at zero frequency and the pole at $f = 1/[2\pi(R_s + R_{in})C_c]$. For the case shown in Figure 5.8(b) the result is the same with R_{bias} replacing R_{in}.

Observe

A good AC interface has its high-frequency pole well below the lower limit, f_{low}, of the signal bandwidth. This condition corresponds to $2\pi R_{in}C_c \gg 1/f_{low}$.

Digital circuits use signals that suddenly change from low to high level or vice versa. In the case of a repetitive sequence of zeros and ones, the signal is a square wave with full amplitude. The given result is like that of Figure 5.10, as predicted by the frequency response equation (5.12). In fact, a high-pass network preserves the sudden changes of a square wave because the step variations correspond to very high frequency components; in contrast, a

COMPUTER EXPERIMENT 5.3

AC Coupling

The coupling between building blocks can be DC or AC. The AC coupling separates the constant voltage level of signals and transfers just the AC part. With this computer experiment you understand the possible effects caused by a simple coupling network and observe results in DC or AC mode. This ElvisLab board contains two equal coupling circuits made up of a capacitor and a biasing resistor. The resulting effect is high-pass. The in/out port has an input resistance of 100 kΩ and input capacitance of 10 pF; the output port replicates the input almost ideally. The result is sent to the scope with AC or DC coupling. The input resistance of the scope is 1 MΩ and the capacitance is negligible (you use active probes). You can:

- set amplitude and frequency of the sine wave and square wave generators;
- control the DC bias of the biasing resistance;
- set the value of coupling capacitor and biasing resistor.

MEASUREMENT SET–UP

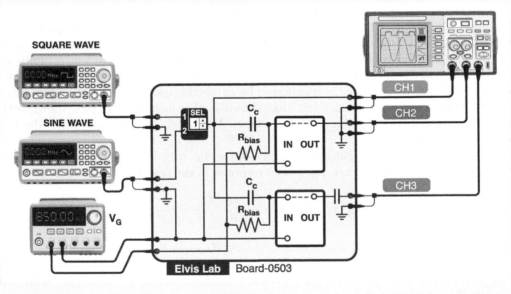

TO DO

- Set the value of coupling capacitor and bias resistance to give rise to a time constant of 100 ns. Use different resistance-capacitance pairs and observe the output at various sine wave frequencies. Explain the outcome.
- Switch to a square wave signal and repeat the experiment. Change frequency and amplitude and explain their behavior.
- Compare waveforms of Channels 2 and 3 with the input signal. Explain possible differences between the traces of Channels 2 and 3.
- Change the DC bias voltage, observe the results, and explain why you see differences in the DC and AC levels.

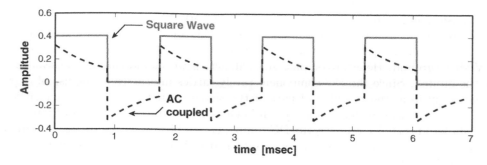

Figure 5.10   Input square wave and its AC coupled equivalent.

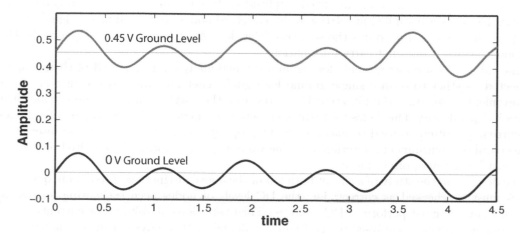

Figure 5.11   The same signals with different ground levels.

steady value, which low-frequency components imply, vanishes exponentially. Note that to obtain negligible decreasing steps in steady values requires periods of the square wave that are much smaller than the time constant $(R_s + R_{in})C_c$.

5.2.5   Ground and Ground for Signal

A voltage signal is the difference between the voltages of two nodes. Often one of the nodes is the reference and many (or all) of the voltage signals use that reference for taking the difference. In those cases we consider the voltage of a node as a signal, and not the voltage difference, because we implicitly assume that the second node is the reference. The reference and its choice are a convention valid for specific circuits or systems. For example, in a car the reference is the minus terminal of the battery, connected to the body of the car itself, and all the electrical apparatus in the car uses it. In the house one wire of the electrical power supply is the reference. That wire may be connected to ground for safety purposes. That precaution gives the name of "ground" to the reference, and, by extension, all the reference nodes used in electronics are called *ground*, even if they are not physically connected to the ground.

Since the choice of ground is a convention, any node of the circuit at a fixed level with respect to supplies can be chosen as the reference. Then all voltages in the system, including the mains, use that ground wire as reference. Consequently we say that supply terminals are 0–1.8 V if ground is the least negative node of the circuit and ±0.9 V if ground is halfway.

The plot of a signal can look like the solid curve of Figure 5.11. It has a varying part that is 0.45 V above ground. Hence, to emphasize small details we should enlarge the scale and not include the ground. Since the DC component rarely carries information, the use of a ground that is far away from the meaningful part of the signal cannot be a convenient choice. It would be more practical to change the convention and use, as in the example of Figure 5.11, a node at 0.45 V. This choice removes the DC component and focuses better on the information carried. Even if the circuit does not have a node at 0.45 V we can assume that it exists and call it *ground for signal*.

As a further step, the concept of ground for signal is extended to any signal of the system. When meanings are changed around fixed amplitudes, the constant terms are disregarded and the variable parts are studied in relation to hypothetical nodes: the grounds for signal. Thus, using ground for signal produces the same result as does AC coupling used to ignore the DC component when it does not carry information.

Since all nodes used as grounds for signal are functionally equivalent, all of them can be connected together to form a single ground for signal. Therefore when studying a circuit it is convenient to separate the quiescent DC parts from the varying parts and determine their effects independently. This helps to distinguish between functions. The fixed part of voltages (or currents) is often entitled to bias the circuit properly and must be stable and insensitive to electrical or environmental interference. The variable part performs the required processing and must be fast and, often, linear.

The design and the study of a circuit both follow the above distinction. First of all, quiescent values must be checked as correct; i.e., the DC levels of nodes need estimating, or, as we normally say, we must perform a DC analysis. Then variations of voltages or currents in the time domain (transient analysis) or the frequency domain (AC analysis) can be studied.

5.2.6   Single-ended and Differential Circuits

In real electronic circuits noise and interference corrupt electrical signals. In order to ensure accurate results, the amplitude of voltages must be much larger than unwanted parts. This requirement is quantified by the SNR defined in Chapter 2. Obtaining large SNRs can be difficult, because the supply voltage and the headroom margin necessary to allow proper operation of the circuit put a limit on the voltage swing of signals. Thus, reducing noise and interference and, when possible, increasing signal level are important topics for electronic design. The use of lower and lower supply voltages, as required by modern submicron and nanometer technologies, worsens the scenario.

One convenient method used for dealing with the above issue is differential or fully differential processing. Instead of handling conventional single-ended signals, which vary with respect to their ground, electronics employs differential signals; i.e., pairs of electrical variables that represent information with the same amplitude but opposite signs. Figure 5.12 shows an example of conventional and differential signals $V_{s+}(t)$ and $V_{s-}(t)$ at input or output of electronics. The subscripts $+$ and $-$, frequently used, are just a convention used to distinguish between signals in the two processing paths.

Differential processing assumes that only the difference between differential signals, and not single components, carries information. So the voltage that matters is

$$V_{s,diff}(t) = V_{s+}(t) - V_{s-}(t) = 2V_{s+}(t), \tag{5.13}$$

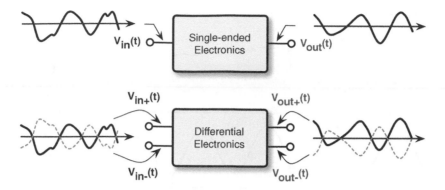

Figure 5.12   Single-ended and differential processing.

which, in the case of opposite values, doubles the signal amplitude.

If noise affects the differential voltages by the additive uncorrelated terms V_{n+} and V_{n-},

$$V_{s+}(t) = \hat{V}_{s+}(t) + V_{n+}(t) \tag{5.14}$$

$$V_{s-}(t) = \hat{V}_{s-}(t) + V_{n-}(t) \tag{5.15}$$

where ˆ indicates the expected signal. The differential signal becomes

$$V_{s,diff}(t) = 2\hat{V}_{s+}(t) + V_{n,diff}(t). \tag{5.16}$$

$V_{n,diff}(t)$ is the superposition of the two noise components that, as we know, must be done by using statistical methods. Since $V_{n+}(t)$ and $V_{n-}(t)$ are uncorrelated, assuming that both have the same mean square value the result is

$$\langle V_{n,diff} \rangle = \sqrt{\langle V_{n+} \rangle^2 + \langle V_{n+} \rangle^2} = \sqrt{2}\langle V_{n+} \rangle. \tag{5.17}$$

This shows that, while differential amplitude increases by 2, the noise augments by only $\sqrt{2}$. So the SNR improves by $\sqrt{2}$, which, in dB, equals 3 dB.

Suppose now that interference alters the differential signals

$$V_{s+}(t) = \hat{V}_{s+}(t) + V_{int+}(t) \tag{5.18}$$

$$V_{s-}(t) = \hat{V}_{s-}(t) + V_{int-}(t), \tag{5.19}$$

where, again ˆ indicates the expected signal. Interferences V_{int+} and V_{int-} depend on coupling the disturbing agent with the nodes carrying V_{s+} and V_{s-}. A proper IC or PCB layout makes the couplings almost match, $V_{int+} \simeq V_{int-}$. Therefore only the mismatch between interferences has an effect on differential signal

$$V_{s,diff}(t) = 2\hat{V}_{s+}(t) + \delta V_{int}(t). \tag{5.20}$$

Hence, the use of differential processing improves the SNR by 3 dB and almost cancels interference, assuming the couplings are matched with lines that carry differential signals.

Figure 5.13 illustrates the benefits of differential processing. Interference of a clock and coupling with a slow signal cause disturbances larger than 0.15 V. The spur is added to the differential component of a signal made by a sine wave with 0.25 V amplitude. The differential

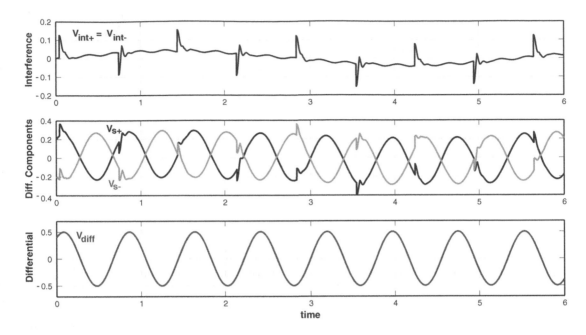

Figure 5.13   Benefit of differential processing: the interference signal (top) is added to the differential components (middle), but the differential signal (bottom) is not affected.

result, assuming that interference equally affects the differential components, returns to being a perfect sine wave with double amplitude, 0.5 V. The amplitude extent of differential components is larger than the expected 0.5 V peak-to-peak (it is about 0.4 V). Thus the electronics must make up for a swing larger than the foreseen maximum in the available dynamic range.

The use of differential processing costs more silicon area and more power consumption. Actually, costs are not doubled because some electronic functions are common to both differential paths. However, a designer accounts for relevant benefits.

Moreover, the signal at input and output of the system cannot be differential but is, as already stated, single ended. Differential processing requires a single-ended to differential converter and, possibly, another block that transforms differential results back into single-ended versions at the end of the differential processing chain.

5.3   INSIDE ANALOG ELECTRONIC BLOCKS

Previous sections have described how electronic blocks are connected and discussed the way electronic circuits properly obtain interfacing and ensure good signal quality. Inside blocks, other electronic circuits carry out the required functions. They are often indicated by a set of equations or statements. However, for electronic realizations it is convenient to translate equations or statements into block diagrams that often report just the relevant interconnections of basic functions.

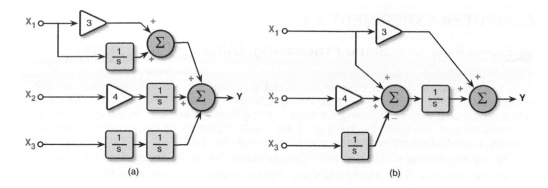

Figure 5.14   Two block diagrams of the same processing function.

Suppose, for example, that an electronic scheme must carry out the following processing function in the s domain:

$$Y(s) = X_1(s)\left[3 + \frac{1}{s}\right] + \frac{4X_2(s)}{s} - \frac{X_3(s)}{s^2}. \tag{5.21}$$

The circuit uses four electrical variables, three for inputs and one for output. They can be voltages or currents. There are various schemes that can carry out what is required by equation (5.21). Figure 5.14 shows two of them, omitting the interfaces, which are supposed ideal. The scheme of Figure 5.14(a), which directly implements equation (5.21), obtains the intermediate terms that are added together. It requires three multipliers by a constant factor, four integrators and two adders, one with two and the other with three inputs. However, other schemes produce the required result, one of which is shown in Figure 5.14(b), used to minimize the number of integrators. That scheme needs two integrators instead of four. The numbers of multipliers and adders remain the same.

Remember

Electronic circuits that carry out processing use intermediate variables. The amplitudes of these intermediate quantities must be carefully controlled to avoid saturation or corruption by noise.

Note that the two schemes of Figure 5.14 use intermediate variables that are different for the two solutions. Obviously, to obtain good results it is necessary to preserve the quality of those variables and, for instance, avoid amplitudes becoming too low or too high. A reduction in levels makes small signals comparable to noise; an increase in amplitudes causes early saturation. Thus it is true that there are many possible solutions, but realization costs and performances must be carefully studied.

5.3.1   Simple Continuous-time Filters

What we discussed in a previous part of this section is a non-specific case, illustrating the link between a processing function and a corresponding block diagram for possible electronic realization. More definite examples, given here, consider simple processing, linear and in the s domain, as required by filters.

COMPUTER EXPERIMENT 5.4

Same Processing, Different Processors

With this experiment you can observe the output generated by two different networks that implement the same processing function. The circuits have two sine waves at input. The input-output relationship of the processing functions is straightforward and can be derived from the schematic diagram. However, the swing of all the circuits implementing any processing function is limited; the output saturates outside a permitted interval. You can:

- set the amplitude of the two sine wave generators and change the phase of the first one;
- set the frequency of the second sine wave. The first frequency is fixes at 2.5 MHz;
- set the upper and lower clipping limits of the range of operation for all the blocks;
- set the position of all the selectors;
- set the value of the processing parameters K_1 and K_2 and the integrators' time constant τ.

MEASUREMENT SETUP

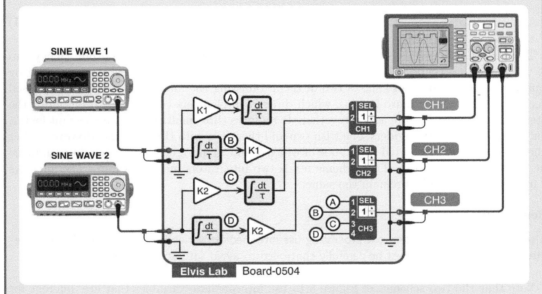

Elvis Lab Board-0504

TO DO

- Set the saturation limits to +5 and -5 and observe the two outputs with small input amplitude at different frequencies and other parameters.
- Reduce the saturation values and increase the sine wave amplitudes until the saturation alters the response. Find a condition at which the outputs are different. Find the block that is the first element responsible for error.
- Repeat the measures with different values of processing parameters.
- Analyze the intermediate responses and try understanding the simpler operations performed. See what happens with asymmetrical clipping values. Notice that the displayed waveforms always start with zero initial conditions.

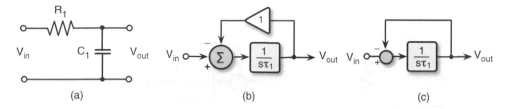

Figure 5.15   (a) Simple low-pass filter; (b) active representation; (c) simplified diagram.

As we know, with a passive network it is possible to obtain elementary filters like the one in Figure 5.15(a). It is a low-pass filter with transfer function

$$V_{out}(s) = V_{in}(s)\frac{1}{1 + sR_1C_1} = V_{in}(s)\frac{1}{1 + s\tau_1}, \tag{5.22}$$

denoting a real pole at $\sigma_1 = -1/\tau_1$ (or $\omega_1 = -1/\tau_1$).

The scheme of Figure 5.15(a) is simple but is often not the right solution, because the interfaces at input and output are poor. The filter is a load for input. The network connected to the output influences performance. Moreover, the resistive link between input and output can cause unwanted interference. The scheme can be improved to make interfaces with electronic circuits before and after the passive network, but the overall cost may be high. A more appropriate solution is to perform the transfer function with active electronic functional blocks, as in Figure 5.15(b). This requires an electronic circuit able to perform integration and another one that subtracts two inputs. The variables involved are voltages, but even an operation in the current domain is a valid solution. By inspection, it results in

$$V_{out}(s) = [V_{in}(s) - V_{out}(s)]\frac{1}{s\tau_1}. \tag{5.23}$$

That, when rearranged, gives the transfer function(5.22).

Observe that the integrator used in Figure 5.15(b) is not a plain $1/s$ but includes a time constant τ_1 to obtain dimensionless operation; i.e, the type of electrical variable at the input and output is unchanged. The schematic of Figure 5.15(b) is normally simplified as shown in Figure 5.15(c) to obtain more compact diagrams. Since output is used at input, it is said that the scheme employs feedback.

The passive RC circuit uses two voltages and one current; the active scheme of Figure 5.15(b) employs only voltages. However, there is a correspondence between variables in the passive and active counterparts. The current of the passive scheme is proportional to the difference between input and output, and this difference is actually what is used at the input of the integrator. Therefore the active circuit uses a voltage to represent the current of a passive network. This is frequent in active circuits: they mimic the operation of passive prototypes to obtain the same input/output behavior in a more reliable and usable way.

Consider a second example, the high-pass scheme of Figure 5.16(a). Current in the RC series and output voltage are

$$I_{in} = V_{in}\frac{sC_1}{1 + sR_1C_1} = V_{in}\frac{sC_1}{1 + s\tau_1}, \quad V_{out} = V_{in}\frac{s\tau_1}{1 + s\tau_1}, \tag{5.24}$$

which corresponds to a high-pass response with a zero at $s = 0$ and a pole on the real axis at $\omega_1 = -1/\tau_1$ ($\tau_1 = R_1C_1$). Since the transfer function is that of the low-pass multiplied

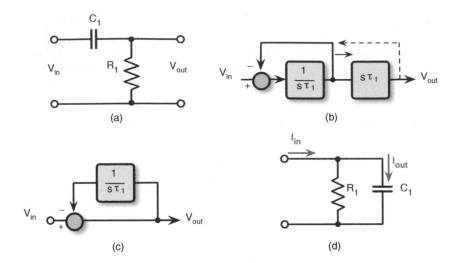

Figure 5.16   (a) Simple high-pass filter; (b) active representation; (c) topological transformation; (d) passive equivalent in the current domain.

by $s\tau_1$, the corresponding active scheme is the one in Figure 5.15(c) plus a derivator, placed indifferently before or after, because the system is linear. The result is shown in Figure 5.16(b).

A topological transformation simplifies the diagram. Let us suppose we move the feedback branch after the derivator, as shown by the dashed line. This is possible if the derivative is multiplied by an integration before feeding back the signal. Moreover, after moving the feedback, the forward path is the cascade of a derivator and an integrator that cancel each other out. The final result is shown in Figure 5.16(c).

It can be observed that the scheme of Figure 5.16(c) is the active representation of the high-pass passive filter of Figure 5.16(d), which uses currents at input and output. Its transfer function is the high-pass

$$I_{out} = I_{in} \frac{sR_1C_1}{1 + sR_1C_1} = I_{in} \frac{s\tau_1}{1 + s\tau_1}, \tag{5.25}$$

that solves the node equation

$$I_{out} = I_{in} - I_{R_1}, \tag{5.26}$$

also represented by

$$I_{out} = I_{in} - \frac{I_{out}}{sR_1C_1} = I_{in} - \frac{I_{out}}{s\tau_1}. \tag{5.27}$$

The diagram in Figure 5.16(d) matches equation (5.27). The only difference is that Figure 5.16(c) uses current while Figure 5.16(d) uses generic electrical variables – voltages in the depicted case.

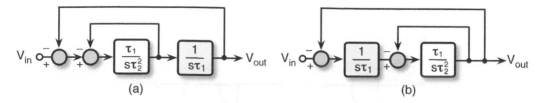

Figure 5.17   Block diagram of a low-pass filter with two poles.

5.3.2   Two-Pole Filters

As a last example, we consider a two-pole low pass filter. The poles are real or complex conjugates. Supposing voltage at input and output, the transfer function has the form

$$\frac{V_{out}}{V_{in}} = \frac{1}{1 + s\tau_1 + s^2\tau_2^2},\tag{5.28}$$

which can be rearranged as

$$V_{out} = (V_{in} - V_{out})\frac{1}{s\tau_1}\frac{1}{1 + s\tau_2(\tau_2/\tau_1)}.\tag{5.29}$$

The scheme in Figure 5.17 performs the function given by equation (5.29). The block that produces $1/[1 + s\tau_2(\tau_2/\tau_1)]$ is like the one in Figure 5.15(c). It can be placed before the plain integrator, as shown in Figure 5.17(a), or after, as depicted in Figure 5.17(b), because the system is linear.

Again, active implementation of the filter uses integrators. That function is frequently used in continuous-time schemes, particularly in high-order filters that obtain multiple zeros and poles.

5.4   CONTINUOUS-TIME LINEAR BASIC FUNCTIONS

Electronic circuits perform linear and non-linear mathematical functions. The analog, continuous-time implementations use passive components and active blocks with proper wire interconnections. The passive components used are resistors, capacitors, and, very rarely, inductors. The active blocks use transistors, consume power, and operate within given voltage and current ranges.

Linear processing functions are:

■ addition of two or more inputs;

■ multiplication by positive or negative constants;

■ integration;

■ obtaining a derivative.

Schemes with passive components, albeit approximately, perform some of the above functions. For others, or when the quality or accuracy are not satisfactory, active elements complement the action of passive elements, to obtain or to improve the functions.

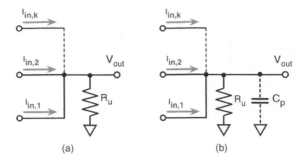

Figure 5.18   (a) Addition of current signals to obtain a voltage signal; (b) scheme with parasitic capacitance accounting for the total parasitic on output node.

Table 5.1   Discrete and integrated resistors

Type	Tolerance	Matching	VC	TC
Carbon composition	±5–20%	n.a.	250 ppm/°C	100 ppm/°C
Carbon film	±1–5%	n.a.	50 ppm/°C	–300 ppm/°C
SMD film	±1–5%	n.a	50 ppm/°C	200 ppm/°C
Metal (thin and thick) film	±1–10%	n.a	1–5 ppm/°C	5–20 ppm/°C
Foil	±0.01%	n.a	1–5 ppm/°C	0.2–50 ppm/°C
Diffusion	±20%	±0.2–0.5%	10000 ppm/°C	1500 ppm/°C
Implant	±5%	±0.1–0.5%	2000 ppm/°C	1000 ppm/°C
Poly	±20%	±0.2%	1500 ppm/°C	1500 ppm/°C
Thin film	±2%	±0.1%	10–100 ppm/°C	10–100 ppm/°C

5.4.1   Addition of Signals

An easy way to add signals is when they are represented by currents and the output wanted is voltage. Multiple currents $I_{in,1}, I_{in,2}, \ldots, I_{in,k}$ injected into a grounded unity resistance, R_u, as shown in Figure 5.18(a), produce an output voltage, V_{out}, given by

$$V_{out} = R_u[I_{in,1} + I_{in,2} + \cdots + I_{in,k}], \tag{5.30}$$

where R_u is 1Ω if input signals are measured in amperes and output in volts. If, on the other hand, inputs are mA or µA and output is volts the resistance is 1 kΩ or 1 MΩ respectively.

Obviously, the unity resistance R_u loads the various inputs, and this influences the input currents, unless they are provided by ideal current sources or are generated by suitable interfaces.

A bit more about resistors

There are different types of resistors, distinguished by the fabrication method. For PCBs they are through-the-hole or surface mounting. In an IC they are fabricated using resistive layers made available by the technology process. The names of different types of resistors are listed in Table 5.1. The values of a real resistor is always inaccurate, and its error affects the output voltage to the same extent. The precision of discrete resistors can be 1% or better for good, expensive components, but, often, for low-cost parts the error is 5% or more. The accuracy of integrated resistors is normally worse than for discrete components; so the expected accuracy is low, and, for precise applications, it is necessary to include trimming or calibration.

The correction of inaccurate responses is an important task, which, instead of resistors, can use components whose value is adjustable and depends on manual or electrical control. The function is carried out by the so-called trimmer or potentiometer with a movable contact. Other possible integrated solutions use the series connection of many equal resistors with intermediate taps. The desired tap is selected electronically.

Table 5.1 provides information on matching, which means the accuracy of the ratio. This parameter obviously makes sense for resistors fabricated together. The use of the same fabrication steps and materials correlates the possible variations against the nominal value and makes the changes of value proportional. This is a valuable feature because often performances of circuits depend much more on the ratio between passive components than the absolute value.

The table also provides the Temperature Coefficient (TC) and Voltage Coefficient (VC). Note that the TC of integrated resistors is much higher than that of discrete types. The reason is that the fabrication steps of an integrated circuit pay attention to transistors and their performance and only indirectly to the features of resistors. The TC of discrete or integrated resistors is normally positive, but for special needs it is possible to find resistors whose TC is negative. Having both positive and negative TC can be a relevant design option.

The accuracy of the result depends on the accuracy of the resistors. There are many limitations that are mainly caused by inaccuracy of the fabrication methods. The most used parameters that quantify the resistor performances are tolerance, matching, and voltage coefficients. These are listed in Table 5.1 for various types of discrete and integrated resistors. Tolerance specifies the maximum difference between the nominal and real values. It is normally relatively high for discrete and integrated components unless one is using expensive parts. Matching refers to the accuracy of the ratio of resistance values. The figures given in Table 5.1 include the temperature coefficient that quantifies how the value changes with temperature (in the column labeled "TC"), and the voltage coefficient (labeled "VC"). Such parameters indicate that the value of the resistance is non-linear. Therefore, harmonic distortion may occur, when the amplitude of the voltage across the resistance becomes high.

Another limit on accuracy is the parasitic capacitance, C_p, between output node and ground (Figure 5.18(b)). This is caused by the unity resistor, wired connections, and the input capacitance of the circuit that utilizes the result of the addition. The output voltage is not a simple addition of inputs but is frequency dependent, because the total current flows in the parallel

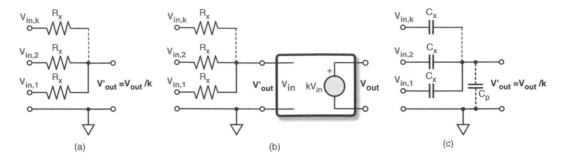

Figure 5.19   (a) Combination of voltage signals to obtain attenuated voltage addition; (b) use of interface with gain; (c) combination of voltages with capacitors.

connection of R_u and C_p and not just in R_u. The real output voltage, V', becomes

$$V'_{out}(s) = \frac{V_{out}}{1 + sR_uC_p},\tag{5.31}$$

where V_{out} is the expected value calculated using equation (5.30). In order to keep the limit low, even at high frequency, it is necessary to keep the parasitic capacitance low. Namely, it is necessary to verify that $C_p \ll 1/(2\pi f_{in}R_u)$. Suppose, for example, that $f_{in} = 100$ MHz and $R_u = 1k\Omega$. The above condition requires $C_p \ll 1.59$ pF, which is a small value difficult to obtain in discrete realizations but certainly possible for integrated solutions.

The method discussed for obtaining addition is for currents at input. With voltages, addition with passive elements is still possible. The scheme that uses equal resistors, depicted in Figure 5.19(a), produces an attenuated version of the expected result. This signal, calculated on the superposition principle, is

$$V'_{out} = \sum_{i=1}^{k} V_{in,i}\frac{R_x}{R_x + R_x(k-1)} = \frac{1}{k}\sum_{i=1}^{k} V_{in,i},\tag{5.32}$$

where k is the number of inputs used. If attenuation by k is not acceptable because the full result is required, the system needs amplification to counterbalance attenuation. Moreover, equation (5.32) assumes that input voltages are provided by ideal voltage sources and that output is measured with a very large resistance (in comparison with R_x). These possible issues are addressed by the interface in Figure 5.19(b), which enlarges the value of R_x, reads the intermediate signal as volt-metric, amplifies the input by k and provides the output in a volt-metric manner.

The accuracy of the Figure 5.19 scheme depends on errors affecting the resistors. Assuming that the ith resistance is $R_x(1 + \delta_i)$, where δ_i is a relative error, equation (5.32) becomes

$$V'_{int} = \frac{1}{k'}\sum_{i=1}^{k} V_{in,i}(1 + \delta_i), \quad k' = k + \sum_{i=1}^{k}\delta_i.\tag{5.33}$$

This denotes an error in attenuation and errors on single inputs. However δ_i refers to the ratio between resistor values and is not an error in the absolute value. So, if resistors match well, the accuracy is also good. Matching features integrated resistors that are realized with the same fabrication steps and materials. The figures given in Table 5.1 show that by using this method in integrated circuits, we can obtain precision as good as 0.1%.

The same results as those in Figure 5.19(a) can be obtained with capacitors instead of resistors, assuming that the capacitors are initially discharged. The scheme of Figure 5.19(c) produces the intermediate voltage independently of the capacitance value, as in the case of resistors:

$$V_{int} = \sum_{i=1}^{k} \frac{V_{in,i}C_x}{kC_x} = \frac{1}{k}\sum_{i=1}^{k}V_{in,i}. \tag{5.34}$$

The difference with respect to the scheme of Figure 5.19(a) is that capacitors drain charge and not current from the input terminals. Therefore the output must be measured with an infinite impedance. Moreover, possible parasitic capacitance between the node joining capacitors and ground for signal attenuates the result. There is also an error that depends on the initial charge (or voltage) of the node joining capacitors.

Self training

Resistors are fabricated by different methods, have various shapes, and withstand different levels of power. Do a search on the Web to find fabrication methods and the key features of the various types listed in Table 5.1. Prepare a short report that expands Table 5.1 with the following information:

- voltage coefficient and its comparison with the values in Table 5.1;

- power dissipation;

- temperature range.

Compare the parameters of Table 5.1 with the ones obtained in your search, and choose the types and fabrication methods that give the two best performances.

5.4.2   The Virtual Ground Concept

Recall the method illustrated in Figure 5.18. It performs the addition in a proper way, without any attenuation, but it needs current at input to produce voltage at output. That mixture of electrical variables is a limitation. However, the scheme can use just voltages at input and output if the inputs are changed by voltage-to-current (V-to-I) converters into currents, as shown in Figure 5.20(a), which uses specific blocks for that function. However, transforming a voltage into a current is not conceptually difficult; a resistance connecting a node to ground does that. This produces a current proportional to the voltage between that node and ground. What is difficult is to use the current obtained, because it flows away to ground. This observation suggests the need for a special node that realizes ground (or, to be precise, ground for signal) and permits the use of the current drained towards it. This task, as will be studied shortly, is achieved by using suitable electronic circuits that include operational amplifiers. Since this node is ground for signal but does not drain current, it is called *virtual ground.*

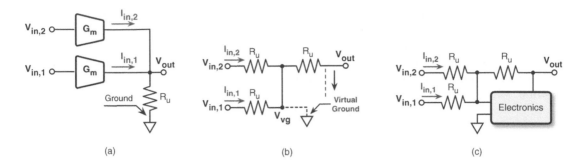

Figure 5.20 (a) Use of voltage-to-current converters for the addition of two voltage signals; (b) transformation into addition of voltage signals by the virtual ground; (c) sustained by electronic circuit.

Figure 5.20(b) illustrates use of the virtual ground concept to obtain the addition of two voltages. The inputs are transformed into current by unity resistors, R_u. Currents through resistors do not disappear to ground but are available to enter another unity resistance, R_u, and give rise to a drop in voltage proportional to the addition of currents. Since the current flow is through the virtual ground voltage, if input voltages are positive the output is negative. Obviously the electronic circuit that enables virtual ground also adjusts the current at the output node. If it is not used, or only partially used, by the output node, it is compensated for through an auxiliary path, as shown in Figure 5.20(c).

Inspection of the circuit shows:

$$I_{in,1} = \frac{V_{in,1} - V_{vg}}{R_u}, \quad I_{in,2} = \frac{V_{in,2} - V_{vg}}{R_u} \tag{5.35}$$

$$V_{vg} - V_{out} = R_u(I_{in,1} + I_{in,1}), \tag{5.36}$$

which yields

$$V_{out} = -(V_{in,1} + V_{in,2}), \tag{5.37}$$

which is the expected addition with an extra minus sign. So the use of virtual ground inverts the addition.

The virtual ground

Virtual ground is widely used in analog processing. Its implementation needs a special analog block, the operational amplifier.

Notice that the result depends on the value of the resistors, but all dependence disappears because they are assumed equal. Since possible errors change the result, if we suppose that the resistors converting voltage into current are $R_{u,1}$ and $R_{u,2}$, then equation (5.37) becomes

$$V_{out} = -\left[V_{in,1}\frac{R_u}{R_{u,1}} + V_{in,2}\frac{R_u}{R_{u,2}}\right], \tag{5.38}$$

showing that good matching between nominally equal resistors ensures accuracy.

The electronic circuit adding two or more inputs is schematically represented by the symbol in Figure 5.21(a), which, for the sake of simplicity, indicates inputs with only one wire.

COMPUTER EXPERIMENT 5.5

Understanding the Virtual Ground

The virtual ground concept is unusual and puzzling. It is a node kept at the ground voltage but the current on it does not flow to ground. This computer experiment uses a black box that is able of producing the virtual ground. The black box drains or sinks current with an extent that makes the voltage of the virtual ground node equal to ground. Actually the board includes two circuits, one that behaves ideally, the other that gives rise to a "weak" virtual ground. "Weak" virtual ground means that the voltage of the virtual ground is not exactly ground but is required to balance the output with an opposite and attenuated value: $V_g = -K V_{out}$.

The control parameters allow you to:

- choose the "balancing" coefficient K. When it is zero the real virtual ground is ideal;
- change frequency and amplitude of the two input signals. For both signals it is possible to change the DC level;
- select the value of resistances R_1 and R_2 and capacitor C_2. Resistances R_u are 1 kΩ.

MEASUREMENT SET–UP

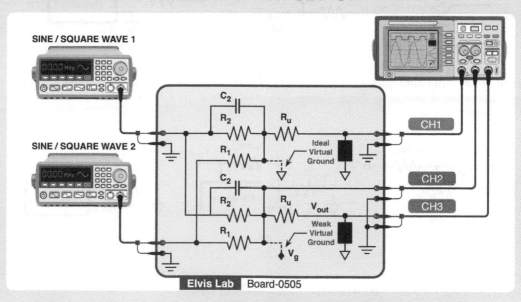

Elvis Lab   Board-0505

TO DO

- Set the amplitude of the upper signal generator to zero. Compare the outputs of ideal and "weak" circuits for different values of K. Estimate the error with constant K and different R_1.

- Set the amplitude of the top generator and change frequencies and components. Observe the "weak" virtual ground voltage with different experimental conditions.

- Set the two signal generators equal and find the equivalent circuit that gives rise to same output but zero amplitude for the second generator.

After practicing on real and ideal virtual ground, do a systematic study of performances. Draft a diagram of error as a function of K and constant values of R_2 and C_2. Use four points per decade.

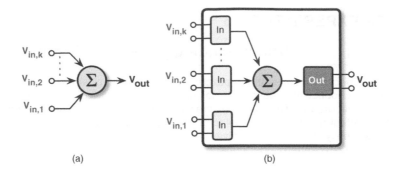

Figure 5.21   (a) Symbol representing addition of signals; (b) symbolic scheme including interfaces.

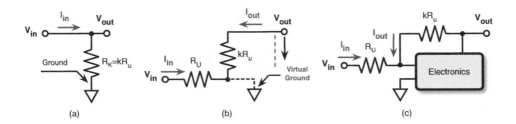

Figure 5.22   (a) Multiplication of current to obtain voltage by the use of non-unity resistance; (b) multiplication of voltage to obtain voltage via virtual ground; (c) scheme using electronics to obtain virtual ground.

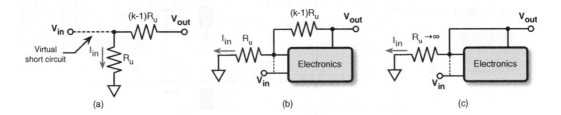

Figure 5.23   (a) Virtual short circuit concept; (b) electronic realization of a multiplier by a positive factor (≥ 1) using the short circuit concept; (c) multiplier by one.

Including interfaces that decouple input sources and output gives rise to Figure 5.21(b). The input and output signals can be either single ended or differential.

5.4.3   Multiplication by a Constant

Multiplication by a constant is straightforward if the electrical variables are current at input and voltage at output. Figure 5.22(a) shows the conceptual scheme. The injection of a current into a non-unity resistor, kR_u, gives rise to multiplication by k:

$$V_{out} = kR_u I_{in}. \tag{5.39}$$

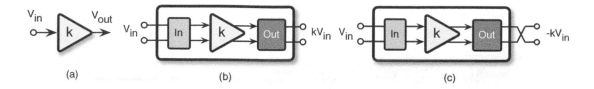

Figure 5.24 (a) Symbol for an amplifier; (b) fully differential scheme including interfaces; (c) changing the amplification sign.

As before, if the current is measured in amperes and output in volts, the unity resistance is 1 Ω; otherwise the unity resistance is a multiple of 1 Ω such as 1 kΩ or 1 MΩ. The absolute accuracy of unity resistance determines the accuracy of the multiplication factor. Since the absolute accuracy of integrated circuits is poor, errors can be as large as 20%, as indicated by Table 5.1. Even the absolute accuracy of discrete elements is low. Only special and expensive parts achieve absolute accuracy lower than 1%. Hence multiplication using the method shown in Figure 5.22(a) can be rather inaccurate.

As with addition, it is possible to use the virtual ground concept to transform the input variable from voltage to current. The resulting scheme is shown in Figure 5.22(b). Since the currents that flow in R_u and kR_u have opposite signs, the result is

$$V_{out} = -kR_u I_{out} = -\frac{kR_u}{R_u}V_{in}. \tag{5.40}$$

The gain, k, is thus negative and depends on the ratio between the resistors. To get positive gain we can use an additional inverting stage or a different method based on a new concept, similar to the virtual ground: the *virtual short circuit*.

Consider now the scheme of Figure 5.23(a). The input voltage is transformed into current by a unity resistance. However, the voltage across its terminal is applied not directly but through a virtual connection that makes the voltage what is needed but does not drain any current. The connection is equivalent to a short circuit, but it is only virtual. Since current $I_{in} = V_{in}/R_u$ does not flow through the virtual short circuit, it comes from the resistance $(1-k)R_u$. So the output voltage is given by

$$V_{out} = V_{in} + (1-k)R_u I_{in} = kV_{in}, \tag{5.41}$$

which has a positive sign or, as we have said, gives non-inverting amplification. The electronics used to obtain virtual short circuits makes the assumed shorted voltages equal, and makes sure the current does not flow out of the output terminal (Figure 5.23(b)). Since the value of any resistance is, obviously, positive, the method admits gain but no attenuation ($k \geq 1$). For unity gain the resistance across the electronics becomes zero. The unity resistance can be any value, even infinite. The scheme of Figure 5.23(c) with $R_u = \infty$ is called a "unity gain buffer."

Having gain controlled by resistor ratios is good for integrated circuits, because resistors fabricated together match well, as reported in Table 5.1. Another interesting feature concerns temperature variation. As is well known, the temperature coefficient primarily describes the change in resistor value due to temperature,

$$R(T) = R(T_0)[1 + \gamma_T(T - T_0)], \tag{5.42}$$

where γ_T is the Temperature Coefficient (TC). The use of resistors made of the same material gives equal temperature coefficients. Moreover, if the circuit or board layout produces equal

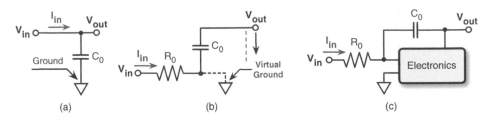

Figure 5.25   (a) Integration of current to obtain voltage; (b) integration by using virtual ground; (c) integrator with electronics to obtain virtual ground.

resistor temperature, the ratio between the resistances cancels out the temperature dependence.

Figure 5.24 shows a schematic representation of a multiplier by a constant. Since the factor is normally larger than 1 the block is called an *amplifier* and the multiplication factor *gain*. Figure 5.24(b) outlines the interfaces with possibly differential signals. With differential processing the gain sign is not a problem. Crossing the output connections, as shown in Figure 5.24(c), transforms an inverting into a non-inverting amplifier.

5.4.4   Integration and Derivative

Integration is easy if the input is current and the output is voltage. The injection of current into a capacitor produces a voltage shown by the equation

$$V_{out}(t) = \frac{1}{C_0} \int I_{in}(t)\, dt \tag{5.43}$$

or, in the s domain,

$$V_{out}(s) = \frac{I_{in}(s)}{sC_0}. \tag{5.44}$$

Hence the simple scheme of Figure 5.25(a) ensures there is integration. Again, it is often preferable to use voltages at both input and output. This is made possible by the virtual ground method, which transforms input voltage into current and enables its use for integration. Figure 5.25(b) shows the operation. The input current, V_{in}/R_0, injected into the bottom plate of the capacitor C_0 makes the output voltage equal to

$$V_{out}(s) = -\frac{V_{in}(s)}{sR_0C_0} = -\frac{V_{in}(s)}{s\tau_0}, \tag{5.45}$$

where τ_0 is the integrator time constant. Note that the denominator is dimensionless, since it is time multiplied by a complex frequency. Thus the dimensions of input and output are both voltages, as expected. Moreover, integration is with a minus sign, or, as explained before, the scheme is an inverting integrator.

About capacitors

The quality and accuracy of capacitors, as already asserted for resistors, determine the quality and accuracy of the circuits that use them. Their features depend on the fabrication process for discrete components, exploiting the conductive insulating materials of the IC technological process for on-chip capacitors.

Integrated capacitors' plates have a very small area, and oxide thicknesses are in the nm range. Because of these small dimensions, absolute accuracy is not good, but, since capacitors are fabricated together, they have good matching accuracy, especially if the capacitors are equal or have a rational ratio made by parallel connections of equal units. The value of integrated capacitors is very low. Elements of just a few pF consume a lot of silicon area, because even with very thin oxides and large dielectric constants the specific capacitance is in the $fF/\mu m^2$ range. Therefore, it is unusual to find capacitors of tens of pF in integrated circuits.

The plate area of discrete components is relatively large, and the capacitance value can be tens of μF or more. Different types of discrete capacitors for PCB surface mounting in a system-in-package are available. They are distinguished by the insulator or fabrication process used.

- **Film** This is made of aluminium plates or tin foils separated by an insulator of polycarbonate, polyester, polypropylene, or another polymer dielectric, wrapped and filled with epoxy. Film capacitors are used in high-performance and precise applications, but the low dielectric constants (from 2.2 to 6) make the capacitive–volume ratio high.

- **Metallized film** This is a variation of film capacitors. Metal foils are replaced by a thin layer of aluminium or zinc deposited onto the insulating film itself. The thickness is in the range of 20 μm to 50 μm. The advantage of these capacitors is their reduced physical size, but their performance can be lower than that of film capacitors.

- **Ceramic** As the name indicates, ceramic materials are the dielectric that separates multilayer electrodes. The value can range from a fraction of a pF to a μF. Components have different shapes depending on their prospective use, for example whether they are to be mounted through holes or on the surface. For miniature products, capacitors as small as $1.0 \times 0.5 \times 0.5$ mm or less, producing a capacitance of 1 μF, can be found on the market.

- **Electrolytic** Very large capacitances are handled by this type of capacitor, which uses an electrolyte as one of its plates. Electrolytic capacitors are polarized and must be biased with the proper DC voltage sign. The dielectric material is the oxide of the anode metal, often aluminium. It is grown during fabrication by a current that flows through the electrolyte. Reverse biasing dissolves the dielectric and gives rise to low resistance between the plates. So, reverse biasing causes a current that destroys the component. This type of capacitor is suitable for low-frequency applications and is often connected in parallel with capacitors with high-frequency performances to enhance the capacitive action over a wide frequency range.

■ **Tantalum** These are electrolytic capacitors with tantalum for the electrodes. This type achieves large capacitance values similar to those of electrolytic capacitors. They are superior to aluminium electrolytic capacitors in their temperature and frequency characteristics. Tantalum capacitors are available for surface mounting with values of fractions of μF.

Possible limits to real capacitors are mainly caused by the parasitics at the two terminals. However, for more accurate analysis it is necessary to account for other effects such as the non-ideal results caused by limited insulator quality. Non-ideal operations are represented by extra elements that make an equivalent circuit similar to the one shown in the diagram below:

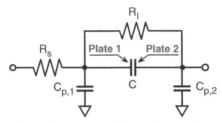

Equivalent circuit of a real capacitor.

The resistance R_I, called the *insulating resistance*, depends on the type of insulator used, on the thickness of the oxide, and on the plate area. For discrete components it is given in $M\Omega \cdot \mu F$ and has a value of several tens or hundreds of $M\Omega \cdot \mu F$. For polyester capacitors the value is in the range 500–2000 $M\Omega \cdot \mu F$. The series resistance R_s is very small and is normally negligible unless we are dealing with very large frequencies. Parasitic capacitance depends on the surroundings defined by the layout.

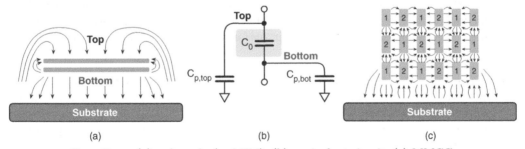

Capacitors: (a) poly–poly (or MIM); (b) equivalent circuit; (c) MMCC.

Integrated capacitors have various forms. Part (a) of the diagram above shows a parallel plate arrangement that uses two poly layers or metals for plates, and silicon dioxide or high-k insulator in between. The metal–metal structure is also called MIM (Metal–Insulator–Metal). The electric field is orthogonal to the plates, as happens with infinite-area electrodes. However, at the boundaries the electric field deviates from the orthogonal direction and gives rise to fringe contributions. If fringing effects are negligible the value of the capacitance is

$$C_0 = \epsilon_0 \epsilon_r W L, \tag{5.46}$$

where W and L are the width and length of the assumed rectangular plate.

The surrounding material and the substrate, both at short distance from the electrodes, determine the parasitics. The top and bottom of the plates should be distinguishable. Accordingly, the equivalent circuit must include top and bottom parasitics, as part (b) above shows. Often the values of $C_{p,top}$ and $C_{p,bot}$ are fractions of C_0. With the substrate very close to the plates, as in poly–poly structures, the parasitics become a significant fraction of C_0, of the order of 2–5%. Moreover, the bottom plate parasitic is larger than the top plate parasitic because the bottom shields the top.

The structure shown in part (c) above is a Metal–Metal Comb Capacitor (MMCC) that uses fingers of parallel metals to make electrodes and exploits the small separation allowed by the technology to obtain a good specific capacitance value (i.e., capacitance per silicon area). The case shown in the figure uses three metal layers, but more layers are possible. The capacitance is given by the fringing contributions in the horizontal and vertical directions. There are also parasitics towards the substrate. The top and bottom parasitics are almost equal and normally very low, if the metals are well separated from the substrate, as happens with modern integrated circuit technologies.

Be aware that the parasitic capacitances that involve the substrate of an integrated circuit are often non-linear because of the presence of diode structures.

Self training

Make a search on the Web and find details of the fabrication methods of discrete capacitors suitable for system-in-package realizations. Prepare a short report giving the following information:

- capacitor dimensions and values;
- voltage and temperature coefficients;
- frequency of operation.

Select the capacitor type suitable for making integrators with time constants of 1 msec, 1 µsec, and 10 psec. The value of the resistance must always be lower than 10 MΩ.

Having discussed real capacitors let us now consider the integrator with current at input, illustrated by Figure 5.25(a), again. This time, though, we'll use a real capacitor such as an integrated MIM capacitor. The circuit is like that shown in Figure 5.26(a). It does not include parasitic resistances but must account for parasitic capacitances. The current goes into the capacitance $(C_0 + C_{p,top})$ rather than only into C_0. Therefore the output voltage is affected by an error that can be expressed as

$$V_0(s) = \frac{I_{in}}{SC_0} \frac{C_0}{C_0 + C_{p,top}} \approx \frac{I_{in}}{sC_0}\left[1 - \frac{C_{p,top}}{C_0}\right] = V_{out,id}(s)(1 - \epsilon_c), \qquad (5.47)$$

where $\epsilon_c \simeq C_{p,top}/C_0$ and $V_{out,id}(s)$ are the expected ideal output. Notice that the use of a top plate on the output node minimizes the error if, as normally happens, $C_{p,top} < C_{p,bot}$. Actually, the problem is not the extent of the error, which is, eventually, comparable with errors caused by inaccuracies in fabrication, but the non-linearity of the error. Often the parasitic coupling involves non-linear effects, as in the case of reverse biased p–n junctions

COMPUTER EXPERIMENT 5.6

Equivalent Circuit of Real Components

There is a certain difference between ideal components and real elements fabricated as a discrete or integrated device. For example, a simple resistance, is intended as a two-terminal component, but in reality has a third terminal: the ground with parasitic components toward it. This computer experiment permits you to extract the equivalent circuit of a "real" resistor. The board of this experiment uses a "real" resistor in series with 10 MΩ resistances and drives the series with two sine wave generators. With a button you can change the "real" resistance, choosing it randomly from a virtual drawer of components. In this experiment you can:

- select frequency and amplitude of the two sine wave input signals;
- change the phase of the sine waves in the 0 to 2π range;
- measure the voltage of nodes, or the voltages across resistances;
- change the resistance under test and repeat the experiment.

MEASUREMENT SET–UP

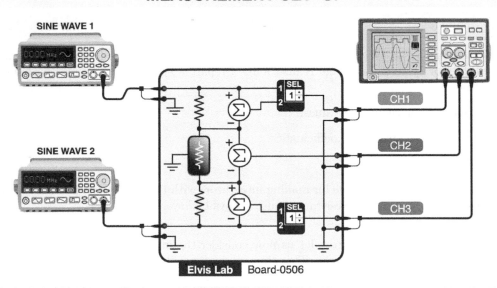

TO DO

- Set a top sine wave amplitude and send the amplitude of the bottom one to zero. Use zero frequency and variable phase to generate a DC voltage. Suppose the 10 MΩ resistances are exact for determining the DC equivalent circuit.
- Use the same frequency for the two sine waves but different phases. Change amplitudes and phase until the phase of the voltage across the test resistance is zero. The scheme enables you to measure the parasitic capacitances of the two input ports.
- Build an equivalent circuit made up of the "real" resistor and two parasitic impedances between the resistance endings and signal ground.

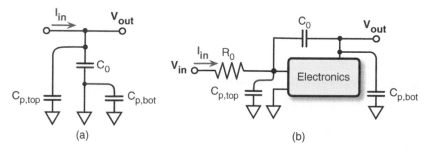

(a) (b)

Figure 5.26 (a) Parasitics of an integrated capacitor; (b) effect of parasitics with possible use of an integrated capacitor.

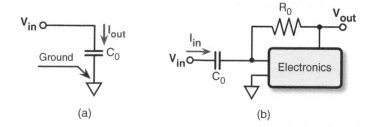

(a) (b)

Figure 5.27 (a) Derivative of input voltage to produce output current; (b) derivative of input voltage that obtains output voltage by using virtual ground.

that give rise to non-linear capacitances. Thus, it is convenient to have small errors because even if they are non-linear the effect is small. We shall not study this point in any more detail here. It is enough to be aware that linear parasitic capacitances distort output voltage and cause harmonic terms.

The use of virtual ground, illustrated again in Figure 5.26(b), resolves the problem of parasitics and gives parasitic independence. Any capacitance loading the virtual ground is at a fixed voltage, the virtual ground level. Therefore those capacitances are not charged or discharged and do not drain any fraction of the signal current away. The input parasitic does not influence the operation of the circuit, which is determined only by the capacitance C_0. The electronic circuit that produces the virtual ground takes care of the parasitic connected to the output. In other words, the electronics of Figure 5.26(b) charges or discharges the bottom-plate parasitic of C_0. Notice that the use of the top plate at the virtual ground side is not strictly essential. However, it is a good precaution because the virtual ground is a critical node.

The derivative operation is the opposite of integration (or its inverse in the s domain). This function is easily obtained with voltage at input and current at output. Notice that the derivative of the voltage across a capacitor is the current that flows into the capacitance itself:

$$I_{out} = \frac{dQ_C}{dt} = C_0 \frac{dV_{in}}{dt}. \tag{5.48}$$

The result is what is obtained by the circuit of Figure 5.27(a).

It is often desirable to have voltages at both input and output. For this it is necessary to transform the current flowing into a capacitor into voltage. This is done by using the virtual ground, as shown in Figure 5.27(b). The output voltage becomes

$$V_{out} = -R_0 I_{in} = -R_0 C_0 \frac{dV_{in}}{dt}, \tag{5.49}$$

or, in the s domain,

$$V_{out}(s) = -R_0 C_0 s V_{in}(s) = -s\tau_0 V_{in}(s), \tag{5.50}$$

which again defines a time constant, given by the product of resistance R_0 and capacitance C_0.

The absolute accuracy of discrete capacitors is rarely better than 1% and is often in the range 10–20%. Since inaccuracy on capacitance adds to the inaccuracy of resistance, the accuracy of the time constant, either of integrators or of derivators, is poor and often not enough for practical applications.

Let us suppose that the values of the resistance and the capacitance deviate from the nominal values R_N and C_N according to the equations

$$R = R_N(1 + \epsilon_R) \tag{5.51}$$

and

$$C = C_N(1 + \delta_C), \tag{5.52}$$

where ϵ_R and ϵ_C are the relative errors with variance σ_R and σ_C respectively. The error, ϵ_τ, affecting the nominal time constant, τ_N, is

$$\tau = \tau_N(1 + \epsilon_\tau). \tag{5.53}$$

The variance is given by

$$\sigma_\tau = \sqrt{\sigma_R^2 + \sigma_C^2}, \tag{5.54}$$

which is dominated by the larger variance of errors if one is dominant. It becomes $\sqrt{2}$ times each variance, if they are equal.

Even for integrated technologies the accuracy of the time constant is a problem. The fabrication steps of resistors and capacitors are uncorrelated. Errors are independent of each other, and even for integrated systems the quadratic superposition of absolute accuracies must be used. However, the errors affecting capacitors and the ones changing resistances fabricated on a single chip go in the same direction and are almost equal. Hence, the time constants generated by the same type of resistors and capacitors on the same IC are affected by errors. The errors may be large, but one is close to the other.

Figure 5.28 shows the symbols for the electronic circuits that produce integration and derivation. The interfaces before and after the blocks cause optimum matching of the input and output electrical variables. When the blocks handle differential signals, the function performed can incorporate a minus sign at zero cost. It is enough to switch the differential inputs or outputs.

Keep in mind . . .

that when errors are random and statistically independent they are superposed quadratically. This is what happens with inaccuracies in resistors and capacitors, because the fabrication steps used for the two components are independent of one another.

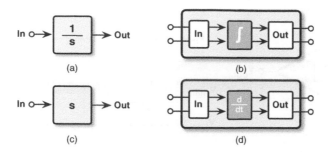

Figure 5.28   Symbols for (a) an integrator and (c) a derivator; (b) and (d) the symbols better specified by adding interfaces that handle differential signals .

Example 5.1

The time constant of an integrated continuous-time integrator is controlled by a resistor and a capacitor. Inaccurate resistivity and oxide thickness determine random errors, ϵ_R and ϵ_C, whose variances are σ_R and σ_C. Moreover, the actual temperature causes variations in the resistance and the capacitance, according to the relationships $R(T) = R(T_0)[1 + \alpha_R(T - T_0)]$ and $C(T) = C(T_0)[1 + \alpha_C(T - T_0)]$ (T_0 is the room temperature; conventionally it is 27°C). Calculate the time constant at $T = T_0$ and $T = T_0 + 47$°C. Use as nominal values of resistance and capacitance, at the nominal temperature $T = T_0$, $R_N = 100$ kΩ and $C_N = 4$ pF. Also use the values $\epsilon_R = 0.23$, $\epsilon_C = 0.14$, $\alpha_R = 0.001$, and $\alpha_C = 0.0002$.

Solution

The nominal value of the time constant at $T = T_0$ is $R_N C_N = \tau_N = 0.4$ μs. The real resistance and capacitance differ from the nominal values because of random errors and temperature coefficients. We obtain

$$R = R_N(1 + \epsilon_R)[1 + \alpha_R(T - T_0)]$$
$$C = C_N(1 + \epsilon_C)[1 + \alpha_C(T - T_0)].$$

The time constant at $T = T_0$ is not correct because of a random fluctuation $\epsilon_{s,\tau}$:

$$\tau = \tau_N(1 + \epsilon_R)(1 + \epsilon_C) = \tau_N(1 + \epsilon_{s,\tau}).$$

The variance of $\epsilon_{s,\tau}$ is $\sigma_{s,\tau} = \sqrt{\sigma_R^2 + \sigma_C^2} = 0.27$. So at "one σ" the value of the time constant ranges between $0.4(1 + 0.27) = 0.508$ μsec and $0.4(1 - 0.27) = 0.292$ μsec. An increase in temperature by 47° augments the resistance by 4.7% and the capacitance by 0.94%. Therefore, because of temperature the time constant increases by 5.68% ($1.047 \times 1.0094 = 1.0568$).

5.5   CONTINUOUS-TIME NON-LINEAR BASIC FUNCTIONS

Analog circuits carry out non-linear functions required by many processing algorithms. A frequently used operation is threshold detection. The output is a logic signal to indicate

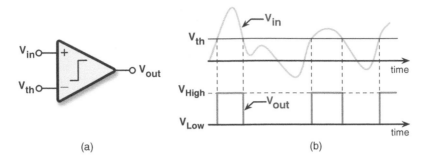

(a)   (b)

Figure 5.29   (a) Symbol used for a continuous-time comparator; (b) example of input and output signals for a time-varying input.

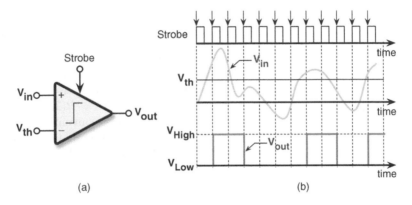

(a)   (b)

Figure 5.30   (a) Symbol used for a sampled-time comparator; (b) example of input and output signals for a time-varying input – the output changes at the rising edge of the strobe signal.

that input is lower or higher than the threshold. Another relevant function is the product of two inputs that, with sine waves, produces two sine waves at the sum and the difference of the input frequencies. Other functions used are the quadratic response, which produces the square of its input, and the logarithmic response, used to compress signals when relevance is at low levels. The logarithmic and quadratic responses exploit the non-linearity of electronic devices in proper regions of operation. In other words, junction diodes produce an exponential response; a MOS transistor in the saturation region enables quadratic response.

5.5.1   Threshold Detection

The electronic circuit used to detect a threshold is called a comparator, because it compares input and reference to perform its function. The signals are normally voltages, but currents are also possible. The output is given by two levels (of voltage or current), one low and the other high. With voltage as output, the lower supply is normally low and the higher supply of the circuit is high. However, low is not a firmly defined voltage; the same holds for high. It is enough is to be able to distinguish clearly the low level from the high.

The comparator can be continuous-time or sampled-data. In the continuous-time case, the ideal output instantly switches from one logic level to the other each time the input crosses the threshold. Figure 5.29 shows the symbol used for a comparator and the waveforms of a possible voltage comparison. As shown, the switching of output from low to high level is, ideally, at the crossing times. The input of Figure 5.29(b) increases or decreases steadily, and the crossing points are well defined. The situation is different with signals affected by noise: when the input is very near the threshold, the noise causes frequent crossing of the threshold and the output switches back and forth many times. To avoid this undesired response, some comparators use a small hysteresis: the upward and downward crossing are slightly different.

The continuous-time comparator is equivalent to a subtraction of input and reference followed by a large amplification (ideally infinite) plus a clipping at low or high logic level. The latter feature is represented by the symbol inside the triangle to remind us of its step-like input/output transfer characteristics.

The sampled-data comparator looks at the difference between input and reference only at sampling times. It also generates the logic output and holds it for an entire sampling period. Figure 5.30 shows the symbol and possible waveforms. Note that the symbol uses an additional input, which is a logic signal whose rising or falling edge determines the operation time. The sampling time is also defined by the rising or falling time of the logic signal. The signal given is a clock also called a *strobe*. Figure 5.30(b) uses a strobe given by the rising time of a logic signal with 50% duty-cycle. The output shown in the figure is immediately after the strobe. This is not realistic, because the circuit needs some time for operation. Only ideal circuits generate the output instantaneously.

5.5.2 Analog Multiplier

Analog multipliers, as indicated by their name, obtain the product of two signals, normally voltages. However, the multiplication produced by real analog circuits is not precise. Their accuracy is acceptable only with small input amplitudes: tens or at most hundreds of mV. Often, the result is not the product but a given function of the inputs. For example, the popular multiplier invented by Barry Gilbert, made from bipolar transistors, obtains

$$V_{out}(t) = k_2 \tanh(k_1 V_{in,1}(t)) \tanh(k_1 V_{in,2}(t)), \qquad (5.55)$$

which is the product of the hyperbolic tangents of its inputs. The coefficients k_1 and k_2 have dimension $[V^{-1}]$, so that the tanh argument is dimensionless. The response of the Gilbert cell can possibly be corrected by non-linear blocks at the inputs with an inverse response, $\tanh^{-1}$, as shown in Figure 5.31. That input pre-distortion compensates for the subsequent opposite distortion, thus extending the linear region of operation.

The analog multiplier is frequently used to move the band of signals to lower or higher frequencies. In this case the multiplier is also called a frequency modulator or a mixer. One input is the signal and the other is a sine wave at the modulating frequency. As we know, the multiplication of two sine waves produces two sinusoids, one at the addition and the other at the subtraction of the frequencies used. When one input is a band-limited signal with spectrum around f_2 in the interval $(f_2 \pm f_B/2)$, its product with a sine wave at f_2 generates two replicas, one around $(f_1 + f_2)$ and the other around $(f_1 - f_2)$. Since the multiplier generates two replicas, the one that is possibly undesired is removed by filtering.

A simple frequency modulator uses a square wave instead of a modulating sine wave. The result obtains the frequency shift but generates multiple spur replicas. Multiplication by a square wave is straightforward. It is obtained, for example, by using switches that reverse the connections of the differential inputs of the other signal, as shown in Figure 5.32. When the

COMPUTER EXPERIMENT 5.7

Understanding Comparators

This computer experiment permits you to do some practice with comparators. The comparator is a block that generates a logic level, 1 or 0, in response to an input that exceeds a reference or not. The ElvisLab board used here includes two comparators: one clocked, the other asynchronous with hysteresis. The reference is zero, the input signal is the addition of two sine waves and noise. This type of input signal corresponds well to real situations, as the noise (that always exists) makes the threshold transition uncertain, when the signal is near the threshold. The hysteresis in the asynchronous comparator serves to avoid output bouncing.
The setting allows you to:

- determine amplitude and frequency of the sine waves and define the noise level;
- set the clock of the clocked comparator by a square wave generator. The sampling of the input signal occurs at the rising edge of the clock, but the result is available at output at the falling clock edge;
- set the hysteresis interval used to establish a different threshold for the asynchronous comparator that switches in opposite directions.

MEASUREMENT SET–UP

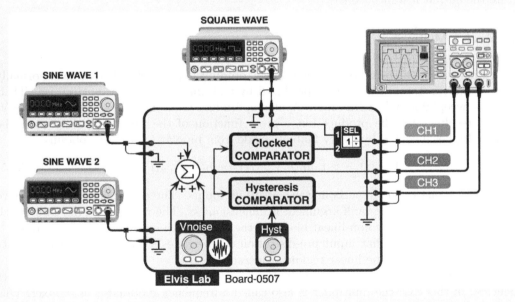

TO DO

- Set the sine waves but zero the noise. Set the clock frequency about four times the fastest sine wave. Observe the difference between the two outputs. Observe that the clocked comparator can generate a positive output when the signal is already negative.
- Add noise and use sine wave frequencies much lower than the clock. Observe what happens near the zero crossings.
- Set the hysteresis of the comparator to the minimum level that avoids output bouncing.

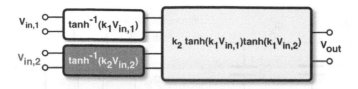

Figure 5.31   Pre-distortion in a multiplier for extending its range of operation.

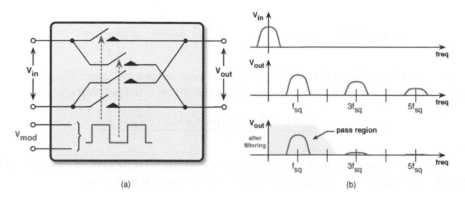

Figure 5.32   (a) Multiplication of the input signal by a square wave; (b) input and output spectra.

input is high the switches direct the signal straight; when the input is low the switches reverse the couplings. The multiplication performed is by $+1$ and -1.

Recall that the Fourier transform of a square wave, $Square(t)$ with periodicity T_{sq} and amplitude 1, is the infinite series of sine wave at multiples of the square wave angular frequency, $2\pi/T_{sq}$:

$$Square(t) = \frac{4}{\pi} \sum_{n=1,3,5}^{\infty} \frac{1}{n} \sin\left[n\frac{2\pi}{T_{sq}}t\right]. \tag{5.56}$$

Therefore, as mentioned before, multiplication by a square wave produces an infinite number of replicas of the input, the replicas being shifted by $\pm kf_1$. The unwanted extra terms can be filtered out by a post-mixer filter, as Figure 5.32(b) schematically shows.

5.6 ANALOG DISCRETE-TIME BASIC OPERATIONS

Analog processing in discrete time, either linear or non-linear, handles samples. Therefore, analog sampled-data processors use a proper way to represent samples in electrical format. Normally the samples are kept for the sampling period, and, when required, they are delayed. As we know, the capacitor can temporarily store signals in the form of charge, provided it is insulated from the rest of the circuit. Therefore, a proper way to process signals in the sampled-data domain is to use capacitors and switches. The simple scheme of Figure 5.33 produces elementary processing of the input signal. The input voltage charges the capacitor C_1 when switch S_1 is closed (phase Φ_1). Then both switches open for a short period (the phases are not overlapped) before switch S_2 connects capacitors C_1 and C_2 in parallel. The sharing

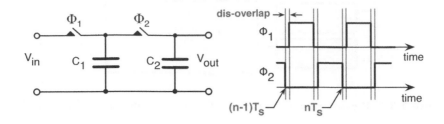

Figure 5.33 Simple switched capacitor processing function.

of the charges on the two capacitors determines the new output voltage. Therefore, at time nT_s, conventionally assumed to be at the end of phase Φ_2, the output voltage is

$$V_{out}(nT_s) = \frac{C_1 V_{in}(nT_s - T_s) + C_2 V_{out}(nT_s - T_s)}{C_1 + C_2}, \tag{5.57}$$

where it is supposed that the input voltage remains unchanged during each clock period.

The amount of charge entering the input node in every clock period is

$$Q_{in}(nT_s) = C_1[V_{out}(nT_s) - V_{in}(nT_s)], \tag{5.58}$$

which is proportional to the voltage difference between output and input. That charge is the same that flows during the sampling period T_s through a resistance R_{eq} connected between input and output, equal to

$$R_{eq} = \frac{T_s}{C_1}. \tag{5.59}$$

Therefore, the capacitor C_1 and the two switches make a structure that, on average, is equivalent to a resistor. Because of this, the structure is called a switched capacitor resistor.

Based on the above, we can conclude that the circuit of Figure 5.33 operates as a sampled-data RC network. The equivalent time constant is $R_{eq}C_2 = T_s C_1/C_2$, proportional to the sampling period, T_s, and a ratio between capacitors. Since a clock period can be very precise and the ratio between capacitance realized in integrated circuits is quite precise, we can obtain an accurate equivalent time-constant.

The switched capacitor resistor of Figure 5.33 depends on the parasitic capacitance between the top plate of C_1 and the other terminal. However, the scheme is not the only configuration capable of producing the function. Other schemes with more switches resolve the problem. We do not consider them here because, at this level of study, it is enough to be aware of the basic idea. It is used together with the virtual ground concept to obtain analog signal processing similar to what we have already studied for continuous-time situations.

Sampled-data processing often requires opening switches and keeping the information stored on a capacitor for a while. The scheme of Figure 5.33 does that for a short period between the two phases; in other situations the storing time can be longer. Suppose that the opening of switches occurs at time nT_s. The charge on the storing capacitor, C_s, is $Q_{in} = C_s V_{in}(nT_s)$. Then the information is supposed frozen. In reality the capacitor is not completely insulated, because switches leak current. The leakage I_{leak} causes a continuous drop of charge on C_s:

$$\Delta Q_{in}(t) = (t - T_s)I_{leak}. \tag{5.60}$$

Therefore, if $I_{leak} = 1$ pA the capacitor loses $1 \cdot 10^{-18}$ Coul in 1 μs, a very small value compared with the charge stored, for example, on a 0.1 pF charged at 1 V (10^{-13} Coul). However, when the sampling frequency is very low or the temperature augments current leakage, the loss of signal can become significant.

5.7   LIMITS IN REAL ANALOG CIRCUITS

Analog processors are not able to produce the expected ideal functions: imperfections of passive components and limits of basic blocks alter their operation. Non-linearities and parasitic components are the key things responsible for flaws. In addition, real active circuits do not realize the virtual ground or the virtual short circuit perfectly; the values of real passive elements differ from the nominal because of inaccuracies in fabrication. Moreover, the noise affecting electronic circuits blurs the outputs. The consequences of those non-idealities are summarized in global parameters such as harmonic distortion, errors in frequency response, and signal-to-noise ratio.

Many active components, such as the operational amplifier or the OTA to be studied shortly, have fairly non-linear responses. Passive components, in particular resistors and capacitors, show limited non-linearities. The designer obtains linear operation of active circuits at a level equal to the passive elements by using feedback. We shall study this technique and verify its ability to reduce the non-linearity of some blocks in a later chapter.

The limits of active performances and errors in the values of passive components cause discrepancies between the ideal expected response, H_{id}, and the one obtained, H_{real}. The deviation of H_{real} from H_{id} depends on the error, ϵ_i, and the sensitivity to that error, $S_{\epsilon_i}^H$, as already discussed in Chapter 4:

$$\delta H_{real} = S_{\epsilon_i}^H \epsilon_i. \tag{5.61}$$

It is important to be aware of the sensitivities because they identify the errors that cause important departures from the expected result. The sensitivity can be estimated by an analytical study or by repeated simulation where the value of the parameter under study is suitably changed. When it is desired to assess the errors in several parameters at the same time, the so-called Monte Carlo analysis is performed. The Monte Carlo analysis is a method that, after selecting the domain of parameters, randomly sets those parameters using a certain specified probability distribution. After performing the simulations, aggregated results show the range of output variations.

The effect of linear limits on the output signal is estimated by the superposition principle. Inaccuracies are modeled by a signal (voltage, current, or value of component), say ϵ_i, applied at some point in the system. The impulse time response, $h_i(t)$, or the frequency response, $H_i(s)$, determine the effect of each input on the output:

$$\epsilon_{i,out}(t) = \int_0^t \epsilon_i(\tau) h_i(t - \tau) d\tau$$

$$\epsilon_{i,out}(s) = \epsilon_i(s) H_i(s). \tag{5.62}$$

The combination of all outputs depends on whether the inputs are correlated or not. Since in the case of noise the sources are uncorrelated, the superposition is quadratic. Therefore, in the s domain,

$$\epsilon_{out}(s) = \sum_i \epsilon_{i,out}(s) = \sum_i \epsilon_{i,in}(s) H_i(s) \tag{5.63}$$

COMPUTER EXPERIMENT 5.8

Multiplier in the Time and Frequency Domains

Analog multipliers are non-linear blocks that, ideally, perform the corresponding mathematical operation. Obviously, since multiplication is a non-linear operation, it gives rise to new spectral components. This computer experiment helps you in understanding signal multiplication. The inputs are two sine waves and a square wave. The ElvisLab board uses two ideal multipliers and permits you to observe waveforms in the time domain. A spectrum analyzer displays the spectra of the output selected by the user. Remember that the spectrum of a sine wave is a delta line at the sine wave frequency. In this experiment you can:

- select frequency and amplitude of input signals. For the first sine wave signal you can also change the phase;
- choose between the two available outputs and see the spectrum;
- set the frequency interval of the spectrum analyzer sweep.

MEASUREMENT SET-UP

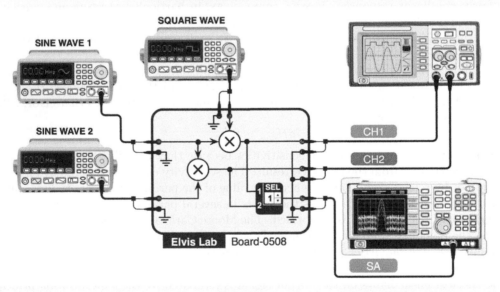

TO DO

- Set the frequency of the first sine wave generator to zero and change the phase. The output is a constant signal whose magnitude is determined by the phase and amplitude. Observe the output voltages for different settings of the other generators and verify the spectra.
- Change the first sine wave frequency to a finite value and see what happens in the time and frequency domains. Change the phase and verify that, although differences in the time plots, the spectrum does not change.
- Use a fixed first sine wave frequency and change the other one from very low to very high with respect to the fixed one. Explain the results.
- Observe the difference between output spectra when one of the inputs of the multiplier is a sine wave or a square wave at the same frequency.

for correlated signals. For noise, the output, $\epsilon_{noise,out}$, is

$$\epsilon_{noise,out}^2(s) = \sum_i \epsilon_{i,noise,out}^2(s) = \sum_i \epsilon_{i,noise}^2(s)|H_{i,noise}(s)|^2. \qquad (5.64)$$

Often, the global error is transformed into an equivalent error applied to the input, because it is convenient to compare that error with the input signal. It is calculated by

$$\epsilon_{in,T}(s) = \frac{\epsilon_{out,T}(s)}{H_{in}(s)}, \qquad (5.65)$$

where $\epsilon_{in,T}(s)$ is the total error due to the superposition of correlated or uncorrelated limits. The name of the transformed error includes the term *input referred*. Thus, we can have *input referred noise, input referred offset*, and so forth.

5.8 CIRCUITS FOR DIGITAL DESIGN

Digital design is important for many modern applications, but, in some sense, it is more straightforward than analog design. Digital circuits, for example, do not use sophisticated notions like the virtual ground or the virtual short circuit; they implement logic or mathematical expressions directly, using basic functions in hierarchical ways. Moreover, they are tolerant with respect to timing and amplitude accuracies of the electrical quantities representing signals. Also, it often happens that obtaining digital functions does not require the development of hardware, because it may be more convenient to employ available software-programmable hardware that can be adapted to user needs by software.

Years ago, digital circuits had a limited number of components. Digital complexity is estimated using a unit called the Gate Equivalent (GE). The size of this unit is, conventionally, that of a two-input NAND cell. A single integrated circuit that contains up to 10 GE is a small-scale integrated circuit. This category is still used for simple needs, but there are many other categories. Thanks to technology evolution digital complexity has increased exponentially, leading to present-day counts of billions of GE. We distinguish between:

- logic gate – 1 GE;

- Small-Scale Integration (SSI) – up to 10 GE;

- Medium-Scale Integration (MSI) – 10–100 GE;

- Large-Scale Integration (LSI) – 100–1000 GE;

- Very Large-Scale Integration (VLSI) – 1000–10 000 GE;

- Ultra Large-Scale Integration (USI) – 10 000–100 million GE;

- Giga-Scale Integration (GSI) – more than 100 million GE.

Digital systems use not just one category but a hybrid of several of the above classes. To produce complex functions the designer uses the last categories massively, with blocks of low complexity that interface macro functions. Then the designer's task is mainly to match the

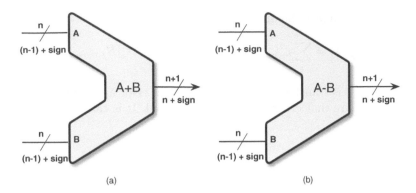

Figure 5.34   (a) Symbol of a digital adder; (b) symbol of a digital subtractor.

inputs and the outputs of the blocks, avoiding spur and noise interference, signal synchronization, speed enhancement, and power optimization.

Since digital processors operate on signals that represent numbers, the basic blocks perform mathematical operations such as addition, subtraction, counting, multiplication, and transfer and storage of data. These actions are described at a high level by the use of symbols or by suitable behavioral language that details logic operations. Then, often, the use of powerful tools automatically enables the designer to translate the high-level description into a design description. Such a tool simulates the behavior of the resulting digital network, accounting for the actual response of real circuits, and verifies that the description is logically correct.

Digital circuits generate logic outputs with a delay that depends on the speed of the electronics. If there is no control on the timing, the circuit is said to be asynchronous. On the other hand, it often happens that a clock signal controls the operation in such a way that every output changes simultaneously (in ideal situations). Such a circuit is called synchronous. In practical situations, different delays of various logic parts determine the clock speed. The longest delay, with a margin that ensures safe operation in the worst case, determines its period.

5.8.1   Symbols of Digital Blocks

Apart from a description using language, a block diagram also represents a digital design. Figure 5.34 shows the symbols for an adder and a subtracter. The difference between the two symbols is only the + or − sign at the input terminals. These are busses with n wires, one of them defining the sign. Since the adder output can be $(n + 1)$ bits wide, the bus at output uses $(n + 1)$ wires plus sign. The subtracter output is n bits plus sign. The number of bits can be any value. However, the word-length should not be more than what is needed to ensure the expected overall accuracy. Long words cost silicon area and power.

The adder is the basis of a counter. Figure 5.35 shows the block diagram for this. It uses an adder with two n-bit inputs and an n-bit output. One of the inputs is a stable 1; the other is a delayed version of the output. The clock that drives the delay block defines the rate of the counter. Therefore at each clock period the output of the adder increases by 1. When it reaches the top of the scale, the next clock period starts again at zero. When necessary the scheme can use a reset that zeros the output when it goes high. What the scheme of Figure 5.35 implements is an *up*-counter. Similarly, it is possible to realize a *down*-counter

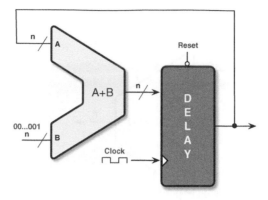

Figure 5.35 Block diagram of a digital counter based on an adder.

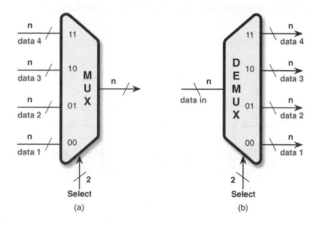

Figure 5.36 (a) Symbol of a digital multiplexer; (b) symbol of a digital demultiplexer.

with a subtracter as its basic block. When its output crosses zero, the next clock period becomes the full scale, decreasing by 1 at each successive clock period.

In modern computers the adder resides in the Arithmetic Logic Unit (ALU), which is a specialized digital circuit where other operations, such as multiplication or division, are also performed. Although adders can be constructed for generic numerical representations, they normally use binary numbering because of the simplicity of the circuit. Using numbering with other bases is possible but requires more complex schemes than the binary ones.

Multiplexing and demultiplexing are functions frequently needed in digital systems. The blocks that do these are also called MUX and DEMUX. A MUX basically switches one input bus to the output depending on the value of a selecting signal that decides which input is to be chosen. Multiplexers do not typically use clocks; the output changes as the select value changes. Figure 5.36 shows the symbol of a four-input MUX and a four-output DEMUX. The select signal has two bits; inputs and outputs have the same number of bits, n.

Multiplication is a complex operation performed by a dedicated circuit. It normally resides in the already mentioned ALU, a key part of the microprocessor. However, it can also be produced with a discrete logic circuit. Different computer arithmetic techniques are used to perform multiplication. Most of them involve the calculation of partial products that are then

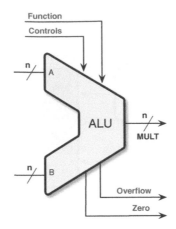

Figure 5.37   Symbol of a digital Arithmetic Logic Unit (ALU).

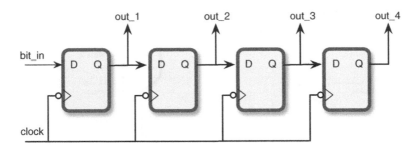

Figure 5.38   Block diagram of a shift register made from cascading flip-flops.

summed together, as in multiplying decimals by hand. The symbol of the multiplier can be the schematic ALU of Figure 5.37, where the multiplication function is selected by a signal at the input. Some block diagrams use more detailed functions to represent the internal operations. For instance, they detail the shift register, one of the functions used by the multiplier. The shift register is a cascade of cells with a single-bit signal at input, schematically shown in Figure 5.38. In hardware a shift register is a sequence of cells where the data shifts from one cell to another at every clock edge. The elementary cell is called a *flip-flop*. Figure 5.38 outlines the input (D), the output (Q) and the clock control. Some cells also provide the complement of the output ($\bar{Q}$).

After digital circuits have been described using symbols or Hardware Description Language (HDL), the result is broken down using basic functions such as logic gates and flip-flops. The process of expanding the design is called synthesis. This, in addition to signal handling, takes care of timings and other details. The operation is normally done not manually but by using special software tools. The input of a synthesis tool is the description code, and the output is the design at transistor or basic logic cell level.

5.8.2 Implementation of Digital Functions

Various options are available for implementing digital functions. The method a particular designer uses depends on overall complexity, personal design expertise, and other conditions such as cost, time-to-market, power budget, and so on. For simple functions it is enough to interconnect a few logic gates normally available in "off-the-shelf" integrated circuits. Simple logic gates also serve for interfacing complex blocks on PCBs. The "off-the-shelf" devices used for those purposes are also called glue logic. Their complexity is low, but their specific design enables the interfacing or adapting of signals at the input and output of complex devices.

Frequently, digital functions are very complicated, and the circuits used for implementing them are also very complex (they are USI or GSI). In those cases the designer is not required to go into the details of the circuit implementations but to work at a high level. Often, knowing the inner operation of complex digital circuits is scarcely relevant because the overall architecture consists of macro-blocks or macro-functions, whose limits are well defined at a high level. They are realized by specific ICs or macro-cells. These are black boxes with detailed input/output descriptions. For high-volume products, instead of assembling ICs and glue logic on a PCB it can be convenient to integrate all of the functions on a dedicated integrated circuit. In this case we have an ASIC – the acronym for an Application-Specific Integrated Circuit. The IC performs a specific function and its use ensures optimal speed, power consumption, and silicon area.

The hard interconnection of macro-blocks (hardware) is increasingly being replaced by "flexible" hardware, whose operation is defined by software used to provide instructions on how the hardware must process digital signals. A second option is to use reconfigurable architecture, with software for establishing the interconnections needed to obtain the architecture. There are three possibilities, as follows.

- **Use of Microcontrollers or Microprocessors** The hardware is pre-designed, and only the interconnections with peripherals and other devices such as memory are needed. The functional operations are defined by software (which runs on a microcontroller or a microprocessor). The hardware executes the programs stored in the memory by fetching instructions, examining them, and executing them one after another, following the program order. The communication with peripherals is via busses. All of the operational steps are controlled by a clock signal, whose period is the processor cycle time. The nano-scale of modern semiconductor technologies enables very high microprocessor clock speeds, measured in GHz or more. The rate depends on delays in transferring instructions and data and on the power consumption.

- **Use of Digital Signal Processors** The method is similar to that of the previous case, being a combination of hardware and software. The difference is that while microprocessors or microcontrollers are for generic logic functions, a DSP is specialized for numerically based operations such as digital filtering or FFTs. Very powerful DSPs are capable of processing at a speed that matches the sampling frequency of analog data converted into digital form, as required by the Nyquist condition. This feature is normally referred to as real-time processing capability.

- **Use of logic devices or Programmable Logic Devices (PLD)s** These are devices that contain large arrays of logic cells, whose correct interconnection produces various digital functions. The interconnections are digitally programmable, and their configuration is stored in memory. This category of device is called a Field Programmable Gate

Array (FPGA). The acronym is also used to refer to devices with different methods of programming. For example, we have the old PAL, Programmable Array Logic, which uses fuses or anti-fuses to obtain one-time programmability, or the GAL, Generic Array Logic, which can be erased and reprogrammed. FPGAs can be extremely complex. The first devices had fewer than 10 000 equivalent gates, but by early 2000 this had risen to several million. Often the designer's task is simply to describe the whole architecture in a suitable language. The design is then translated into the FPGA configuration by a software tool. The use of programmable devices is either for prototyping or for medium-volume applications.

In summary, electronic circuits for digital design, or hardware, are strictly linked to software development. Often the major effort of the designer is on the software side. However, it is important to be aware, even at a high level, of how digital circuits work, so as to identify optimal solutions to specific signal processing problems.

PROBLEMS

5.1   The input port of an integrated processing block is equivalent to a 10 kΩ resistance in parallel to 0.2 pF. The output port corresponds to a voltage generator with in series 1 kΩ. The input signal is provided by a signal generator with 100 Ω output resistance. The load of the processor is 7.8 kΩ with in parallel 2 pF. What is the ratio between the expected output and the actual amplitude at 330 MHz?

5.2   The trace of a PCB can be modeled by the cascade of four equal RC cells made by $R = 0.1\ \Omega$, $C = 0.2$ pF. With that trace you connect a voltage sine wave generator and an amplifier. The output resistance R_{out} of the generator is 50 Ω. The input resistance R_{out} of the amplifier is 100 Ω. Estimate the signal at input of the amplifier with a sine wave of 1 V at 1 GHz.

5.3   Design the interface that gives rise at the input of the last block in the figure below to a voltage V_{in} equal to $i_s \cdot 10^5$. Realize the interface with only resistances and $VCVS$ or $CCCS$ with unity gain.

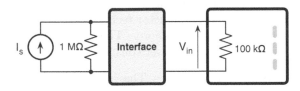

5.4   You have four loudspeakers. Two of them have 8 Ω input impedance, and the other two 16 Ω. You can use two power audio amplifiers, both with 100 W output power (with impedance matching). One of the amplifiers has 8 Ω output impedance, and the other 16 Ω. What is the best interconnection of one (and only one) amplifier with the speakers so as to give rise to the maximum audio power? (Efficiency is not important; what matters is the loudness of the music.)

5.5   The output buffer of a digital integrated circuit must drive a 10 pF load. The output resistance of the buffer is 100 Ω and the maximum current that the buffer can deliver is 200 mA. Determine the time required to charge the load capacitance at 50% and 75%

of the logic transition. The supply voltage is 0–1.2 V. How long does the output buffer stay in the slewing condition?

5.6 The equivalent circuit of an output stage is a current generator I_o and, in parallel, 100 Ω and 5 pF. The spectrum of the current signal I_o spans from 10 MHz to 120 MHz. Determine the value of the AC coupling capacitor, C_x, that connects the output with a grounded load resistance $R_L = 100$ kΩ. The maximum signal loss across R_L must be 1 dB in the entire signal frequency range.

5.7 A single-ended processor generates the required output voltage V_{out} but adds a random noise $V_{n,s}$. Remember that the random noise voltage is the square root of the integral of the power spectrum $v_{n,s}^2$ over the signal band. The noise causes 89 dB SNR. The processor is transformed into differential form with differential outputs given by

$$V_{out,dp} = V_{out} + V_{n,s} + V_{d,p}$$
$$V_{out,dn} = -V_{out} + V_{n,s} - V_{d,n},$$

where $V_{d,p}$ and $V_{d,n}$ are generated by noise spectra equal to that of the single-ended noise $v_{n,s}$. Estimate the SNR of the fully differential signal. Repeat the estimations, assuming that the signal band doubles and all the noise spectra are white.

5.8 Draft the functional diagram of at least two processors that generate each of the transfer functions below. Use adders, subtractors, multipliers by a constant, s, and $1/s$ as building blocks.

$$Y(t) = \int \left\{ \left[\int X_1 \, dt \right] + 3X_2 - 2\frac{dX_3}{dt} \right\} dt$$

$$Y(t) = \int \left\{ \int [3X_1 + 4X_2] \, dt - 2\frac{dX_3^2}{dt^2} \right\} dt$$

determine in the time and the s-domain the relationship between the inputs and the intermediate nodes of the architecture designed. Calculate the signal amplitude in every node at $f = 0.01$. Change the design so that the frequency response scales according the rule $f \to kf$.

5.9 An electronic circuit realizes a virtual short circuit between nodes A and B of the figure below. The voltages of A and B are equal, but there is no current flowing through the virtual short connection. The operation of the black boxes sustains the virtual short circuit. Determine the output voltage.

(a)   (b)

5.10 Repeat Problem (5.9), but suppose that the black box is not able to give rise to a perfect virtual short circuit. The operation causes a fixed difference (or offset) by V_{os} between node A and node B.

5.11   Repeat Problem 5.9 with a capacitor C_2 replacing the resistance R_2 and the parallel connection of R_0 and C_0 between virtual short circuit and output. Determine the output voltage and the currents in the s-domain. Suppose that the capacitor C_2 is initially charged at V_3. What is the effect of V_3 on the output voltage? Estimate the current through C_2 with two sine waves at the frequencies f_1 and f_2 at inputs. What is the difference caused by the initial charge?

5.12   Consider again the schemes of Problem 5.9, and assume a parasitic capacitance equal to $C_0/25$ from node A and ground, from node B and ground, and from node A and node B. Determine the output voltages.

5.13   A real electronic circuit is not able to give rise to a perfect virtual ground. The operation of the black box of the scheme shown in (a) below can be described by the equivalent circuit shown in (b). The resistance R_0 is divided into two parts, R_0' and R_0''; $R_0 = R_0' + R_0''$. The black box realizes the virtual ground at the junction A' instead of A. Determine the output voltage in the ideal case ($R_0' = 0$) and when $R_0' = R_0''/500$.

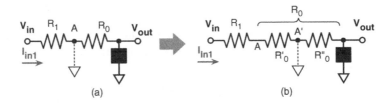

(a)   (b)

5.14   Repeat Problem 5.13 but account for a parasitic capacitance C_p from node A and ground. Determine the amplitude and phase errors for various value of k being $R_0' = R_0''/k$. Plot the Bode diagram of the output voltage and estimate the value of C_p that gives rise to the first pole at 100 MHz. Use the following numerical values: $R_0 = 40 \cdot R_1 = 867$ kΩ, $k = 100, 500, 1000$.

5.15   Determine the transfer functions of the schemes in the diagrams below. The black boxes give rise to zero voltage (virtual ground) at the joining points. What are the input currents? Suppose that R_0 changes by $\pm 20\%$. Estimate the values of other circuit elements that mantain the two transfer functions unchanged.

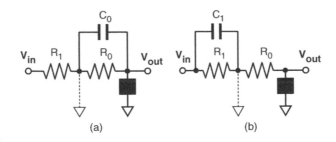

(a)   (b)

5.16   Use the circuit diagrams of Problem 5.15 and suppose that the inaccuracies of all components are random, with the relative variances of resistors and capacitors given by

$$\sigma_R^2 = \frac{\delta R^2}{R^2} = 0.043, \quad \sigma_C^2 = \frac{\delta C^2}{C^2} = 0.027.$$

Determine the variance of the frequency gain and the pole (or zero) of the transfer functions.

5.17   Sketch a generic bipolar (positive and negative values) waveform applied at the input of a comparator with zero threshold. Determine the output signal. Include noise (just roughness added to the waveform) and evaluate the period of time during which the comparator bounces back and forth. Estimate the extent of the comparator hysteresis that avoids bouncing.

5.18   One of the inputs of an analog multiplier is the superposition of two sine waves with amplitudes 0.3 V and 0.58 V, and frequencies 234 MHz and 432 MHz, respectively. The other input is a square wave with amplitudes switching from 0 V to 1 V and frequency 244 MHz. What is the spectrum at the output of the multiplier? Where is the tone at lower frequency? What happens if the second input becomes a sine wave at 333 MHz?

5.19   Draft the circuit diagram of a three-bit multiplexer and a three-bit demultiplexer using analog switches. The input data is the parallel representation of two-bit words. Plot possible waveforms for input and output and those used for the control of the switches.

CHAPTER 6

ANALOG PROCESSING BLOCKS

This chapter studies key components used in linear and non-linear analog processing. These are active parts to be interconnected with passive components such as resistors, capacitors, and, rarely, inductors. We shall study the relevant features of these analog elements so as to learn how to use them for implementing various functions, such as amplification, addition, subtraction, and modification of spectral features. We shall also consider non-linear basic processing blocks. The chapter focuses on the properties, limits, and use of analog building blocks but does not go down to the component level. We remain at the level of black boxes.

6.1 INTRODUCTION

Only a few years ago, analog circuits were made up of many discrete elements: many resistors and capacitors and a few inductors and transistors, assembled on a large, single-layer, PCB. Since then the goal of continuous miniaturization and the demand for increased effectiveness have reduced the analog system to the interconnection of functional blocks and a small number of passive elements.

Another limitation of past solutions was that a large number of components required many solder joints. Also, error in component placement is more likely when a large number of elements are used. All of those weaknesses worsen the yield. Furthermore, since PCBs were

Understanding Microelectronics: A Top-Down Approach, First Edition. Franco Maloberti.
© 2012 John Wiley & Sons, Ltd. Published 2012 by John Wiley & Sons, Ltd.

too large, it was necessary to change from single- to double-sided boards and to accommodate components on both sides, which was made possible by new assembly methods.

The desire to increase the yield, shorten production time, and reduce costs moved electronic systems towards integration and volume reduction. On the assembly front, modern PCB fabrication techniques make multiple connective layers with dedicated ground plane and power plane possible. The thickness and separation of copper lines match the needs of the designer; the so-called strip lines method routes high-frequency signals, and shielding layers protect sensitive nodes. Placement of components and soldering is done automatically by passing the board over a ripple, or wave, of molten solder in automatic soldering machines.

Continuous improvement in fabrication of ICs, which can now house an entire system on a single integrated circuit (SoC, System-on-Chip) or in a single package (SiP, System-in-Package), and new assembly technologies produce small PCBs with high yields and low cost. These boards accommodate a few functional analog and digital blocks, already tested and qualified.

Important analog blocks are the Operational Amplifier (often called, for short, the *op-amp*), the Operational Transconductance Amplifier, or OTA, and the comparator. These functional blocks are available as discrete parts, single or multiple elements, housed in packages with a suitable number of pins. They provide input and output ports used to interconnect the function to the rest of the circuit and to ensure biasing or functional controls.

In some cases analog blocks are fabricated together with other functions as part of a large integrated system. To enable modular design they are treated as distinct parts with their own inputs, outputs, and bias terminals. The schemes are predefined and functionally verified by prefabrication of the part and its experimental test and characterization. Then the analog blocks are made available in a library database. Blocks of this kind, often called standard cells, satisfy most design needs. Only in demanding cases, when library cells are not completely adequate, does the system designer develop ad hoc cells.

With discrete parts or standard cells it is essential to choose the right solution and verify its correspondence with what is needed. The necessary information is given in data sheets in the form of numbers, tables, or diagrams. The tables typically provide the range of parameter variations for nominal, minimum, and maximum conditions. The diagrams plot changes in performance with supply voltage, temperature, signal amplitude, frequency, and other quantities. In addition to data sheets the designer uses application notes, which provide possible solutions to typical design problems, the architecture of corresponding circuits, and recommendations on the parts and their optimal use in the system.

Data sheet

Carefully read the data sheet of whatever you plan to use in your electronic system. Data sheets provide much information to assist in proper use. Numbers, diagrams, and written notes greatly help the designer in avoiding mistakes and misunderstandings.

Often, the data sheets have legal value: the supplier guarantees the customer that the purchased part ensures the stated performance within given limits, if the part is used as specified. The user relies on these numbers for the design, and, when needed, takes action to compensate for possible non-idealities. If the part does not meet the specifications, the entire system does not work in the expected manner, causing possible economic damage to the customer. In some cases, after purchasing the parts, the user wants to verify, for quality assurance, that performances correspond to what is expected and guaranteed. The data sheet provides information on how to perform verification measurements, indicates the procedures to follow, and recommends operational conditions.

Self training

Download from the Web the data sheets of two operational amplifiers and two comparators fabricated by two different suppliers. Look at the organization of these documents and make notes on:

- the type of package and mechanical description;

- the key features and maximum ratings;

- the type of static and dynamic characteristics;

- the kind of information given in tabular or graphical format and the scales used;

- recommendations, if any, for assembling the part on a PCB.

Compare the documents and outline the possible different strategies in providing information. Write a short note on the differences.

6.2 CHOOSING THE PART

Selecting the right parts is important for designing an effective system. The choice requires a knowledge, at least at the functional level, of the circuit topology of the blocks and their operation, but it also demands a thorough knowledge of the overall system specifications.

Parametric catalogs facilitate the selection of analog blocks. Many parametric catalogs, available on the websites of integrated circuit providers, start with a classification of products on the basis of functional and key performances. For instance, operational amplifiers are categorized as high speed, instrumentation, audio, video, programmable, and so forth. Then, step by step, the relevant features of each category are better depicted. Instrumentation amplifiers, for example, can be divided into low power, low noise, high voltage, high output current, and other. Elements that can be relevant for the choice are the availability of samples

Self training

Search on the Web for a parametric catalog of operational amplifiers or comparators, and list the categories used. Select one category, and expand the menu to examine key performances. Become familiar with the use of parametric catalogs by making a few trial selections. Download the data sheet of one part and the application notes provided.

Let us suppose we are interested in a part for low power and 10 MHz speed. Compare possible alternatives for these assumed needs, and understand the limitations and benefits of the possible choices.

Write a short report on your search.

for experimental verifications or of macro-models for computer simulations. The package is also important. Size and pin-configuration assist in estimating overall system dimensions. For portable applications volume is important, and it is often essential to have micro-packages or to have devices available as a bare die.

6.3   OPERATIONAL AMPLIFIER

The most important analog block is the operational amplifier. Its name indicates its main functions. Since the block performs operations, the first word is "operational." Since the scheme achieves amplification, the second word is "amplifier." Operational amplifiers, or op-amps for short, are designed using bipolar, Complementary MOS (CMOS), or BiCMOS (with bipolar and CMOS transistors on the same chip) technologies. The supply voltage ranges from small (1 V or less) to large (12 V or more) values, depending on the expected use but also constrained by technology. Normally, bipolar or BiCMOS technologies sustain higher voltages than does CMOS. The power consumed mainly depends on the speed. If speed is high, power is also high. A typical op-amp consumes a fraction of a mW (or even less) at low speed (a few MHz); at very high speeds power can go up to several tens of mW. Since op-amps for portable applications impose a careful use of power, it is necessary to envisage low-speed analog processing, for optimal use of power. Often an op-amp for portable systems consumes a few µW.

The circuit scheme of an op-amp is mainly an interconnection of transistors. A few capacitors and resistors can be part of the scheme. The circuit complexity critically depends on the gain needed. Simple schemes obtain low gain; high gain requires complicated architectures. Another relevant parameter is noise. This depends on the current flowing in transistors and on their size. Normally obtaining low noise costs power and silicon area.

6.3.1   Ideal Operation

Ideally, an op-amp is described as a voltage-controlled voltage amplifier with an infinite differential gain. The relevant terminals of the block are the differential inputs and output. It is supposed that the frequency of operation of an ideal op-amp is unlimited (infinite band). Moreover, the input and output interfaces are ideal: input impedance is infinite and output impedance is zero. In summary, the relevant features of ideal operational amplifiers are:

- infinite differential voltage gain;

- infinite bandwidth;

- infinite input impedance;

- zero output impedance.

Figure 6.1(a) shows the op-amp symbol: a triangle with two terminals at the input and one at the output. Input and output voltages are measured with respect to a reference level, not shown in the symbol. The voltage difference between input terminals is the differential input. Since the gain is ideally infinite, a finite output voltage requires a zero (or infinitesimal) differential input. This, obviously, cannot be verified in practical situations. However, with very large gains the differential input is very low.

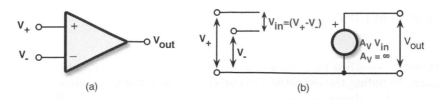

Figure 6.1 (a) Symbol for an operational amplifier; (b) equivalent circuit of an ideal operational amplifier.

The symbol shown in Figure 6.1(a) just uses the terminals that are necessary to describe the behavioral operation. Real op-amps allow for additional terminals used to connect the part to the supply generators and to the reference level; in some cases, also, extra terminals are used to adjust important features, such as the offset.

As with all electronic circuits, it is customary to represent a block with its equivalent scheme. Figure 6.1(b) shows the equivalent circuit of an ideal op-amp consisting of input and output terminals and a voltage source. The impedances between inputs and common reference, as well as the impedance between differential inputs, are infinite (open circuits). The output impedance is zero. The voltage source is a Voltage-Controlled Voltage Source (VCVS) whose output is

$$V_{out} = A_V(V_+ - V_-),\qquad(6.1)$$

with $A_V = \infty$. Notice that the names of voltages are given with capital letters. This, conventionally, shows a large signal operation; i.e., it supposes that the amplitude of the input signals does not affect the linearity of the response. Real circuits do not provide that feature, because at given input levels the output voltage saturates.

Since the output depends on the differential input, it is worth representing the inputs as

$$V_+ = V_{CM} + \frac{V_d}{2}\qquad(6.2)$$

$$V_- = V_{CM} - \frac{V_d}{2},\qquad(6.3)$$

which defines V_{CM}, the common-mode voltage: i.e., the mean value of the inputs. V_d, the differential input, is a symmetrical signal added to the common-mode level. Notice that, since the output depends on V_d only, the value of V_{CM} is irrelevant, whether it is constant or changes in time. Therefore, we can state that the common-mode gain, defined by $A_{CM} = V_{out}/V_{CM}$, is zero. Obviously, that is for ideal cases. We shall see that the output of real op-amps is slightly influenced by the common-mode input level.

Describing an op-amp as an ideal block is useful for initial and high-level studies to verify the signal flow and determine the values of the passive components needed. After that, more precise estimations are necessary, and detailed descriptions of the op-amp must be used. The effect of non-idealities can be studied at the transistor and passive components level or, more efficiently, by the use of behavioral models based on linear and non-linear equations or equivalent circuits. Obviously, the reliability of results depends on the accuracy of models. In turn, complicated models require considerable computation effort. Therefore, very large systems require a suitable trade-off between accuracy of results and simulation time.

6.4   OP-AMP DESCRIPTION

A real op-amp is specified by parameters (or specifications) that account for non-idealities and limitations, including the expected inaccuracies occurring during fabrication. These parameters, listed in data sheets, provide a common understanding of the circuit's behavior. The parameters and data sheets of a certain part mainly highlight the features that are relevant for that part. For instance, the data sheet of a high-frequency op-amp describes in detail its frequency response and speed parameters. On the other hand, an operational amplifier for instrumentation focuses on high gain and low offset.

The numerical values given in the data sheets are not precise figures. They result from extensive measuring of different samples that gives rise to an average and statistical distribution, typically Gaussian. The value reported on the data sheet is usually the average. When the expected average is zero, the data sheet provides the variance, σ. Remember that if the variance of a Gaussian distribution is σ, this means that 68.3% of the devices have an error within the $\pm\sigma$ limits, 95.4% within $\pm 2\sigma$ and 99.7% within $\pm 3\sigma$. Wider ranges around the average lead to higher percentages of samples.

Figure 6.2 shows the statistical offset distribution of a hypothetical op-amp. This histogram denotes very good numbers, typical of a precise operational amplifier.

In addition to typical values, some data sheets also give a minimum and a maximum, defining a range guaranteed by the supplier. This can be ensured by testing and selecting parts before delivery. In many cases, the minimum and the maximum are the 6σ limits; the fraction of samples that fall outside the 6σ limit is extremely small – zero, in practice.

The organization of a data sheet is in sections. Normally we have four sections: the general description of the part, tables giving absolute maximum ratings, tables and diagrams showing electrical characteristics, and information on packaging and board assembly.

6.4.1   General Description

The general description, as the name implies, provides a first introduction to the circuit and a quick glance at its key characteristics. Often the general description is the first page of the data sheet, with a list of the relevant features and suitable applications.

In addition, the first page can include the scheme of a significant application, including details on interconnections with passive components and their values.

6.4.2   Absolute Maximum Ratings and Operating Rating

The absolute maximum ratings give operating conditions that can impair the device. The given values hold under any conditions of use, either for service or for testing. Maximum ratings concern, for instance, the supply voltage and the voltage at input terminals that can cause aging of the circuit. These rates are lower than the ones corresponding to the over-biasing limits (voltages or currents) that give rise to catastrophic consequences. Observe that these maximum rating limits are non-disruptive but must be avoided anyway, because they degrade the device. When they are exceeded the specifications are not met and so are no longer guaranteed.

Table 6.1 gives an example of absolute maximum ratings. The maximum supply voltage is 3.6 V, twice the nominal supply, which is 1.8 V. Obviously the input voltages are limited

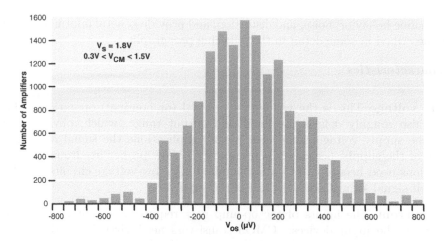

Figure 6.2   Statistical distribution of the offset of an op-amp for precise applications.

Table 6.1   Absolute maximum ratings – example

Parameter	Rating
Supply voltage	3.6 V
Input voltage	$V_{SS} - 0.4$ V to $V_{DD} + 0.4$ V
Differential input voltage	± 3.6 V
Output short-circuit duration to GND	indefinite
Storage temperature range	-65°C to $+150^\circ$C
Operating temperature range	-40°C to $+125^\circ$C
Junction temperature range	-65°C to $+150^\circ$C
Lead temperature (soldering, 60 seconds)	300°C
Power dissipation	800 mW

within similar boundaries. The maximum temperatures are given for normal use, storage, and soldering of lead to the board. These are recommended because high temperatures cause a diffusion of dopant and a consequent degradation of junctions. This, as will be studied in a later chapter, changes the electrical behavior of transistors and consequently alters op-amp performance. Even if the output short circuit can be indefinite, the power dissipation must be under control because excessive power can cause self heating and high temperatures. On the other hand, too low temperatures can freeze the package and cause possible micro-cracks. Afterwards, humidity and other corroding elements may enter through the cracks and deteriorate the silicon protecting layers and the wire connections.

Some data sheets list the conditions of operation for which the device is intended to be functional. These conditions exceed the normal expected settings; therefore specified performances are not guaranteed, but the device is not influenced. They are just operational conditions that do not impair the device permanently.

6.4.3   Electrical Characteristics

The electrical characteristics give information about the electrical behavior and use of the device in circuits and systems. They are shown using tables or diagrams. They concern the static and dynamic behavior, noise, and distortion, and provide general information. The most relevant are described in some detail in the following paragraphs.

Static characteristics

- **Supply voltage** This is the voltage for optimal (or nominal) operation. The supply voltage also roughly defines the input and output range, which, obviously, must be within the supply voltage range. For precise applications the signal range must be a fraction of the supply, because it is necessary to allow proper headroom. Since precise applications need headroom of several hundred mV, low-voltage circuits (1.2 V or less) can admit signal swings as low as half of the supply.

- **Input current** The inputs of the op-amp may require a certain bias current. This depends on the input devices. CMOS transistors need virtually zero input currents (nanometer technologies have a small but not completely negligible input current); bipolar elements require some input current, whose value depends on schematic and operational conditions. The input bias current can be of concern when the input source has very high impedance, because that current causes undesired voltage drops.

- **Input offset voltage, V_{os}** This is the voltage between inverting and non-inverting input necessary to balance mismatches associated with asymmetries and fabrication errors. It is represented by a voltage source, V_{os}, in series with one of the input terminals. Figure 6.3 accounts for it. Since the differential voltage at the input of the ideal op-amp is

$$V_d = V_+ - V_- = V_{in+} - V_{in-} - V_{os} = V_{d,in} - V_{os}, \qquad (6.4)$$

the output becomes zero for $V_d = 0$. Therefore, the offset is also the voltage across the effective input terminals V_{in-} and V_{in+} needed to compensate for mismatches and fabrication errors.

Some operational amplifiers provide for electrical control of the offset. The adjustment is with an external potentiometer that regulates the DC voltage of a pin, whose value is used internally to adjust the offset.

- **Input offset drift** This is a parameter that shows how much the offset changes with temperature. A possible value is 10 µV/°C. If the circuit has to operate in a 100°C temperature interval, the input offset changes by 1 mV over the temperature range of operation.

- **Large-signal voltage gain, A_{V_0}** The large-signal gain derives from the input–output transfer characteristic, a curve that plots the output voltage versus the differential DC signal. A typical curve is a straight high slope line around $V_d = 0$ that bends when output voltage gets close to the supply limits. Figure 6.4(a) is an example of input–output transfer characteristics. The large signal gain is defined as the ratio between the output and input differential voltages,

$$A_{V_0} = \frac{V_{out}}{V_d}. \qquad (6.5)$$

The value of A_{V_0} is large with small inputs but drops when the input level brings the output near to saturation level. Therefore, as shown in Figure 6.4(b), the large signal

COMPUTER EXPERIMENT 6.1

Op-Amp Offset, Gain, and PSRR

There are many specifications that feature a real operational amplifier. With this computer experiment you can analyze the input–output transfer characteristic, estimate the offset, and observe that the output also depends on signal at input as well as on spurs corrupting the supply. The use of a sawtooth signal at one of the input terminals estimates the input–output transfer curve, if the other terminal is at the average level of the sawtooth. The sawtooth value, for which the output equals a defined reference, gives the offset. The use of a sine wave (with very small amplitude) at input or added to the supply voltage gives rise to A_V and PSRR. For that you can:

- set the DC level of the sawtooth signal and its peak-to-peak amplitude. Remember that the frequency must be very low when determining "static" parameters;
- set the voltage of the non-inverting terminal and that of the positive supply. Notice that the input common mode should be around half of the positive supply;
- define amplitude and frequency of the sine wave generator and its connection.

MEASUREMENT SET–UP

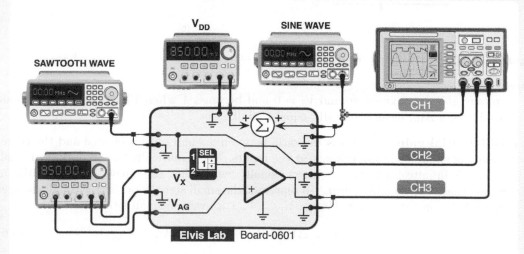

TO DO

- Select the sawtooth as input signal and set the DC levels of inverting and non–inverting terminals properly. Use a small amplitude and low frequency to determine the input–output transfer response. Assume that the nominal supply voltage of the op-amp is 1.8 V and that the output saturates 0.3 V near to the supply rails.
- Use a small sine wave at the input, adjust the offset and measure the small signal gain at various frequencies. Plot the Bode diagram of the module (use a logarithmic frequency axis).
- Add the sine wave signal to the supply voltage and determine the power supply gain. Use the result to draw the plot in dB of the PSRR with logarithmic frequency axis.
- Increase the amplitude of the sine wave sent to the input terminal and slightly change the offset correction. Observe the output voltage.

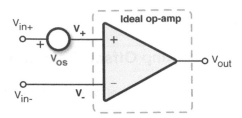

Figure 6.3   Equivalent circuit of operational amplifier with offset.

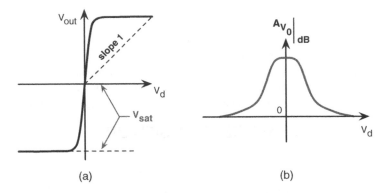

Figure 6.4   (a) Input–output transfer characteristic of an op-amp; (b) large-signal voltage gain.

voltage gain is maximal around $V_d = 0$ and becomes 1 when $V_d = \pm V_{sat}$. This input level makes A_{V_0}, measured in dB, equal to 0 dB.

- **Differential gain, A_d** This is the ratio between a small output change and the corresponding small differential change at the inputs. Mathematically the amplitude of small signals is infinitesimal. For engineers "small" means that the output changes within a region that is reasonably linear. To represent the small signals, or, better, small changes around the operating points $(V_{+,op}, V_{-,op}, V_{out,op})$ we use small letters:

$$v_d = v_+ - v_- = \Delta V_d = (V_+ - V_{+,op}) - (V_- - V_{-,op}) \tag{6.6}$$

$$v_{out} = \Delta V_{out} = V_{out} - V_{out,op}. \tag{6.7}$$

The differential gain is

$$A_d = \frac{v_{out}}{v_d} = \frac{\Delta V_{out}}{\Delta V_d}, \tag{6.8}$$

which, for an infinitesimal differential signal, becomes the derivative of the input–output transfer characteristic.

For an ideal op-amp A_d is infinite. For a real scheme the differential gain is very large but not infinite. Indeed, the gain needed by a system is never infinite but depends on the accuracy required. It can be verified that, at first approximation, the error caused by non-infinite gain is inversely proportional to the gain itself. Therefore some applications can admit 80 dB of gain or even less. On the other hand, precise applications, like the large amplification of tiny signals generated by sensors, need op-amps with very high gain, 120 dB or even more.

Ideally the differential gain does not depend on the common-mode input. This is not completely verified by real circuits; for many of them it is enough that the input common mode is within a suitable range. When inputs get too close to supplies, operation can be affected and the gain drops. When the common-mode input goes too close to the supply, it is necessary to use a special kind of op-amp: one capable of rail-to-rail input.

Notice that small signal gain (as well as large signal gain) is a DC property, being obtained from the input–output static characteristics. We shall see shortly that gain depends on frequency. Typically gain diminishes with frequency because of the limited speed capabilities of the basic components used. The effect is described by specifications to be defined shortly.

- **Common-mode gain, A_{cm}** The output of the operational amplifier changes when the common-mode voltage varies. Common-mode gain is the ratio of output voltage to the common-mode input. Accounting for both differential, A_d, and common-mode, A_{cm}, gain, the output voltage is

$$v_{out} = A_d v_d + A_{cm} v_{cm}. \tag{6.9}$$

- **Common-mode rejection ratio (CMRR)** This defines the op-amp's capability to distinguish between differential signals and their common-mode component. It is the ratio

$$\text{CMRR} = \frac{A_d}{A_{cm}}. \tag{6.10}$$

The value of the CMRR depends on the op-amp architecture and frequency. Careful design can obtain a CMRR as good as 80–100 dB at low frequency. However, as the frequency increases, the CMRR drops significantly.

- **Total power dissipation, P_D** This is the total DC power supplied to the device minus any power delivered from the device to a load. Without load it is the product of the supply voltage and the DC current.

Observe

Differential gain, common-mode gain and CMRR are defined for small signals. They assume that the input signal is small enough to keep the output swing in a linear region.

Dynamic characteristics

The specification of small signal parameters can extend from static to frequency dependent. Small signal differential gain is also regarded as a dynamic quantity, $A_d(s)$. Its amplitude and phase are described by Bode diagrams like the ones shown in Figure 6.5, which are typical plots of a good op-amp. The low-frequency gain is almost 100 dB. Phase at low frequency is not zero but 180°, because it is customary to represent $-A_d(s)$ to account for an extra 180° phase shift. The initial phase in the figure is slightly less than 180° because the frequency axis begins at 100 Hz, not far from the dominant pole that already caused some gain drop and phase shift. Then the gain plot line rolls off by 20 dB/dec. At the pole frequency the phase shifts by an extra 45°, so that it is 135°. Far away from the first pole the phase becomes 90°. The first pole is largely separated from the second one, which occurs at more than 100 MHz. Therefore, it dominates the Bode diagram plot lines. In fact, the amplitude diagram rolls off by 20 dB/dec

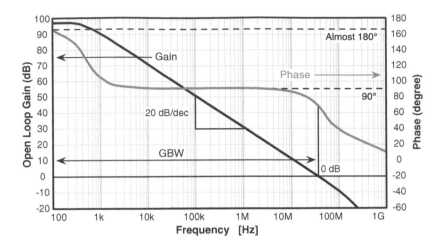

Figure 6.5 Frequency response of an op-amp: modulus and phase Bode diagrams.

with unchanged pace, and the phase remains constant at 90°. For this reason the first pole is called the *dominant pole* of the frequency response. When the frequency approaches the second pole (called the non-dominant one) the modulus diagram starts to increase its bending, and this becomes a 40 dB/dec well after the second pole. The phase diminishes again and reaches 0° at high frequency. Possible additional non-dominant poles (assumed to be in the left-hand half of the s plane) augment the roll-off and diminish the phase.

Bode diagrams like the ones of Figure 6.5, which correspond to op-amps for precise use with high gain and relatively low speed, are useful to explain the definitions of some of the following dynamic parameters.

- **Gain bandwidth product (GBW)** This is the product of the low-frequency gain and the frequency of the first or dominant pole. Typically the non-dominant pole is at high frequency, and the Bode diagram rolls off by 20 dB/dec until and above the GBW.

- **Unity gain frequency, f_T** This is the frequency at which the gain crosses the unity gain (0 dB) axis. To ensure stability, it normally requires a -20 dB/dec slope until it crosses the 0 dB axis, which means having the non-dominant pole after the crossing. Therefore, if significant extra bending caused by the non-dominant pole occurs after the crossing then f_T almost equals GBW.

- **Phase margin, Φ_m** This is the phase of $A_d(f)$ at f_T. Usually a phase margin of at least 60° is desirable. In order to obtain the condition it is necessary to have at f_T a mere 30° phase shift at all the non-dominant poles. If the response has just one non-dominant pole its frequency must be higher than $1.73\, f_T$.

- **Slew-rate** This feature describes the driving capability of the op-amp. It has already been discussed in an earlier chapter that many active circuits have a maximum deliverable current. This also holds for op-amps, and their maximum current in given operating conditions determines the maximum change in time of the output voltage $\frac{\Delta V_{out}}{\Delta t}|_{max}$. The slew-rate is typically measured in V/μsec. Figure 6.6 shows the step response of an op-amp that jumps from -1 V to 1 V. The plot line of the real response performs the

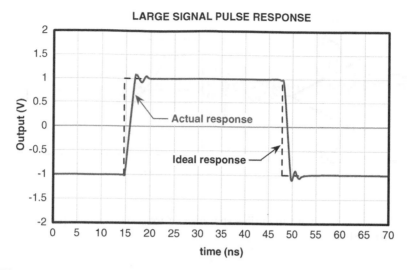

Figure 6.6 Typical large-signal step response of a fast operational amplifier.

2 V transition (including a ringing transient) in less than 1 ns. Therefore that op-amp has a good slew-rate: more than 3000 V/μsec.

- **Rise (or fall) time, T_r (or T_f)** This is the time required for an output voltage step to change (going up or down) between 10% and 90% of its final value. These parameters are obviously related to the slew-rate of the op-amp.

- **Power supply rejection ratio (PSRR)** Any undesired signals affecting the power supply inevitably appear at output. The power supply rejection ratio denotes the ability to reject those undesired signals. Figure 6.7(a) describes possible spurs on the supply lines with two signal generators, v_{ps+} and v_{ps-}, in series with the supply connections. The output is

$$v_{out} = v_{ps+}A_{ps+} + v_{ps-}A_{ps-}, \tag{6.11}$$

which defines two PSRRs. They are the ratio between differential gain and power supply gains, A_{ps+} or A_{ps-} respectively:

$$\mathrm{PSRR}_+ = \frac{A_d}{A_{ps+}}, \quad \mathrm{PSRR}_- = \frac{A_d}{A_{ps-}}. \tag{6.12}$$

Figure 6.7(b) shows a typical PSRR plot. The rejection is good at low frequency but at high frequency the ability to block spurs becomes weaker and weaker. Having a poor high-frequency PSRR is often a problem at the system level. Its solution requires special care in design and always demands clean supply voltages over a wide frequency range. A circuit method capable of limiting the injection of spurs from the power supply is to process signals using a fully differential scheme.

Noise and distortion response

The quality of a signal processor depends on its ability to distinguish between useful signal and undesired components. The unwanted spurs are internally generated noise (whose wide-band

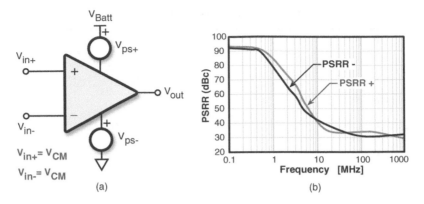

Figure 6.7   (a) Equivalent circuit to outline spur coming from the power supply; (b) typical plot of positive and negative PSRRs.

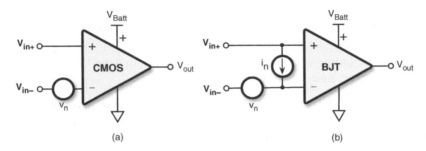

Figure 6.8   (a) Equivalent noise source for a CMOS op-amp; (b) equivalent noise source for a Bipolar Junction Transistor (BJT) or a JFET op-amp.

spectrum corrupts the signal spectrum), harmonic components caused by non-linear terms, and tones determined by interference. Some parameters specified in data sheets feature these limitations. The most relevant of them are the following.

■ **Input noise generators** One or two noise generators connected at input of the op-amp describe the overall noise of internal active and passive devices. For CMOS operational amplifiers a single voltage source is enough to represent noise. Bipolar or JFET op-amps need two generators, one producing voltage and one delivering current, as shown in Figure 6.8. A $1/f$ region that vanishes at the corner frequency f_c features the spectra. After the corner frequency the spectrum is white because white contributions become dominant. The data sheet gives only the noise voltage or current in the white region together with the corner frequency value. For example, a typical low-noise, medium-power part can show the following noise spectra:

$$\text{Input noise voltage} \quad f \geq 2 \text{ KHz} \quad 1.82 \text{ nV}/\sqrt{\text{Hz}}$$
$$\text{Input noise current} \quad f \geq 2 \text{ KHz} \quad 3.1 \text{ pA}/\sqrt{\text{Hz}}.$$

That example corresponds to a relatively low value of corner frequency. Typically it is in the 1–100 kHz range.

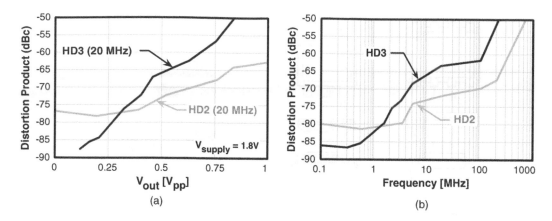

Figure 6.9   (a) HD2 and HD3 of a hypothetical op-amp versus the signal amplitude; (b) harmonic terms for a fixed input.

The noise voltage (or noise current) results from the noise spectra. It is the square root of the integral of the spectrum over a defined frequency interval.

- **Second harmonic distortion (HD2)** This parameter and the HD3 that follows serve when the op-amp is intended for high linearity use. Let us suppose we have a defined configuration with an input signal at a specified frequency, and measure the output amplitude at twice the input frequency. The HD2 is the ratio between amplitudes of the second harmonic and the fundamental. Normally the op-amp configuration produces a given gain, say 1 or 10. The harmonic distortion increases with frequency; the data sheet, normally, gives it for a defined amplitude and two signal frequencies. The information given looks like:

$$\text{Second harmonic distortion} \quad V_{out} = 1\ V_{pp} \quad f = 20\ \text{MHz} \quad \text{HD2} = -86\ \text{dB}_c$$
$$\text{Second harmonic distortion} \quad V_{out} = 1\ V_{pp} \quad f = 70\ \text{MHz} \quad \text{HD2} = -68\ \text{dB}_c.$$

- **Third harmonic distortion (HD3)** The third harmonic is also caused by non-linearity. The parameter is provided in conjunction with the HD2, assuming the same configuration and gain. The levels of distortion of third and second harmonics are normally much larger than those for other harmonic terms. Figure 6.9 shows the plot of HD2 and HD3 of a hypothetical op-amp. The performances, given in dBc, are as a function of input signal amplitude and frequency. Notice that HD3 is better than HD2 over a given range of amplitude and frequency. The figure shows that with a supply voltage of 1.8 V the distortion products become significant for an input amplitude of just half the supply.

Miscellaneous specifications

This data-sheet section provides various information such as the following.

- **Rail-to-rail operation** Input rail-to-rail means that the input terminals can have any value within the supply voltage interval. Therefore, it allows any input common-mode voltage, V_{CM}. Output rail-to-rail is the same for the output voltage. In real cases this feature means that the required headroom voltages are very small.

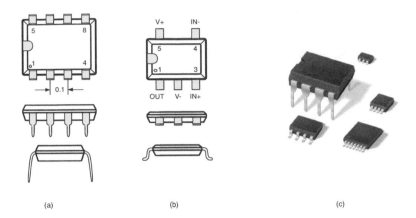

Figure 6.10   Typical packages used for operational amplifiers: (a) an eight-pin package for through-hole mounting; (b) a five-pin package for surface mounting; (c) various plastic packages.

- **Shutdown mode** This is a specification that indicates the possibility of switching the op-amp off with a logic signal. This function places the output node of the amplifier in a high impedance state. Moreover, the quiescent current goes to zero or becomes very low.

- **Wake-up time, t_{ON}** The power-on of any electronic circuit is not immediately followed by full operation. Often it is necessary to wait for some time to charge some capacitance or to settle some reference voltage. The wake-up time is the time necessary, after the op-amp is turned on, to obtain full operation.

6.4.4   Packaging and Board Assembly

This section of the data sheet provides information on package, physical dimensions, and pin configurations. Some integrated circuits house two or four op-amps on a single package. They share the supply terminals to allow (at lower cost) functions that need more than a single op-amp.

The same part can be available with different packaging options. Assembly can be through a hole or alternatively by surface mounting. The through-hole solution facilitates manual positioning on the board and provides high power dissipation, but parasitic capacitances are large. Accordingly, through-hole packages are suitable for low speed and relatively high power consumption. Chips for surface mounting have very little lead or no lead at all. They are often recommended when parasitic capacitances are a concern. Depending on needs, the designer can choose, for example, an eight-lead Plastic Dual In-Line (PDIL) or a surface-mount package as shown in Figure 6.10(a) and (b). Figure 6.10(c) illustrates various plastic packages.

The assembly of the op-amp on the PCB together with other active and passive components is important. Optimal performances of architectures depend on the board layout and component selection. In the light of this, the specifications provide suggestions on how to design the PCB. It should have a low-inductance ground plane and well-bypassed supply lines. Critical traces must be short. Critical components and supply bypass capacitors should be placed very near the amplifier. These and other recommendations are also given in application notes.

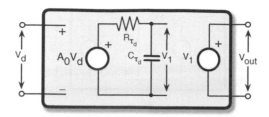

Figure 6.11   Equivalent circuits of op-amp with dominant pole.

6.4.5 Small-signal Equivalent Circuit

The small-signal operation of a circuit that includes an operational amplifier is approximately derived by using op-amp models. Finite gain and the position of poles of small-signal differential gain describe the op-amp's behavior. If the description accounts for the dominant pole only, the differential gain is

$$A_d = \frac{A_0}{1 + s/\omega_D},\qquad(6.13)$$

with dominant pole at the angular frequency ω_D.

A simple low-pass RC filter realizes the denominator of equation (6.13) provided that the product of resistance and capacitance gives the time constant $\tau_d = 1/\omega_D$. Thus any arbitrary resistor R_{τ_d} and a capacitance C_{τ_d} equal to τ_d/R_{τ_d} realize the denominator.

The equivalent circuit of Figure 6.11 uses two voltage sources; one obtains the amplified differential, and the other generates the output voltage. The first voltage source drives the filter. The second one decouples the output of the filter from the possible op-amp load. Therefore the input and output resistances are infinite and zero respectively.

A more accurate model should include all of the other poles, located at high frequency. Typically they are after the GBW because they must not affect the frequency response before f_T. This, again, is why the high-frequency poles are called non-dominant. If small-signal differential gain has many poles it becomes

$$A(f) = \frac{A_0}{(1 + s\tau_d)(1 + s\tau_2)(1 + s\tau_3)\cdots},\qquad \tau_d \ll \tau_2, \tau_3 \cdots .\qquad(6.14)$$

Normally it is supposed that the op-amp has only one second high-frequency pole (non-dominant). As said, this causes an additional roll-off by 20 dB/dec, making the magnitude slope 40 dB/dec. At the second pole frequency there is an extra phase shift of $-45°$.

Be aware

The equivalent circuit of ideal op-amps is useful only for initial approximate studies. Never rely on results obtained without accounting for non-idealities and fabrication limits.

The equivalent circuit of Figure 6.12 describes the op-amp with two poles. Each of them is real and contributes with a transfer function like an RC network. The scheme uses an intermediate variable V_1 that outlines the first pole term. Supposing that the time constant of the non-dominant pole is τ_2, the values of resistance R_2 and capacitance C_2 must satisfy the condition $R_2C_2 = \tau_2$. The third VCVS of the equivalent circuit decouples the output of the second filter from the output of the op-amp. It provides ideal output zero impedance.

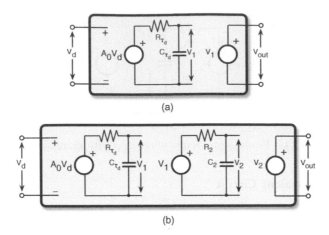

(a)

(b)

Figure 6.12   Equivalent circuits of op-amp with two poles.

For more than two poles the model must expand into further sections, each of them emulating the transfer function of a single pole. Remember that what matters is the product of resistance and capacitance and not their individual values.

Example 6.1

The figure below shows the amplitude Bode response of an op-amp. The non-dominant pole, f_2, and the extrapolated f_T are both at 430 MHz. The low-frequency gain is 92 dB. Determine the position of the dominant pole and estimate the component values of the equivalent circuit for small signals.

Solution
The Bode diagram confirms that a 20 dB/dec roll-off starting from the dominant pole crosses the 0 dB axis at the frequency f_2, the non-dominant pole. Therefore, the separation of the

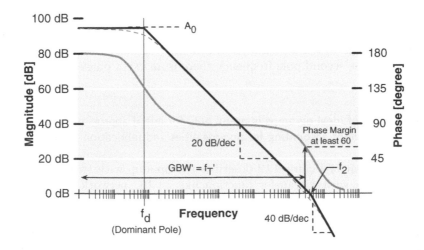

dominant and non-dominant poles is 82/20 decades: $10^{4.6} = 39\,811$. That value, other than being the low-frequency gain, A_0, produces f_d, given by $f_2/A_0 = 10.8$ kHz. Therefore the time constants of the dominant and non-dominant poles become

$$\tau_d = 1/(2\pi \cdot 34.156 \cdot 10^3) = 14.73 \text{ µsec}, \quad \tau_2 = 1/(2\pi \cdot 430 \cdot 10^6) = 370 \text{ psec}. \qquad (6.15)$$

The small-signal differential gain is

$$A_d = \frac{39,811}{(1 + s \cdot 14.73 \cdot 10^{-6})(1 + s \cdot 370 \cdot 10^{-12})}. \qquad (6.16)$$

The equivalent circuit with two poles is provided in Figure 6.12. Assuming we use $R_{\tau_d} = 10$ kΩ and $R_2 = 10$ Ω, the values of the capacitances become

$$C_{\tau_d} = 1.47 \text{ nF}, \quad C_2 = 37 \text{ pF}. \qquad (6.17)$$

6.5 USE OF OPERATIONAL AMPLIFIERS

The key use of operational amplifiers is for realizing virtual ground or obtaining a virtual short circuit between the input terminals. We get these functions because the op-amp has very large differential gain and infinite input impedances. Thanks to the large gain, a finite voltage at output gives a very small voltage (ideally zero) at input. Therefore with finite output the inputs are virtually shorted, with one (the non-inverting one) connected to ground and the other at virtual ground. The infinite input impedance ensures that no current flows into the input terminals, but that it flows through the external connection.

Notice that we obtain virtual ground or a virtual short-circuit only if the output voltage is finite. With outputs infinite or, more realistically, saturated to one of the supplies, the differential input can have any value. Therefore it is necessary to make sure that the output voltage is constrained, and this is done by a suitable network connecting output and inverting input, as shown in Figure 6.13. That kind of network, to be studied in detail shortly, establishes *negative feedback*. Feedback is an important concept in control theory. It is a way of stabilizing the output, because if the value tries to increase with respect to the correct amplitude there is a fed back signal that opposes the increase, and, vice versa, when output tries to decrease the feedback favors an increase. Feedback will be studied in detail in its own chapter; for now the intuitive description is enough to explain the concept.

Implementing the virtual ground or obtaining a virtual short circuit with an op-amp is what is needed to realize the conceptual schemes studied in the previous chapter. Therefore, with op-amps and passive components we obtain various analog linear operations. The circuits that realize the most important of these are studied in the following subsections.

6.5.1 Inverting Amplifier

The scheme of Figure 6.14 provides inverting amplification. The circuit uses an ideal op-amp and two resistors. One of these, R_1, is connected between input source (assuming an ideal voltage generator) and inverting input; the other resistance is between the same inverting node

COMPUTER EXPERIMENT 6.2

Understanding the Inverting Amplifier

The simplest use of an op-amp is for inverting amplification. The scheme of this experimental board uses two resistors. The ratio between these components almost determines the inverting amplification, because bandwidth and DC gain of the op-amp are large enough. With this virtual experiment you can observe the response with a sine or square wave input signal. Input amplitudes are variable but the frequency is fixed. The supply voltage of the op-amp is 1.5 V. The differential gain of the op-amp is very large but, when the output approaches the rails, performances worsen. Since the voltage of the inverting terminal gives you relevant information on operation, it is worth observing this. In this experiment you can:

- set the amplitude of sine wave and square wave. The frequency of the two signals is fixed and holds 1 kHz;

- define the finite gain of the op-amp. Since the used part has a large bandwidth, at 1 kHz you do not see any speed limitations;

- select the type of input signal and set the non-inverting voltage V_{NI}. Set R_1 and R_2.

MEASUREMENT SET–UP

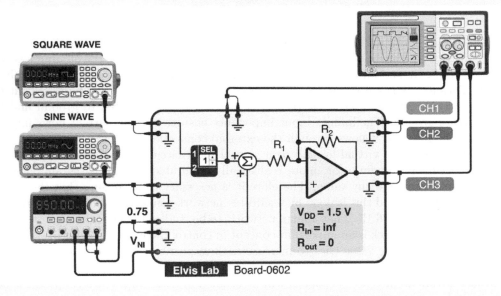

TO DO

- Set the value of both resistances to 1 kΩ and the non-inverting input voltage at 0.75 V. Set the gain at 100.000: a value that almost corresponds to an ideal operation. Observe the output for the two possible inputs (amplitude 2 mV) and notice the inverting gain by -1.

- Increase the value of the resistance connected in feedback up to 500 kΩ. Observe the waveforms at output. Notice that at 0.25 V and 1.25 V there is clipping of the output signal. Change the voltage of the non-inverting input and see what happens for different values of the gain.

- Change the value of the op-amp gain. Measure the error with respect to a nominal amplification of -10, -100 and -500 and observe the virtual ground.

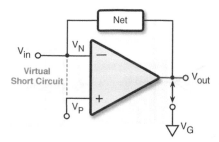

Figure 6.13   Use of feedback to obtain virtual ground.

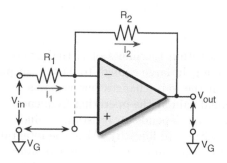

Figure 6.14   Use of the operational amplifier to obtain inverting amplification.

and the output. The connection established by R_2 produces negative feedback because any output variation with respect to the expected value causes an opposite change at the input capable of stabilizing the output. All voltages are measured with respect to the analog ground.

Since the input impedance of the op-amp is infinite, the current entering the negative input terminal is zero. This gives

$$I_1 = I_2. \tag{6.18}$$

Moreover, since the virtual ground concept makes inverting and non-inverting terminals at the same voltage, V_G, inspection of the circuit gives

$$I_1 = \frac{V_{in} - V_G}{R_1}, \tag{6.19}$$

$$I_2 = \frac{V_G - V_{out}}{R_2}, \tag{6.20}$$

which yields

$$V_{out} - V_G = -\frac{R_2}{R_1} \cdot (V_{in} - V_G). \tag{6.21}$$

If the information on the ground level is omitted (which means assuming $V_G = 0$), the output is the amplification of the input by G_V:

$$G_V = \frac{V_{out}}{V_{in}} = -\frac{R_2}{R_1}. \tag{6.22}$$

This shows that the gain is negative and is controlled by the ratio between the resistors.

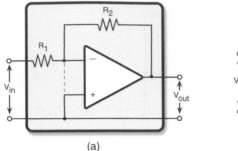
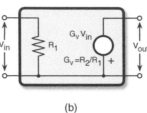

(a)   (b)

Figure 6.15   Inverting amplification: (a) circuit schematic; (b) equivalent circuit.

Notice that this result holds within the limits of virtual ground approximation. In many situations it is enough to have a gain large enough to ensure, in practice, the virtual ground concept. It does not matter if the gain changes but remains large enough. Therefore the requirement is to ensure minimal gain at the operating frequency.

Equation (6.22) shows that gain depends on resistance values. We have already observed that resistors are components with good linearity. Also, the integrated implementations made from the same material and structure produce accurate matching. Therefore, equation (6.22) predicts linear and accurate gains for real circuits, provided that the op-amp satisfies the virtual ground approximation.

The scheme operates for any waveform. If the input is a sine wave

$$V_{in}(t) = \bar{V}_{in} \cdot \sin(\omega_0 t + \varphi_0), \tag{6.23}$$

the output is

$$V_{out}(t) = \bar{V}_{in} \frac{R_2}{R_1} \sin(\omega_0 t + \varphi_0 + \pi), \tag{6.24}$$

a sine wave with just the amplitude changed. Moreover, inverting causes a phase shift by π (or 180°). That is for ideal op-amps. In real cases, the operation holds until given low frequencies are reached. Because of op-amp speed limitations the response starts lagging at high frequency. As we already know, the gain of the op-amp diminishes with constant pace after the dominant pole. The drop makes the virtual ground approximation weaker and weaker, especially for high gain, G_V. Because of this, with real op-amps, the amplitude and phase of G_V become frequency dependent. Therefore, the described operation of the inverting amplifier works properly only within a given frequency range. As a rule of thumb, the range is $f_T/(k \cdot G_V)$ with k between 10 and 100.

Figure 6.15 shows the equivalent circuit of an inverting amplifier built with an ideal op-amp. Since the scheme establishes virtual ground at the non-inverting terminal, the resistance R_1 is virtually connected to ground. Therefore the input impedance is R_1. The generator is a Voltage-Controlled Voltage Source (VCVS) whose gain depends on the ratio between resistors.

Notice the degree of freedom in choosing resistor values. R_1 can be large for establishing a large input impedance but $R_2 = -G_V R_1$ becomes even larger with large gains. Moreover, observe that the zero output impedance established by the ideal op-amp can deliver any current to feed R_2 and the possible load. This is, obviously, a limit of the model.

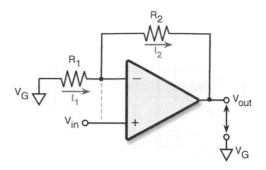

Figure 6.16   Use of operational amplifier to obtain non-inverting amplification.

6.5.2   Non-inverting Amplifier

The scheme of Figure 6.16 is non-inverting. This is a slight modification of the inverting scheme: the input signal is at the positive terminal while the resistance R_1 is connected to ground. The negative feedback established by R_2 determines a virtual short circuit between positive and negative input. Thus across R_1 we have the input signal. Moreover, the infinite input impedance of the op-amp leads to

$$I_1 = I_2. \tag{6.25}$$

Using the ground voltage V_G as reference ($V_G = 0$), we obtain

$$I_1 = -\frac{V_{in}}{R_1} \tag{6.26}$$

$$I_2 = \frac{V_{in} - V_{out}}{R_2}, \tag{6.27}$$

which, resolved, gives

$$V_{out} = V_{in}\left(1 + \frac{R_2}{R_1}\right), \tag{6.28}$$

denoting a positive gain, larger than 1 by the resistive ratio R_2/R_1.

The equivalent circuit of the non-inverting amplifier is shown in Figure 6.17(a). The main difference from the one in Figure 6.15 is that input resistance is infinite as no current flows through the non-inverting terminal. For this scheme, too, the output resistance is zero, being determined by the ideal op-amp operation.

Again, since resistors are linear elements, the gain of the inverting amplifier (as will result for other schemes we study) is linear when the op-amp reasonably meets the virtual short circuit approximation. With linear schemes we can use the superposition principle. Therefore, the input superposition of sine waves produces an amplified addition of sine waves. However, the gain can depend on frequency. As discussed above, real op-amps give rise to gains that lag at high frequencies. Therefore the output spectrum may not be just a shift up on the dB axis but can display differences at high frequency. Because of this, it is advisable to use operational amplifiers whose f_T is large enough in the signal band.

Observe that the op-amp input terminals are both at the input voltage. Therefore the op-amp must ensure an input common mode able to accommodate DC and signal swing of V_{in}. In some cases an input rail-to-rail operation may be necessary.

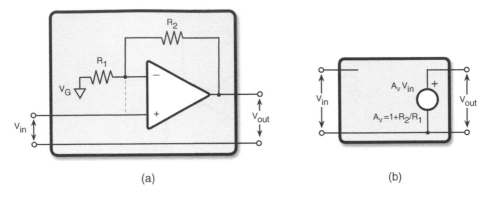

Figure 6.17   Non-inverting amplification: (a) circuit schematic; (b) equivalent circuit

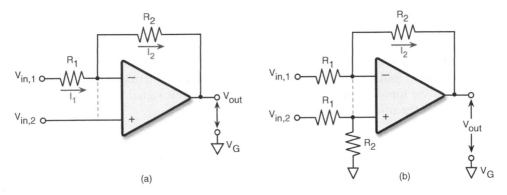

Figure 6.18   Superposition of inverting and non-inverting amplifier: (a) different gain; (b) equalized gain.

6.5.3   Superposing Inverting and Non-inverting Amplification

Since the operation of inverting and non-inverting schemes is linear, we can superpose their functions and obtain both tasks with a single op-amp, as shown in Figure 6.18(a). The output voltage is

$$V_{out} = V_{in,2}\left(1 + \frac{R_2}{R_1}\right) - V_{in,1}\frac{R_2}{R_1}, \tag{6.29}$$

which is the amplifications, one inverting the other non-inverting, of $V_{in,1}$ and $V_{in,2}$. The two gains differ. The amplitude of the non-inverting gain is higher by one than that of the inverting one. Moreover, the loads established at the two inputs are different. The positive terminal has infinite resistance because the resistances between the input terminals and between a single terminal and ground are supposed infinite; the current through the negative terminal depends on both input voltages:

$$I_1 = \frac{V_{in,1} - V_{in,2}}{R_1}, \tag{6.30}$$

denoting an equivalent resistance, R_1, across the two inputs but with the current that flows only through the $V_{in,1}$ generator. $V_{in,2}$ can be a voltage source with any output resistance. $V_{in,1}$ must have a low output resistance to deliver the required current properly.

Having dissimilar inverting and non-inverting gains can be a problem. They can be made equal by attenuating $V_{in,2}$ by $R_2/(R_1 + R_2)$ before its use. Figure 6.18(b) shows the corresponding scheme. The voltage at the inputs becomes

$$V_+ = V_- = V_{in,2}\frac{R_2}{R_1 + R_2}, \tag{6.31}$$

which, amplified by the non-inverting gain $(R_1 + R_2)/R_1$, gives rise to R_2/R_1, as expected.

The $(R_1 - R_2)$ attenuator is a load for the $V_{in,2}$ generator. The resistance at the non-inverting input becomes $(R_1 + R_2)$.

Again, the input common-mode terminal swings with $V_{in,2}$. Real schemes need to use op-amps with common-mode input range capable of accommodating the DC level and the swing of the signal at the non-inverting terminal.

Example 6.2

Realize, with a single op-amp, the processing function $V_{out} = -5\,V_{in,1} + 11\,V_{in,2}$. Use an ideal op-amp and minimum resistance 1 kΩ.

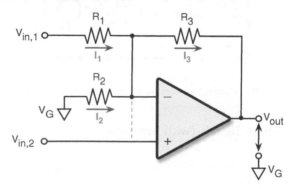

Solution

The scheme of Figure 6.18(b) reduces the non-inverting gain. This example asks for augmentation of the gain. Therefore it is necessary to increase the current due to $V_{in,2}$ and to inject the result into the virtual ground. Since the virtual short circuit is at $V_{in,2}$, a resistance between the inverting node and ground produces the extra current needed. The scheme given here, which uses an extra resistance between the inverting terminal and ground, achieves this aim. For determining the resistor ratios we use the superposition principle. If $V_{in,2} = 0$ and we apply only $V_{in,1}$, the inverting input is ground and no current flows into R_2 regardless of its value. Therefore resistor R_2 does not affect the inverting gain. The inverting gain requires

$$R_3 = 5\,R_1.$$

Now set $V_{in,1} = 0$. Since the parallel of R_1 and R_2 is connected to ground, the non-inverting gain becomes

$$G_{V+} = 1 + \frac{R_3(R_1 + R_2)}{R_1 \cdot R_2}.$$

To obtain a non-inverting gain of 11, R_3 being $= 5\,R_1$, the parallel of R_1, R_2 must be $\frac{1}{2}R_1$, leading to $R_2 = R_1$. Both resistors R_1 and R_2 have the minimum required value 1 kΩ; $R_3 = 5$ kΩ.

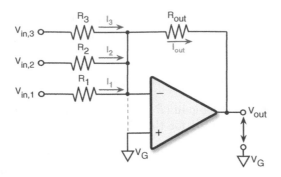

Figure 6.19   Addition of multiple signals.

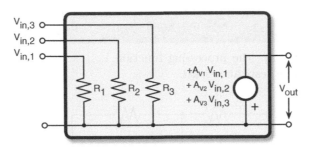

Figure 6.20   Equivalent circuit of the scheme in Figure 6.19.

6.5.4   Weighted Addition of Signals (with Inversion)

A single operational amplifier can create inversion of multiple signals. They can also be weighted by different gain factors and realize

$$V_{out}(t) = -[k_1 \cdot V_{in,1}(t) + k_2 \cdot V_{in,2}(t) + k_3 \cdot V_{in,3}(t) + \cdots]. \tag{6.32}$$

Combining multiple signals is used, for example, to mix the outputs of several microphones. The sensitivities or the signal levels can differ, and for their equalization different amplifications are required. It may also be necessary to adjust the gains in different frequency bands. Here we study only the mixing with frequency-independent gains. More complex processing is normally done with multiple op-amps or using digital solutions.

The scheme of Figure 6.19 produces what equation (6.32) needs. Each input is connected by a resistance to virtual ground, to transform the input voltage $V_{in,i}(t)$ into a current $I_i(t)$. All currents are then summed up to flow through the feedback resistor R_{out}. The voltage drop across R_{out} gives the output voltage.

The values of the resistors and coefficients in equation (6.32) are related by

$$k_1 = \frac{R_{out}}{R_1}, \quad k_2 = \frac{R_{out}}{R_2}, \quad k_3 = \frac{R_{out}}{R_3}, \quad \cdots, \tag{6.33}$$

which use R_{out} as the normalizing element. Its choice is a degree of design freedom that results because the gain depends on a ratio of resistors and not on absolute values.

Remember that each input resistance loads its input signal generator. Figure 6.20 shows the equivalent circuit of Figure 6.19. Each ith input establishes a load resistance, R_i. If the

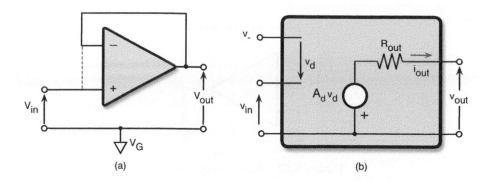

Figure 6.21   (a) Use of operational amplifier to produce unity gain buffer; (b) equivalent circuit with op-amp finite output resistance.

value is not suitable for the input source it may be necessary to use decoupling interfaces with additional op-amps.

6.5.5   Unity Gain Buffer

A voltage decoupling interface is an amplifier with unity gain, infinite input impedance and zero output impedance. That function is obtained by the non-inverting scheme of Figure 6.21(a). This corresponds to the circuit of Figure 6.16 with zero feedback resistance and infinite resistance between inverting input and ground. The zero feedback resistance makes input and output equal. Indeed, even the current in the feedback path is zero. Notice that even a non-zero resistance would make input and output equal. We have

$$G_V \cong 1. \tag{6.34}$$

The input resistance is high because it is given by the input resistance of the op-amp. The output resistance is zero for an equivalent reason. Indeed, the output resistance is zero even if the op-amp has infinite gain and finite output resistance. This results from a property of feedback that will be studied in a later chapter. However, since the verification is easy we can already study this simple case.

Let us consider the equivalent circuit of Figure 6.21(b) and suppose that the differential gain $A_d \to \infty$. By inspection we obtain

$$(v_{out} - v_{in})A_d - R_{out}i_{out} = v_{out} \tag{6.35}$$

which gives the output voltage, as a function of input voltage and output current,

$$v_{out} = v_{in}\frac{A_d}{A_d + 1} - \frac{R_{out}}{A_d + 1}i_{out}, \tag{6.36}$$

denoting a reduction of the output resistance by $(A_d + 1)$. Therefore, for infinite differential gain the output resistance goes to zero. The result is also evident by inspection. Since the virtual short circuit makes input and output equal, an infinitesimal change of the differential input compensates for any possible drop voltage across R_{out}.

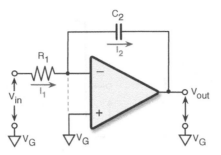

Figure 6.22 Active integrator.

A mystifying result?

It may be confusing to learn that the output resistance of a block goes to (almost) zero thanks to a circuit connection. This result is because of feedback. This and other features will be studied in a later chapter.

This scheme is also called a unity gain buffer. "Unity gain" outlines the gain of the scheme. "Buffer" indicates the output-to-input insulating action of the circuit.

Observe that, since the swings of input and output are equal, the real op-amp must ensure a wide swing both at input and at output. The op-amps that satisfy this requirement use so-called "rail-to-rail" input stages.

6.5.6 Integration and Derivative

Conceptual schemes studied in Chapter 5 produce integration and derivative. Those functions are implemented with operational amplifiers, as accomplished by the inverting integrator of Figure 6.22. The diagram uses capital letters to indicate voltages or currents. Since that means using large signals, the scheme assumes that the components are always linear: i.e., they respond linearly even with large signals. In reality the op-amp is not perfectly linear, and this holds in just one given region of operation. Therefore, both the generated voltage and the input can be "large," but only under certain conditions.

The operation of the circuit is as follows: the resistor R_1 transforms the input voltage into a current, I_1:

$$I_1 = \frac{V_{in}}{R_1}. \tag{6.37}$$

The current is integrated on the feedback capacitor to give, in the time domain,

$$V_{out}(t) = V_{out}(t_0) - \int_{t_0}^{t} \frac{I_1(\tau)d\tau}{C_2}. \tag{6.38}$$

So in the s domain this becomes

$$V_{out}(s) = -\frac{1}{sR_1C_2}V_{in}(s) = -\frac{V_{in}(s)}{s\tau_1}, \tag{6.39}$$

where τ_1, the time constant, is R_1C_2.

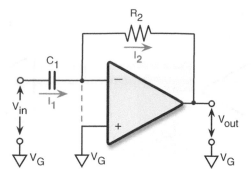

Figure 6.23   Active derivator.

As was just mentioned, the time constant depends on the proper choice of both resistor and capacitor. Thus integrators also have a degree of freedom. R_1 can increase so as to reduce C_2, or vice versa.

The accuracy of the time constant depends on resistor and capacitor accuracy. The fabrication steps of different components are different, and this is also true in integrated circuits. Therefore errors in resistors and capacitors do not track each other. Since the errors are uncorrelated, again, we have

$$\left(\frac{\delta \tau_1}{\tau_1}\right)^2 = \left(\frac{\delta R_1}{R_1}\right)^2 + \left(\frac{\delta C_2}{C_2}\right)^2, \tag{6.40}$$

giving time constant inaccuracy in real circuits.

Often the time constant value must be precise. Its accuracy influences, for example, the performance of filters that, often, require precision to the fraction of a % or less in their response. For such needs the time constant must be adjusted by trimming. The changed component can be the resistor or the capacitor. Its adjustment is possibly done with a digital control.

Figure 6.23 gives the derivative. The input element is a capacitor, while a resistor establishes the feedback. Study of the derivator scheme is similar to the one just made for the integrator. By inspection, the current in capacitance C_1 is

$$I_1 = C_1 \frac{dV_{in}}{dt}, \tag{6.41}$$

which is injected into the virtual ground to obtain

$$V_{out} = -R_2 C_1 \frac{dV_{in}}{dt}. \tag{6.42}$$

The equation, in the s domain, becomes

$$V_{out}(s) = -s R_2 C_1 V_{in}(s) = -s \tau_1 V_{in}(s), \tag{6.43}$$

giving, with a minus sign, the expected derivative.

For the derivative also, the time constant is the product of a resistance and a capacitance. Therefore its accuracy is the same, and possibly as problematic, as for the integrator. When the accuracy required exceeds the precision achievable with the technology used, it is necessary to

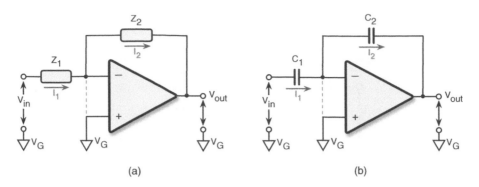

(a)   (b)

Figure 6.24   (a) Amplifier that obtains a generalized response; (b) generalized amplifier with capacitors instead of resistors.

adjust the time constant value. Moreover, since a change of temperature alters the component value, the time constant can also be temperature dependent.

It is worth remembering that these configurations, as for the previous ones, rely on the implementation of virtual ground. Non-infinite gain and speed limits affecting the op-amp alter the transfer function in low- and high-frequency regions.

6.5.7   Generalized Amplifier

We generalize the inverting (or non-inverting) amplifier by replacing the two resistances with two impedances, as shown in Figure 6.24(a). Assuming the virtual ground approximation to have been verified, the output voltage is given by

$$V_{out}(s) = -\frac{Z_2(s)}{Z_1(s)} V_{in}(s), \tag{6.44}$$

showing a frequency-dependent gain, whose zeros and poles are a possibly generalized amplifier given by impedances. Suppose that they are a simple series of resistor R_s and capacitor C_s, or the parallel connection of resistor R_p and capacitor C_p. The impedance values are

$$Z_s = \frac{1 + sR_sC_s}{sC_s}, \quad Z_p = \frac{R_p}{1 + sR_pC_p}. \tag{6.45}$$

Thus, with those types of impedances it is possible to realize responses with one or two zeros and one or two poles, all of them real. However, if poles and zeros are coincident they cancel each other out, and the response is frequency independent. This special case occurs for the scheme in Figure 6.24(b), where impedances are both made by simple capacitors. Each impedance gives a pole at $f = 0$ and the gain becomes

$$G_v = -\frac{C_1}{C_2}, \tag{6.46}$$

which is frequency independent like the inverting amplifier studied before. However, the circuit shows an input impedance that is capacitive (and not resistive like the configuration of Figure 6.14). Therefore, the input impedance is frequency dependent and very large at low frequency.

The scheme in Figure 6.24(b) is conceptually feasible. However, it lacks good control of DC voltages. If the op-amp input impedance is infinite (as in CMOS schemes), possible charges can be trapped in the virtual ground node, and this shifts the DC output voltage value. With zero input we expect to have zero output. Instead, a charge Q_t trapped in the virtual ground node makes the output non-zero. The virtual ground action charges and discharges C_1, but all the charge Q_t stays in C_2, making the output voltage with zero input signal

$$V_{out} = -\frac{Q_t}{C_2}. \tag{6.47}$$

Therefore, to make sure that the limit caused by trapped charges does not shift the output into regions that impair the operation, it is advisable to reset the capacitors of Figure 6.24(b) periodically (remember that a reset switch leaks current).

6.6   OPERATION WITH REAL OP-AMPS

The above study assumed the operational amplifier to be ideal. However, as indicated by data sheets, real op-amps feature many non-idealities. It is important to know possible effects on performance so as to make the best selection of the part or to improve the circuit design by a possible compensation for limitations. Here we consider the most relevant of them. In particular, we study the effect of input offset, finite gain, non-infinite input impedance, and non-zero output impedance. In addition, we shall study the effects of finite bandwidth and limited slew-rate. Noise is important and will also be examined. Often other limits, which must be kept in mind anyway, are relevant for specific uses of the op-amp.

6.6.1   Input Offset

As defined by the specifications, input offset accounts for the mismatch between elementary components of the op-amp. It is described by a DC voltage generator in series with one input terminal. Normally this is the non-inverting terminal. Obviously, the offset can be moved from one input to the other. Figure 6.25 shows how to represent the offset in the two cases. The schemes are equivalent. In the first case the positive terminal is at the offset voltage, and, thanks to the virtual ground concept, the point joining R_1 and R_2 is at the offset voltage, V_{os}. In the second case the inverting terminal is at ground, and the offset generator makes the voltage of the joint equal to V_{os}.

Since the study performed assumes that the system is linear (or the circuit operates in a region that, to a good approximation, can be assumed linear), we can use the superposition principle and notice that the input offset is amplified at output by the non-inverting gain. For the schemes in both Figure 6.14 and Figure 6.16 the output accounts for the additional term

$$\Delta V_{out} = V_{os}\left[1 + \frac{R_2}{R_1}\right]. \tag{6.48}$$

The offset changes the output level by a constant shift proportional to offset and voltage gain. Thus amplifiers that use large gains, like the ones for sensors, should ensure a very low offset. Since offset depends on the technological limits that cause mismatch, it cannot be reduced to zero even with careful design and fabrication. It can be canceled or compensated

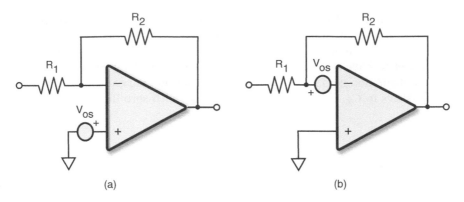

Figure 6.25   (a) Offset as an input-referred generator at the non-inverting terminal; (b) scheme with the offset source moved to the inverting terminal.

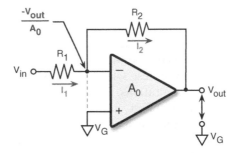

Figure 6.26   Inverting amplifier that uses an op-amp with finite gain A_0.

for with trimming methods or by special techniques, called *auto-zero* and *stabilized chopper*. With these methods it is possible to obtain offsets of a few μV.

6.6.2   Finite Gain

If the output voltage of an op-amp with finite gain A_0 is V_{out}, its differential input is not zero but $-V_{out}/A_0$. For normal cases the error is small but is output dependent. It influences the behavior of the scheme by an extent that depends on the scheme itself and the vales of passive components. For example, the finite gain in the inverting amplifier of Figure 6.26 determines the condition

$$\frac{V_{in} + V_{out}/A_0}{R_1} = \frac{-V_{out}/A_0 - V_{out}}{R_2}. \tag{6.49}$$

That condition also implies $I_1 = I_2$. The solution of equation (6.49) is

$$\frac{V_{out}}{V_{in}} = -\frac{R_2}{R_1 + (R_1 + R_2)/A_0} = -\frac{R_2}{R_1} \frac{1}{1 + (1 + R_2/R_1)/A_0}, \tag{6.50}$$

which, obviously, becomes the ideal result when $A_0 \to \infty$.

Equation (6.50) includes the inverting gain $-G = -R_2/R_1$ with ideal op-amp. More markedly

$$\frac{V_{out}}{V_{in}} = -G' = \frac{-G}{1 + (G+1)/A_0} \simeq -G[1 - (G+1)/A_0] = -G(1 - \epsilon_G), \quad (6.51)$$

which defines a relative gain error, $\epsilon_G = (G+1)/A_0$.

If, for example, an inverting amplifier with gain -20 needs an error ϵ_G lower than 0.1%, the finite gain A_0 should verify the condition $((G+1)/A_0) < 0.001$. Therefore, the required minimum gain is $A_0 > 21\,000$ (~ 86 dB), a value that is achievable but not easy to sustain for a large interval of the output voltage.

The above result shows that A_0 must be high when the expected signal amplification is high. However, since the signal gain is the ratio between resistors, the accuracy of that ratio depends on the matching accuracy. The errors caused by finite gain and matching are summed, and it may be that the second one is dominant. However, the mismatch error is static while the finite gain error, which depends on A_0, can be frequency dependent. Therefore, the choice of the op-amp must make sure that gain error is negligible or comparable to other sources such as that due to matching resistors. Very large op-amp gains (such as 100 dB or more) are needed in a few special situations. For normal cases a gain of 80–90 dB is enough to meet specifications.

Example 6.3

The signal gain of an inverting amplifier is -50 (34 dB). The discrete resistors used have 0.5% accuracy. Estimate the minimum finite gain of the op-amp that gives rise to a gain error equal to half the one caused by resistor mismatch.

Solution
Since resistors are discrete components the matching accuracy is no better than the absolute value. Assuming that the fabrication steps of the two resistors determine uncorrelated errors, the matching is the quadratic superposition of errors

$$\epsilon_R = \sqrt{2}\epsilon_R = 7.07 \cdot 10^{-3}. \quad (6.52)$$

In order to secure a finite gain error lower than $(0.707/2)\%$ it is necessary to use an op-amp with gain

$$A_0 > \frac{(G+1)}{\epsilon_R/2} = \frac{51}{3.53 \cdot 10^{-3}} = 14.45 \cdot 10^3 \rightarrow 83.2 \text{ dB}, \quad (6.53)$$

which is not difficult to find in the specifications of many commercial op-amps.

6.6.3 Non-ideal Input and Output Impedances

Impedances between inputs and ground in ideal op-amps are infinite. In real circuits they can have a large but finite value. This is due to the input stage of the op-amp. However, for CMOS operational amplifiers the input is often a terminal (the gate) that is separated from the rest of the circuit by a thin layer of silicon dioxide. In this case the input resistance is

COMPUTER EXPERIMENT 6.3

Inverting Approximate Integrator

The use of the operational amplifier enables you to perform any analog linear processing function. This Computer Experiment studies inverting integration. The scheme replaces the feedback resistor of the previous Computer Experiment with a capacitor. Possibly, a resistance in parallel to the capacitor limits the maximum gain. The circuit of the board uses an op-amp with a supply range equal to 0-1.5 V. The output starts saturating at 0.3 V and 1.2 V and has a smooth transition that reaches hard minimum and maximum outputs at 0.2 V and 1.3 V, respectively. The input signal is a sine or a square wave with controllable amplitude and frequency. Finite gain and finite bandwidth are features of the op-amp response. You can:

- set the frequency of sine wave and square wave and choose the type of input signal;
- set gain and unity gain frequency of the op-amp;
- control the voltage of the non-inverting input, to compensate for possible op-amp offset;
- choose the values of the input and feedback resistances and integrating capacitor.

MEASUREMENT SETUP

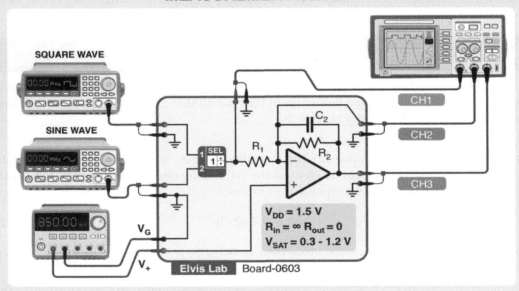

SQUARE WAVE

SINE WAVE

CH1

C_2

CH2

SEL

R_1

R_2

CH3

$V_{DD} = 1.5$ V
$R_{in} = \infty$ $R_{out} = 0$
$V_{SAT} = 0.3 - 1.2$ V

V_G

V_+

Elvis Lab   Board-0603

TO DO

- Set the input resistance to 1 kΩ and the feedback resistor to 1 MΩ, gain at 100 dB, unity gain frequency 1 GHz and non-inverting input at 0.75V. Use a value of feedback capacitor that gives 20 dB gain at 100 kHz, 1 MHz and 10 MHz. Determine the frequencies at which the gain becomes -1.
- Use a square wave at the input and observe the output. Increase the input amplitude until clipping arises. Change the voltage of the non-inverting input.
- Reduce the input frequency and use very low amplitudes. Get to the point where, even if frequency diminishes, the output amplitude does not significantly change.
- Set the bandwidth of the op-amp to 200 MHz and observe how it behaves with very high input frequencies for different values of the integrating capacitance.

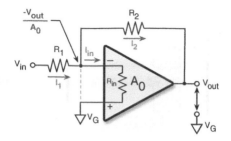

Figure 6.27   Inverting amplifier with finite gain and input resistance.

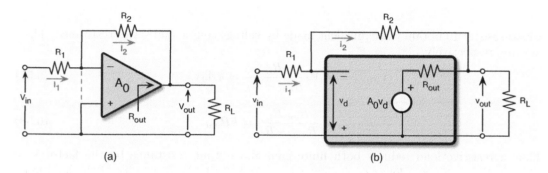

Figure 6.28   (a) Inverting amplifier with finite gain and output resistance; (b) equivalent circuit.

almost infinite. For a bipolar input stage the input resistance can be relatively low (a few kΩ), but with specific circuit techniques it is possible to obtain much higher input resistances.

The input impedance from each terminal and ground is mainly a load to the input source. The impedance between the two inputs possibly affects circuit operation. However, its effect is attenuated because that impedance is connected across nodes almost at the same voltage, V_{out}/A_0. For the inverting amplifier of Figure 6.27, for example, finite gain and input resistance cause a current I_{in} that flows into the inverting terminal. Therefore

$$I_1 = I_2 - I_{in}, \tag{6.54}$$

which gives rise to

$$\frac{V_{in} + (V_{out}/A_0)}{R_1} = -\frac{V_{out}}{A_0 R_{in}} - \frac{(V_{out}/A_0) + V_0}{R_2}, \tag{6.55}$$

which, solved, gives

$$\frac{V_0}{V_{in}} = -\frac{R_2}{R_1 + (R_1 + R_2 + R_{in})/A_0}. \tag{6.56}$$

Thus a finite input resistance determines an amplification error, similar to the one caused by the finite gain. The error is negligible if R_{in} is much larger than R_1 and R_2.

The equivalent output stage of ideal op-amps uses only a Voltage-Controlled Voltage Source (VCVS). A resistor, R_{out}, added in series to the VCVS, models the finite output resistance, as shown in Figure 6.28. The resistances R_1 and R_2 make an inverting gain stage with ideal gain $-R_2/R_1$. The diagram also includes a load at output, R_L.

The Thévenin equivalent of the output stage explains the effect of the resistive load on circuit performance. That equivalent circuit, as shown in Figure 6.29, combines real output generator

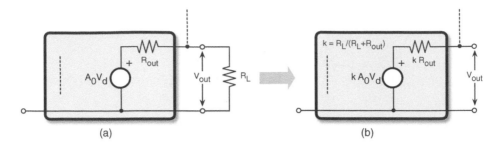

Figure 6.29 (a) Output stage with output resistance; (b) its Thévenin equivalent circuit.

and load resistor to obtain the equivalent made by voltage source and series resistance. Their values are, respectively,

$$V_{Th\acute{e}v} = A_0 V_d \frac{R_L}{R_{out} + R_L} = k A_0 V_d \tag{6.57}$$

$$R_{Th\acute{e}v} = \frac{R_{out} R_L}{R_{out} + R_L} = k R_{out}. \tag{6.58}$$

Thus a resistive load reduces both finite gain and output resistance by the factor $k = R_L/(R_L + R_{out})$. Since a lower equivalent gain is undesired, the resistance of the load must be much higher than the output resistance.

The equation $I_1 = I_2$ for the scheme including the Thévenin equivalent circuit yields

$$\frac{V_{in} + V_d}{R_1} = \frac{-V_d - V_{Th\acute{e}v}}{R_2 + R_{Th\acute{e}v}}, \tag{6.59}$$

which, using equations (6.57) and (6.58), becomes

$$\frac{V_{in} + V_d}{R_1} = -V_d \frac{1 + k A_0}{R_2 + k R_{out}}; \tag{6.60}$$

therefore, the differential input is

$$V_d = -V_{in} \frac{R_2 + k R_{out}}{R_2 + k R_{out} + R_1(1 + k A_0)}. \tag{6.61}$$

Moreover, remembering that

$$\frac{V_{in} + V_d}{R_1} = -\frac{V_d + V_{out}}{R_2}, \tag{6.62}$$

the voltage gain is

$$\frac{V_{out}}{V_{in}} = \frac{k R_{out} - R_2 k A_0}{R_2 + k R_{out} + R_1(1 + k A_0)}, \tag{6.63}$$

showing that the non-zero output resistance combined with the finite op-amp gain and resistive load causes two sources of error in the inverting gain; one is due to the resistive load that causes an equivalent attenuation of the finite op-amp gain, and the other depends on the output resistance.

COMPUTER EXPERIMENT 6.4

Understanding Op–Amp Limitation

Real op-amps limit the operation of analog circuits, especially at high speed. With this virtual experiment you can verify how real performances affect results. This ElvisLab board allows you to verify the limits of finite bandwidth and slew rate. Both parameters depend on the bias current being changed by a control on the board. An input sine wave is the optimum for studying the bandwidth limit. A square wave investigates the slew-rate restraints better. Since the results depend on expected amplification, the resistances are variable. Notice that the small signal gain A_0 is constant and holds 60 dB.

In this experiment you can:

- set amplitude and frequency of sine wave and square wave. The amplitudes of both signals are symmetrical with respect to the analog ground given by $V_{DD}/2 = 0.9$ V;

- set the values of resistances used for the inverting gain configuration;

- move the dominant angular frequency and slew rate (interrelated) by setting the value of the bias current.

MEASUREMENT SET–UP

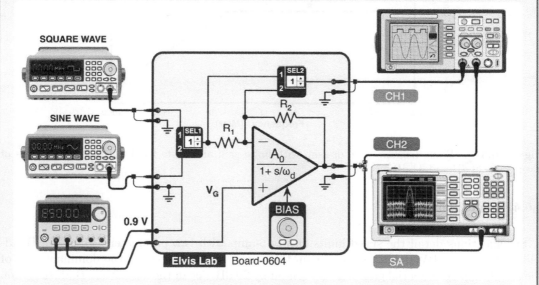

TO DO

- Set the bias current to maximum value and use the values of resistances that determine an expected DC gain of 37 dB. Observe waveforms with sine wave and square wave at input but at various frequencies.

- Reduce the bias current and estimate the gain over a logarithmic span (three points per decade) until the gain is -20 dB. Estimate the op-amp bandwidth. Be informed that the capacitive load, established by oscilloscope and spectrum analyzer, does not affect the experiment because of a "strong" output op-amp stage.

- Use the square wave and observe the output waveform in the time and frequency domain. Change the voltage of the non-inverting terminal to optimize the output waveform.

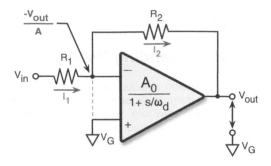

Figure 6.30 Inverting amplifier with single-pole op-amp.

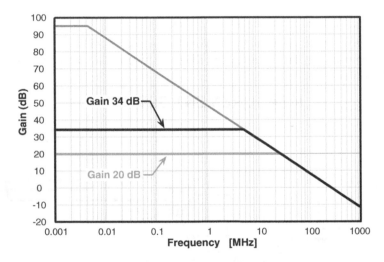

Figure 6.31 Bode diagram of an op-amp with finite gain and finite bandwidth; gain responses of inverter at 20 dB and 34 dB.

6.6.4 Finite Bandwidth

Specifications detail the speed limits of an op-amp with two parameters: the Gain Bandwidth Product (GBW), and the unity gain frequency, f_T. As discussed, the dominant pole of the small signal differential gain causes a roll-off by 20 dB/dec in the magnitude Bode diagram. The slope remains almost constant until it crosses 0 dB. Other poles located at higher frequencies just slightly bend the Bode diagram (and change the phase) near the GBW. Since well below f_T the effect of non-dominant poles is negligible, this study uses a single-pole model,

$$A_d(f) = \frac{A_0}{1 + s\tau_d}, \tag{6.64}$$

where, as usual, $\tau_d = 1/(2\pi f_d)$ and A_0 is the low frequency gain.

Let us consider the inverting amplifier in Figure 6.30. Its inverting input is at $-v_{out}/A_d$. The condition that current I_1 equals that on the feedback resistor, I_2, leads to

$$\frac{V_{in} + V_{out}/A_d}{R_1} = \frac{-V_{out}/A_d - V_{out}}{R_2}; \tag{6.65}$$

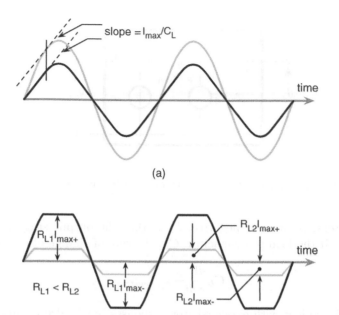

Figure 6.32   Effects of limited output driving capability: (a) limited output slew-rate; (b) clipping of the output voltage.

that, like equation (6.51), leads to

$$\frac{V_{out}}{V_{in}} = -\frac{R_2}{R_1} \cdot \frac{A_d}{A_d + 1 + R_1/R_2}. \tag{6.66}$$

The use of equation (6.64) in equation (6.66) yields

$$\frac{V_{out}}{V_{in}} = -\frac{R_2}{R_1} \cdot \frac{A_0}{A_0 + 1 + R_1/R_2} \cdot \frac{1}{1 + s\tau_G}, \tag{6.67}$$

which is the gain of an inverting amplifier with finite gain A_0 and frequency limitation (Figure 6.31) given by a pole at ω_G/τ_G. τ_G and ω_G are, respectively,

$$\tau_G = \frac{\tau_d R_2/R_1}{A_0 + 1 + R_1/R_2}, \quad \omega_G \simeq \frac{\omega_d \cdot A_0}{R_2/R_1}. \tag{6.68}$$

Remembering that $\omega_d \cdot A_0/(2\pi)$ is the unity gain frequency, f_T, the pole is at f_T/G, which is close to f_T if the gain is not very large.

6.6.5   Slew-rate Output Clipping and Non-linear Gain

Various non-linearities affect the operational amplifier. They are described in several specifications. The most important of them are the slew-rate and the harmonic distortion coefficients. The slew-rate caused by limited driving capabilities has already been discussed. It is due to

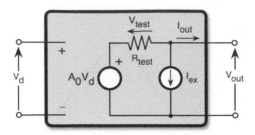

Figure 6.33   Equivalent circuit suitable for modeling limited driving capability.

the maximum current, positive or negative, I_{max}, that the output stage of the op-amp can deliver to the load. If the load is a capacitor, C_L, the rate of change of output voltage is

$$C_L \frac{dV_{out}}{dt} = I_{max}. \tag{6.69}$$

The limitation is evident with expected step responses but it also occurs with sine waves. For example, Figure 6.32(a) illustrates the difference between an ideal sine wave at output and the generated waveform with a strong limited driving capability. The output is not a sine wave, especially at the zero crossing where the expected slope is maximum. Near zero we have constant slope $\pm I_{max}/C_L$ until the required rate of change falls below the limit. Then the output follows sine wave behavior.

If the load is a resistor, R_L, the maximum current does not affect dynamic behavior but maximizes the output voltage

$$V_{out,max} = R_L I_{max}. \tag{6.70}$$

Since the output cannot exceed this level there is a sharp clipping of the output waveform that depends on the load. Figure 6.32(b) shows the possible outputs of an amplifier using an op-amp with limited positive and negative driving capability. Clipping with a small load resistance, R_1, occurs at a relatively high voltage. A lower load R_2 causes significant clipping even if the expected output is smaller than in the other case. Notice that Figure 6.32(b) shows it is possible to have different positive or negative output driving capabilities.

A limited output current can be modeled by a non-linear equivalent circuit. Figure 6.33 shows a possible scheme. It is a bit more complex than the circuits used so far but much simpler than the transistor-level equivalent. It uses a test resistance R_{test} to measure current. If the value exceeds the limit I_{max}, the non-linear current source, I_{ex}, drains the surplus toward ground so that the output current is I_{max}. The excess current is

$$I_{ex} = u(V_{test}/R_{test} - I_{max})(V_{test}/R_{test} - I_{max}), \tag{6.71}$$

where $u(x)$ is the step function, which is equal to zero if $x \le 0$, and set to one when $x > 0$. Equation (6.71) can be embedded into the model using suitable behavioral language.

Take note

The effects of slew-rate, clipping, and other non-linear limits are attenuated until the virtual ground concept is almost verified.

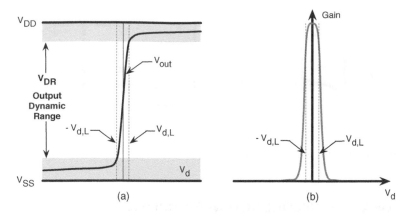

Figure 6.34   (a) Input/output response of a real op-amp; (b) small-signal gain.

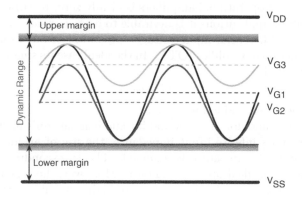

Figure 6.35   The wider output sine wave with different analog grounds.

In addition to limited driving capabilities that cause non-linearity, there is the shortcoming of non-linear input/output characteristics. Strong non-linearity occurs when output voltage approaches the rail limits. Figure 6.34(a) shows a typical input/output response. The slope is high and almost constant within a small range of input differential voltage ($\pm V_{d,L}$). Then, near the positive and negative supply voltages, V_{DD} and V_{SS}, it drops. Since the regions where the gain becomes low are unusable it is necessary to know the extent of the output dynamic range, V_{DR}, outlined in Figure 6.34(a). The slope of the input/output response, which is, actually, the small-signal differential gain, is high in only a small range of the input differential voltage. It has to be higher than a defined amount, $A_{d,min}$, to ensure good performance. Figure 6.34(b) shows the slope of the curve and delimits the input differential range for which the small-signal gain is higher than $A_{d,min}$.

The level of analog ground is a designer choice. It is set to the midpoint of the output dynamic range (V_{G1} in Figure 6.35) to ensure symmetrical swings. Values higher or lower (V_{G2} or V_{G3}) expand downward or upward – the permitted unilateral swing. However, for analog ground above the midpoint the maximum sine wave that an op-amp can generate without distortion diminishes, because the positive swing enters the non-linear region before the negative one, and vice versa for analog ground below the midpoint.

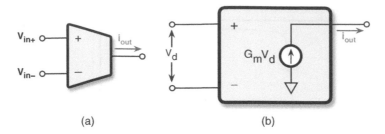

(a) (b)

Figure 6.36 (a) Symbol for an OTA; (b) small-signal equivalent circuit.

6.7 OPERATIONAL TRANSCONDUCTANCE AMPLIFIER

The output impedance of operational amplifiers is ideally zero, because it generates voltage at output. For a number of applications it is desirable to have, instead, current at output with value proportional to the differential input voltage. This kind of function is provided by the Operational Transconductance Amplifier (OTA). In the ideal case its small-signal input/output relationship is approximated by

$$i_{out} = G_m v_{d,in}, \tag{6.72}$$

where G_m represents the transconductance gain.

Figure 6.36 shows the symbol for an OTA and its small-signal equivalent circuit in the ideal case. The input stage shows infinite input impedance; the use of an ideal current source at output is because the output impedance is assumed. The transconductance gain is a finite value and not, as happens for ideal op-amps, infinite. Moreover, the scheme in Figure 6.36(b) supposes infinite output resistance. Real circuits have high but finite output impedance.

There are two key differences between the OTA and the conventional op-amp. First, the output impedance of the OTA is high, while the op-amp has very low output impedance. Second, the op-amp needs a feedback network that reduces sensitivity to non-idealities and produces finite gain. In contrast, the OTA can be used without feedback. Its transconductance gain is a design parameter necessary to obtain the desired functions.

OTAs are rarely available as discrete parts because they are normally used inside integrated circuits. The few OTAs commercially available are specified by parameters similar to the ones used for operational amplifiers. An important and useful specific feature offered by many OTA schemes is that the transconductance gain is electrically controllable. Often, just varying a bias current changes G_m. Thus, the user can obtain electrically tunable responses.

The use of OTAs is often advantageous because the speed achieved is higher and the power consumed is lower than corresponding op-amps. However, the G_m obtained is not constant but depends in a non-linear way on input voltage and the output terminal voltage. Since having non-linear responses is problematic for many applications, the choice between op-amp and OTA is the result of a benefit/limitation analysis.

6.7.1 Use of the OTA

Like op-amps, operational transconductance amplifiers give key analog processing functions. Here we discuss some of them. Consider, for example, the scheme in Figure 6.37(a).

COMPUTER EXPERIMENT 6.5

Understanding OTAs

The OTA (Operational Transconductance Amplifier) generates at output a current proportional to the input differential voltage. OTAs can make amplifiers and integrators, with which more complex processing functions can be constructed. The ElvisLab board contains two OTAs with a bias current control that regulates transconductance gains. The output current flows into a resistance or a capacitor (with the output resistance in parallel). The input is a sine wave symmetrical with ground. The supply voltage is bipolar: ±0.9 V. The voltage of the inverting terminal can vary in the positive or negative direction. In this experiment you can:

- set amplitude and frequency of the input sine wave;
- change resistance and capacitor in the two OTA schemes. The output resistance is constant and holds 1 MΩ;
- decide whether the measured output is done with the capacitive load of the instruments (20 pF) or through active probes that give rise to a negligible capacitive load.

MEASUREMENT SET–UP

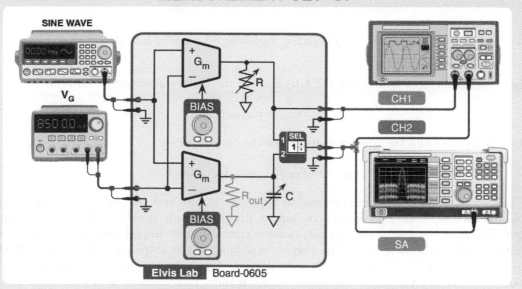

TO DO

- Set the sine wave amplitude to zero and trim the bias of the inverting terminal until the output of the integrator is around zero. The reset button on the ElvisLab console enables you to discharge the integrating capacitor. Notice that the two OTAs have different input–referred offsets (that also depend on the board used with this experiment).
- Set the value of the resistance to 10 kΩ and measure gain versus bias current at different frequencies. Change the amplitude of the sine wave to avoid output saturation.
- Focus on the integrator and estimate frequency response with different bias currents. Determine the maximum low frequency and achieved gain versus bias frequency.
- Observe the spectrum of the output voltage with inadequate bias conditions. Change the voltage of the non-inverting terminal and observe what happens.

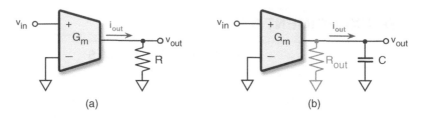

Figure 6.37   (a) Voltage amplifier with OTA; (b) integrator with OTA.

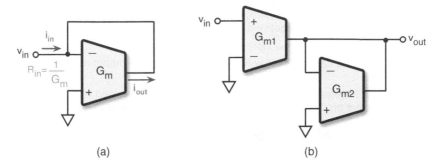

Figure 6.38   (a) Equivalent resistance; (b) amplifier with OTAs.

The transconductance gain of the OTA transforms the input voltage into a current

$$i_{out} = v_{in} G_m,$$ (6.73)

which is injected into the resistor R to obtain the output voltage

$$v_{out} = v_{in} G_m R \quad \rightarrow \quad G_v = G_m R.$$ (6.74)

The result shows a gain proportional to a fixed resistance and G_m. The value can change if G_m is tunable. The gain is positive, but an inverting result is obtained by just reversing the OTA input terminals. We can also observe that the input impedance equals that of the OTA: infinite in the ideal case. The output impedance of the amplifier is determined by the resistance R. This holds if the OTA output is ideal (a Voltage-Controlled Current Source (VCIS)). If the OTA's output impedance is not infinite, that element is in parallel to R. In this case, both gain and output impedance change. Since a load at the output terminals gives rise to a similar effect, a negligible error on G_m requires a value for R that is much smaller than the OTA output resistance and load to be chosen.

Figure 6.37(b) is an integrator. As you already know, a current injected into a capacitor integrates the current. Since in the scheme the current is proportional to input voltage, the output voltage is

$$v_{out} = v_{in} \frac{G_m}{sC},$$ (6.75)

an integrator with time constant C/G_m. The use of OTAs with adjustable G_m possibly regulates the position of the pole.

For this scheme, too, input impedance equals the impedance of the OTA. Moreover, a finite output resistance (and any load) is in parallel with C. Thus the current of the scheme outlined

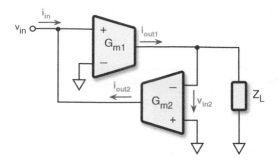

Figure 6.39   Impedance transformation with gyrator.

in Figure 6.37(b) flows in the parallel connection of capacitor and resistor, leading to

$$v_{out} = v_{in} \frac{R_{out} G_m}{1 + s C R_{out}}. \tag{6.76}$$

The pole is not at zero frequency but at $1/(2\pi C R_{out})$. Notice that at low frequency the gain, v_{out}/v_{in}, becomes $R_{out} G_m$, a value that is expected to be large. Remembering that the equivalent limit caused by the finite gain of the op-amp causes low-frequency gain equal to A_0, the product $R_{out} G_m$ must be as large as the gain of an op-amp counterpart.

Figure 6.38 shows two other possible uses of the OTA. The first scheme uses feedback between output and inverting input. Since the input impedance is infinite, the input current is

$$i_{in} = -i_{out} = v_{in} G_m, \tag{6.77}$$

which establishes a simple relationship between input voltage and current. The input resistance is

$$R_{in} = \frac{1}{G_m}. \tag{6.78}$$

The use of the scheme in Figure 6.38(a) leads to an amplifier made with two OTAs, as shown in Figure 6.38(b). This is the scheme of Figure 6.37(a) but with the resistance R_{out} implemented by the diagram in Figure 6.38(a). The first OTA generates a transconductance current proportional to the input voltage. The second OTA makes a resistance that transforms current into voltage.

An interesting use of OTAs is shown in Figure 6.39. This uses two OTAs in a feedback loop to establish a given impedance from input to ground. By inspection, we have

$$v_{in2} = -v_{in} G_{m1} Z_L \tag{6.79}$$

$$i_{in} = -i_{out2} = v_{in} G_{m1} G_{m2} Z_L, \tag{6.80}$$

which produces the input impedance

$$Z_{in} = \frac{1}{G_{m1} G_{m2} Z_L}. \tag{6.81}$$

Thus the scheme can give the equivalent of an inductor with a capacitance, and vice versa. Because of this feature the scheme is called a gyrator. Realizing the operation of an inductor with OTAs and a capacitor is a relevant result, because inductors are bulky elements with

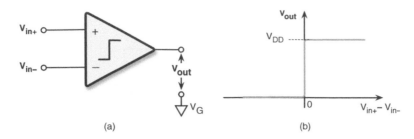

Figure 6.40 (a) Symbol for a comparator; (b) input–output transfer characteristic.

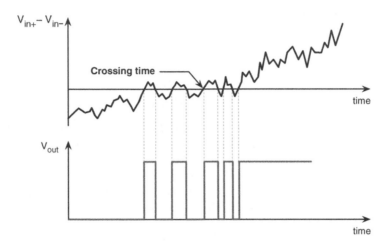

Figure 6.41 Response of a comparator with noisy inputs.

performances far away from that of an the ideal component. The scheme can be conveniently used in filters operating at low to medium frequencies.

There are several other circuit configurations that produce analog processing functions with OTAs. They are studied in other courses, such as those on analog filtering.

6.8 COMPARATOR

Another relevant analog block is the comparator. It transforms a magnitude relationship between analog voltages or analog currents into a logic signal, often represented by voltage. Comparators can be used for polarity identification, threshold detection, square/triangular-wave generation, and many other needs. A comparator typically has two inputs and one output. Some comparators use multiple inputs. In those cases the output depends on a defined relationship between input magnitudes. The output has two values: low – close to ground or the negative voltage supply – and high, close to the positive rail.

As indicated by the name, the output of a comparator gives a comparison between the inputs. If the first input (let us say this is applied to the positive terminal) exceeds the level on the negative terminal, then output is high; otherwise it is low. In principle, any high-gain amplifier does this. In reality there are differences between devices designed as op-amps and

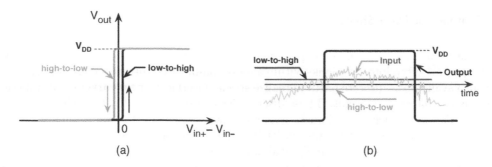

Figure 6.42   (a) Input–output response with hysteresis; (b) response of comparator with hysteresis and noisy inputs.

those designed as comparators. For example, op-amps aim to operate in the linear region, and use very small amplitudes as differential inputs. By contrast, the input of comparators can be large, and large inputs bring the comparator into non-linear saturated regions, from which it must be able to recover in a very short time.

Figure 6.40(a) shows the comparator symbol. Like that for operational amplifiers, it is a triangle with two inputs and one output. The step inside the triangle indicates the transition that occurs at output when inputs cross one another. Figure 6.40(b) shows the input–output transfer characteristic. In the ideal case it is a step that jumps from zero to V_{DD} when the difference $(V_{in+} - V_{in-})$ crosses zero. In real cases the response is not so steep but shows a finite slope around zero.

Ideally the output of a comparator switches immediately in response to input. In reality circuits take some time to give output. Typically the delay depends on how fast the zero crossing is and the extent of the signal before and after the crossing. An immediate response seems to be a good feature. However, is some situations high switching speed is problematic because the input differential signal does not cross the zero in a definite way. Because of noise or interference, the crossing often looks like the signal in Figure 6.41. The output is zero well before and one well after crossing, but it jumps back and forth many times because of multiple zero crossings caused by small, uncertain changes. This behavior is unwanted, because it is normally desirable to have a well-defined transition.

A possible solution is to have hysteresis in the response, as shown in Figure 6.42(a). The transition from low to high occurs for a small positive input. Switching in the other direction needs a slightly negative signal. Therefore if the output is high it remains high, even if input becomes negative but higher than the high-to-low threshold. Only when input crosses the high-to-low threshold does the output swing to zero. It remains at zero until the next crossing of the low-to-high threshold. Figure 6.42(b) shows a possible time-domain response of a comparator with hysteresis. The signal is noisy and fluctuates around an average waveform. It crosses zero many times, but the output changes only twice to record a solid trend of input signal. The time of switching does not exactly indicate that inputs are equal, but the small error is an acceptable compromise.

Hysteresis is not a feature normally offered by commercial components. An external network made up of passive elements (capacitors or, more often, resistors) carries out the function by establishing what is called positive feedback. The value of the resistors (or capacitors) determines the width of the hysteresis interval.

6.8.1 Comparator Data Sheet

Comparators are normally used to perform complex functions, typically data conversion. Therefore they are often internal blocks within a more complicated integrated circuit. However, they are also available as discrete parts to realize special interfaces or for carrying out functions not given by the integrated circuit. Those discrete parts are fabricated with CMOS or bipolar technology. The IC, which often houses four comparators sharing the same pins for supply biasing, satisfies requirements for high sensitivity, high speed or low power.

Comparator data sheets are normally available on the websites of providers to give all the relevant information for use. They are structured in the same manner as op-amp data sheets, whose organization was discussed before in detail. An initial general description gives the overall information on the parts of a given family of devices. Often the general description also includes a schematic diagram of a typical application. Then the data sheet gives warnings on absolute maximum ratings, tables and diagrams with electrical static and dynamic characteristics, and, at the end, information on packaging and board assembly.

Many of the parameters given are similar to those of the operational amplifier. Parameters that are specific to comparators are as follows.

- **Large signal voltage gain, $A_{v,d}$**, indicates the sensitivity of the comparator. The action of a comparator is, indeed, amplification of the input difference and saturation of the voltage generated within the supply rails. The value of $A_{v,d}$ is normally given in [V/mV].

- **Static sensitivity, Δ_s**, is the minimum input imbalance (excluding offset) that gives a well-defined logic signal at output. It depends on $A_{v,d}$ and the supply voltage.

- **Logic "1" and logic "0" output voltages** are the minimum guaranteed voltage (close to the positive rail) generated to provide a logic one and the maximum guaranteed voltage (close to the negative supply) generated to provide a logic zero.

- **Propagation delay, t_p**, is the time required to obtain a logic level at output after a step with a given amplitude is applied at input. The amplitude before the step change, specified in the data sheet, is typically ± 100 mV. The signal after the step change, called overdrive, has the opposite sign and an amplitude higher than Δ_s. The response depends on the input amplitude and, obviously, is faster for higher inputs, as shown in Figure 6.43 for a low-power comparator.

- **Large-signal response time** is the response time with large overdrive, as it results from a logic input. It is much smaller than the propagation delay.

Some values for electrical characteristics are given in Table 6.2. These refer to hypothetical comparators and give typical, minimum, and maximum values. An offset of 1 mV or less indicates a very good design. Even the drift of offset is remarkable. The propagation delays are for low-power and high-speed parts. The differences (more than two orders of magnitude) are significant. This is because when the design focuses on low power it must sacrifice other kinds of performance. Since speed is power hungry, related performances such as propagation delay are greatly affected.

The specification describes through diagrams how parameters change with electrical or physical variables. For example, the offset changes with common-mode input voltage. It is a

COMPUTER EXPERIMENT 6.6

Understanding Clocked Comparators

The cascade of a sample-and-hold, a preamplifier and a latch makes a clocked comparator. In some cases, as in this ElvisLab experiment, the input is differential. The board uses a single-ended to differential (S/D) block that transforms the addition of two input signals into a differential pair whose values cross at half of the supply voltage. The selectors of this computer experiment enable you to test relevant internal points like one of the outputs of the S&H and the differential outputs of the preamplifier. Suitable buffers at the test points (not shown in the diagram) avoid the load that would influence normal operation of the circuit.

In this experiment you can:

- set amplitude and frequency of sine wave and triangular wave. You can also set the phase of the sine wave;
- set the frequency of the clock. The rising edge activates sampling, while the fall time starts the latch regenerating.

MEASUREMENT SET–UP

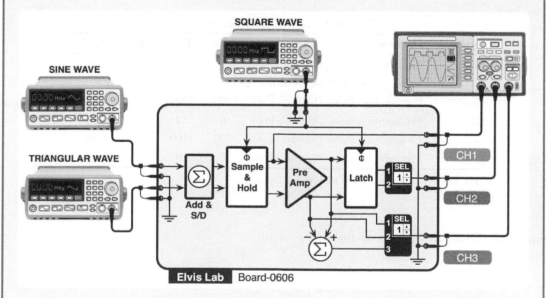

TO DO

- Set the triangular waveform amplitude to zero. Use zero frequency and change the phase of the sine wave to give a constant input voltage. Observe the latch waveforms.
- Set amplitude and frequency of both sine and triangular waves. Choose a frequency of the clock that gives rise to a demanding sequence of samples, for example, a large positive followed by a large negative, a large positive followed by a small negative and vice-versa.
- Observe the outputs of the preamplifier and their difference. Measure the low–frequency gain of the preamplifier and observe what happens when the sampling frequency increases.
- Use very high sampling frequency and increase the frequency of the input signal until the preamplifier starts lagging behind and the signal at the latch input becomes inadequate.

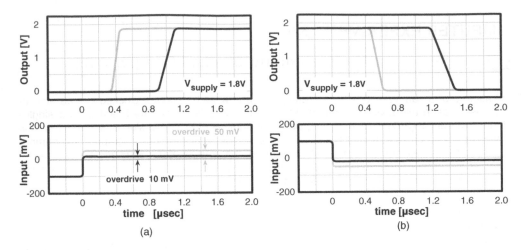

Figure 6.43   Propagation delay of a low-power comparator with two different overdrives: (a) low to high; (b) high to low.

Table 6.2   Electrical characteristics

Symbol	Parameter	Min	Typ	Max	Unit
V_{OS}	Input offset voltage		0.2	1	mV
$TC\,V_{OS}$	Input offset temperature drift		15		μV/C$^{\circ}$
t^{*}_{PHL}	Propagation delay (high-to-low)	800	1200		ns
t^{**}_{PHL}	Propagation delay (high-to-low)	6	12		ns
t^{*}_{PLH}	Propagation delay (low-to-high)	700	1100		ns
t^{**}_{PLH}	Propagation delay (low-to-high)	5	10		ns

*For low-power parts. ** For high-speed parts.

parameter similar to what was defined for op-amps:

$$V_{CM} = \frac{V_{in+} + V_{in-}}{2}. \tag{6.82}$$

Figure 6.44(a) shows possible behavior. The supply voltage is 1.8 V and the common-mode voltage covers the entire supply range; therefore, the input performance is rail-to-rail. The offset is for three different temperatures: room temperature (27°C) and the two limits of industrial temperature range, −40°C and 85°C (the commercial temperature range is 0°C to 90°C). The offset is 0 mV for input common mode at half the supply voltage and room temperature, and it becomes positive or negative, depending on the conditions.

Figure 6.44(b) gives another possible diagram. It depicts propagation delay versus input overdrive. As expected, the diagram shows that the parameter drops when overdrive increases. The specified numerical values are small, indicating a high-speed comparator that, probably, consumes power.

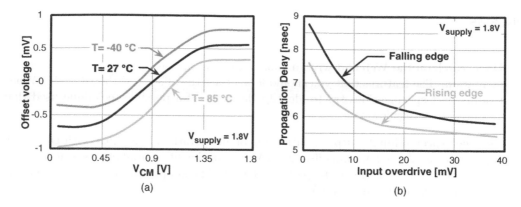

Figure 6.44   (a) Offset of hypothetical comparator as function of input common-mode voltage at different temperatures; (b) propagation delay with rising and falling edges versus overdrive.

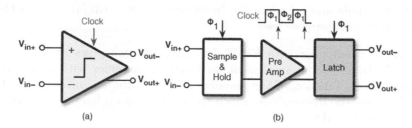

Figure 6.45   (a) Schematic of a clocked comparator; (b) details of the high-level architecture of clocked comparator: sampling lasts for the entire phase Φ_1, and latch becomes active at the falling edge of Φ_1.

6.8.2   Clocked Comparator

The comparator is often used in discrete-time systems. The required function is to compare two inputs at discrete times, defined by the clock (or strobe). We have already briefly discussed this type of comparator in Chapter 5. Their output refers to the value of inputs at the sampling times and remains unchanged for an entire sampling period. Moreover, the operation is not affected by the inputs at times other than what is defined by the discrete system. Normally the output of clocked comparators is available for half or an entire clock period after sampling. The output can be a single logic signal or a pair of complementary signals.

Figure 6.45 shows the symbol for, and a possible architecture of, the clocked comparator. The square wave making the clock identifies two phases, Φ_1 and Φ_2. The first determines the sampling of input and its hold. The end of Φ_1 is the time at which input is sampled. Then the input held during Φ_2 is amplified by the block called *PreAmp*. The amplified result is the input of another block, the latch, activated by the phase Φ_1. The latch is a special circuit that generates a logic signal based on the sign for differential input at latch time. The circuit produces output by augmenting the difference of inputs in a regenerative manner. This is obtained by using positive feedback. However, the latch needs a minimum amplitude difference between inputs to operate properly. This is why Figure 6.45 uses the pre-amplifier before the latch.

PROBLEMS

6.1   Do a search on the Web and write a short report on operational amplifiers for instrumentation. The features that you should look at are the offset and the DC gain. When reading part specifications try to identify the name of a possible technique used that gives rise to a very low offset.

6.2   Plot the expected input/output relationship of an operational amplifier with DC small-signal gain equal to 80 dB and offset 2.3 mV. One of the differential inputs is at constant voltage. Use that voltage as the plot parameter. The supply voltage is 1.2 V and the common-mode input range is 0.5–0.9 V.

6.3   The small-signal frequency response of an operational amplifier features 76 dB DC gain and two poles, at $f_1 = 1$ kHz and $f_2 = 10$ MHz. Write the transfer function using the complex variable s, and estimate the phase margin at f_2. Extrapolate the Bode plot after the first pole and calculate the frequency at which $|A(f)| = 1$. Is the op-amp suitable for general-purpose use?

6.4   An operational amplifier with 83 dB DC gain and two poles at $f_1 = 1$ kHz and $f_2 = 10$ MHz is used to implement an inverting amplifier with gain -10. Determine the transfer function of the amplifier and draft the magnitude and phase Bode plots. Estimate the frequency at which the Bode diagram loses 3 dB with respect to the low-frequency value.

6.5   The small-signal equivalent circuit of a two-stage op-amp has the configuration shown in the figure below.

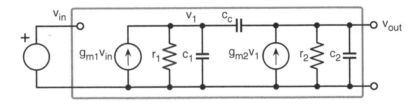

Determine the transfer function, and plot poles and zeros on the s plane, for $r_1 = r_2 = 10$ kΩ, $c_1 = c_2 = 0.2$ pF, $g_{m1} = g_{m2} = 10$ mA/V.

6.6   Determine the gain of the amplifier shown in the figure below. Estimate the gain at very low frequency and very high frequency. Suppose $R_2 = 10$ kΩ, $R_1 = 100$ kΩ, $C_1 = 10$ pF. The op-amp is ideal. Plot the Bode diagrams and determine the gain (modulus and phase) at $f = 100$ MHz.

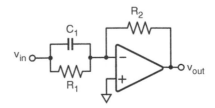

6.7   The circuit schematic of the figure below uses $R_2 = 20R_1 = 5R_3 = 250$ kΩ, $C_2 = 10$ pF, $C_3 = 5$ pF. Determine the frequency response and plot the approximate Bode diagram of the modulus.

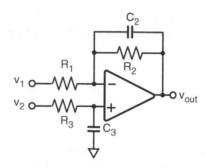

6.8   Determine the input currents of the schemes of Problems 6.6 and 6.7. Plot on log–log diagrams the current amplitude versus frequency using 1 mA and 1 Hz as the current and frequency references.

6.9   The transfer function of the operational amplifier used in Figure 6.18(b) is

$$A(s) = \frac{A_0}{(1 + s\tau_d)(1 + s\tau_{nd})},$$

where A_0 is 78 dB, $\tau_d = 10^{-5}$ sec, and $\tau_{nd} = 10^{-9}$ sec. Suppose $V_{in,2} = 0$. Determine the open loop unity gain frequency, the gain of the amplifier and the phase at 100 MHz. Calculate the frequency at which the phase is 330°. Use $R_1 = 100$ kΩ, $R_2 = 1$ MΩ.

6.10   A unity gain buffer uses an operational amplifier with finite gain equal to 2300 and $f_T = 830$ MHz. Determine the value of the gain at $f_1 = 1$ MHz and $f_2 = 830$ MHz. Suppose that fabrication advantages or inaccuracies change the op-amp gain in the range 35–90dB. What is the effect on the low-frequency gain? Plot the result on a dB–log diagram.

6.11   Design an active circuit performing the linear operation $V_{out} = 3 \cdot V_{in,1} - 5 \cdot V_{in,2}$. Use an operational amplifier with finite gain 9860 and $f_T = 1.1$ GHz. Estimate the low-frequency error and determine the response at $f_1 = 800$ MHz and $f_2 = 1.4$ GHz.

6.12   A generalized amplifier like the one of Figure 6.24(a) uses the following impedances:

$$z_1 = (10^4 + j \cdot 10^{11}/f)[\Omega]$$

$$z_2 = \frac{10^5}{1 + jf \cdot 10^{-7}}[\Omega].$$

Determine the small-signal response with an ideal op-amp and with an op-amp whose DC gain is 10^3 and has a dominant pole at 10^4 Hz.

6.13   An analog integrator and a derivator (Figure 6.22 and Figure 6.23) both use an operational amplifier whose small-signal equivalent circuit is made by a transconductance generator and a parallel RC. The DC gain is 7300, the dominant pole is at 300 Hz, and the output resistance is 500 Ω. Determine the frequency responses with $R_1 = R_2 = 10$k Ω and $C_1 = C_2 = 10$ pF. Plot the amplitude Bode diagram.

6.14   The offset of an operational amplifier is 8mV. What is the waveform at the output of an integrator and a derivator with time constant $\tau = 2 \cdot 10^{-6}$ s? Assume the use at input of a 10 mV sine wave at 180 kHz and 10 MHz.

6.15   You have to design an amplifier with gain -100. The resistors available have a matching accuracy of 0.3%. What is the minimum DC gain that gives rise to an error lower than the one caused by the worst-case resistor mismatch? Repeat the calculation for a -1000 gain.

6.16   An inverting amplifier with gain -10 has at input a sine wave and a DC component. The circuit uses an op-amp with 0.3–1.5 V as output dynamic range. Draw a plot of the sine wave amplitude versus the DC level that avoids output saturation.

6.17   Determine the gain of a circuit that uses two equal OTAs with $G_m = 2$ mA/V. The input voltage generator drives the first OTA, whose output current flows into an impedance made by the series of $R = 100$ kΩ and $C = 2$ pF. The voltage across the capacitor is the input of the second OTA, whose output current flows into the parallel connection of $R = 180$ kΩ and $C = 1$ pF. The output is the voltage across the parallel load.

6.18   A transconductor with $G_m = 10.2$ mA/V and input 2 mV drives a latch whose parasitic capacitance is 0.2 pF. The latch requires for a safe comparison at least 25 mV at its input. What is the maximum possible speed of the clock, supposing that half of the period is for the pre-amplification and half for the regenerative operation of the latch?

CHAPTER 7

DATA CONVERTERS

This chapter deals with features, algorithms and design techniques for analog-to-digital and digital-to-analog conversion. Analog and digital signals have already been studied. Moreover, the benefits and the limitations of signal processing in the analog and digital domains have been analyzed. As well as that, it is necessary to know how to obtain, using electronic circuits, the transition from one format to the other. This is actually what this chapter does. We shall start with the features and specifications of data converters; then we shall learn some popular conversion algorithms and their high-level circuit implementations. The benefits and limitations of key architectures are also examined.

7.1 INTRODUCTION

An Analog-to-Digital (A/D) Converter (ADC) transforms analog signals into digital codes at the interface between analog and digital processing. In the opposite direction, a Digital-to-Analog (D/A) Converter (DAC) generates an analog signal under the control of a digital code. All signals involved in these processes are discrete-time signals and have continuous amplitude on the analog side and discrete amplitude on the digital side. A number of electrical signals, carried by wires, represent the digital code. If the code uses N bits, there are N logic signals whose amplitude is low or high to represent two possible values: one or zero.

Understanding Microelectronics: A Top-Down Approach, First Edition. Franco Maloberti.
© 2012 John Wiley & Sons, Ltd. Published 2012 by John Wiley & Sons, Ltd.

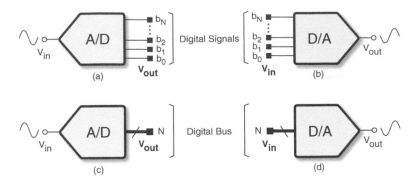

Figure 7.1 (a) Symbol for an A/D converter; (b) symbol for a D/A converter; (c) simplified symbol for A/D with a single wire representing digital bus; (d) simplified symbol for D/A, also with single wire representing bus.

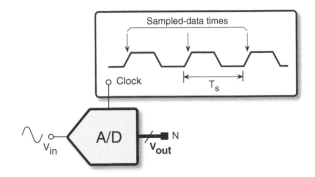

Figure 7.2 Symbol for an A/D converter, outlining the clock – its waveform gives information on conversion speed and sampled data timing.

Figure 7.1 shows the symbols for an A/D and a D/A converter. The signal in all of the diagrams in this figure flows from left to right. There is only one wire on the analog terminals, and there are multiple wires on the digital sides. The symbols in Figure 7.1(c) and (d) group the N wires into a bus, which in some cases also provides information on the number of wires making up the bus, as shown in Figure 7.1(c) and (d).

Obviously the converter requires a timing control to establish the rate of input sampling and the rate at which the output is generated. This is established by a specific signal, the clock, carried by a wire that, in some symbols, is not shown. Some implementations do not need a clock from outside because it is generated internally. The clock, which can be obtained from a sinusoidal waveform, is a square wave with defined rise and fall times, as shown in Figure 7.2. The clock period establishes the conversion frequency with its rise or fall front defining the sampling instants. Thus the clock transitions should be sharp.

Real circuits do not generate their output instantaneously. Architectures that use one or more data converters must account for that. Only a few schemes make available the converted result immediately after one clock period. Often the converter needs more time for the operation. Moreover, the conversion period can differ from the time elapsed between input and output. In such a case the converter is said to have a *latency time*. Some schemes an algorithm for which the latency time can be many clock periods. That is not a serious limitation;

however, the issue must be kept in mind, because having delays (analog or digital) can be problematic in real-time processors with feedback.

Data converters are tailored to diversified applications that require different levels of accuracy, speed, and power consumption. For example, the conditioning of signals from sensors, such as strain gauges, flowmeters, or load cells, can require as many as 24 bits, with more than 19 bits of signal-to-noise ratio with low signal bands. Data conversion for wireless communications is also very demanding. The very high bandwidth, the need to increase data rates, and the requirement to improve spectral efficiency lead to challenging specifications. The trend of these systems is to perform almost all the signal processing in the digital domain, with data converters at the ends of the processing chain. The performance required implies conversion speeds of hundred of MHz or even GHz and accuracies of 10–12 bits or more.

7.2   TYPES AND SPECIFICATIONS

A data converter is normally named after the conversion algorithm. For example, we have flash, successive approximation, pipeline or Sigma–Delta converters. Besides that, data converters are also classified into two big categories: Nyquist-rate and oversampled converters. The Nyquist-rate kind uses sampling frequencies near the minimum prescribed by the Nyquist theorem, which, remember, requires a sampling rate at least twice the signal band, f_B. On the other hand, oversampled data converters use much higher sampling rates. The Oversampling Ratio, OSR $= f_s/(2f_B)$, distinguishes the Nyquist-rate from the oversampling architectures. With an OSR equal to eight or more we consider the converter to be oversampled.

> **Notice**
>
> Oversampling means using electronic circuits that work at frequencies higher than their Nyquist-rate counterparts. The extra effort spent in handling high-frequency signals should be repaid by corresponding benefits.

In addition to type, number of bits, and sampling frequency, the designer obtains technical information and functional descriptions from the data sheet. This document is organized into four sections:

- general features, applications and description;
- electrical characteristics;
- electrical dynamic specifications;
- digital and switching data.

7.2.1   General Features

The general features section always provides a description of the data converter together with other miscellaneous information. The type of converter, its key features, and lists of possible applications are reported. A short description outlines the benefits of the part, mentions the technology used, and highlights useful characteristics such as pin-to-pin compatibility with other parts or low-power operation.

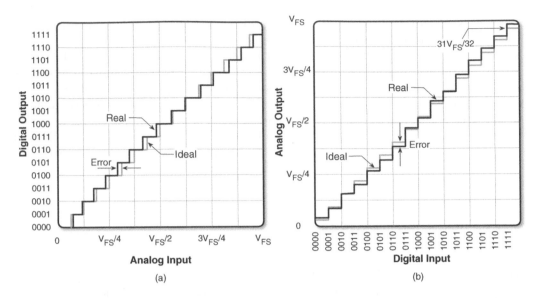

Figure 7.3   Input–output static characteristics: (a) of a four-bit ADC; (b) of a four-bit DAC.

Other data gives the number of pins, and their configuration and function. Often integrated circuits use multiple pins for the supply voltages. This ensures a low series inductance that possibly causes drop voltages in the presence of quick variations of the current. Moreover, some integrated circuits use separate pins for the supply of analog and digital sections. Information on how to assemble the part on the PCB are also provided.

7.2.2   Electrical Static Specifications

As implied by the name, the electrical static specifications concern the circuit behavior with a constant signal at input. The most relevant of them is the number of bits. It assumes a binary format with N bits to denote the digital code. The number of discrete analog levels is 2^N. Other static specifications result from the so-called input–output transfer characteristic. For an ADC this gives the output code as a function of the input analog signal. In a DAC, on the other hand, the transfer curve displays the analog output versus the digital input code.

Figure 7.3 shows a typical input–output transfer characteristic of a four-bit A/D converter (or ADC) and a four-bit D/A converter (or DAC). The ideal response of Figure 7.3(a) distinguishes 16 equal parts in the ADC's input range $(0, V_{FS})$. The code 0000 denotes the first interval $(0, V_{FS}/16)$. The second interval is associated with 0001, and so forth until 1111, which symbolizes the last interval $(V_{FS} \cdot 15/16, V_{FS})$. A staircase with equal steps plots the static curve. Figure 7.3(a) also displays a real input–output response. The transitions from one code to the next do not occur exactly where they might be expected, because of positive or negative errors that alter the transition points. Thus the steps have equal risers but different treads.

Figure 7.3(b) shows the DAC response. Steps equal to $V_{FS}/16$ feature the difference between the analog signals generated in response to digital codes. The first amplitude is inside the first analog interval (or step). Any value from that interval is acceptable. Normally it is the beginning, the middle point, or the end of the interval. Figure 7.3(b) uses $V_{FS}/32$, half of

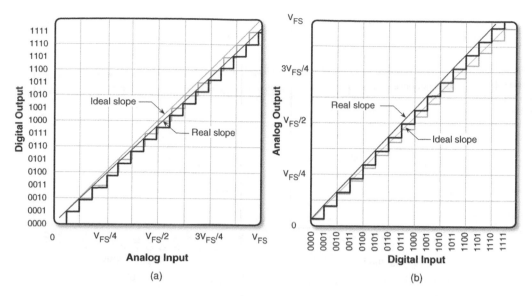

Figure 7.4   Gain error in the static characteristics: (a) of a four-bit ADC; (b) of a four-bit DAC.

the step; accordingly, the full analog scale becomes $31/32$ V_{FS}. Figure 7.3(b) shows a real transfer characteristic in addition to the ideal curve. The amplitude of its steps differs from the expected $V_{FS}/16$ in a random manner because of errors in fabrication; the value of the last generated voltage incidentally exceeds the correct value.

> **Observe**
>
> The input–output transfer curve uses a discrete analog quantity on one axis and discrete integer numbers on the other. The (Least Significant Bit) LSB is the difference between adjacent analog quantities.

The two real transfer characteristics of Figure 7.3 show disparities with respect to the ideal curves. The differences are due, among other things, to gain error, offset error, and monotonicity error. The gain error, ϵ_G, of a D/A converter denotes analog outputs equal to the expected values multiplied by $(1 + \epsilon_G)$. Having a ϵ_G gain error in A/D converters means that the full digital scale is for input analog voltages in the interval

$$V_{FS}(1 + \epsilon_G)\frac{2^N - 1}{2^N}, \ldots, V_{FS}(1 + \epsilon_G). \tag{7.1}$$

Figure 7.4 shows the transfer curves of hypothetical A/D and D/A converters with gain error. Straight lines with slopes given by the ratio of the maximum code and the highest analog voltage (or vice versa) interpolate the staircases. The difference between the slopes defines the gain error. The plot of Figure 7.4(a) reaches the full digital scale almost at V_{FS}, making the last channel very small. It may also happen that the last channels vanish. For the D/A converter the gain error of the figure gives rise to an output voltage representing code 1110 almost equal to what we expect for the full-scale code 1111. However, in both pictured cases the gain error causes full-scale deviations of less than one LSB (Least Significant Bit).

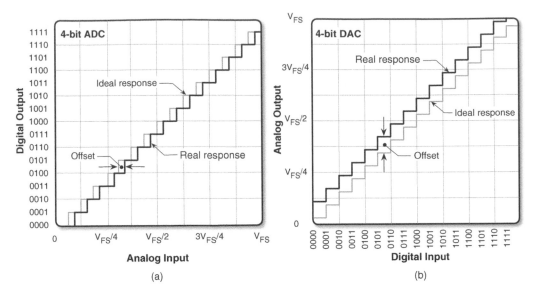

Figure 7.5   Offset error in the static characteristic: (a) of a four-bit ADC; (b) of a four-bit DAC.

The offset is a constant shift of the static response. If an analog input $V_{in,1}$ is expected to generate a digital output Y_1, the offset is such that

$$V_{in,1} - V_{os} \rightarrow Y_1, \tag{7.2}$$

or, for a digital-to-analog converter, a digital input X_1 generates the output voltage

$$X_1 \rightarrow V_{out,1} + V_{os}. \tag{7.3}$$

Figure 7.5 illustrates input–output responses with offset. They show a shift of the staircases in the horizontal or the vertical direction.

Other limitations identified by the transfer curve are non-linearity, missing codes, and non-monotonicity. Figure 7.6 illustrates them. As we know, the input–output transfer characteristic is supposed to be a staircase that rises with the input signal. Gradients in the value of the components determine non-linearities similar to the one shown in Figure 7.6(b); the curvature can also be on the other side, upwards, or the non-linearity can give an S-shaped deviation. A missing code occurs when no analog input is able to generate a given output code; Figure 7.6(c) shows the corresponding transfer curve. The response is non-monotonic, as shown in Figure 7.6(d), when an increase in the input amplitude determines a decrease in the digital output.

The transfer curve also helps in defining and estimating other parameters such as the Differential and Integral Non-Linearities (DNL and INL), which are determined by inaccurate quantization steps for different codes. Neither these nor other static parameters are discussed here in detail. You can refer to their definitions and descriptions in data sheets or specialized books.

The static specification section of the data sheet also gives self-explanatory information about supply voltage, consumed current, power, and other static parameters of the data converter.

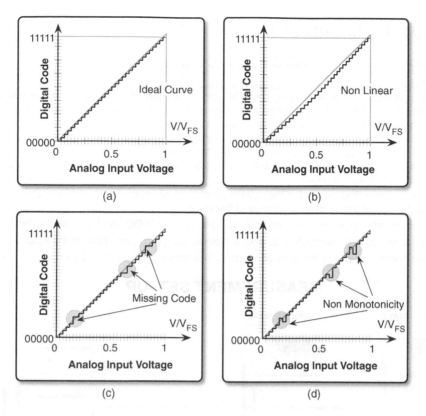

Figure 7.6   (a) Ideal static response; (b) non-linear ADC response; (c) response with missing codes; (d) non-monotonic ADC response.

7.2.3   Electrical Dynamic Specifications

Dynamic specifications describe the time effectiveness of data converters. Some of them are similar to what we already discussed for the operational amplifier and the comparator; others are specific to the data converter. For an ADC we have:

- **Aperture delay** This describes the speed of response to a sampling command. A digital "encode" signal sends the command for sampling, but the effective sampling time is not immediate. The delay may depend on signal amplitude and its derivative.

- **Aperture uncertainty (jitter)** This shows that the sampling times are not exactly what is expected, but that small fluctuations affect them. The clock is the main source of jitter inaccuracies. It must be very precisely generated and carefully handled in the network that distributes and regenerates it on the board or inside the IC. For high-resolution and high-speed A/D converters, ensuring low jitter is the key to obtaining good performance.

- **Signal-to-Noise Ratio (SNR)** You will learn at the end of this chapter that, under some conditions, the dynamic effect of quantization is equivalent to the addition of noise corrupting the signal spectrum. The assumption that the quantization error can be

COMPUTER EXPERIMENT 7.1

ADC Input–Output Characteristics

The static specifications of a Nyquist-rate ADC are generated (or verified) by the input–output static characteristic. This is the plot of analog input versus digital code at the output. Since using digital numbers is not very illustrative, it is convenient to transform the digital code into an analog equivalent and to plot the results in the analog domain. In this computer experiment, after A/D conversion, a precise D/A converter returns the signal back to analog to drive the oscilloscope display. The board also uses an interface that drives a seven-segment display showing the digital result. In this experiment you can:

- control a slow ramp generator that starts synchronously with the clock generator. You can set minimum and maximum amplitude of the ramp;
- set voltage references of ADC and DAC and set a possible DC input for static tests;
- set the three control parameters to alter the offset, gain error, and DNL of the ADC;
- set the number of bits of the data converters. The maximum possible is 12 bits.

MEASUREMENT SET–UP

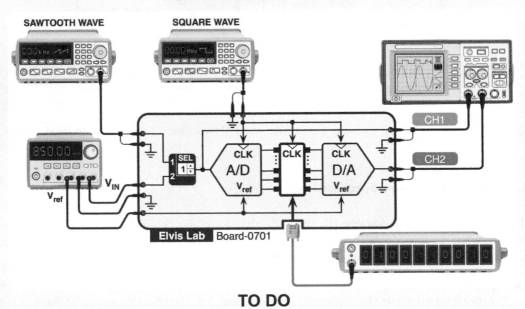

TO DO

- Set the reference voltage to 1 V and resolution to six bits. Use a 64 kHz clock frequency and set the sawtooth repetition rate to 100 Hz. This nominally leads to 10 samples per digital code. Choose sawtooth limits of 0 V and 1 V and set references to 1 V. Observe the results.

- Increase the converter resolution and use smaller sawtooth ranges. Add non–idealities and observe the results. Verify gain and offset error, and estimate the INL.

- Increase the value of INL until you notice non–monotonicity. Push the button that changes the ADC part and make a statistical analysis of offset and gain error.

- Use a fixed DC voltage at input and change it manually. Notice that the DC voltage has superposed some noise. At critical input amplitudes you would notice a fluctuation in the seven-segment display (it may be it is not visible on the scope trace).

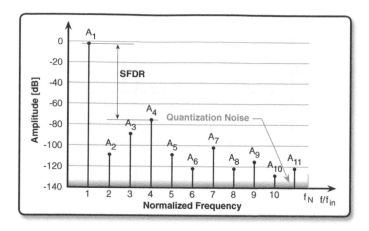

Figure 7.7 Spectrum of the signal at the output of an ADC with an analog sine wave at input.

treated as noise enables us to adopt the signal-to-noise ratio, defined in Chapter 2, as a feature of data converter performance. As you know, more bits means better accuracy; therefore, we have a higher SNR with more bits. An equation quantifies that relationship. The link between SNR measured in dB and the number of bits is calculated by

$$\text{SNR}_{\text{dB}} = 6.02 N_{bit} + 1.76, \tag{7.4}$$

where the constant factor 1.76 assumes the use of input sine waves.

- **Total Harmonic Distortion (THD)** This is caused by the non-linearity of the input–output transfer response. An input sine wave at f_{in} with amplitude A_1 produces tones at $k f_{in}$ with amplitude A_k (see Figure 7.7). The total harmonic distortion, in dB, is

$$\text{THD}_{\text{dB}} = 10 \log \frac{\sum_2^{10} A_k^2}{A_1^2}, \tag{7.5}$$

showing that the THD is normally estimated by accounting for the first ten tones. Notice that because of aliasing these can be folded. Figure 7.7 uses a low frequency of the test signal that keeps unfolded the sequence of tones in the Nyquist interval.

- **Spurious Free Dynamic Range (SFDR)** For many communication applications it is important to keep the amplitude of tones low. The reason is that tones caused by unwanted signals can fall close to the frequency range of the signal. These spurs blur the signal and degrade the associated information. It is therefore necessary that the highest spur is less than a certain level. The SFDR gives the ratio between the amplitude of a test sine wave and the largest spur generated. It is the distance in dB between the signal amplitude and the highest tone, as shown in Figure 7.7. In the case shown in Figure 7.7 the SFDR is a mediocre 74 dB.

Many other dynamic parameters of DACs are almost equivalent to those of the op-amp. A few are specific to digital-to-analog conversion. One of these is for DACs used for video applications: the glitch power. It happens that with some critical code transitions where, for instance, many bits change simultaneously from one to zero (or zero to one), the switching

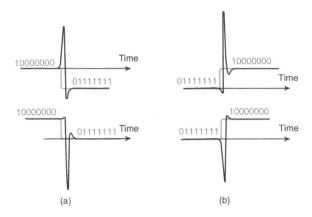

Figure 7.8   Typical glitch waveforms: (a) for downward crossing of the midpoint; (b) for upward crossing.

on and *off* of many bits can occur with short periods of overlap or disoverlap. The output becomes something like the waveforms of Figure 7.8. The downward crossing of the midpoint (Figure 7.8(a)) occurs with a delay in the switching off of the top waveforms, or the opposite happens with zeros that turn off before. The same may happen in the upwards crossing. The resulting waveforms change the average signal only slightly but contain high-frequency components. Glitches are undesirable for video applications because the eye is very sensitive to the derivative of light variation. The limit is quantified by the glitch power, which, in reality, is not power but the time integral of the positive or negative glitch measured in volts or amperes, $[V \cdot \sec]$ or $[A \cdot \sec]$.

Keep in mind

The importance of static and dynamic specifications depends on the data converter. The conversion of a signal coming from a sensor requires a very low offset. A data converter for communication needs very high linearity and ultra-high SFDR. Audio applications need at the same time very good resolution and linearity.

This subsection has examined the most important dynamic parameters. Other features are important for some applications. When using a data converter whose dynamic behavior is important it is therefore necessary to read the data sheets and application notes available on the Web carefully, so as completely to understand the dynamic operation and its effect on the system to be designed or used.

7.2.4   Digital and Switching Data

The digital and switching data section of data sheets helps in the proper handling of digital signals and gives instructions on generation and clock control. Proper sequences of digital signals are used at the input and output ports. Moreover, the algorithm implemented by the data converter can need special logic patterns. This section specifies these and defines the correct logic levels and timing. When relevant, the data sheet specifies delay, rise period, fall period, and synchronization for exchanging data between the data converter and external circuitry.

7.3   FILTERS FOR DATA CONVERSION

The use of analog or digital filters accompanies and sustains the operation of data converters. Analog filters pre-condition the signal before the A/D conversion or post-process it after the D/A conversion. Digital filters support an increase or a reduction in sampling frequency or improve the performance of the digital results. This section studies the types of analog and digital filters frequently used with data converters, and looks at their features.

7.3.1   Anti-aliasing and Reconstruction Filters

We have already studied aliasing in Chapter 2. It is the folding in the Base-Band (BB) of frequency components from high Nyquist bands. The base-band $(0, f_{CK}/2)$, in addition to its own spectrum, can contain terms folded from even Nyquist intervals, and terms shifted from odd Nyquist intervals. Obviously aliasing is undesirable because mixing spectra corrupts the signal.

The so-called anti-aliasing filter obtains a substantial reduction in the spur components that would be aliased in the band of interest. Notice that the filter is necessary even when the input is band limited, because, superposed to the signal, we always have high-frequency spurs. The anti-aliasing filter is, indeed, not for the signal that is normally band limited, but mainly used to reject spurs before sampling.

Specifying the anti-aliasing filter is part of the designer's job. Let us suppose that f_B is the band of interest and $f_N = f_s/2$ is the Nyquist interval. It is evident that $(0, f_B)$ is the frequency range that needs protection against aliasing. However, since f_B is just a fraction of the Nyquist interval, what falls out of band (f_B, f_N) does not alter the information carried by the signal. Possible spurs aliased in the out-of-band interval are obviously unwanted, but they can be filtered out afterwards in the digital domain.

Figure 7.9 identifies the first frequency region, indicated by "AL", that aliases in the base-band. Other folded bands will come from higher Nyquist zones. Since the "AL" frequency interval is problematic, the designer defines the anti-aliasing filter mask accordingly. The bandpass region is the signal band. The stopband starts at $f_s - f_B$ and the gray region is in between, as shown in Figure 7.9.

Remark

Architectures with data converters always include analog filters. They are necessary before sampling in A/D, and are used in D/A for reconstruction. In some cases the filter is not an electronic circuit but is obtained by a parasitic or an actuator. A loudspeaker, for instance, operates as low-pass filter.

The filter complexity depends on the requirements on the pass-band, the rejection in the stopband, and the width of the gray region. If the gray interval is large, filter specifications relax. Thus the relatively low sampling frequency of Figure 7.9(a) generates more difficult anti-aliasing specifications than the situation of Figure 7.9(b).

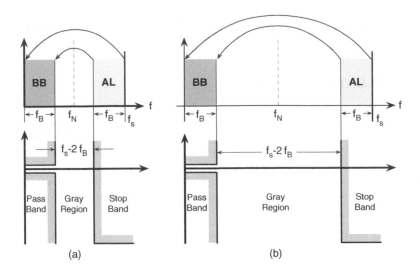

Figure 7.9 Frequency region that causes aliasing and anti-aliasing filter mask: (a) low sampling frequency; (b) high sampling frequency.

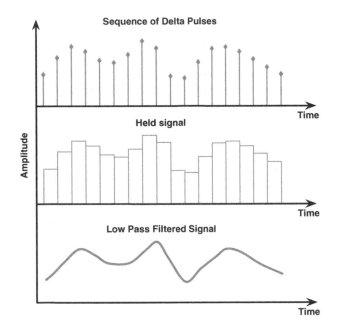

Figure 7.10 Steps of the reconstruction process: sequence of delta pulses (top), held signal (middle), and reconstructed result (bottom).

The anti-aliasing is a pre-filter that is always necessary before sampling. On the other side the digital-to-analog converter uses post-filters to obtain reconstruction. The conceptual output of a D/A converter is a sequence of deltas (or short pulses), whose amplitude is the analog representation of the input code. A sequence of deltas is not a convenient analog waveform because it contains large high-frequency components. A first improvement toward

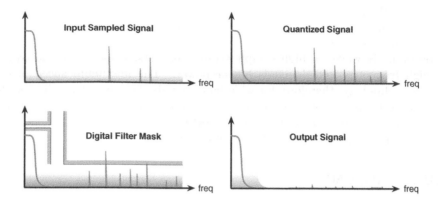

Figure 7.11   Steps for the improvement of the output resolution by digital filtering.

continuous time is obtained by holding the amplitude of the deltas for an entire sampling period as shown in Figure 7.10. The use of a low-pass filter further improves the spectrum. The name of this filter is the reconstruction filter because it serves to attenuate harmonic bands, as shown in Figure 7.10, and reconstruct a continuous-time waveform. Indeed, the result is not what we strictly desire, because any real reconstruction filter is not able to completely remove the high-frequency band. A real filter can only attenuate signal components. The complexity of the reconstruction filter depends on the separation of the base-band and the first image, for similar reasons concerning the anti-aliasing specifications. Close images make the reconstruction filter specifications more complex.

7.3.2   Oversampling and Digital Filters

We have just observed that the use of large oversampling (the ratio between the Nyquist limit and the band of the signal, OSR $= f_s/(2f_B)$) relaxes the specification of analog filters in a data conversion system. The benefit of oversampling is more general because errors causing frequency spurs outside the signal band can be removed digitally. Suppose that the quantization error gives rise to out-of-band spurs or noise-like components. A digital filter removes those terms and improves the quality of the result. Figure 7.11 resumes the steps of improvement. Interfering tones and noise corrupt the input signal. A digital filter with a suitable mask cleans up the out-of-band spectrum.

Notice that digital filtering involves additions and multiplications. The word length of the result can be larger than that of the input. Suppose, for example, that we use the simple digital transfer function

$$H(z) = 1 + 2z^{-1} + z^{-2}, \tag{7.6}$$

i.e., the superposition of the input with its weighted delayed versions by one and two clock periods. The full-scale output is the multiplication by four of the full-scale input. So its number of bits is two more than those of the input. This does not mean that the resolution increases by two bits, because what matters is the number of meaningful bits. We have to discard bits that are just numerical noise. Therefore, if we know that the use of the filter improves the accuracy by k bits we have to consider only k bits more than the input resolution. This is convenient even with intermediate results during the digital calculation. It is worthwhile to truncate partial results because they do not carry any information.

The above study shows that oversampling and digital filtering can improve the data converter's resolution. However, a high sampling-rate has costs, because the electronics must operate at high frequency and demands higher power. The maximum speed of operation depends on the technology; therefore, for wide-band signals there is not much room for oversampling. Moreover, storing or transmitting digital data with a high sampling rate is not convenient. Often an extra processing step, supported by digital filtering, called *decimation* reduces the sampling frequency to near the Nyquist limit.

7.4 NYQUIST-RATE DAC

Nyquist-rate digital-to-analog converters generate an analog signal using a reference quantity as the full scale. The sampling rate, as implied by the name, is a little more than twice the signal band.

There are many architectures that provide the DAC function. Often they attenuate the reference variable. Since the accuracy of the reference influences the accuracy of the final result, having a good reference, stable in time and temperature, and free from noise, spurs and interference, is the key to good designs. The reference generator, in addition to having accurate features, must be able to drive the circuit making the converter. As will be studied shortly, ADCs often use networks that change the load depending on the input code. This can complicate the design because of the request that within a conversion period voltages or currents settle with an accuracy better than the resolution requires.

Reference quantities can be generated inside the circuit or provided from the outside. In the latter case the connections and the PCB layout must avoid drops in voltage, leaks of current, and mismatches at a level that degrades results.

Both D/A and A/D converters use a clock to control the conversion timing. Any error in the clock causes errors in the final accuracy. Often there is a master clock generated internally or provided outside the circuit as a sine wave or a square wave with sharp rise and fall transitions. Inside the circuit a logic section uses the main clock to generate all the signals required.

For high speed a sine wave as the main external clock is more convenient than a square wave, because it is easier to have a pure sine wave. Then, inside the chip, the sine wave is squared to become the master clock.

7.4.1 Resistor-based Architectures

The simplest form of D/A converter is the resistive (or Kelvin) divider. It is the series connection of unity resistors biased by the voltage reference. Figure 7.12 shows the circuit configuration for three bits. The resistive divider generates eight voltages, including the analog ground. All of the voltages can be selected by closing one (and only one) switch.

Suppose that we close the ith switch; then

$$V_{out} = \frac{iR}{8R}V_{Ref}. \tag{7.7}$$

Thus the simple scheme of Figure 7.12 generates all of the possible quantized outputs with quantization step equal to $V_{Ref}/8$. The clock and the digital input control the switches. If the

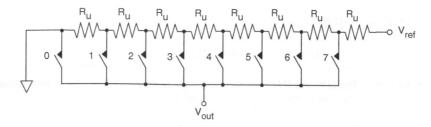

Figure 7.12   Three-bit DAC made with a resistive (Kelvin) divider.

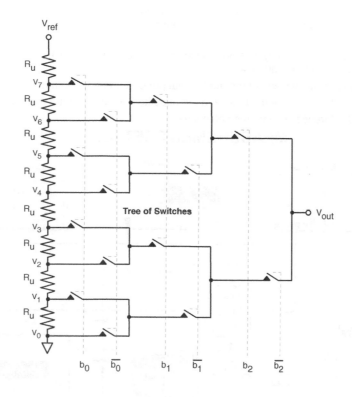

Figure 7.13   Tree of switches for the selection of output voltages.

selected switch remains closed for an entire clock period, the output voltage provides a held version of the DAC output.

The logic closes the switch 0 when the three-bit input is 000; the switch 1 closes for 001 at the input, and so forth. Therefore the maximum voltage, for 111, is $(7/8)V_{Ref}$. Some DACs use only seven resistors to obtain V_{Ref} as at the full digital scale.

A decoder controls the switches. It is made by logic circuits or by a tree of switches, as shown in Figure 7.13. The scheme uses three switches, controlled by a bit or its inverse (as denoted by the over-bar), from each intermediate node to output. With 000 at input, the three switches at the bottom of the diagram close, leading the analog ground voltage to the output. The selection of the Kelvin divider voltages just needs the inputs and their inverse with a level able to operate switches. A digital decoder, on the other hand, requires a more complex logic

COMPUTER EXPERIMENT 7.2

Understanding the Kelvin Divider DAC

This experiment uses Nyquist-rate DAC architecture with a Kelvin divider. As is well known, a string of unity resistors makes all the discrete output voltages available and an array of digitally controlled switches makes the selected voltage available at output. This computer experiment studies the mismatch between nominally equal resistances and observes the limit of noise over the reference generator. The experimental setup uses a multi–format data generator that gives, at output, the data pattern sequence and its corresponding analog waveform. You can:

- set the sequence of bits generated by the multi-format generator. The sequence is made of up to 16 words of up to 12 bits each. The sequence is cyclically repeated at fixed rate (1 kHz);
- set the conversion frequency. It is recommended to use a multiple of 1 kHz multiplied by the number of words of the sequence;
- regulate the full scale voltage of the multiformat generator analog output;
- define the number of bits of the DAC and the expected accuracy of the resistors;
- set the DAC reference voltage and noise level.

MEASUREMENT SET–UP

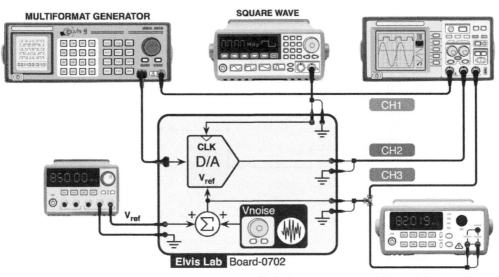

TO DO

- Set noise and mismatch of unity resistances to zero. Use six-bit and data pattern that alternates from 000000 to 001111. Select a sampling frequency that gives rise to eight samples per code. Observe outputs and trim the D/A reference for making the two staircases equal.
- Change the mismatch of resistors and observe waveforms. Increase the number of bits. Observe the staircases at the beginning, middle and end of the response. Draft the INL.
- Add noise and observe that the response changes every measure. Expand one region of the input-output curve and use a sampling frequency that gives 128 points per code. Determine the actual width of that code for a given resistor mismatch.
- The multimeter displays the average voltage every data sweep. Notice how the noise changes in time with the average reference.

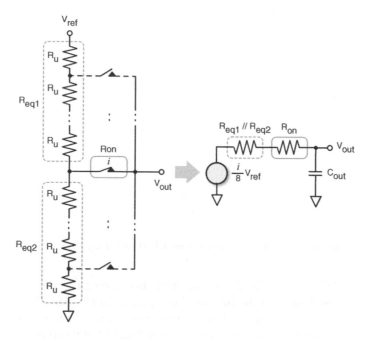

Figure 7.14   Equivalent circuit of the Kelvin divider DAC.

that processes the digital input to operate one of the switches of the basic scheme shown in Figure 7.12.

A drawback of the resistive divider is that the output voltage is a source generator with variable output resistance. Thévenin's theorem, with the ith switch closed, gives the equivalent circuit of Figure 7.14. The voltage is what is required; the equivalent resistance is

$$R_{eq} = \frac{i(8-i)}{8} R_u + R_{on}, \tag{7.8}$$

where R_{on} is the on-resistance of switches in the *on* state.

The scheme also accounts for the capacitance C_{out}. This is the sum of all parasitics and the actual capacitor at the output node. The result is a low-pass RC filter that limits speed. A change of the input code produces not a voltage step at output but an exponential transition with time constant $\tau = (R_{eq} + 3R_{on})C_{out}$. A limited conversion period leads to a final voltage affected by an error that depends on the time constant. Since R_{eq} is code dependent, different time constants give rise to non-linear errors as a source of harmonic distortion. It is therefore necessary to use low values of unity elements and, also, to employ switches with an *on*-resistance that is low and that have a small voltage sensitivity.

A practical limitation of the Kelvin divider comes from the exponential increase of the number of resistors with the number of bits. For n bits it requires 2^n resistors. When the resolution goes to 10 bits, the number of unity resistors is 1024, doubling for each additional bit. Thus resolutions larger than 12 bits or more lead to schemes that are not practical to realize. The increase of the unity resistors adds to the complexity of selection logic. A tree of switches becomes impractical for two reasons. The *on*-resistance of a single switch, multiplied by the number of bits, becomes significant in limiting τ. The number of switches in the tree is $(2^n - 1)$, which almost doubles with every additional bit. The use of digital logic with a

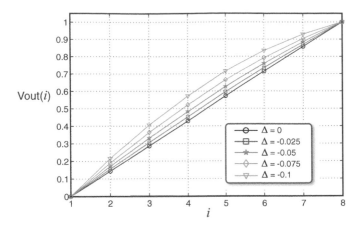

Figure 7.15   Input–output response of a resistor-based DAC with a gradient in resistor values.

scheme like the one of Figure 7.12 is also problematic because of the increase of the number of wires carrying logic controls and the increased complexity of the decoding functions.

The Kelvin divider is normally adopted for medium resolutions. However, it is used for high resolutions because of its intrinsic monotonicity. The voltage of any tap is always higher than that of the previous one, making it impossible that increasing the code reduces the output voltage. However, a gradient in the resistors' value can give rise to large INLs. Thus, when using the method for high resolutions, it is necessary to design INL correction methods.

To understand why INL depends on resistive gradients, suppose that we use unity resistance fabricated with an integrated circuit process. The value of the resistors is

$$R_u(i) = R_u(0) + i\Delta R. \tag{7.9}$$

The ith code selects the voltage

$$V_{out}(i) = \frac{\sum_{j=1}^{i}(R_u + j\Delta R)}{\sum_{j=1}^{2^N}(R_u + j\Delta R)}V_{ref}, \tag{7.10}$$

where at the denominator we have the total resistance of the string. The numerator is zero with the $000\cdots0$ code and almost the total resistance for the $111\cdots1$ code, making the error at those two points either zero or very small. On the other hand, as can be easily verified using the equation, at the midpoint (code $1000\cdots0$) the error is maximal. Figure 7.15 shows the input–output response for $N = 10$ and different $\Delta R/R_u$. As expected, the transfer characteristic deviates from the linear interpolation, with highest shift at the mid-code. This limitation implies non-linearity and, therefore, harmonic distortion.

As was studied above, the equivalent circuit of the Kelvin scheme is a voltage source and an equivalent output resistance, calculated by an equation such as (7.8), which is valid for a three-bit scheme. In order to avoid non-linear drop voltages the load must be higher than the output resistance. Often, to avoid possible errors, a unity gain buffer follows the DAC scheme to ensure good operation with any load.

The resistive divider becomes impractical for a large number of bits. A solution that avoids the problem at the cost of non-intrinsic monotonicity is the R–$2R$ ladder shown in Figure 7.16. This is a cascade of equal cells, each made up of a resistance R_u and a grounded resistance

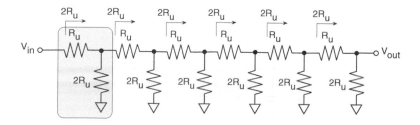

Figure 7.16   Resistive R–$2R$ ladder network.

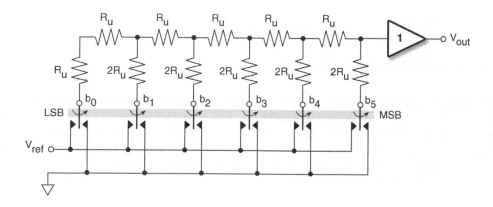

Figure 7.17   DAC converter realized with an R–$2R$ ladder network.

$2R_u$. The interesting property of the ladder is that the resistance seen at the right-hand side of every input is $2R_u$. The result is that the voltage at the right of each resistance R_u is half the value at its left. Moreover, the current flowing out of resistance R_u splits into two equal parts, one flowing toward ground through $2R_u$, and the other flowing to the right.

This property of a recursive division by two is interesting for using binary weighted parts. The network realizes a DAC with the scheme of Figure 7.17. Switches, under the control of bits $(b_0, b_1, \ldots, b_{N-1})$, connect the $2R_u$ resistors to ground or to the reference voltage. The unity gain buffer that replicates the voltage of the last right-hand node provides the output. It can be verified that the Most Significant Bit (MSB) switch determines $V_{Ref}/2$. The one at the left, connected to V_{Ref}, produces $V_{Ref}/4$, and so forth. Therefore,

$$V_{out} = \sum_{i=0}^{N-1} b_i \frac{V_{Ref}}{2^{(N+1)}}, \tag{7.11}$$

which is the needed DAC output, with the benefit of simple selection logic and the need for only a small number of unity resistors: $(3n + 2)$.

Other architectures also use resistors. For the scope of this book it is enough to know just two of them, which are the basis of other schemes. The reader can refer to other solutions in specialized books or application notes.

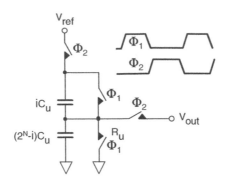

Figure 7.18 Conceptual scheme of a DAC based on capacitances.

7.4.2 Capacitance-based Architectures

The resistor-based DAC obtains its output as a digital controlled fraction of the reference voltage. The same can be done with capacitors. Consider the scheme of Figure 7.18. The switches discharge two capacitors, iC_u and $(2^N - i)C_u$, during the phase Φ_1. C_u is a unity capacitance of any reasonable value. "Reasonable" means ensuring the required matching accuracy and complying with the noise requirements. During phase Φ_2, the capacitors make a capacitive divider whose output voltage is

$$V_{out} = \frac{iC_u}{2^N C_u} V_{Ref} = \frac{i}{2^N} V_{Ref}, \tag{7.12}$$

which is the required DAC conversion if i corresponds to the input code.

The output voltage is valid only during Φ_2. Moreover, the scheme does not admit any resistive loads that would discharge the array. Another limitation comes from parasitic capacitance on the output node that alters the value of the generated voltage. The change is just an attenuation if the parasitic or the load are initially discharged to analog ground; having attenuation is not a serious limit because it is a mere gain error. However, if the parasitics are non-linear, the attenuation depends on the output amplitude giving a non-linear response.

Normally the scheme of Figure 7.18 includes a unity gain buffer with infinite input impedance.

The capacitor-based DAC needs a simple selection network – much simpler than the resistor-based counterpart. Figure 7.19 shows two possible schemes for capacitor-based DACs, both easy to control. The schemes are for four bits. Figure 7.19(a) uses binary weighted capacitors. An array of 16 unity elements is made by the parallel connections of $C_u, C_u, 2C_u, 4C_u$, and $8C_u$. During phase Φ_{RES} all capacitors are discharged: the bottom switches are all in the right position. On the conversion phase Φ_2, the bit b_0 controls the switch connecting C_u, the bit b_1 drives the switch associated with $2C_u$, and so forth. The task of the logic is simple: to connect the bottom plate to analog ground if the bit is zero and to V_{Ref} when the bit is 1. It is therefore just necessary to buffer the logic input to drive a switch toward V_{Ref} and to use its inverse so as to obtain the complementary operation.

The scheme of Figure 7.19(b) uses 16 unity capacitances. One of them, as in the other scheme, is always connected to analog ground. The conversion of a digital code corresponding to number i requires i elements connected to V_{Ref}. The scheme needs $2^N - 1$ control signals (15 for a four-bit DAC) to control the 15 switches. Since they are normally operated sequentially,

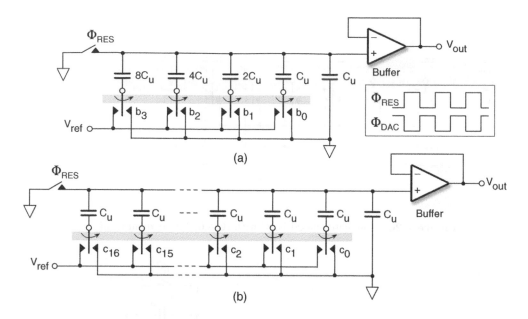

Figure 7.19   (a) Capacitor-based DAC with binary weighted capacitor; (b) scheme with unary elements.

the control set is called thermometric. For example, the input binary code 0101 produces the thermometric code 000000000011111, i.e., five ones and ten zeros.

The selection logic for the scheme of Figure 7.19(b), which uses unity elements, is more complex than what Figure 7.19(a) needs. However, Figure 7.19(b) ensures intrinsic monotonicity. If the input code increases by one, for example from five to six, the logic adds the sixth element to the ones already connected to V_{Ref}. Thus the generated voltage is certainly higher than the previous one. On the other hand, for the circuit of Figure 7.19(a) an increase by one of the input can require it to switch many elements *off* and others *on*. For example, at the midpoint transition (from seven to eight for the four-bit converter) C_u, $2C_u$, and $4C_u$ switch off and only $8C_u$ switches on. The operation is critical with a large number of bits. It may happen that, because of fabrication inaccuracies, the set of elements switched off has a value bigger than that of the capacitor that is turned on, even if the sum is expected to be smaller. This circumstance causes non-monotonicity.

Another possible problem with the scheme of Figure 7.19(a) is a consequence of the inaccuracies of switching times. It may happen that the switching off is not exactly at the same time as the switching on or vice versa. Thus having capacitors on or off simultaneously for a short period of time generates a glitch. To summarize, the scheme of Figure 7.19(a) needs simple selection logic. The scheme of Figure 7.19(b) ensures intrinsic monotonicity and gives rise to small glitches.

A good compromise between benefits and costs is to use a mixture of the two solutions. The technique is called segmentation. The N bits of the input digital word are divided into two sections (or three for more sophisticated architectures), N_1 grouping the low-resolution bits and the remaining $N - N_1$ consisting of the more significant bits. The first segment uses a binary weighted scheme because a low number of unity capacitors, C_u, does not cause non-monotonicity. The second segment uses the unary method. If the matching accuracy between

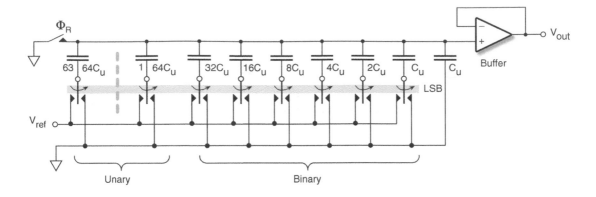

Figure 7.20   Capacitor-based DAC with $(6+6)$-bit segmentation.

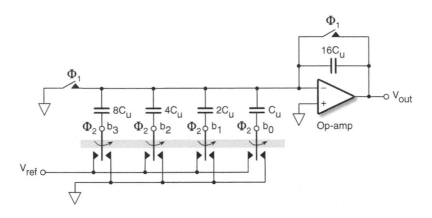

Figure 7.21   Multiplying DAC with capacitors.

unity elements ensures monotonicity with seven or eight bits, a segmented 12-bit converter can use a $6+6$ segmentation with a binary array made up of $C_u, C_u, 2C_u, 4C_u, 8C_u, 16C_u$, and $32C_u$ elements and 63 groups of $64C_u$ unary elements. The thermometric decoder must serve only six bits. Figure 7.20 depicts the schematic diagram. The reset phase, Φ_R, closes the top switch on the top plates while all the bottom plates are connected to analog ground. Then the top switch opens and the control phases connect some of the bottom capacitances to V_{Ref}.

7.4.3   Parasitic Insensitivity

The schemes studied in the previous section obtain, during the conversion phase, the output voltage on the top plate of the capacitor array. The parasitic capacitance of that node is not negligible and can be non-linear. Thus the sensitivity of the scheme to parasitics can result in a limitation. Modifications of the architecture ensure insensitivity to parasitics. The solutions use an op-amp, like the scheme of Figure 7.20. Therefore they do not require additional active blocks and power.

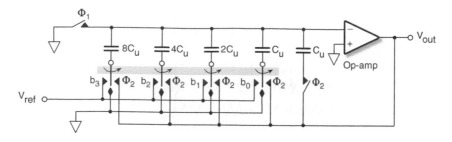

Figure 7.22   Flip-around DAC.

There are two possible techniques: the Multiplying DAC (MDAC) and the flip-around DAC. The first architecture is an inverting amplifier, a circuit already studied in the previous chapter. The gain is digitally controlled by the input code. Figure 7.21 shows the MDAC circuit diagram for four bits. The circuit uses binary weighted capacitors because of the low resolution. It is possible to realize the same function with a unary scheme and a more complex selecting network.

Phase Φ_1 and controlled switches discharge the input array and the $16C_u$ capacitance in feedback around the op-amp. During the conversion phase Φ_2, the input bits that are equal to one connect the controlled capacitors to V_{Ref}; the others remain at analog ground. The charge of the capacitors connected to V_{Ref} comes from the virtual ground and the feedback element. It results in

$$V_{out} = -\frac{V_{Ref}}{16C_u}[b_0 C_u + b_1 2C_u + b_2 4C_u + b_3 8C_u], \tag{7.13}$$

which is the inversion of the expected value. The inversion is not a problem because reversing the reference voltage, or another minus in subsequent processing, cancels the limit. The scheme is parasitic insensitive because the top plates of the input array are at analog ground during phase Φ_1 and at the virtual ground during Φ_2. Since analog ground and virtual ground are almost coincident, the parasitic capacitances do not need charge when switching from the reset to the conversion phase.

The architecture of Figure 7.21 uses 2^N unity capacitors in the input array and 2^N unity capacitors in feedback, thus leading to doubled capacitances with respect to the ones used by the DAC of Figure 7.19. The flip-around technique, illustrated in Figure 7.22, resolves the limitation. The schematic, again, uses a binary weighted array, but the method also applies to unary configurations. Three switches can connect each bottom plate of the capacitors either to analog ground, to the reference voltage, or to the output of the operational amplifier. During phase Φ_1, instead of a reset the circuit pre-charges fraction $i/2^N$ of the entire array to the reference voltage. The digital input defines i. The total charge on the array is

$$Q_T = C_u(b_0 + 2b_1 + 4b_2 + 8b_3)V_{Ref}. \tag{7.14}$$

The controls during phase Φ_2 connect all the bottom plates together. The entire array connected across the op-amp stores Q_T. Therefore, the output voltage becomes

$$V_{out} = \frac{Q_T}{16C_u} = \frac{b_0 + 2b_1 + 4b_2 + 8b_3}{16}V_{Ref}, \tag{7.15}$$

which is the four-bit conversion of the digital input. The scheme of Figure 7.22, used to illustrate the method with only four bits, can obviously be extended to any number of bits.

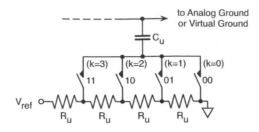

Figure 7.23   Use of a resistor-based DAC to increase the resolution of a capacitor-based scheme.

Even the flip-around solution is parasitic insensitive. The top plate of the array switches from ground to virtual ground, which are almost equal voltages. Since during the conversion the virtual ground node is floating, the charge on it remains unchanged. That is the charge that determines the output voltage. The output node of the operational amplifier provides the charge needed by the bottom plate parasitic capacitances.

7.4.4   Hybrid Resistive–capacitive Architectures

The number of unity capacitances used in the previously studied schemes increases exponentially with the number of bits. Combined resistor–capacitor schemes obtain more a effective solution. Figure 7.23, for example, uses a resistive divider DAC to generate a fraction of the reference voltage to charge the unity capacitance. If the voltage of the top plate is ground or virtual ground, the charge stored on the bottom plate is the fraction $k/4$ of V_{Ref}, depending on a two-bit word. The stored charge is the same as that of a two-bit array with a quarter of C_u as the elements used. Thus the simple circuit of Figure 7.23 would obtain two more bits with just four more unity capacitors.

The concept of a mixed resistive and capacitive divider leads to hybrid architectures with the number of bits divided into two segments, one for the resistive and the other for capacitive division. The segmentation used also reduces the complexity of the selection logic. The optimal division of bits between the resistive and the capacitive array depends on the area of unity resistors and capacitors and that of the selection logic.

There are two options for the hybrid DAC: use the MSBs for the capacitive array or use them for the resistive array. Moreover, the method can be used for any architecture that admits segmentation. Figure 7.24 shows the two hybrid implementations of a $(6 + 6)$-bit flip-around converter. The circuit of Figure 7.24(a) uses the LSBs in the resistive section. The capacitive section uses the same LSBs in Figure 7.24(b). The resistive-based DACs provide one output for the first scheme and two outputs for the second configuration. They are the top and the bottom voltages generated across each capacitor of the string. These voltages are then finely divided by the capacitive division.

The schemes of Figure 7.24 look complicated – much more so than those that have been studied up to now. However, don't be shy. You have the basic knowledge to read them and to understand the roles of various components. Some details, such as the phases of switches and the names or values of some components, are not given. That is what normally happens with complex schemes. They assume that missing information can be deducted or is not relevant at that stage of the description.

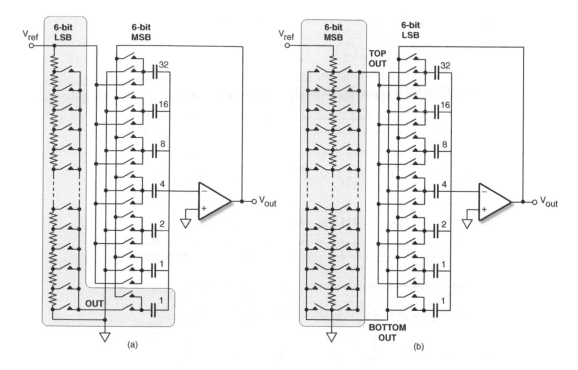

Figure 7.24   Mixed resistive–capacitive DACs with resistors used to convert the MSBs: (a) the MSBs control the capacitive section; (b) the MSBs drive the resistive section.

7.4.5   Current-based Architectures

The use of equal current generators, as distinct (or, as is usually said, unary) elements or as parallel connections of binary weighted elements, produces a DAC with current as the output variable. If necessary, the circuit obtains voltage by injecting the output current into an output resistance. That method enables, for high-speed applications, the matching of the output resistance and the impedance of coaxial cables or strip lines (I am not sure you know what a strip line or matching are. It is enough to be aware that coaxial cables and strip lines are suitable media for transferring very high-frequency signals and that matching obtains an optimal transfer of the signal).

Consider the schemes of Figure 7.25. The currents are directed toward the output node or a dummy point that discards unused currents. The sum of the selected currents generates the output variable. Notice that the switches used in this architecture carry currents. On the other hand, the switches of previous schemes with resistors or capacitors transfer a voltage. The logic control that steers the current sources attains a thermometric or a binary function. Some architectures use segmentation control with a mix of binary weighted and unary current sources. The three diagrams of Figure 7.25 show the schemes (p is the MSB segment).

Notice that control of the current generators in Figure 7.25 does not open the connections but deviates the current from the output terminal to a dummy node. This strategy serves to avoid opened current sources, as would happen with a single switch breaking off unused currents. Obviously it is impossible to open the connection of an ideal current source. However,

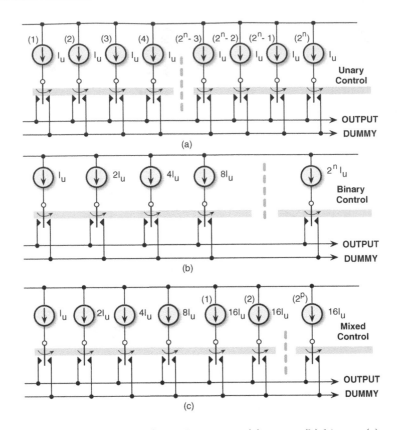

Figure 7.25 Current-steering DAC architectures: (a) unary; (b) binary; (c) segmented.

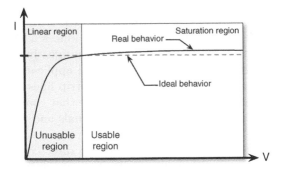

Figure 7.26 Voltage/current relationship of real current sources.

real current sources realized with electronic components can do that because the input–output characteristic looks like the plot of Figure 7.26. When the output current is zero, the voltage is zero as well. However, even if it is possible, the operation is not advisable because having current and voltage sent to zero is problematic. The parasitic capacitance of the output node get discharged, and turning the current on again requires time. A first drawback is that

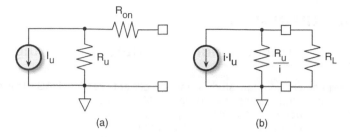

Figure 7.27   Behavioral model of a switched real current source and equivalent circuit of a current-steering DAC with i current sources toward output.

transients limit the speed. Moreover, the integral over time of transients can be non-linear because parasitic capacitances are often non-linear. The use of a dummy drain, indicated in Figure 7.25, avoids switching off the unused current sources.

The architecture of Figure 7.25 uses ideal current generators and assumes zero on-resistance for the switches. Real circuits, as we are reminded by the response of Figure 7.26, do not deliver the same current at different output voltages. Moreover, they need a voltage across that is larger than the saturation limit. If the current sources used operate in the saturation region we can study the problem with the simplified equivalent circuit of Figure 7.27(a). This represents a real unity cell by its output resistance R_u and, for the sake of completeness, with the *on* resistance of the switch, R_{on}. However, in practical cases, R_{on} is much lower than R_u and can be ignored. The parallel connection of i current sources leads to the scheme of Figure 7.27(b), leading to a code-dependent output resistance, R_u/i. The voltage at output becomes

$$V_{out}(i) = i\,I_u\frac{R_L R_u}{iR_L + R_u},\tag{7.16}$$

which describes a non-linear deviation from the expected ideal input–output characteristic, given by

$$V_{out}(i) = i\,I_u R_L.\tag{7.17}$$

In addition to non-linearity, equation (7.16) implies gain error as the full-scale output is not $(2^N - 1)R_L I_u$ when i is at full scale $(2^N - 1)$. The gain error is

$$\varepsilon_G = \frac{R_u}{2^N R_L + R_u},\tag{7.18}$$

showing the need for an output resistance R_u larger than $2^N R_L$. A gain error is normally acceptable; it may be compensated for by slightly increasing the value of unity current.

The above study, even if it is felt to be too detailed, serves to make you aware of the limitations produced by simple non-idealities as the finite output resistance of a current source. It is possible to adjust the gain error with a constant increase of unity currents. On the other hand, non-linearity is difficult to correct and can be problematic for circuit operations.

Another limit that the designer accounts for is the matching between unity current generators. They are assumed to be equal, but real implementations produce

$$I_u(k) = I_u(1 + \varepsilon_k),\tag{7.19}$$

where ε_k is the mismatch error.

COMPUTER EXPERIMENT 7.3

Understanding the Current Steering DAC

With this virtual experiment you can study a four-bit current–steering DAC. The setup uses a custom-made instrument that allows you to select 16 current sources with 16 buttons. The instrument can define a sequence of 12 different patterns repeated sequentially at a programmed speed. An active circuit transforms an input voltage into the reference current. The transconductance gain is 10 mA/V. The 32 unity current generators are not perfect replicas of the reference but feature random and systematic mismatches. The output currents flow through two 600 Ω loads. The input code drives a precise D/A converter to generate the reference analog signal. In this computer experiment you can:

- define the custom selection of current sources repeated every 12 conversion periods. The custom selection allows a thermometric, binary, or random choice of elements;
- set the supply voltage, the noise, and the reference voltage (then transformed into current);
- set the clock frequency of the converter;
- set systematic and random mismatch of a matrix array of 4×4 current generators. The systematic mismatch in the x direction is twice the one in the y direction.

MEASUREMENT SET–UP

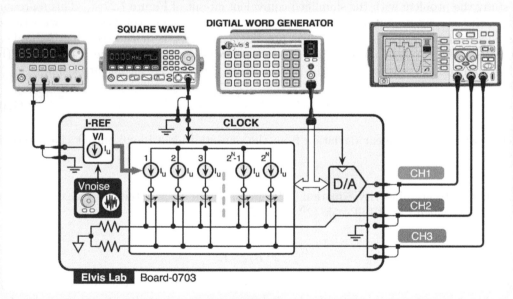

TO DO

- Set reference current (zero noise) to give rise to a 1 V full scale at output of the DAC.
- Choose a random mismatch value and construct the input–output characteristic. Repeat the experiment with systematic mismatch. Determine the DNL and the INL in both cases.
- Add noise to the current reference and observe the output pattern for various sequences.
- Compare the responses with thermometric and binary selections when the data pattern crosses half-range. Determine the unity element selection that, with an equal input sequence, minimizes error.
- Increase the clock frequency and observe possible speed limits.

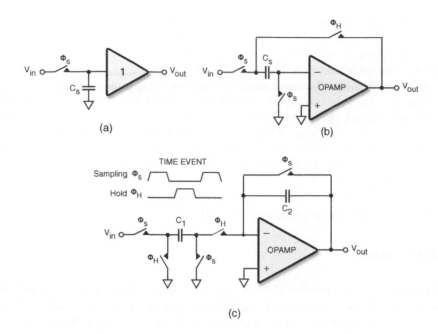

Figure 7.28 (a) Simple sample-and-hold; (b) flip-around sample-and-hold; (c) switched-capacitor sample-and-hold.

The selection of the first i elements gives rise to an actual output current given by

$$\sum_{k=1}^{i} I_u(k) = I_u\left(1 + \sum_{k=1}^{i} \varepsilon_k\right). \tag{7.20}$$

The errors ε_k include the random errors and the correlated terms caused by a gradient in the fabrication parameters. The addition of all the errors gives rise to a difference in the full-scale current, separately accounted for with a gain error. The piling up of errors causes integral non-linearity that becomes large in the presence of correlated terms.

7.5 NYQUIST-RATE ADC

Coding analog signals into digital formats requires, as the first step of the processing stage, input sampling. This happens at times nT_s (T_s is the sampling period) under the control of the clock. The sampler accuracy, both in time and amplitude, must preserve the input signal-to-noise ratio. However, unwanted components always corrupt real signals, and the SNR quantifies this. Thus, as happens in all electronic circuits, samplers add noise but its amplitude should be a fraction of what already affects the input.

Since the data converter needs time to perform the translation into digital format, the sampler, in reality, is a sample-and-hold (S&H) that, after sampling, keeps the signal unchanged for the time needed. The storage element is a capacitor, and the snap-shot trigger is a switch.

Figure 7.28 shows some schemes used for sample-and-hold. They operate under the control of the clock, which generates one or two phases driving switches. The phases are disoverlapped to avoid having switches closed at the same time. The scheme of Figure 7.28(a) uses a passive sampling, occurring when the phase Φ_1 opens the switch. Actually, when the switch is *on* the output voltage, supplied by a unity gain buffer, follows the input. For this reason the real operation of this scheme is that of a Track-and-Hold (T&H).

Figure 7.28(b) also samples the input signal at the end of the phase Φ_1. Then the capacitor C_s goes in feedback around the op-amp to generate the held version of the input during Φ_2. The output is valid only during this phase, because the input capacitance during Φ_1 samples the input voltage. The scheme of Figure 7.28(c) shows a switched-capacitor circuit that samples the input during Φ_1 and injects the charge stored on C_1 into C_2 during Φ_2. If $C_1 = C_2$, the output is a replica of the input. If the ratio of the two capacitors differs from unity, the scheme provides attenuation or amplification.

After sampling, an A/D circuit realizes the conversion algorithm. The following subsections describe the methods and circuits mostly used for that.

7.5.1   Flash Converter

The flash architecture obtains the ADC function by using a "brute force" method. Exploiting the fact that an N-bit A/D converter divides the quantization range into 2^N intervals, the flash uses $2^N - 1$ comparators to compare the input with all of the threshold transitions between quantization intervals. Each comparator decides whether input is higher or lower than each threshold, giving altogether a set of $2^N - 1$ logic outputs. These are a digital one until a given level is reached, and they turn into a digital zero above that. The result is the thermometric view of the digital output. After that, suitable logic transforms the $(2^N - 1)$ set of digits into a binary code.

Figure 7.29 shows a four-bit flash converter. The circuit uses 15 comparators and 15 threshold references. As is commonly done, a Kelvin divider obtains the references whose voltages are

$$V_{Th}(i) = \frac{V_{Ref}}{16} i. \tag{7.21}$$

If the input is below the first threshold, all the comparators give zero at output, producing the code 0000. With a voltage falling in the second interval only the first comparator switches; a thermometric output made by a single logic one and 14 zeros gives the binary 0001; and so forth. The thermometric to binary coder is a combinatory logic system with low complexity if there are few bits.

The time needed for the conversion is small. The scheme requires just the time for sampling and comparison: a clock semi-period can be enough for sampling and another semi-period for comparison. Thus the flash conversion needs only one clock period. The digital circuit performing the thermometric to binary conversion also requires time. This operation does not affect the next sample and is normally done afterwards.

The advantage of flash architectures, also implied by the name of the method, is their speed of operation. With modern technologies it is possible to design comparators running at speeds measured in GHz. Sampling with moderate accuracy is also very fast. Therefore the flash converter is the right solution for applications that require low or medium accuracy but where the speed must be as fast as possible – at the limit allowed by the technology.

This technique, suitable for medium resolutions, is not appropriate for a large number of bits because the exponential increase of number of comparators with number of bits is problematic. The limitations that occur with many bits are as follows.

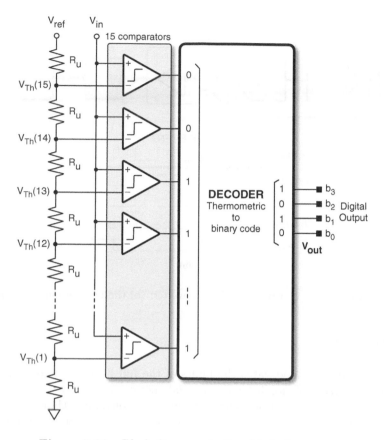

Figure 7.29   Block diagram of a four-bit flash converter.

- The consumed power increases as 2^N or even more, because it is necessary to increase the number of comparators exponentially. Moreover, more power is required to a single comparator because the sensitivity must increase since the threshold separation is lower.

- The silicon area increases because we have more comparators and threshold references to generate. There are many connections to route and the thermometric to binary converter strongly increases its circuit complexity with the number of bits.

- The resistance of the Kelvin divider becomes a limit to overall speed. The parasitic capacitance of the input terminals of comparators distributed along the resistive string increases in a non-linear manner the settling time of possible transients.

- The load established by comparators becomes too large for the sample-and-hold. It becomes more and more difficult to drive large capacitors with the required accuracy.

In summary, the straightforward flash converter serves for applications that use few bits, normally eight bits or fewer, and high conversion speed. When the required resolution increases there are more convenient solutions, even if high speed is also required.

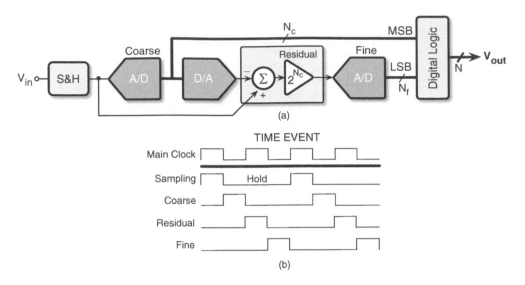

Figure 7.30   (a) Architecture of a two-step flash converter; (b) timing diagram of controlling phases.

7.5.2   Two-step Flash

The two-step architecture moderates the limits of the full flash at medium resolutions. It obtains analog-to-digital conversion in two sequential operations. The first determines a coarse conversion of input; then the resolution improves by a fine conversion that appends other bits to the previous result. If the bits of the coarse conversion are N_c and the fine are N_f, the overall resolution becomes

$$N = N_c + N_f. \tag{7.22}$$

The architecture of Figure 7.30(a) gives the result. Actually it is a combination of low-resolution data converters to obtain a higher resolution. The first block is the sample-and-hold used to make sure that the signal at input of both coarse and fine converters is the same. The first low-resolution ADC determines N_c. The digital output of the coarse flash forms the input of a DAC of the same resolution that generates an analog signal subtracted from the held input. This quantity is the quantization error, with information on missing accuracy:

$$V_Q = V_{in} - V_{DAC}. \tag{7.23}$$

The fine A/D converter receives (7.23) at input to generate the fine conversion with N_f LSBs. Recall that the quantization error, V_Q, is at most the quantization step, and much less than the reference voltage

$$V_Q < \Delta = \frac{V_{Ref}}{2^{N_c}}; \tag{7.24}$$

the fine flash must be able to handle such a small signal, or, as is done in some architectures, the level is increased by amplification. If the gain of the interstage amplifier is 2^{N_c}, the maximum amplitude is V_{Ref} as the input. The same reference can be used in both A/D converters. The amplified quantization error is often called "residual."

The time required by the two-step flash is longer than for the flash architecture. As shown in Figure 7.30(b), it uses four time-slots: sampling, coarse conversion, generation of the residual,

COMPUTER EXPERIMENT 7.4

Two-Step Flash A/D Converter

A two-step flash converter determines the digital output in two successive phases. The first, as you certainly know, estimates the (N1) MSBs which, in this Computer Experiment, range from 2 to 6. The generated residual is the input of the second A/D converter that gives rise to N2 LSBs (also variable from 2 to 6). For your convenience the MSBs are transformed into analog signal. A selector allows you to observe MSBs and residual on channel 3 of the scope. The plot of the residual shows a pattern that looks like noise or a fixed pattern depending on the input frequency. All the blocks of this experiment work ideally, because the aim is to understand the algorithm and not to explore limits. The MSBs and the LSBs are properly combined and transformed into analog for the display of the output conversion. You can:

- change amplitude, DC level and frequency of the sine wave used as test signal;
- define the number of N1 and N2 bits of the MSB and LSB sections;
- set the clock frequency and choose between MSB and residual analog signals. The full scale is the same as input, because of the proper amplification of the residual generator.

MEASUREMENT SETUP

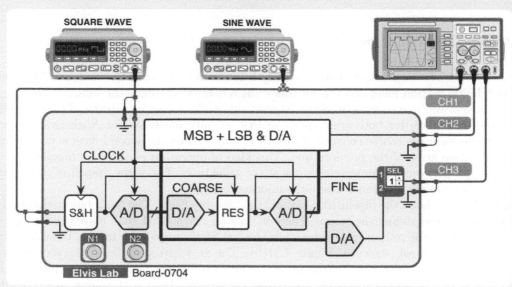

TO DO

- Set the amplitude of the input sine wave. Use a clock frequency at least 10 times the input sine wave. Set the number of MSB bits to 6. Observe the effect of amplitude and input frequency at different sampling frequencies. The scope repetition rate is relatively slow to observe the waveforms better.

- Display the residual at different sampling frequencies (go down until twice the sine wave frequency). Notice that the residual becomes a repetitive pattern when the ratio between the sampling and the signal frequency is an integer.

- Set the value of LSBs to 6 and look at the quantized output.

- Reduce the number of bits of the MSB section and observe the residual. Use small input amplitude and change the input offset until the residual is a replica of the input.

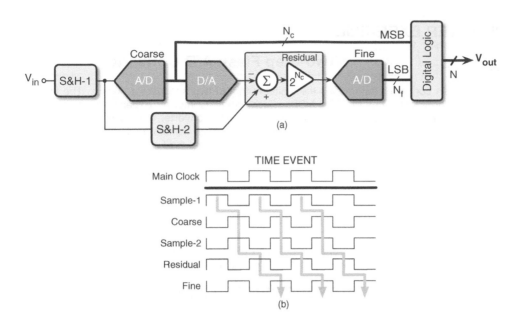

Figure 7.31   (a) Pipeline version of a two-step flash converter; (b) phases for pipeline operation.

and fine conversion. Notice that there is a delay between the setting of the coarse and the fine bits; the digital section must account for this using a delay that aligns the bits.

Since coarse and fine conversion occur sequentially, we have two architectural possibilities: to use the same flash for both conversions so that we save hardware, or, with two distinct flashes, to perform the coarse conversion of a new sample while the second flash is converting the quantization error of the previous one. This kind of operation is called a "pipeline;" i.e., the two stages perform two processing steps at the same time. The name "pipeline" describes in a figurative manner a channel where the information flows.

Figure 7.31(a) shows the pipelined version of the two-step flash. The basic blocks that build up the diagram are almost the same. There is only an extra S&H that makes the sampled signal available for the residual generator while the first S&H samples the input. The difference is in the timing control, shown in Figure 7.31(b). The gray arrows indicate the sequence of conversion steps. Sampler 1 stores the input at the end of the sampling phase and holds it during the next half period to enable the coarse conversion. During the same time-slot the second sampler stores the input value that is combined with the DAC output to generate the quantization error. That time-slot also includes the amplification that generates the residual. The last phase is for the fine conversion.

The conversion requires two clock periods, like the two-step scheme of Figure 7.31(a), but the sampling of input and the output speed doubles. Thus we have a latency time because the delay between input and output is longer than the conversion period.

The two-step flash is more complicated than the flash architecture because it requires estimation of the quantization error or its amplified version (called "residual"). However, the number of comparators needed is less than for the full flash. If $N_c = N_f$, each flash uses $(2^{N_c} - 1)$ comparators. Thus, for example, if $N_c = N_f = 4$, the total number of comparators needed by the two-step flash is 30 while the full flash counterpart uses 255 comparators. Moreover, the number of reference voltages is much less.

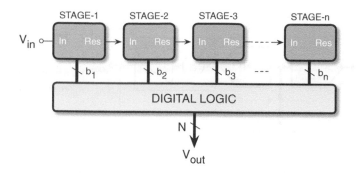

Figure 7.32   Architecture of a pipeline A/D converter.

The two-step flash is less robust than the full flash because of its more complex architecture. There are several limits to accuracy. These mainly come from the generation of the residual. The errors of the DAC and the amplifier give rise to an overall error that affects the generation of the LSBs. That error must be less than the quantization step of the fine converter.

7.5.3   Pipeline Converters

The previous subsection studied the pipeline version of a two-step flash converter. It is possible to extend the method to more stages to obtain the pipeline architecture. Its block diagram is shown in Figure 7.32. It is the cascade of a number of cells, each of them generating a N_i-bit digital signal and a residual. The residual is the input to the next stage, which adds new bits to the already determined ones, together with a new residual. This continues until the end of the chain, where the last stage does not generate any residual because it is not needed anymore. The total number of bits of a k-stage pipeline is

$$N = \sum_1^k N_i. \tag{7.25}$$

Figure 7.33 shows a typical scheme for a pipeline stage. It includes a sample-and-hold followed by an A/D converter that provides i bits at the digital output. The DAC, the subtracting cell, plus the amplifier by 2^i generate the residual. These three functions are often realized with a single electronic circuit made from a capacitive array and an operational amplifier.

The number of bits at each stage is a designer choice. We can have, for example, a 10-bit converter made up of five stages with two bits per stage, a four-stage pipeline with $2+2+3+3$ bits, or other configurations with any combination of bits per stage. For the choice of the bit allocation there are trade-offs because, for instance, having a small number of bits requires low inter-stage gains but to obtain a given resolution it is necessary to use many stages.

The digital section of the pipeline must account for different resolutions and timings. It gives at output the full resolution with a given latency that, obviously, depends on the pipeline length. The complexity is low because the functions needed are just shifts and delays.

The pipeline architecture, like the two-stage one, relies for its accuracy on the precision of the residual generated by each stage. Since the first stage determines N_1 bits, after it the required accuracy is $N - N_1$, which is the number of bits to be estimated. The accuracy after

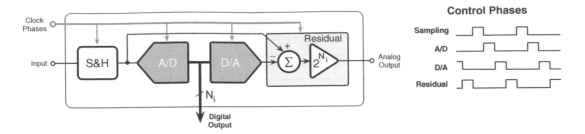

Figure 7.33   Block diagram of a stage of a pipeline converter. The digital output provides N_i bits, and the full scale of the residual equals that of the input.

the second cell must be $N - N_1 - N_2$, and so forth. Thus the precision needed in the generation of the residual diminishes along the pipeline. This is the reason why the first stages of the architecture are critical while further along the pipeline the required precision is relaxed.

With the pipeline method it is possible to obtain resolutions as good as 14 bits with conversion rates of 200 MS/s or even more. For this level of accuracy and speed it is necessary to use special technologies and design methods. In particular, it is necessary to compensate for fabrication errors. This is also done by making use of part of the resolution given by the stage. The method is called digital correction because it corrects analog inaccuracies by using a special scheme and simple digital processing. Such methods, and other design techniques, are beyond the scope of this book.

7.5.4   Slow Converters

The architectures described in the previous subsections favor conversion speed. The accuracy of a flash or a two-step flash is low or moderate. It is also moderate for the pipeline, though by the use of special design methods high accuracy values can be achieved. Power consumption is not very much considered.

Slow converters, on the other hand, value the number of bits and the power consumption. The conversion speed is not an issue because they are intended for applications with low or very low signal bandwidth.

There are many slow converter architectures. An example is the ramp converter, based on the conceptual scheme of Figure 7.34(a). A ramp generator starts under the control of digital logic. The logic also starts a counter of the clock generated by the circuit itself, as shown by Figure 7.34(b). The comparator detects the crossing of the ramp with the input voltage and produces a stop signal. The input amplitude is proportional to the time interval between start and stop. Thus the method performs a voltage-to-time conversion. Figure 7.34(c) gives a possible circuit implementation. An integrator with time constant $\tau = R_1 C_1$ provides the ramp. Its output changes in time as

$$V_{out}(t) = V_{Ref} \frac{t}{R_1 C_1}. \tag{7.26}$$

When the output voltage crosses V_{in}, the comparator detects the event at t_{stop}. At that time

$$V_{in} = V_{Ref} \frac{t_s}{R_1 C_1}. \tag{7.27}$$

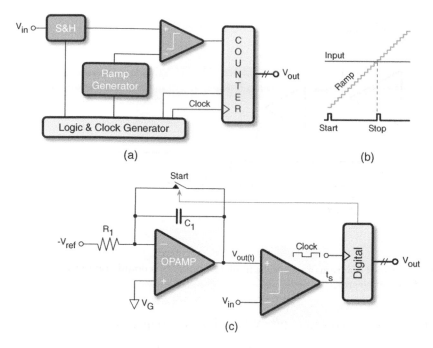

Figure 7.34 (a) Conceptual scheme of a ramp converter; (b) waveforms at the input of the comparators; (c) block diagram of a possible implementation of a ramp converter.

If the input is at full scale V_{Ref}, the stop time must equal the time constant R_1C_1. Moreover, if it is desired to have N-bit resolution, it is necessary to use a clock such that

$$2^N t_{CK} = R_1 C_1. \tag{7.28}$$

The circuit is simple and requires just an operational amplifier and a comparator. The accuracy depends on the linearity of the ramp converter, the control of the time constant, the precision of start and stop signals, and their synchronization with the clock. Moreover, the frequency of the clock generator must be stable with temperature.

An alternative solution that compensates for the inaccuracy of the time constant is the dual ramp or the dual-slope architecture. Figure 7.35(a) shows its simplified scheme. The conversion takes place in three phases:

- reset of integrator;

- integration of the input;

- integration of the inverse of the reference voltage.

The reset phase discharges the capacitor across the op-amp C_1. The second phase integrates the input signal for 2^N clock periods. The voltage at output of the op-amp becomes

$$V_{out} = -\frac{V_{in} 2^N t_{CK}}{R_1 C_1}. \tag{7.29}$$

Phase three switches the input resistance from V_{in} to $-V_{Ref}$. This discharges the integrator toward zero with a slope proportional to V_{Ref}. Since $V_{Ref} > V_{in}$, the output crosses zero in a

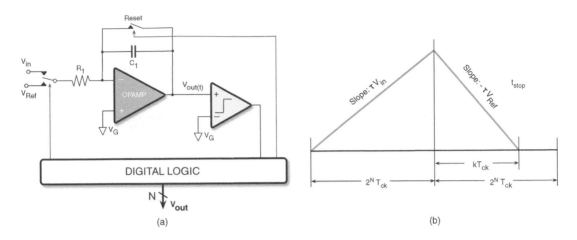

Figure 7.35 (a) Block diagram of a possible implementation of a double ramp converter; (b) waveform at the output of the integrator.

number of clock periods lower than 2^N. When crossing occurs, we have

$$V_{out}(t_{stop}) = -\frac{V_{in}2^N t_{ck}}{R_1 C_1} + \frac{V_{Ref}k\,t_{ck}}{R_1 C_1} \simeq 0, \qquad (7.30)$$

which cancels the dependence on $R_1 C_1$. The result is

$$V_{in} \simeq V_{Ref}\frac{k}{2^N}, \qquad (7.31)$$

which is the A/D conversion with N bits of accuracy.

The algorithm needs 2^N clock periods for phase two and another 2^N clock periods for phase three. Thus it needs twice the conversion time of the single-ramp method. In return for this speed cost, operation independent of $\tau_1 = R_1 C_1$ is obtained.

The method is frequently used for very high resolutions. In these cases it is necessary to design the circuit carefully so as to address some potential problems:

- leakage current;

- residual charge after the reset;

- non-linear behavior of components;

- frequency limitation of the operational amplifier;

- non-linearity of resistor and capacitor.

Another popular slow converter uses the successive approximation method. The algorithm obtains N-bit resolution with $N+1$ clock periods. One serves for sampling and the others are one per bit. The successive approximation method achieves the final digital value with the help of a DAC that generates an analog quantity that approximates the input better and better. The DAC output generates, in successive steps of approximation, the input quantity. Thus the DAC control is the result of the A/D being closer and closer to the input.

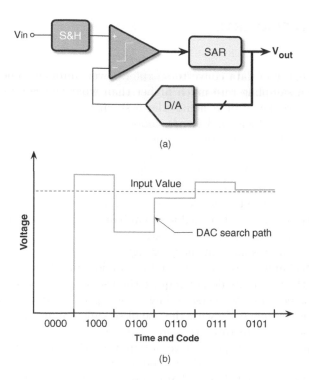

(a)

(b)

Figure 7.36   (a) Simplified block diagram of a SAR converter; (b) waveform of the search path at output of the DAC.

The digital output of the DAC obtains an analog search path by the action of logic called a Successive Approximation Register (SAR). Figure 7.36(a) shows a simplified scheme for the converter. The sample-and-hold stores the input for the entire conversion. At the beginning the SAR drives the DAC with one as the MSB and zero for the other bits, to generate $V_{Ref}/2$. The comparator decides whether the input is bigger or smaller than $V_{Ref}/2$. Assuming that, as shown in Figure 7.36, the input is lower, the logic information provided to the SAR indicates that the MSB is not one but zero. The next attempt uses 01 as the two most significant bits and leaves the others as zero. The DAC generates $V_{Ref}/4$ for the new comparison with the input voltage. The comparator reveals, for example, that the input is higher than $V_{Ref}/4$, confirming the pair 01 of bits. In the next clock period the SAR drives the DAC to a voltage that is expected to be closer to the input ($3/8\ V_{Ref}$, equivalent to 011 followed by zeros). This determines the third bit. The search continues by adding one bit at each clock cycle until the full resolution of N bits is reached.

Successive approximation converters normally consume little power. They typically use for the DAC a binary weighted capacitive array to attenuate the reference voltage. Its output can drive the input of the comparator directly without the need for an op-amp. Moreover, the power used by the comparator is low. Therefore, the overall power is only what the comparator uses with, in addition, the power consumed by the logic and what is necessary to charge and discharge the capacitive array. At very low speeds a SAR converter of 10 or more bits can require just a few μW.

7.6 OVERSAMPLED CONVERTER

The second big category of data converters exploits oversampling. The algorithms of this kind of converter use a sampling rate much higher than what the sampling theorem recommends. They obtain two kinds of benefit: the first ensures relaxed specifications for the anti-aliasing filter in A/D systems, and favors smooth requirements for the reconstruction filter in D/A applications. The second aims at obtaining extra bits to add to the ones given by the A/D used. This result is possible for given types of input signals.

Using high sampling frequencies is expensive: the power of active blocks is higher than that of their Nyquist-rate counterparts. Also, oversampling indirectly limits the signal bandwidth. Indeed, for a given technology and available power the gain requirements determine a maximum speed for op-amps or comparators. The usable signal band is that maximum frequency divided by twice the OverSampling Ratio (OSR). In addition, storing or transmitting data at the oversampled frequency requires more memory or signal bandwidth.

The oversampling technique was initially used for audio-band data conversion. The advantage exploited is that the extra bits do not require the use of very precise analog components. This, as is often said, is because the oversampling trades speed against resolution. The benefit of oversampling increases remarkably with the use of so-called Sigma–Delta ($\Sigma\Delta$) techniques. The pay-back with those architectures is more than the costs of the higher sampling rate. Indeed, with $\Sigma\Delta$ architectures it is possible to obtain as much as 16–20 bits for a signal bandwidth of about 20 kHz as required by MPEG audio coders and other more demanding audio standards. An audio signal (20 kHz band) that can be sampled at something more than 40 kHz is processed by a 3 MHz sampling rate or even more.

Oversampled converters, mainly continuous-time or sampled-data $\Sigma\Delta$, are now used for a wide range of applications. It is possible to obtain 22+ bits for very precise needs; $\Sigma\Delta$ converters can also achieve medium–high resolutions with many MHz of signal band, as required by communication architectures.

Understanding oversampled data converters and the Sigma–Delta ($\Sigma\Delta$) category is difficult and involves a new concept: one must assume, under certain conditions, that the quantization error is equivalent to noise. Therefore, before analyzing oversampled converters, we recall this important issue.

7.6.1 Quantization Error and Quantization Noise

We know that quantization changes the amplitude of a signal from any value to one of the discrete levels of the set

$$\overline{x}(nT) = k_n \Delta, \tag{7.32}$$

where k_n is an integer number representing a digital N-bit code ($k_n = 0, \ldots, 2^N - 1$) and Δ is the quantization interval. Also, if the full-scale voltage is V_{Ref}, the quantization step is

$$\Delta = \frac{V_{Ref}}{2^N - 1}. \tag{7.33}$$

The operation of the quantizer that transforms an analog signal x into the closer discrete level $\overline{x}$ is equivalent to the addition of a quantization error ε_Q,

$$\varepsilon_Q = \overline{x} - x, \tag{7.34}$$

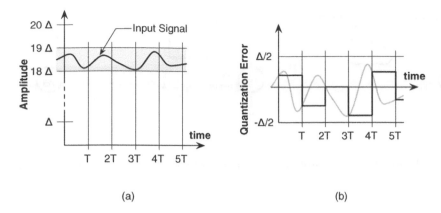

(a)   (b)

Figure 7.37   (a) Input signal with a small change of amplitude; (b) its quantization errors.

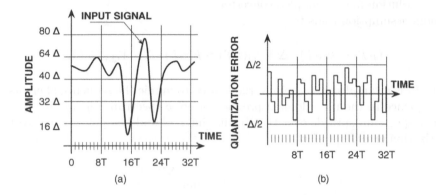

(a)   (b)

Figure 7.38   (a) Input signal with a large change of amplitude; (b) its quantization errors.

whose amplitude is within the limits

$$-\frac{\Delta}{2} < \varepsilon_Q < \frac{\Delta}{2}. \tag{7.35}$$

The actual value of ε_Q depends on the input amplitude. It is zero if the signal is at the midpoint of the quantization interval and near $\Delta/2$, positive or negative, if the input is close to the quantization step limits. When the input amplitude changes in a busy way, as happens with many real signals, successive samples fall anywhere inside one of the quantization intervals, and the sampled-data error looks like a random variable in the $\pm\Delta/2$ range. This feature, as will be better specified shortly, is essential for exploiting the benefits of oversampling.

In order to understand the point, let us first analyze when the quantization error is not suitable for augmenting the number of bits. A constant input gives rise to a constant sequence of errors that, obviously, is not a random variable. A signal with small changes that remains within one quantization interval gives rise to an error that is just a replica of the sampled signal itself, as shown in Figure 7.37(b). Even this does not lead to good behavior.

On the other hand, a busy signal like the one of Figure 7.38(a), spanning a large number of quantization intervals, gives rise to the quantization error shown in Figure 7.38(b). Actually,

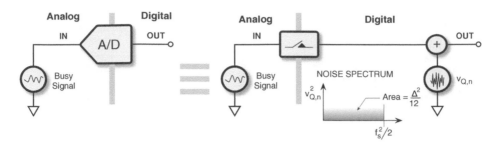

Figure 7.39 Pictorial equivalence of the quantization of a busy signal with the addition of quantization noise.

this looks like sampled-data noise, because it is difficult to predict one error on the basis of the previous ones and there is no resemblance to the input signal. This is what we need to secure extra resolution in oversampled converters.

The random assumption leads to

$$\bar{x}(nT) = k(nT) \cdot \Delta + \varepsilon_Q(nT) \simeq k(nT) \cdot \Delta + V_{Q,n}(nT), \tag{7.36}$$

where $V_{Q,n}(nT)$ is a sampled-data noise variable used to mimic the sequence of errors $\varepsilon_Q(nT)$.

A noise sequence in the time domain produces a noise spectrum $v_{Q,n}(f)$ in the frequency domain. That spectrum spreads over the Nyquist interval and has power equal to the power of the quantization error, $\varepsilon_Q(nT)$:

$$V_{Q,n}^2 = \int_0^{f_N} v_{Q,n}^2(f) \, df = \lim_{i \to \infty} \frac{\sum_1^n \varepsilon_Q^2(iT)}{i}. \tag{7.37}$$

A study not reported here has shown that, under some conditions typically verified by busy signals, the spectrum of $v_{Q,n}$ is white. Therefore,

$$V_{Q,n}^2 = v_{Q,n}^2 f_N = v_{Q,n}^2 \frac{f_s}{2}, \tag{7.38}$$

showing that increasing the sampling frequency diminishes the noise spectrum, $v_{Q,n}^2$. Moreover, the assumption that the statistical distribution of ε_Q is uniform, together with equation (7.37), gives for the noise power

$$V_{Q,n}^2 = \frac{\Delta^2}{12}; \tag{7.39}$$

this states that the quantization noise power and, consequently, its spectrum diminishes when the number of bits increases.

To summarize, even if it is not sustained by a solid theoretical basis, we have learned what Figure 7.39 pictorially represents: the use of suitable busy signals as input of the quantizer gives rise to a quantization error resembling an added noise. The noise power is proportional to the square of the quantization step divided by 12. It is spread uniformly over the Nyquist interval with a white spectrum.

COMPUTER EXPERIMENT 7.5

Understanding Quantization Noise

This computer experiment allows you to understand the quantization error. It is the error occurring when the continuous amplitude of a sampled-data signal becomes discrete. The quantization error is the difference between the actual analog signal and its quantized version. Under special conditions, the quantization error looks like noise, i.e. a signal that changes in an unpredictable manner. The ElvisLab board permits you to verify if and when the noise approximation holds. It includes two channels that generate the quantization error of the same input signal with same number of bits but different sampling frequencies. You can also study the limits caused by signal amplitude and converter resolution. In this virtual experiment you can:

- change amplitude, frequency and DC level of the input sine wave. You can also change the frequencies of the two clocks;
- change the equal resolution of two converters ($N1$ bits);
- set the time division of the horizontal axis of the oscilloscope. The trigger is established by the input signal.

MEASUREMENT SET–UP

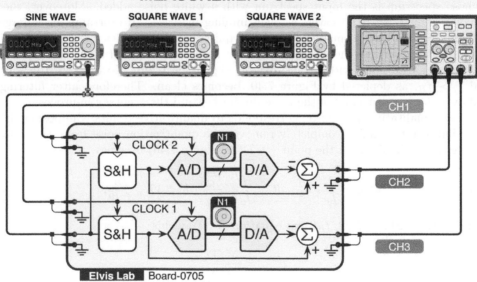

TO DO

- Set the frequency of the sine wave at 1 kHz and both resolutions at eight bits. Use a sampling frequency equal to 8 kHz for one converter and 16 kHz for the second one. Observe the behavior of the two outputs and explain the results. Change the second frequency by 1 kHz steps.
- Increase the time span of the scope to accommodate 10 periods of the input sine wave, and observe the outputs of both channels. Increase the resolution.
- Change the input sine wave frequency and observe the effect on quantization error.
- Go back to 1 kHz at input and change the sampling frequency of one of the converters to a fractional number, for example 13.27 kHz. Observe the resulting quantization error and try to find some correlation with the input sine wave (if any).

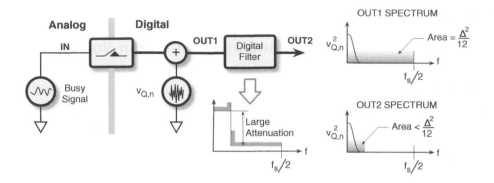

Figure 7.40   Improvement of the SNR in oversampled data converters.

7.6.2   Benefit of the Noise View

We have learned that an A/D converter with a given number of bits determines a digital code whose spectrum is the input spectrum with a white noise added. Moreover, the floor of the noise $v_{Q,n}^2 = \Delta^2/(6f_s)$ lowers as oversampling increases. If oversampling is large, the signal band occupies only a small part of the Nyquist interval, thus leaving a good fraction of the noise spectrum outside the region of interest. Therefore, we could filter out with a digital filter this fraction of noise without altering the signal spectrum. If we do so, the spectrum interval $f > f_B$, as depicted by Figure 7.40, becomes clean. Therefore, after filtering, the integral over the Nyquist band of the noise diminishes and the accuracy improves.

The above qualitative description corresponds to quantitative results. Supposing that the low-pass digital filter almost completely removes the quantization noise from f_B to f_N, the noise power after the filter (at the point $OUT2$ in Figure 7.40) becomes

$$v_{n,B}^2 = \int_0^{f_B} v_{Q,n}^2 \, df = \int_0^{f_B} \frac{V_{Q,n}^2}{f_s/2} \, df = V_{Q,n}^2 \frac{f_B}{f_s/2}, \qquad (7.40)$$

which, using the OSR $(= f_s/2/f_B)$ and equation (7.39), yields

$$v_{n,B}^2 = \frac{\Delta^2/12}{\text{OSR}}. \qquad (7.41)$$

The above predicts an equivalent reduction of Δ by $\sqrt{\text{OSR}}$; thus, since reducing the value of Δ by two means adding one bit to the resolution, increasing the OSR by four secures an extra bit after perfect digital filtering.

Observe that the result is not a simple description of a benefit of oversampling but is also a link between number of bits and noise or, better, between number of bits and Signal-to-Noise Ratio (SNR). Higher resolution means less quantization noise. In turn, lower noise gives a higher SNR. The relationship between number of bits and SNR is

$$\text{SNR}|_{\text{dB}} \cong 6.02 \, N, \qquad (7.42)$$

where 6.02 is the quantity $20 \log_{10} 2$. The result is approximated because $V_{Q,n}^2$ slightly depends on the input waveform. A sine wave at input produces about two more dB in the SNR.

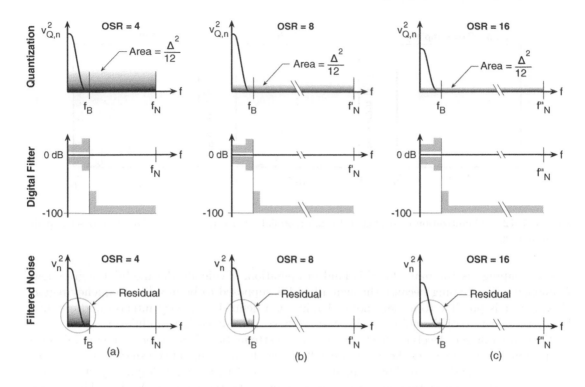

Figure 7.41   Noise spectrum and its in-band reduction with different oversampling factors: (a) OSR = 4; (b) OSR = 8; (c) OSR = 16.

The factor 6.02 in equation (7.42) indicates that increasing the number of bits by one augments the SNR by about 6 dB. Thus a 15-bit converter gives about 92 dB of SNR (with sine wave at input). The SNR becomes 98 dB (or 16 bits) with OSR = 4 and digital filtering of the quantization noise over the frequency interval $f_B - f_s/2$. An OSR = 16 secures two more bits. With OSR = 64 the resolution increases by three bits, and so forth. The rule is that every doubling of the Oversampling Ratio can increase the resolution by half a bit. Figure 7.41 shows the spectrum of a signal after its conversion, with the same number of bits but with different oversampling. The use of a sharp digital filter that leaves the signal band unchanged and strongly removes the out-of-band noise gives the residual noise spectra shown in the bottom diagrams. It is evident that the areas diminish with oversampling.

7.6.3   Sigma–Delta Modulators

What we have studied in the previous section is the basis for understanding Sigma–Delta architectures. They are continuous-time or sampled-data schemes that process the analog input together with the D/A conversion of the digital output. The name Sigma–Delta comes from accumulation (or integral), Σ, and difference, Δ. Indeed, a proper use of those two functions builds so-called $\Sigma\Delta$ modulators, which are able to enhance the oversampling benefit.

Resolution increases because the $\Sigma\Delta$ action attenuates the quantization noise spectrum in the signal band, while at high frequencies, near the Nyquist limit, the noise is augmented.

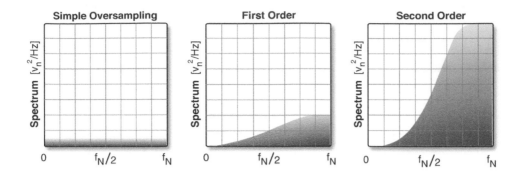

Figure 7.42   Quantization noise shaped by $\Sigma\Delta$ modulators of various orders. Order zero is the plain oversampling.

"Noise shaping" is the name for this kind of operation. Typically the modulator gives low-frequency noise shaping because the signal band is supposed to be in a low-frequency region. However, it is possible to achieve noise shaping in a defined frequency interval and obtain a bandpass rejection of the noise. In this case we have a bandpass $\Sigma\Delta$ modulator.

There are different levels of effectiveness in noise attenuation. The order of the modulator is the parameter that features them. Figure 7.42 shows the spectrum of the shaped quantization noise for a first- and a second-order modulator. Zero order means no shaping (just oversampling); i.e., the quantization spectral noise is white with total power $\Delta^2/12$. The shaped spectra of first- and second-order modulators, as shown in the figure, experience a significant growth of noise at high frequency, but near zero, within a fraction of the frequency interval, the noise level is significantly attenuated. The low-frequency reduction of the noise spectrum improves with the order at the cost of increased high-frequency values. However, having lot of noise at high frequency is not so important because a digital filter can remove it in the digital domain.

Designed $\Sigma\Delta$ modulators obtain noise spectra shaped in the form

$$v_{n,k}^2(f) = v_{n,0}^2 |H_{n,k}(f)|^2 = v_{n,0}^2(f) G_k^2 [2\sin(\pi f/f_s)]^{2k}, \tag{7.43}$$

where k is the order and $H_{n,k}(f)$ is the noise transfer function. The amplification factor G_k is one for low orders but can become larger than one for orders higher than two, depending on the architecture. For a first-order modulator the spectrum at $f = f_s/2$ is four times the plain oversampled amplitude. For higher orders, equation (7.43) predicts a better shaping at low frequency, but there is a significant spectrum amplification at the Nyquist frequency because of the 2^{2k} term plus the possible G_k that amplifies the quantization noise.

Observe that for a low value of the argument we can approximate the sin function as

$$\sin(x) \simeq x; \tag{7.44}$$

therefore, supposing $G_k = 1$, the approximate low-frequency $|H_{n,k}(f)|^2$ of a k-order modulator becomes

$$|H_{n,k}(f)|^2 = 2k \sin^{2k}(\pi f/f_s) \simeq 2k(f/f_s)^{2k}, \quad \text{for } f/f_s \ll 1, \tag{7.45}$$

and its integral from zero to f_B is

$$\int_0^{f_B} |H_{n,k}(f)|^2 \simeq 2k \frac{(f_B/f_s)^{2k+1}}{2k+1}, \quad \text{for } f_B/f_s \ll 1. \tag{7.46}$$

This, remembering that $f_s/(2f_B)$ is the Oversampling Ratio (OSR), yields

$$\int_0^{f_B} |H_{n,k}(f)|^2 \simeq K \cdot \text{OSR}^{-(2k+1)} \tag{7.47}$$

where K is a suitable constant.

Equation (7.47) marks a key feature of $\Sigma\Delta$ modulators: the dependence of the in-band power noise on oversampling. It shows that doubling the OSR of a first-order modulator reduces the noise power by $8 = 2^3$, of a second-order modulator the reduction is by $32 = 2^5$, and so forth. Since the reduction concerns the power, the corresponding benefits measured in bits are as follows.

First-order	$\times 8$ less noise power	1.5 bits per doubling,
Second-order	$\times 32$ less noise power	2.5 bits per doubling,
Third-order	$\times 128$ less noise power	3.5 bits per doubling.
$\cdots$	$\cdots$	$\cdots$

The noise-shaping benefit enables a reduction in the number of bits of the quantizer. At the limit it is possible to reduce the resolution of the ADC to just one bit, using only one comparator. Even the DAC needed for returning to the analog domain as required by the modulator processing is very simple. This is the solution adopted by single-bit $\Sigma\Delta$ modulators in contrast with the multi-bit $\Sigma\Delta$ architectures.

Obviously, the task of the modulator is to shape the quantization noise but not to alter the spectrum of the signal. Since signal occupies a fraction of the Nyquist interval, it is therefore necessary to leave it unchanged at least in the frequency interval $(0, f_B)$; i.e., the signal transfer function $H_s(f)$ of the $\Sigma\Delta$ must be flat and equal to one in the entire signal band.

The architectures used to secure noise shaping are not analyzed here, being the topic of more advanced studies. It is enough to know that obtaining a given order requires to use a number of analog accumulators (or integrators) equal to the order. For that function the schemes employ op-amps or OTAs. Limitations caused by circuit schematic, supply voltage, power budget, and fabrication inaccuracies affect the performance of modulators. These make it difficult to obtain very high resolutions with high-order modulators. Indeed, high-order shaping is problematic because the operation often involves stability issues. The designer solves the problem with schemes that give rise to an extra gain factor G_k, which partially reduces the shaping effectiveness.

The complete scheme of a $\Sigma\Delta$ converter includes, in addition to the modulator, a digital filter needed to reject the out-of-band shaped noise. The design of the filter is relevant for the overall performance because the total noise outside the signal band must be a fraction of what is inside. Moreover, having very little noise in the signal band because of high-order shaping means that a severe specification for the digital filter is required.

Figure 7.43 shows the block diagram of a sampled-data and a continuous-time $\Sigma\Delta$ converter. The modulator has two inputs: the signal under conversion and the output, converted into analog format by a DAC. Notice that the transition from continuous time to sampled data occurs at different points. It is at the input of the modulator for the discrete-time version and before the A/D for the sampled-data scheme. This difference is important for the circuit implementation and for anti-aliasing requirements. The sampled-data modulator uses discrete-time analog accumulators realized with op-amps. The continuous-time one employs integrators obtainable with op-amps or OTAs. Having sampling after the modulator incorporates in the anti-aliasing function the filtering action of the modulator itself.

COMPUTER EXPERIMENT 7.6

Understanding Noise Shaping

Noise shaping is a technique used to reduce the quantization noise in the band of interest for increasing the Signal-to-Noise Ratio (SNR) and, consequently, the effective number of bits. The method is very popular for audio signals and quality music. With this computer experiment you can understand what noise shaping means. The ElvisLab board contains a signal quantizer made of an ADC and a DAC. The subtraction block generates the quantization error that, under proper conditions, is equivalent to noise. An A/D converter with high resolution transforms the quantization error into digital for performing "noise shaping" in the digital domain. A digital filter allows you to add extra filtering. Then, the results return to analog by D/A conversion for a spectral analysis. In this experiment you can:

- select the number of bits of the quantizer and of the quantization error converter. The frequencies used establish the conditions under which the quantization error looks like white noise;

- define the specifications of the digital filter. Below fL the gain is 0 dB, whereas above fH the attenuation gain is given by the selected value. The transition between fL and fH is linear;

- change the order k of the noise shaper given by $(1-z^{-1})^k$.

MEASUREMENT SET–UP

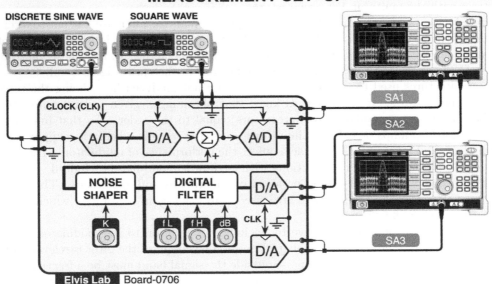

Elvis Lab Board-0706

TO DO

- Set the order of the noise shaper to zero and observe the spectra with different settings of the post-processing filter.

- Choose the order of the noise shaping and set the specifications of the digital filter. Observe the spectra at outputs and notice the low–frequency benefit and the cost to pay at high frequency.

- Use the available controls of the experimental board to find an answer to the possible questions that you have. Explain why and in which cases the effective number of bits increases.

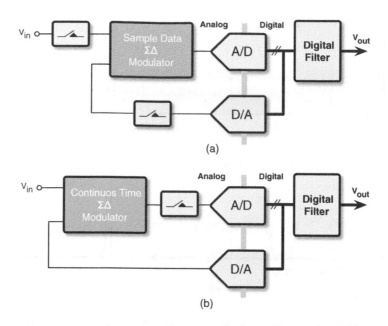

Figure 7.43 Schematic diagram of: (a) a sampled-data $\Sigma\Delta$ converter; (b) a continuous-time $\Sigma\Delta$ converter.

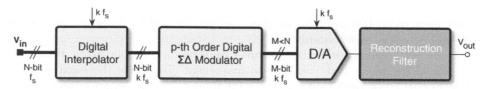

Figure 7.44 Schematic diagram of a $\Sigma\Delta$ DAC.

The digital filter gives at output a number of bits higher than the number for the A/D, because it includes the one granted by noise shaping. The data rate at input is obviously the oversampled frequency of the modulator. The one at output is the same or, better, a lower value, as the next section discusses in some detail.

> **Important remark**
>
> The output of an oversampled converter is not the representation of the analog input sampled at a well-defined time, but gives rise to a waveform with same spectral features of the input, whose signal-to-noise ratio is good because of the oversampling benefit. Use Nyquist-rate architectures to convert well-defined sequences of samples into digital form.

The method described here for the A/D conversion is also used for the D/A needs. A digital circuit, called an interpolator, increases the data rate of the digital input by k to obtain an oversampled data flow. This signal enters a digital Sigma–Delta architecture to reduce the number of bits. A high-speed DAC uses that result, as illustrated in the scheme of Figure 7.44. The benefit of the use of the $\Sigma\Delta$ technique is a reduction of the number of bits of the DAC. The cost is operation at a high conversion rate. In addition to a simpler DAC, the system also

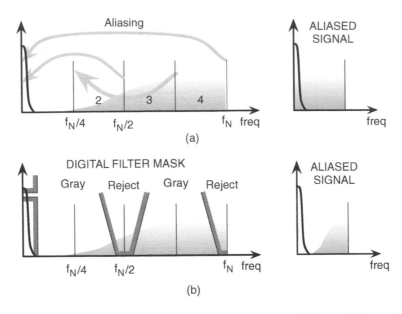

Figure 7.45   (a) Aliasing of the signal at the output of a Sigma–Delta modulator; (b) mask of the digital filter needed to protect the signal band.

uses an elementary reconstruction filter because the distance between the base-band and the first image is made large by oversampling.

ΣΔ DACs are very common in audio consumer applications. A solid-state memory stores the digital data of the music that you hear so frequently. To enable the maximum storage the sampling frequency of the signals is just a bit more than the Nyquist limit. The data produced by oversampling and modulation become sound via a ΣΔ DAC that directly drives the earphones. A reconstruction filter is not necessary because the mechanical low-pass filtering of earphones obtains the function.

The above is just a quick description of the features of a large category of data converters. The aim was to give you the flavor of the methods without going into the details of the architectures and the circuit implementation. I believe that this is enough to enable you to use ΣΔ converters. Obviously, for designing them much more is necessary.

7.7   DECIMATION AND INTERPOLATION

The use of oversampling provides advantages in terms of resolution. We have seen that the number of bits increases in A/D schemes and becomes lower for D/A applications. The cost of oversampling is the use of circuits operating at frequencies much higher than their Nyquist-rate counterparts. That is affordable for many applications, but having oversampled digital data is not suitable for storing and transmitting the information. We know that the optimal sampling rate is a little bit more than what is prescribed by the Nyquist criterion. For example, an audio signal whose band is 20 kHz is conveniently represented by a 48 kHz sampling frequency. Having a 3 MS/s (Mega-Sample per Second) data flow is unsuitable and a waste of resources. Therefore, in addition to the oversampled processing it is necessary to

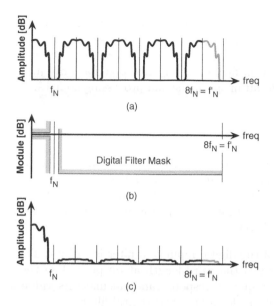

Figure 7.46 (a) Spectrum of sampled-data signal made by the band-base and its replicas; (b) response of a possible digital filter used to attenuate replicas; (c) spectrum after filtering.

reduce the output data rate of oversampled A/D converters and to augment the input data rate of oversampled D/A architectures.

Decimation means reduction of the sampling rate; interpolation increases the data rate. Decimation, as already mentioned, is conceptually a very simple operation: it is enough to discard $(k-1)$ samples out of k to obtain a data set decimated by k. However, before doing that it is necessary to digitally pre-condition the signal. Decimation under-samples the input and that operation can alter the information content.

A possible signal degradation comes from aliasing. Consider for example the decimation by four of the signal generated by a $\Sigma\Delta$ A/D. Figure 7.45(a) shows the initial spectrum. The Nyquist interval divided into four equal parts identifies the bands folded in the base-band. The bands #2 and #4 are folded an odd number of times. The #3 flips an even number of times. Figure 7.45(b) shows the superposition of the folded bands. Evidently, we have a noise-significant worsening in the signal band that undoes the noise-shaping benefit. The figure outlines the need for a filter before decimation that removes the noise from the out-of-band region. Actually not all the out-of-band noise falls in the signal band. Thus we can use a cheap solution like the one depicted in Figure 7.45(b): a digital filter removes the noise only from critical regions and admits a mild attenuation in the others. The noise after filtering and folding, as shown by the right-hand spectrum of Figure 7.45(b), is low up to f_B but increases above that limit. What is illustrated is an intermediate step before further filtering at a lower sampling rate.

The final topic of this exhausting chapter is interpolation. It is used in front of oversampled DACs to augment the sampling digital rate. Interpolation is an easy operation. It is enough to add zeros in between successive samples to obtain high-frequency data flows. However, the new Nyquist interval contains the original input spectrum and its replica. Figure 7.46(a) shows the result for an interpolation factor equal to eight. What is necessary is to keep the signal band but to reject images. This is done by the interpolation filter. The mask is normally difficult since the transition region is small because of the limited distance between the signal band and f_N. Figure 7.46(b) illustrates a possible mask, and Figure 7.46(c) shows the result achieved. The use of computer tools facilitates the design. Often it is a cascade of sections with distributed interpolation factors. Even this topic has only just been mentioned here. It will be studied in more detail in specific signal-processing courses.

PROBLEMS

7.1   The input of an A/D converter is a sine wave at 1 MHz with 1 V amplitude. The spectrum of the digital output shows a 0.7 V_{FS} component at 1 MHz. Other sinusoidal components are at 3 MHz, 4 MHz, 6 MHz and 9 MHz with amplitudes 20 μV_{FS}, 10 μV_{FS} 8 μV_{FS}, and 12 μV_{FS} respectively. Determine the reference voltage used by the ADC and the THD.

7.2   The word-length of the digital input of a filter is 14 bits. The digital transfer function is $(1 - z^1)^4$. What is the word-length at output? Use the $z \rightarrow j\omega$ transformation to determine the frequency response and calculate its value at DC, $F_S/4$ and $F_S/2$. Repeat the estimations for a $(1 + z^1)^4$ digital filter.

7.3   The bandwidth of a continuous-time signal at the input of a data converter is 5 MHz. The sampling frequency is 26 MHz. Sketch the mask of an anti-aliasing filter that attenuates by 64 dB the possible high-frequency spurs that fall into the signal band. Assuming that the anti-aliasing filter has only real poles, estimate its order.

7.4   An integrated DAC is based on a resistive Kelvin divider. It uses 256 unity resistors arranged in a 16×16 two-dimensional array connected in series in a serpentine fashion. The distance between the unity resistors is $d_u = 5$ μm. The value of a generic resistance in the (x, y) position is

$$R(i, j) = R_u(1 + 0.001x/d_u)(1 - 0.0014y/d_u),$$

where R_u is the value of the resistor $R(1, 1)$ at $x = 0$ and $y = 0$. Determine the input–output characteristic of the DAC and, using the help of a computer program, estimate the INL and the DNL.

7.5   A three-bit Kelvin divider uses eight equal resistances of 100 kΩ. The selection of the intermediate voltages is done by eight switches connecting intermediate points to a unity gain buffer with 10 MΩ input resistance. The resistance of the switch is 1000 Ω in the on state and 1 MΩ in the off state. Determine the voltages generated by the buffer for the eight different input codes. (Hint: write a computer program, or assume the result is known and determine the known reference voltage.)

7.6   Study the mismatch limit in an R–$2R$ ladder network used in a 10-bit DAC. The reference is voltage, equal to 1 V. Focus on the first two cells and suppose that the next eight are ideal. Estimate the DNL error at the crossing of one-quarter, half and three-quarters of the full scale. What is the required accuracy in order to have, in the worst case, a DNL smaller than 1 LSB?

7.7   Sketch the circuit schematic of a 12-bit monotonic DAC made by a mixture of an R–$2R$ ladder architecture and a Kelvin divider. The mismatch of resistors is such that an eight-bit R–$2R$ ladder ensures monotonicity. The last resistance of the ladder is the series connection of multiple unity resistances that realize the Kelvin divider. What is the total number of unity resistances?

7.8 The spread of the capacitances in a capacitor-based DAC of the type shown in Figure 7.19(b) can be reduced by using C–$2C$ ladder cells in a similar fashion to an R–$2R$ scheme. Draft the architecture of a 10-bit DAC that uses three ladder cells for the LSBs. How many unity capacitors are saved? What is the limit determined by parasitic capacitances?

7.9 The op-amp used in the sample-and-hold of Figure 7.28(b) and (c) features a 57 dB finite gain and has a 3 mV offset. A random generator with 1 mV variance describes the noise performance. Estimate for both schemes the output voltage and noise with 0.657 V at input. In the second scheme $C_1 = C_2$.

7.10 A Kelvin divider generates the reference voltages needed by a flash ADC. The unity elements are arranged as a folded array with half of the resistors running from bottom to top and half from top to bottom. Due to the resistivity gradient, two neighboring resistor differ by $\epsilon_R R_u$. What is the minimum value of ϵ_R that makes the INL lower than 1 LSB for a six-bit and an eight-bit flash?

7.11 Suppose that the occurrence probability $p(\epsilon_Q)$ of the quantization error is constant in the interval $-\Delta/2 < \epsilon_Q < \Delta/2$ and goes to zero outside it. Use the statistical equation

$$\langle \epsilon_Q \rangle^2 = \int_{-\infty}^{\infty} p(\epsilon_Q) \epsilon_Q^2 \, d\epsilon_Q,$$

to estimate the average power of the quantization error. Notice that the integral of the occurrence probability is, obviously, equal to one.

7.12 The noise shaping of a pth order $\Sigma\Delta$ modulator is $(1 - z^{-1})^p$. Use the relationship $z \to e^{sT}$ that links z and the complex frequency s to determine the frequency at which the amplification becomes equal to one, for $p = 1, 2, 3, 4$. Remember that $1/T$ is the sampling frequency.

CHAPTER 8

DIGITAL PROCESSING CIRCUITS

This chapter introduces you to the electronic circuits for digital processing. These very complex circuits are made up of millions or even billions of transistors. The complexity involves many levels of hierarchy and requires the use of macro-functions. Often circuits are software controlled or software reconfigurable. The goal of this chapter is not to describe the circuits in detail but to help you in understanding their high-level features and in getting the flavor of typical circuit tasks – what is commonly called "hardware design." We shall study how to interface digital blocks, how to transfer digital data and how macro-functions are implemented. Finally we shall discuss the types and features of memories.

8.1 INTRODUCTION

Digital circuits handle numbers. The word "digital" comes from "digit," which means one of a set of symbols representing integers or real numbers. The Latin word *digita* (which refers to the fingers of the two hands) corresponds to 10, the basis of the 10-number decimal system. However, although the word "digital" therefore implies 10 symbols, the numerical calculation performed by digital circuits uses two digits, one ("1") and zero ("0"), to represent numbers using binary methods. Therefore, in electronics disciplines the word "digit" serves to distinguish not 10 but just two symbols, also called *bits*.

Understanding Microelectronics: A Top-Down Approach, First Edition. Franco Maloberti.
© 2012 John Wiley & Sons, Ltd. Published 2012 by John Wiley & Sons, Ltd.

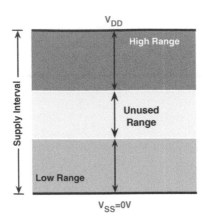

Figure 8.1   Voltage ranges with admitted levels to represent the two logic signals, and an unused interval.

The use of two symbols is required so as to distinguish between two electrical states. Typically, low voltage (or current) represents the logic zero and high voltage (or current) depicts the logic one. Low and high are obviously defined with respect to the supply interval. Moreover, the logic levels are within the supply range. If the bias of the circuit is 0–1.8 V, logic zero is equal to or slightly above 0 V and logic one is somewhat below 1.8 V. To be more precise, we divide the supply interval into three regions: a low range, a high range, and an intermediate interval not defined by any symbol, as shown in Figure 8.1. The amplitudes of the low and high intervals ensure a safety margin for representing the given symbol; the separation between high and low intervals guarantees discernibility between symbols.

> **Notice**
>
> The big advantage of digital electronics is that there are relatively wide margins in the definition of logic levels. Even in the presence of noise that dynamically alters the signal value the digital information does not change.

Having two symbols gives rise to binary representations. As in the case of decimal numbers, the position from right to left indicates the weight of the symbol. This is an increasing power of the base of the numbering system. For the decimal system, the weights are 1, 10, 100, 1000, and so forth; for the binary system the weights are 1, 2, 4, 8, 16, and so forth. For example, the binary number 101001 corresponds to 41 ($4 \cdot 10^1 + 1 \cdot 10^0$) in the decimal numbering system, being given by ($1 \cdot 2^5 + 1 \cdot 2^3 + 1 \cdot 2^0$).

Other methods use bits in a different manner. Alternative numbering systems serve to optimize specific digital algorithms. We do not discuss those systems in more detail, because for our purposes it is enough to study the operation in pure binary terms.

8.2   DIGITAL WAVEFORMS

Digital signals often use sampled data. They change their logic values at a given rate, defined by a clock. Normal schemes usually employ a single clock. However, complex systems,

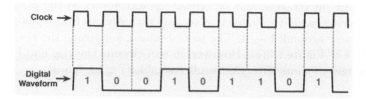

Figure 8.2 Clock and bit-stream waveform corresponding to the sequence 100101101.

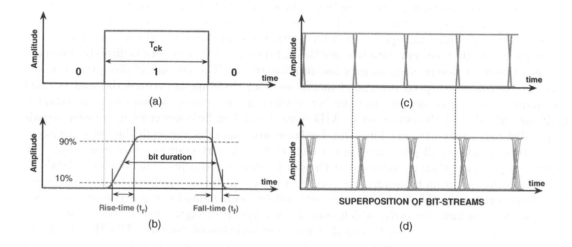

Figure 8.3 (a) Ideal waveform representing the bit sequence 010; (b) possible waveform in real situation with different loads for $0 \to 1$ and $1 \to 0$ transitions; (c) superposition of multiple waveform generated by generic bit sequences; (d) more realistic waveforms for case (c).

or architectures that include analog sections, can use multiple clocks whose frequencies are, preferably, related to one another by integer ratios. The clocks of digital processors determine the speed of mathematical operations; algorithms employ a given number of clock cycles, making the processing speed equal to the clock speed divided by that number.

The bits may change value at the clock's rising or falling edge. In the case of $0 \to 1$ or $1 \to 0$ transitions we ideally expect sharp steps synchronous with the clock. For example, the sequence of bits 100101101 would give rise to the digital waveform of Figure 8.2. The output changes at the rising edge of the clock and remains unchanged for the entire clock period. However, in real cases switching from "1" to "0" or vice versa is not instantaneous but occurs with some delay because of the finite speed of electronic circuits, which means that the rising and falling times are not of zero length. Depending on design and load the 010 ideal sequence of Figure 8.3(a) can look like the diagram in Figure 8.3(b). The rise time, defined as the interval needed to go from 10% to 90% of the full amplitude, is low because the circuit is assumed to be driving a capacitive load, which limits the driving capability of the electronic circuit taking the signal from "0" to "1". The fall time shown is faster, probably indicating a lower load. The result is that the waveform duration in the "1" state, measured by considering the crossings of the 50% amplitude, does not match the clock period.

Now consider many bit streams, and superpose the waveforms on the same generated diagram. Each clock period can have a value that repeats the previous bit, depicted as a transition $0 \to 1$ or, the other way around, $1 \to 0$. The plot obtained, supposing the rise or fall times are small, ideally looks like Figure 8.3(c). However, in real circuits the rise and fall times depend on load, which changes dynamically. The result resembles Figure 8.3(d).

8.2.1 Data Transfer and Data Communication

Data transfer or data communication, which consists of the exchange of digital messages between electronic apparatus, moves data within a variety of electronic systems ranging from computers to integrated circuits. For data communication the distance over which the data flows can be very high, and special data transformations are needed, which make long-range communication effective. For data transfer the distance is small, and data is directly transmitted as a two-level electrical signal over low-resistive traces. The amount of data, the distance and the transmission speed are the elements used for deciding between serial and parallel communication and for defining the number of wires on the busses. When the information is multi-bit, like that at the output of an A/D converter, it can be convenient to transfer signals in parallel using one wire per bit. But for large word lengths the multi-bit information is better broken into small parts sent in parallel, with each part transferred sequentially. The individual parts are then reassembled at the destination to recompose the multi-bit data.

The information from transmitter to receiver can be sent by an unrelated sequence of single bits; however, frequently, many bits are encapsulated into multi-bit unitary sequences. For example, we can have the *byte*, which consists of a group of eight bits; a larger number of bits can be grouped into *packets*, together with the addition of "service" bits that facilitate communication and enable correction of possible transmission errors (measured by the so-called Bit-Error-Rate (BER)), caused by noise, interference, or circuit malfunctioning.

The data flow in the cases discussed previously is in one direction. This is the simplest way to transfer digital information, and is called *simplex*. There is also bidirectional communication, or *duplex*. Duplex communication is further divided into two subcategories: *half duplex* and *full duplex*. Half duplex uses only one wire (or, more usually, one communication channel); this is used either for transmission or reception, but not both at the same time. The blocks at the interfaces, as shown in Figure 8.4, operate as either transmitter or receiver. Full duplex is in reality two simplex channels, as it uses two wires, one to transmit and the other to receive data at the same time.

Be aware that . . .

data transfer and data communication are not just a "communication" matter, but also something to consider carefully when designing digital circuits.

As shown by Figure 8.3(d), timing accuracy is relevant for data transfer. It is worth discussing the issue in some more detail, especially for cases that use very high data rates. Indeed, modern digital systems transmit data by the use of extremely high clock frequencies, making it difficult to maintain data integrity (a concept that will be discussed in a later chapter) either for transferring data through wire connections or when information flows through optical or wireless channels (data communication).

The diagram showing the superposition of multiple data streams displayed over a single clock period can be worse than what Figure 8.3(d) indicates. What is necessary is to make the

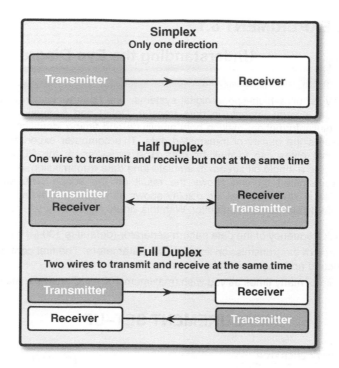

Figure 8.4 Simplex and duplex (half and full) communication schemes.

high level, "1", and the low level, "0", distinguishable in amplitude and time, with transitions occurring within fractions of the clock period. Since the diagram we get looks like an eye, the figure is called an *eye diagram* – a valuable tool used to assess the quality of data transfer.

Digital waveforms at very high transmission rates depend on the bit pattern. For example, we have different responses for repeated sequences made by 111000 or 101010 because the persistence of three "1"s leaves more time to bring the waveform closer to full amplitude. The effect of the bit sequence on received waveforms makes the curves on the eye diagram different. Critical cases affected by poor synchronization, by noise, and by interference lead to quite different responses, which occupy a significant part of the eye – or, as we say, the eye closes. Figure 8.5 shows two possible situations. Having the eye too closed prevents reliable bit identification, with consequent possible errors. Therefore the goal of the circuit designer is correct operation at the transmission and receiving sides to improve reliability or, as we frequently say, to open the eye.

Board-to-board and chip-to-chip communication often use serial data transfer because it limits the number of wires. A bus handles many bits in parallel until the data transfer bottlenecks. Then the data converges into a single channel and is transmitted at a rate equal to that of parallel communication multiplied by the number of wires in the bus, $f_{ser} = N f_{par}$.

The device that performs this operation is called a *serializer* or, for short, *ser*. At the receiver side there is a complementary electronic circuit that performs the *deserialization* or, for short, *des*. The blocks devoted to the function are normally called *SerDes*. The schematic diagram in Figure 8.6 is a representation of the two blocks. The communication channel, especially at very high frequencies (tens of GHz, in some cases) attenuates high-frequency

COMPUTER EXPERIMENT 8.1

Understanding the Eye Diagram

Data transfer is key for ultra–high–speed digital systems. The communication channel must be inexpensive but must ensure integrity in data transfer. Delay and frequency dependence of transmission lines blur information, making both amplitude and timing identification difficult. The eye diagram measures the quality of transmitted data. This computer experiment uses a data pattern generator that gives rise to multi-bit data. The serializer on the PCB transforms the multichannel signal into a single bit stream to emulate long data transmission. A button triggers a snapshot of the transmitted bit and shows the result on the scope. The same snapshot, converted into digital, is stored so that the display shows all snapshots taken, to build, step by step, an eye diagram displayed on the monitor. With this virtual experiment you can:

- select the clock frequency of the data pattern generator (default is 100 MHz);
- set the response of the transmission line with two parameters. The first controls rise and fall times. The second adds a jitter in delay times;
- push the button as many times as you wish (maximum 100) and reset the memory.

MEASUREMENT SET–UP

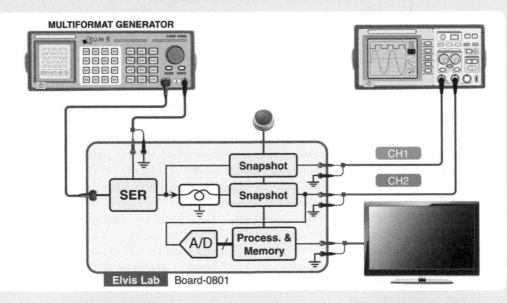

TO DO

- Set to near zero the transmission line parameters, and zero the jitter, to emulate an ideal situation. Push the button many times and generate a well formed eye diagram.
- Reset the memory and set the delay line parameters, including jitter. Construct the eye diagram for a clock frequency around 500 MHz.
- Increase the clock frequency and observe how the eye diagram worsens. Compare the cases with and without jitter.

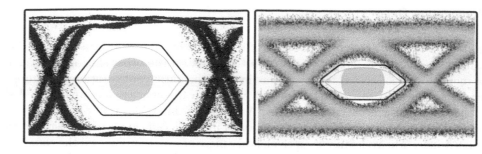

Figure 8.5 Eye diagrams for two typical cases. For the right-hand measurement the eye is more closed than for the one on the left.

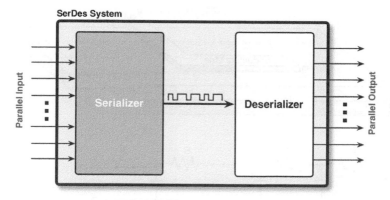

Figure 8.6 Block diagram of "SerDes" architecture for parallel serial communications.

components much more than the low-frequency terms. That action closes the eye. In order to compensate for the limitation, the circuit uses a filter (often a Finite Impulse Response (FIR) filter) in front of the transmit driver, and an equalizer on the receive side. The clock must be synchronous with the data. In some cases the clock is directly extracted from the transmitted data by the time-recovery section.

Self training

Search the Web for different types of architectures for SerDes communication. Look at features of at least three integrated circuits for that function that are available on the market. Make a list of:

- the type of equalization and pre-emphasis;
- maximum transmission rate;
- possible applications;
- required supply voltage and power consumption.

Sketch the block diagram of one of the three selected integrated circuits.

8.2.2 Propagation Delay

The trace on a PCB or the interconnection of integrated circuits is a metal deposited onto an insulating material. At a short distance from the trace, on the other side of the insulator, there is a metal plate connected to ground, as in the schematic configuration of Figure 8.7. The structure, when used to transfer signals at very high frequencies, has well-defined sizes and dielectric constant. It is called a *strip-line*. A convenient model for that physical structure is illustrated in the same figure: a series connection of multiple cells made of a resistance and a parallel capacitance. The resistance is an elementary piece of the strip-line and the capacitance is the corresponding parasitic quantity. The model is a good approximation of reality but it

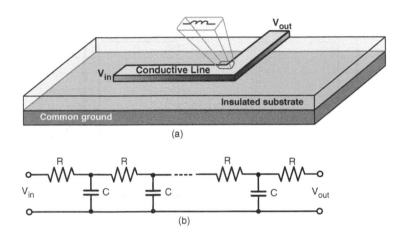

Figure 8.7   (a) Pictorial representation of metal trace on an insulating substrate; (b) its equivalent circuit.

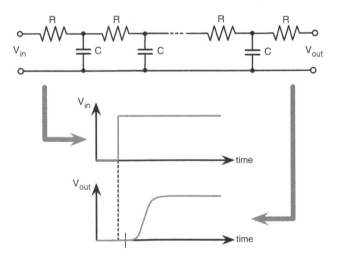

Figure 8.8   Response of resistive transmission line with input step signal.

is not exact. For example, a sharp curve on the conductive line gives rise to an inductance not accounted for in the circuit. Increasing the width of the strip reduces the resistance, but, at the same time, it increases the parasitic capacitance. The two effects compensate for each other and do not change the frequency response.

The equivalent circuit of metal connections predicts a step response like the one in Figure 8.8. The result depends on the RC product and the distance traveled. There is a time delay and some smoothness in the amplitude transition. The time delay can be reduced by diminishing the elementary cell time constant. The smoothness is often reduced by intermediate active buffers placed along lines to reconstruct the step waveform.

As well as the existence of delay at interconnections, note that no real electronic circuit responds instantaneously. This introduces a delay with respect to input transitions, and determines logic signals with a delay and given rise and fall times. The delay caused by a logic cell is called *propagation delay*. It typically increases with chip temperature and supply voltage.

Example 8.1

The delay of interconnections, also contributing to the propagation delay, depends on the speed of the digital waveforms along the transmission media. For a metal connection on an integrated circuit or a trace on a PCB, the speed is about 0.7 c, where c is the speed of light. Determine the delay caused by a metal connection of 1 mm. What is the mismatch, measured as fraction of the period, determined by two PCB traces of 5 mm and 26 mm for a clock at 2.4 GHz?

Solution

Simple calculations give the results. If the signal travels for a distance d and v is the speed, the delay is obviously $\Delta t = d/v$. Since 0.7 $c \simeq 2.1 \cdot 10^{11}$ mm/s, one millimeter line causes a delay of 4.75 ps. The 21 mm difference between the two PCB traces gives rise to a delay mismatch of about 100 ps. Since the period of a 2.4 GHz clock is 416 ps the mismatch in two transmitted clocks is as large as 24% of the period.

Propagation delay causes problems for high-speed digital processors as well, because of different propagation delays in signals. Bit streams can reach the inputs of a subsequent block after traveling through different paths. Even if signals start at the same time, they arrive at mismatched times, thus causing synchronization problems.

8.2.3   Asynchronous and Synchronous Operation

A digital system is made up of both combinational and sequential parts. The task of combinational logic is just to relate signals at input and generate defined logic signals at output. Obviously there is some delay in generating output, but that issue is not the focus of the operation. On the other hand, sequential circuits do care about timing. Circuits of that kind use one or more outputs generated by the logic itself. As implied by the name, a sequential system relies on a sequence of actions that are required to occur in the right order.

Sequential digital circuits can be asynchronous or synchronous. In the former case the data carried by digital waveforms flows with the only constraint established by the delay in the cells performing sequential operations. Flow correctness therefore depends on input that changes levels and on accuracy in propagation delays.

Asynchronous operation serves for various specific processes that do not require the use of a clock. Obviously, since mismatches between delays can impair the results, it is necessary to study timing carefully and to compensate for mismatches by artificially adding delays at critical points. A more reliable alternative operation is when timing is controlled by a clock. In this case synchronous functioning is made possible by digital circuits whose output changes only when the clock changes. Since the clock serves to align slow and fast paths, the period must be such as to accommodate the most time-consuming processing section. Often the architecture is divided into sections where processing is performed in successive clock periods.

> **Be advised**
>
> The use of a clock does not ensure total safety. The propagation delay along different paths can differ but not by very much. Clock distribution always causes clock misalignment.

8.3 COMBINATIONAL AND SEQUENTIAL CIRCUITS

Logic circuits that process digital data can be combinational or sequential. The output of combinational circuits depends only on input. On the other hand, sequential circuits produce output on the basis of both input and output. Since output depends on past input, it is the sequence over time of input that determines the results. For example, a memory circuit is essentially sequential, because its output depends on the input that occurred in the past, held over time: output equals itself unless there is an erase or a rewrite input. On the other hand, the addition of two inputs is combinational because the result just depends on changing inputs. Even the serializer or deserializer circuits discussed above are combinational. The data rates at input and output are different, but the output depends only on the inputs.

Figure 8.9 shows the difference between combinational and sequential block diagrams. The combinational outputs are, as can be seen, a given function of the inputs. Sequential schemes, which include a combinational logic circuit, use inputs and returning signals. Feedback is part of what is delivered to the output – or signals specifically generated to depict the output state. These circuits may use a clock to update the status of feedback controls.

8.3.1 Combinational Circuits

From a mathematical point of view, a combinational circuit implements a given mapping between input and output. At the lowest level there is the comparison between just two bits to give rise to one or more outputs depending on the specific rules used to perform the comparison. Often, combinational circuits implement Boolean logic functions, the rules for combining bits developed by the mathematician George Boole (1815–1864). You probably know about basic Boolean operators like AND and OR. The AND, as implied by its name, gives "1" at output if its inputs are both "1". For the OR it is enough to have one of the inputs "high" to generate "1" at output. Other simple logic operators are NAND and NOR, which give rise to outputs opposite to the ones given by AND and OR. There is also the NOT, which, applied to a single bit, generates the inverse at output. Boolean logic and how to handle complex functions by decomposing operations or simplifying them are not studied

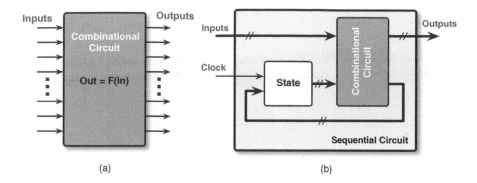

(a) (b)

Figure 8.9 (a) High-level representation of combinational circuit; (b) sequential circuit including combinational circuit.

here, as they form part of a specific course. At this level it is enough to know a few basic rules and to be aware that complicated combinations can be transformed into an optimal sequential combination of basic Boolean functions.

A custom network of elementary digital cells, like the ones to be studied shortly, implements any combinational function. The network gives rise to outputs within a single clock period or is made by a cascade of sections operating in successive clock periods, thus producing outputs after a given latency time. Instead of using a circuit that tailors output to comply with given rules, it can be convenient to use a Look-Up Table (LUT), a storage circuit that utilizes input data as the address of a memory and generates, at output, the information stored at that address. The storage circuit contains all the combinational results in a permanent manner or can be re-written. The storage (or memory) element is called "look-up" because the circuit simply looks up what the output for a specific input is. Therefore combinational schemes generate the result every time the input changes; the look-up solution already stores all the possible outputs, and fetches the one needed in response to the input command.

Choosing between a combinational circuit and a LUT depends on the complexity of functions to be implemented and the speed required. If the function is simple, it is obviously convenient to interconnect logic cells. However, for complex functions or when it is necessary to reconfigure the hardware, the use of a reprogrammable LUT ensures flexibility and better speed.

Self training

Find an example of a combinational circuit. Write the set of rules that the digital processor must implement, and verify that the operation does not need to know past history. Sketch the block diagram of one of these processing systems, and define the number of clock periods needed to generate outputs.

An example of a combinational circuit is the burglar alarm system that you may have at home. There are door and window sensors and motion detectors. They send logic signals to a console via connections that may be directly wired or wireless. A control panel enables you to select zones or activate extra sensors. A siren emitting a very loud noise hopefully causes burglars to give up and leave the house. Alternatively, or in addition, there is an automatic

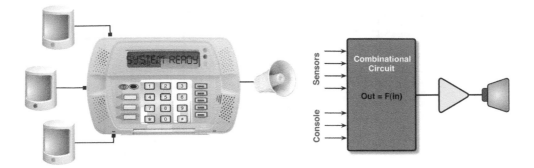

Figure 8.10 Example of system using combinational circuit: a burglar alarm and its high-level schematic block diagram.

call to an emergency telephone number when the alarm is triggered. The combinational circuit receives logic signals from sensors and, depending on the setting, switches on the siren and/or makes the emergency call, when one of the sensors of activated zones indicates an intrusion. Figure 8.10 shows a possible arrangement and the schematic block diagram. The output is a single logic signal, which needs a power amplifier to drive the siren, perhaps for a given period of time.

8.3.2 Sequential Circuits

Sequential circuits, as shown in Figure 8.9(b), establish feedback because they use output signals as part of their input. There are two possible sequential architectures: synchronous and asynchronous. The former uses a clock to establish the update times of the feedback signals. The latter does not have any time control and sends back outputs continuously as soon they change value. Both synchronous and asynchronous architectures enable higher-level processing than combinational schemes but can give rise to problems, especially in the asynchronous case, because of possible instability. Certain output statuses can induce alternating changes of output, regardless of input value. What happens is that the output, which returns immediately in an asynchronous configuration, can cause a sudden inversion of some output bits. Those changed bits returning immediately at input give rise to further change and, because of this, the output oscillates. Instability persists until a different input configuration takes the output out of the unstable conditions. The oscillation frequency depends on time delay along the loop. For modern technologies the bouncing frequency can be extremely high.

Another critical circumstance occurring with asynchronous operation is when two or more inputs change almost simultaneously. If the order in which inputs change status does not affect the final state, the situation is not critical. Otherwise, the so-called *race condition* impairs the architecture. The exact time at which the input bits change depends on dynamic factors, and it is not possible to know a priori which bit will win the "race." Since it is the winner that determines output status, an unpredictable final value will result.

Using a clock avoids the above problems, because feedback signals are univocally determined if update occurs when inputs and outputs are well stabilized. The synchronous scheme uses storage elements to hold the feedback signals for a complete clock period. Since the memory stores the status of the system, a sequential circuit is also called a *state machine*.

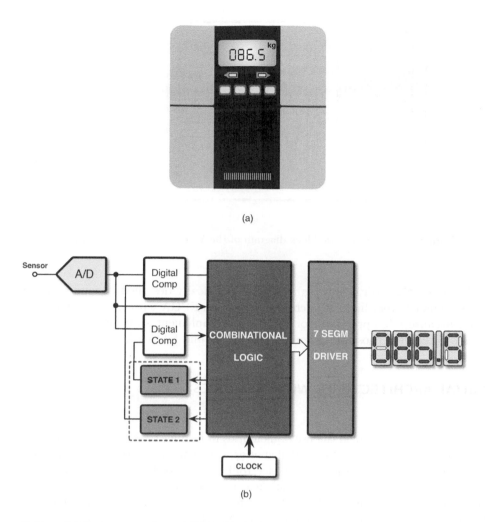

Figure 8.11   (a) Bathroom scale and (b) its block diagram, intended as a sequential logic circuit. The state of previous measurements serves to validate the final measured value.

We have many examples of sequential circuits in real-life situations. For example, consider the digital bathroom scale, used almost every day by people who hope to lose weight. The scale has a strain-gauge sensor that generates an analog voltage proportional to the weight. The analog output becomes digital thanks to the action of an A/D converter. Notice that the measurement is not immediately valid, but it is necessary to wait for a stable output with its value within an interval lower than the accuracyof the scale. This is done automatically by an electronic circuit using a sequence of scale measures. A possible solution employs two digital comparators, whose set values depend on the previous states. Suitable combinational logic, as shown in Figure 8.11, decides for every measurement whether the value is staying inside the accuracy range. Only after a defined number of valid results does the logic acknowledge the measurement and generate the control for the seven-segment drivers lighting the display.

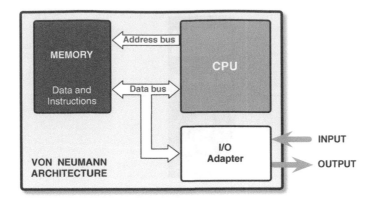

Figure 8.12   Schematic block diagram of the Von Neumann architecture.

Normally sequential logic is not realized with specifically designed circuits, unless the expected sales volume is very high. A microprocessor or an FPGA with external memory realizes the system, because microprocessors or FPGAs can implement state machines at costs that are normally lower than those of designing, debugging, and fabricating custom solutions.

8.4   DIGITAL ARCHITECTURES WITH MEMORIES

The previous section outlined the use of memories in feeding back the state of logic circuits. I suppose you know enough about digital memories, which are a means of storing and accessing binary data. The logic information sets an electronic circuit so that it can distinguish between two levels, "1" and "0". In addition to storing and receiving data, it is necessary to know where these circuits are located – that is to say, to know the address of each memory cell.

There are two different types of information stored in a memory: data or instructions. Data is pure information; an instruction, as the name implies, indicates the manner in which electronic circuits process data. How electronic circuits do that can be complicated, and, actually, it is not necessary to know in detail how memories control circuits, at this level of study. What is important is to be aware that digital circuits are often controlled by an instruction set stored in a memory, retrieved for execution when necessary.

Suppose you want to multiply two numbers deposited somewhere in the memory. It is necessary to fetch the two numbers and instructions for the multiplication procedure. The numbers are temporarily stored in a register and entered in the so-called Arithmetic Logic Unit (ALU), which performs the multiplication. Then the result is possibly sent back to the memory for future use.

The above introduces the concept of general digital architectures, whose high-level schemes depend on the nature of the task to be implemented. Figure 8.12 shows the simplified diagram of what is known as the Von Neumann architecture (after the mathematician John Von Neumann). It is made of two main blocks, the memory and a special type of circuitry called the Central Processing Unit (CPU). The CPU, in addition to the capability of performing data processing, includes temporary memories where data is stored until it completes the

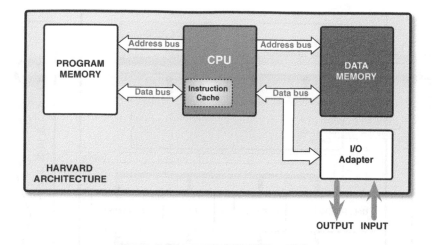

Figure 8.13 Schematic block diagram of a Harvard architecture.

tasks. Communication channels connect memory and CPU, to transfer data and provide the addresses of data. Moreover, there is a block for the input and output of data.

The Von Neumann architecture typically performs the following steps in a repeated manner:

- fetching an instruction from the memory at the address given by the program counter;

- updating the instruction address in the program counter;

- decoding the instruction and performing some operations.

In many cases there is another step for managing interrupts, which are used during lengthy tasks to allow you to do other things while the system waits for output from time-consuming computational sections.

The advantage of the architecture is mainly in its relative simplicity. Indeed, sharing the busses for data and instructions requires less hardware. However, it may happen that, by mistake, instructions are treated as data and data interpreted as instructions, resulting in a system crash. Moreover, the execution process is relatively slow, because it requires two memory accesses: one to fetch data and the other to acquire the instructions.

A specialized solution for heavy mathematical loads is the Harvard architecture, which uses separate memories for data and instructions. Instead of instructions and data flowing through the same bus, the Harvard solution employs separate busses for data and instructions, thus providing high speed. There are different versions of Harvard architecture that have been developed over the years: Figure 8.13 shows an example of a typical block diagram. The scheme includes the so-called *instruction cache* in the CPU. The cache is an auxiliary memory used by the CPU to store instructions that are repeatedly used. The result is improved overall system speed.

The single-memory Von Neumann architecture gave rise to the computer and its core part, the microprocessor (μP). The Harvard architecture is suitable for applications with a heavy computational load, as provided by Digital Signal Processors (DSPs). Both μP and DSP are extremely complex devices with hundreds of millions or even billions of transistors. We do not go into further details on their architectures here, nor study in detail the benefits that some of

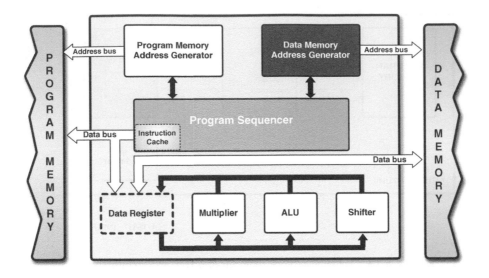

Figure 8.14   Schematic block diagram of the CPU of a signal processor architecture.

them have with respect to others. However, just to give you an idea about what is inside, let us expand a little bit a possible CPU of a DSP to produce, as an example, the block diagram in Figure 8.14. The scheme shows blocks such as the data register, multiplier, Arithmetic Logic Unit (ALU), and shifter; there are blocks dedicated to mathematical operations and to data handling. The diagram also shows functions such as memory address generators that are used by the architecture to exchange data and instructions with memories.

It is obviously necessary to go further and to analyze what is inside key components. We shall do that shortly for basic combinatorial, sequential functions and for memories, being now aware that memories are used to control digital circuits in addition to their well-known and common use for data storage.

8.5   LOGIC AND ARITHMETIC FUNCTIONS

After a short discussion of high-level architectures, we consider here a few key functions employed for logical and arithmetic operations. We shall analyze addition, multiplication, and data handling of multiple-bit numerical quantities. This study, however, remains at a relatively high level; in a later chapter we shall consider schemes and transistor-level details that perform the elementary operations.

8.5.1   Adder and Subtracter

Addition is a well-known mathematical function. It is relevant by itself but is also important because addition is the basis of multiplication. Since data processing uses multipliers heavily, it is important to have effective adders to secure high computational speed. The addition of binary numbers is like the addition of decimal numbers. The sum of two numbers with

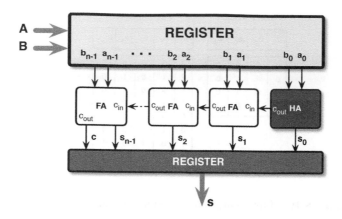

Figure 8.15 Ripple-carry adder made from half-adders and full-adders.

10 as base produces one of the 10 symbols of the base plus a carry over, equal to 1, if the result equals or exceeds the base. For a binary system we do the same with base 2. However, having a carry is much more frequent than for additions in the decimal system. It is therefore necessary to handle carries properly so as to optimize the operation.

Figure 8.15 shows a possible configuration of a multi-bit adder. Two n-bit signals A and B are temporarily stored in a register until the addition is performed. The two LSBs, a_0 and b_0, feed a block called a half-adder, to generate the LSB of the addition and a carry. This enters the next block, FA, together with the bits a_1 and b_1 to generate bit s_1 of the addition and a new carry. The block FA is called a full-adder. The operation continues until all of the bits are summed.

Figure 8.16 shows the input–output relationships of the half- and the full-adder. This kind of table is called a *truth table*. They consist of a number of columns, one for each input variable, and other columns giving the results. The truth tables for half- and full-adders are easily understood. There is a carry when the two inputs (or three for the full-adder) equal "1". The sum is "1" for just a single "1" among the inputs or when all three inputs equal "1". Otherwise the sum bit is "0". Sum and carry are the outputs.

The scheme in Figure 8.15, called a *ripple-carry adder*, generates the digital result sequentially because in order to generate the output any full-adder needs the carry at input. There is a time delay in the addition process, mainly given by the carry propagation delay, t_c. It is defined as the time delay needed to generate the carry-out after the carry-in is applied, assuming that the inputs are already present. If inputs enter simultaneously in the scheme of Figure 8.15 and t_s is the time needed for the sum, the worst case addition delay is

$$t_{add} = (n - 1)t_c + t_s, \tag{8.1}$$

assuming that the first half-adder generates the carry at the same time as the sum.

The propagation delay of the ripple-adder scheme is not optimal. To reduce that delay, the designer often uses other architectures. For example, we have the carry-look-ahead adder, which calculates the carry signals in advance, based on input signals. The method exploits the fact that carry occurs in two cases: when both the input bits are "1", or when one of the two bits is "1" and the carry-in is "1". The observation, applied iteratively to the cells of the adder, gives rise to a logic circuit that eliminates linear dependence of the delay on the

HALF ADDER

a_0	b_0	s_0	c_1
0	0	0	0
0	1	1	0
1	0	1	0
1	1	0	1

FULL ADDER

a_i	b_i	c_i	s_i	c_{i+1}
0	0	0	0	0
0	1	0	1	0
1	0	0	1	0
1	1	0	0	1
0	0	1	0	1
0	1	1	0	1
1	0	1	0	1
1	1	1	1	1

Figure 8.16 Input–output relationships of half- and full-adder.

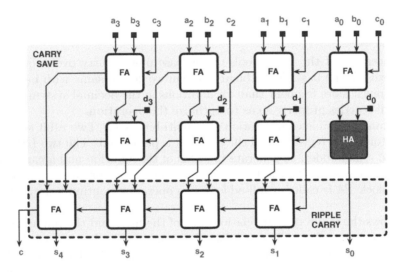

Figure 8.17 Carry-save adder used to add four four-bit numbers.

number of bits. The method, not studied here in detail, requires a carry logic that becomes quite complicated for adders with more than four bits. For that reason, the adder with the carry-look-ahead technique is made of four-bit modules. These modules are then used in a hierarchical structure, whose input word-length is a multiple of four bits.

There are cases where it is necessary to add more than two numbers. This is what happens, for example, in multipliers, to be studied shortly. In those cases instead of adding the first two numbers and then adding the sum to the next and so forth, it is more convenient to add the bits with equal weight all together, and to compute sum and carry separately. Then the results are summed up with a ripple-carry adder to produce the result. This method is called *carry save*. The delay of n-bit addition can become proportional to $\log(n)$ instead of n if this technique is used.

Figure 8.17 shows an example of adding four four-bit numbers. Four full-adders add the bits of three of the inputs, generating partial sums and carries. The partial results enter the second line of adders to give rise to sums and new carries constructing the carry-save section.

Then a ripple-carry adder sums them together with a carry from the first addition that was left behind. The propagation delay is

$$t_{add} = 3t_c + 3t_s, \tag{8.2}$$

assuming again that the adders generate the carry at the same time as the sum. On the other hand, a cascade of just two ripple-carries produces a longer propagation delay: $6t_c + 2t_s$.

Subtraction of binary numbers is also an important mathematical operation. Even though it can be done by following the definition, digital processors actually convert subtraction into addition by using the complement method. There are two complement systems: the ones' complement and the two's complement. They are almost equivalent because the two's complement is the ones' complement with the addition of one. The ones' complement of a number is formed by reversing all bits. For example, the eight-bit number 01011011 has ones' complement 10100100. Notice that for an n-bit number, B_n, the ones' complement is

$$\bar{B}_n = 2^n - 1 - B_n, \quad \text{with } B_n = b_{n-1}b_{n-2} \cdots b_1 b_0. \tag{8.3}$$

Therefore, the subtraction $A_n - B_n$ can be obtained by the addition

$$A_n + \bar{B}_n + 1 = A_n - B_n + 2^n, \quad \text{with } A_n = a_{n-1}a_{n-2} \cdots a_1 a_0 \tag{8.4}$$

which gives rise to an overflow equal to 2^n if the result is positive. Otherwise the result indicates a negative number. It may be that the bit-set obtained must be properly corrected, depending on how the numbering system specifies the sign.

The two complements directly generate the number to be added to the minuend, A, to give rise to the result. A convenient method to obtain the two's complement is to start at the least significant bit and move to the left. Each bit remains the same until the first "1" is reached. The successive bits reverse. For example:

$$\text{two's complement of } B_8 = 10110100 \rightarrow \mathbf{01001100}, \tag{8.5}$$

an operation that can be easily performed by a simple combinatorial network.

In summary, subtraction is performed by a preliminary processing of the subtrahend, an addition, and a possible post-processing of negative results.

8.5.2 Multiplier

The multiplication of two digital words is an indispensable mathematical operation for implementing many processing algorithms. Multipliers enable relevant functions such as the Fast Fourier Transform (FFT) and digital filtering. The result is that the efficiency and speed of digital multipliers is crucial, especially with numbers of bits as large as 32, 64, or even more.

Multiplication is essentially a sequence of repeated additions of partial products. Figure 8.18 shows an example of binary multiplication. The first row is the multiplicand; the second row is the multiplier. With digital numbers the partial product of two bits is "1" if both bits are "1", and otherwise it is zero – a function determined by a logic AND. The single bits that provide the partial products are then arranged onto an addition matrix like the scheme in Figure 8.18. The row arrangement ensures that elements with same weight are in the same column.

Notice that, for the numerical example of the table at the right-hand side of Figure 8.18, the partial products of the second row are all "0" because b_1 is zero. Moreover, the partial

MULTIPLICAND				a_4	a_3	a_2	a_1	a_0
MULTIPLIER					b_3	b_2	b_1	b_0
PARTIAL PRODUCTS				$a_4 b_0$	$a_3 b_0$	$a_2 b_0$	$a_1 b_0$	$a_0 b_0$
			$a_4 b_1$	$a_3 b_1$	$a_2 b_1$	$a_1 b_1$	$a_0 b_1$	
		$a_4 b_2$	$a_3 b_2$	$a_2 b_2$	$a_1 b_2$	$a_0 b_2$		
	$a_4 b_3$	$a_3 b_3$	$a_2 b_3$	$a_1 b_3$	$a_0 b_3$			
m_8	m_7	m_6	m_5	m_4	m_3	m_2	m_1	m_0

				1	1	0	0	1
					1	1	0	1
				1	1	0	0	1
			0	0	0	0	0	
		1	1	0	0	1		
	1	1	0	0	1			
1	0	0	1	1	0	1	0	1

Figure 8.18 Example of binary multiplication as addition of partial products.

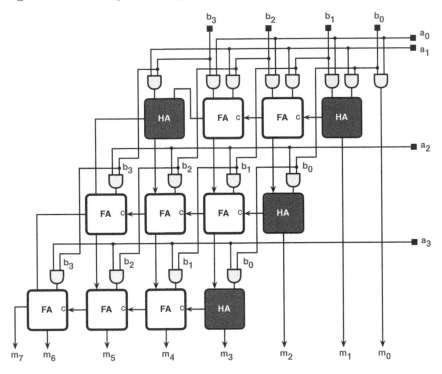

Figure 8.19 Block diagram of matrix multiplier.

products of other rows are simply shifted versions of the multiplicand. The third line is shifted by two positions and the fourth line by three. Adding the bit matrix, column by column, produces the result.

The simplest circuit implementation of the multiplier is a matrix of half- and full-adders. Figure 8.19 shows an architecture for two numbers of four bits each. The bits of the multiplicand, A, run horizontally and those of the multiplier, B, flow vertically. The symbols used at input of the adders represent the circuit realization of the AND. The two inputs give rise to the one-bit partial product. Since the weight of the first column is "1" (2^0), only the $a_0 b_0$ product contributes to that weight. The next column, with weight $2 = 2^1$, has two terms, $a_0 b_1$ and $a_1 b_0$. The next has three terms, and so forth, until the number of terms equals the number of bits.

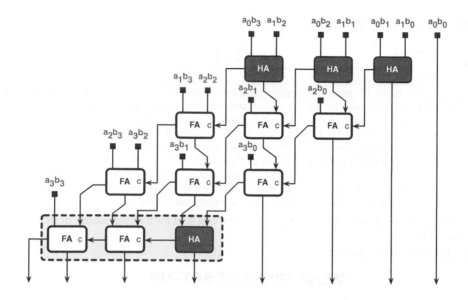

Figure 8.20   Carry-saver (4×4)-bit multiplier.

The computation time needed by the multiplier of Figure 8.19 is long because sequential addition entails multiple and long propagation paths. The sum time is proportional to the number of rows; the carry time depends on the number of columns. Therefore, for architectures like the scheme of Figure 8.19 and multiplications of two numbers with N and M bits, the equation comes out as

$$t_{mult} = (N + M - 3)t_c + (N - 1)t_s + t_{and}, \tag{8.6}$$

where t_{and} is the time required for the initial AND operations.

Often it is necessary to use architectures faster than the scheme described above. The carry-save method used for adders is also effective for multipliers. Figure 8.20 illustrates an architecture of a carry-saver (4×4)-bit multiplier. The row that gives rise to m_1 in the table of Figure 8.18 contains two terms. A half-adder produces the sum. The value of m_2 depends on three bits and the carry that comes from the first adder. Since a full-adder generates the carry almost at the same time as the sum, the carry-save architecture uses a half-adder to sum two terms and a full-adder to sum the result, the third term, and the carry of the previous row. In this manner the delay is only that needed to make the two sums. The next column receives two carries that are summed to one of the terms, as in the previous case – with partial sum and one of the terms of that column. The last row is a ripple-carry adder that sums the carry and sum of the previous row plus the last term of the multiplication.

The computation time is reduced, being given by

$$t_{mult} = 3t_c + (N - 1)t_s + t_{and}. \tag{8.7}$$

An effective mathematical method that improves the multiplier architecture is to represent the multiplier by so-called Booth coding. The aim is to reduce the number of ones with additions and subtractions instead of just additions. The Booth transformation consists of re-coding each "1" as $1 + 2 - 1$. That means adding a "1" in the next column and -1 in that

COMPUTER EXPERIMENT 8.2

4 x 4 Digital Multiplier

A multiplier is the basis of many processing functions. This computer experiment uses a simple matrix multiplier with four bits for the multiplier and four bits for the multiplicand. A data pattern generator builds three words of eight bits. The user defines the first and last words while the middle word is made with 0 for all bits. The 48 bits are transmitted over six channels to the ElvisLab board. A SerDes gives rise to a burst of two words at the input of the multiplier. A square wave generator defines the operational speed of the circuit. The display driver serves to store the results of multiplication and generates video signals for the monitor used to visualize the digital output waveform. The experiment allows you to study timing responses featuring switching from "0" to "1" at different clock frequencies. In this virtual experiment you can:

- define with the pattern generator the bits of the first and third digital words of multiplicand and multiplier;
- change the clock frequency.

MEASUREMENT SET–UP

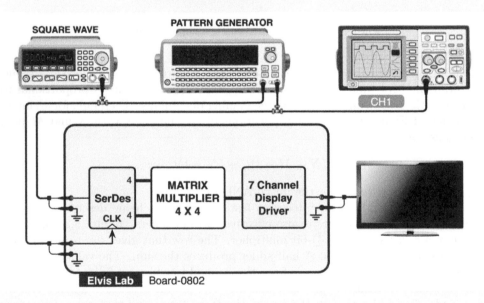

TO DO

- Set for both multiplier and multiplicand of the two digital words the same sequence (0001). Observe the product result, made by a single logic one at output. Determine the maximum speed of operation under these conditions.

- Change the code of the two words and observe the different time delays at output. Notice that the delays depend on the input code.

- Find the input sequence that produces the longest delay. Identify the critical path in the detailed scheme of the matrix multiplier given in this chapter.

- Estimate the maximum clock frequency that can be used by this circuit. Allow a suitable time margin.

column. Consider for example the eight-bit multiplier 01111110. The transformation gives rise to

$$
\begin{array}{cccccccc}
0 & 1 & 1 & 1 & 1 & 1 & 1 & 0 \\
\hline
 & & & & & 1 & -1 & 0 \\
 & & & & 1 & -1 & & \\
 & & & 1 & -1 & & & \\
 & & 1 & -1 & & & & \\
 & 1 & -1 & & & & & \\
1 & -1 & & & & & & \\
\hline
1 & 0 & 0 & 0 & 0 & 0 & -1 & 0,
\end{array}
\tag{8.8}
$$

which reduces the number of ones from six to a pair: $1, -1$.

Example 8.2

Determine the multiplication table of the decimal product 14×21 with normal and Booth coding. Use six-bit binary without a sign.

Solution

The two numbers to be multiplied using six-bit binary are 01110 and 10101. The choice of different multiplicands and multipliers gives rise, for normal matrix representation, to

$$
\begin{array}{ccccc}
0 & 1 & 1 & 1 & 0 \\
1 & 0 & 1 & 0 & 1 \\
\end{array}
\qquad
\begin{array}{ccccc}
1 & 0 & 1 & 0 & 1 \\
0 & 1 & 1 & 1 & 0 \\
\end{array}
$$

$$
\begin{array}{ccccccccc}
 & & & & 0 & 1 & 1 & 1 & 0 \\
 & & & 0 & 0 & 0 & 0 & 0 & \\
 & & 0 & 1 & 1 & 1 & 0 & & \\
 & 0 & 0 & 0 & 0 & 0 & & & \\
0 & 1 & 1 & 1 & 0 & & & & \\
\hline
0 & 1 & 0 & 0 & 1 & 0 & 0 & 1 & 1 & 0
\end{array}
\qquad
\begin{array}{ccccccccc}
 & & & & 0 & 0 & 0 & 0 & 0 \\
 & & & 1 & 0 & 1 & 0 & 1 & \\
 & & 1 & 0 & 1 & 0 & 1 & & \\
 & 1 & 0 & 1 & 0 & 1 & & & \\
0 & 0 & 0 & 0 & 0 & & & & \\
\hline
0 & 1 & 0 & 0 & 1 & 0 & 0 & 1 & 1 & 0
\end{array}
$$

For Booth coding it is more suitable to use 01110 as the multiplier, because it has a group of three "1"s. The Booth coding is 10010, where the bold "1" indicates negative. The resulting addition matrix is

$$
\begin{array}{ccccc}
1 & 0 & 1 & 0 & 1 \\
1 & 0 & 0 & 1 & 0 \\
\hline
\end{array}
$$

$$
\begin{array}{ccccccccc}
 & & & & 0 & 0 & 0 & 0 & 0 \\
 & & & \mathbf{1} & \mathbf{0} & \mathbf{1} & \mathbf{0} & \mathbf{1} & \\
 & & 0 & 0 & 0 & 0 & 0 & & \\
 & 0 & 0 & 0 & 0 & 0 & & & \\
1 & 0 & 1 & & 1 & & & & \\
\hline
0 & 1 & 0 & 0 & 1 & 0 & 0 & 1 & 1 & 0
\end{array}
$$

where the terms of the bold line must be subtracted rather than added.

The result is obviously the same but the number of elementary operations diminishes.

Table 8.1 Modified Booth coding table

Pair $b_n b_{n-1}$	Pair $b_{n-1} b_{n-2}$	Code	Why?
00	00	0	Nothing is at input
00	01	1	Receiving a 1 from the next pair of bits
01	10	1	This is a 1 and nothing from next pair of bits
01	11	2	This is a 1 plus a receiving 1
10	00	-2	The one sent to forward causes a -2
10	01	-1	There is the -2 for forwarded bit receiving a 1
11	10	-1	-2 for forwarded bit, holding a 1
11	11	0	-2 for forwarded bit, holding a 1 receiving a 1

This method potentially reduces the number of "1"'s but does not improve speed because it does not eliminate critical paths. Moreover, the area increases, because of the higher complexity of subtractors as opposed to adders.

A refinement of Booth encoding is the modified Booth method, which requires adders and subtractors with radix 2. The addition can be by one or two, as the subtraction can be by one or two as well. Thus the modified Booth encoder uses five symbols: $\{-2, -1, 0, 1, 2\}$. The idea is to look behind a pair of bits and to account for that. If the first bit of a pair is "1", according to normal Booth encoding it will generate a "1" in the next bit position and -1 in the current position (weight 2 for the pair). If the first bit of the pair is "0", it does not have any effect. A bit following the pair equal to "1" indicates that the pair will receive a "1". Therefore, pairs of bits are coded according to Table 8.1. The encoder divides the bits of the multiplier, which are presumed to be even, into groups of two, and uses, at the input, a pair and the first bit of the next one. For the LSBs group it is necessary to add a following zero.

Consider, for example, the modified Booth coding of the 12-bit number 011011110010. The steps followed are

$$
\begin{array}{c|c|c|c|c|c}
01 & 10 & 11 & 11 & 00 & 10 \\
011 & 101 & 111 & 111 & 001 & 100 \\
+2 & -1 & 0 & 0 & +1 & -1,
\end{array}
\tag{8.9}
$$

where the first line is the input number, the second line gives the overlapped three-bit groups, and the third line is the encoding according to the rule of Table 8.1. Notice that multiplying a binary multiplicand by two is not a problem, as it is just a shift of bits and the addition of a zero at the LSB position. The reduced number of multiplying coefficients reduces circuit complexity and computation time, features verified in more advanced courses.

The architectures studied in this subsection can be implemented in a pipeline fashion. Every line of the schemes performs the function in successive clock periods, with intermediate data temporarily stored in memory cells, called *latches*. The generic configuration of a pipelined multiplier is like the one in Figure 8.21(a). The algorithm that defines the connections inside gray boxes, and the number of input and output wires, depends on any one of the already studied parallel architectures realized as a pipeline scheme.

Another technique that builds multiplication along multiple clock periods is the *shift and add* method. This approach adds partial products in successive steps by reusing the hardware needed by sequential operations. The algorithm can be the simple one that mimics the paper

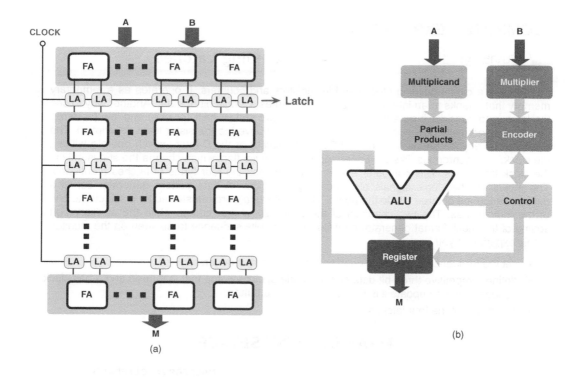

Figure 8.21 (a) Pipeline configuration of multiplier; (b) shift and add multiplier architecture.

and pen process, shown in Figure 8.19, or a much more complex algorithm with encoded versions of the multiplier and additions and subtractions of terms. Figure 8.21(b) shows the conceptual diagram of a possible architecture. The block that performs arithmetic operations is the Arithmetic Logic Unit (ALU). The data generated by the ALU is suitably stored, shifted, and partially sent back to the ALU itself because bits of the result may already be solidly defined and do not need to pass through the arithmetic unit anymore.

8.5.3 Registers and Counters

Multipliers or, more usually, digital processors need to store data or move it around. More-over, signal processing often needs counting. The digital block that enables those functions is the latch. It is an electronic circuit that stores a single one-bit word and makes it available at output. A simple latch configuration is the bi-stable scheme featuring two stable conditions representing "1" and "0". The scheme is volatile (needing power to maintain its memory) and very fast. In the form of a D-latch (its symbol is shown in Figure 8.22(a)) it uses the D input and a signal, E, that enables the circuit to "read" D when E is high, and replicate the input at output Q. The latch also provides the inverse output signal $\bar{Q}$.

There are latches, like the set–reset (S–R) latch of Figure 8.22(b), that do not use a clock. As indicated by the symbols, they have two inputs: S, set, and R, reset. When one of them goes high, the outputs respond accordingly. However, having both S and R equal to "1" must be avoided because both outputs would become zero or go into undefined states. That input

COMPUTER EXPERIMENT 8.3

Understanding Latches

The latch is a digital building block used in registers and counters. It operates as a temporary memory that, thanks to an internal feedback path, stores a single bit of data. Another way to see the latch is to consider it as a bi-stable system with two states: high (or one) and low (or zero). The latch is normally built with basic gates, one level above the transistor level description. The latch can be asynchronous, like the Set-Reset (SR-latch) or the D-latch, which operate without the clock, or synchronous, like the edge-triggered or the flip-flop, which change the status only at the clock transitions. This computer experiment examines the four types of latches mentioned. The inputs are two logic signals plus, possibly, the clock. A data pattern generator sends a defined sequence of three bits to the board. The input signals are transferred by a selector to one of the four latches. The two channels of the scope show the outputs while the auxiliary port setting of the multi-format generator enables the three bits sequence to be seen on the monitor. In this experiment you can:

- set the frequency of the clock used by the multi-format generator;
- define a repetitive three-bit data pattern made of eight words that last three and a half clock periods. For their update it is necessary to push a button;
- select one of the four latches.

MEASUREMENT SET–UP

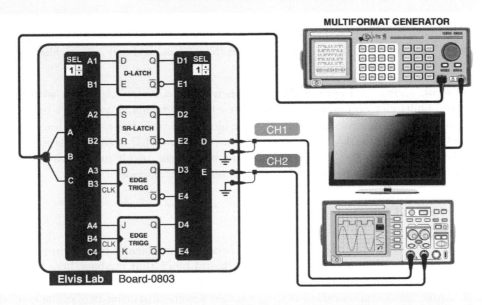

TO DO

- Set the data pattern and look at it on the monitor to verify waveforms and timing. Push the update button and see the outputs.
- Set an input pattern for the SR-latch that gives rise to an illegal state at output. Observe the output.
- Study the operation of other latches with various data patterns.

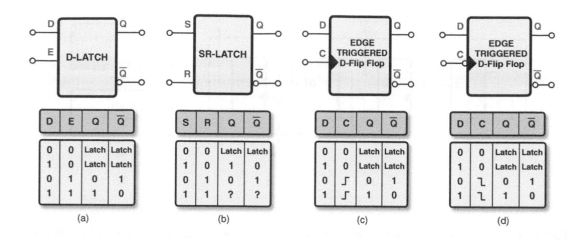

Figure 8.22   (a) Symbol for a D-latch (top) and its truth table (bottom); (b) the same for a S–R latch – notice that, if both S and R are "1", the outputs are undefined (or both zero); (c) positive and (d) negative edge-triggered D-flip-flops with truth tables.

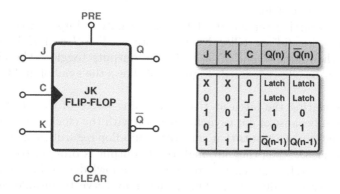

Figure 8.23   Symbol for a JK-latch and its truth table. The block has asynchronous inputs used to set or clear the output regardless of clock status.

is called an invalid or illegal state for the S–R latch. Making $S = 0$ and $R = 1$ brings Q to "0" and $\bar{Q}$ to "1". Otherwise, making $S = 1$ and $R = 0$ "sets" the circuit so that $Q = 1$ and $\bar{Q} = 0$.

Another type of latch is the edge-triggered version or *flip-flop*. The input controls operation only when the clock switches from one state to the other. There are two different types, one positive, which changes status when the clock goes from low to high, and the other negative, which is sensitive to high-to-low transitions. Their symbols and truth tables are shown in Figure 8.22(c) and (d). Notice that the symbols use a small bubble to denote the inverting role of a terminal. The input of the D-edge-triggered circuit determines the output whenever D is high; the outputs define precisely the time at which input passes and changes the output. Edge-triggered circuits serve to synchronize multiple data items that are thus observed at the

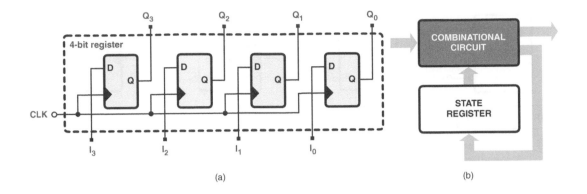

Figure 8.24 (a) Four-bit register used to store data at the rising edge of the clock; (b) possible use of a register in sequential logic.

same time. There are also schemes with set and reset inputs that switch in the same way as D-type edge-triggered circuits do.

The JK-flip-flop is a further version of a latch. Essentially, it improves on the S–R flip-flop because it avoids the "invalid" or "illegal" output state. The J and K inputs operate like the S and R, but when both J and K are "1" the outputs toggle. The circuit can work asynchronously or can be edge-triggered. Figure 8.23 shows the symbol and truth table of the JK-flip-flop.

In addition to activating the latch at rising or falling edges of the clock, which denotes synchronous operation because of data synchronization with the clock signal transitions, some latches use extra asynchronous inputs that control the flip-flop regardless of clock status. These are called preset and clear; when they are high the Q output is brought to "1" or "0".

Some asynchronous latches use extra circuitry to acknowledge stable output status by a signal flowing in the reverse direction. This signal may serve to disable input during the output transitions. This corresponds to the so-called handshaking technique, used in complex asynchronous circuits together with completion logic to identify separate events.

With latches it is possible to realize various functions that need to store digital information temporarily. These kinds of circuits are called *registers*. For example, the register of Figure 8.24(a) stores four bits at the rising edge of the clock. The register can be used in sequential logic, as already discussed, like the example in Figure 8.24(b).

Other relevant examples are the shift register and the rotate register shown in Figure 8.25. Since the cells are positive edge triggered, a new input (also called a *token*) enters at each rising of the clock, and the content of the latches shifts right by one position. The N latches need N clock periods to go through the entire register. Thus the stored information is a picture looking back in time by N clock periods. The register can have parallel or sequential output, depending on needs.

The rotate register is the same as the shift register, but the output goes back to input. At each rising of the clock the token stored in one of the latches moves to the right. The token in the last latch returns to the first one. The "preset" asynchronous controls enable us to load data into the rotate register in parallel. The tokens go on rotating until a new preset occurs.

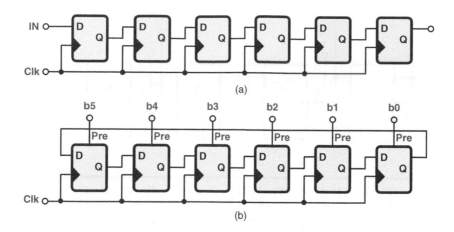

Figure 8.25 (a) Shift register and (b) rotate register made from edge-triggered D-latches.

Table 8.2 Sequences produced by three-bit up- and down-counter

Clock	Up	Down
1	000	111
2	001	110
3	010	101
4	011	100
5	100	011
6	101	010
7	110	001
8	111	000
9	000	111

A binary up-counter is a logic circuit that provides, at each clock period, one more of an increasing sequence of digital numbers from zero to a maximum (when all bits are "1"). Then the bits become all zero and the count starts again. The circuit is called a down-counter when the sequence of bits decreases instead of increasing. For example, a three-bit up- or down-counter gives rise to the sequences shown in Table 8.2.

There are various methods for realizing a binary counter; an easy way is to divide the frequency of the square wave used as the clock. The logic block suited to that function is the JK-flip flop. Figure 8.26 shows the architecture of a four-bit counter. Four clock-edge flip-flops (FFs), one cascading to the next, make up the circuit. The first FF is positive clock edged. The other three are sensitive to the falling edge. The positive output of each FF is the clock of the next one. All of the J and K inputs are "1". The clock, as shown in the figure, is not necessarily a square wave, because the first FF senses the rise transition, no matter when the fall transition occurs.

There are other types of counter. Just to give you an example of the many possible demands on digital systems, we consider here the modulo-N counter, which produces N bits at output, whose values increase sequentially until the full scale is reached. Then the outputs reset to zero. For modulo-3 and modulo-5 counters we have three bits and five bits that change as shown in Table 8.3.

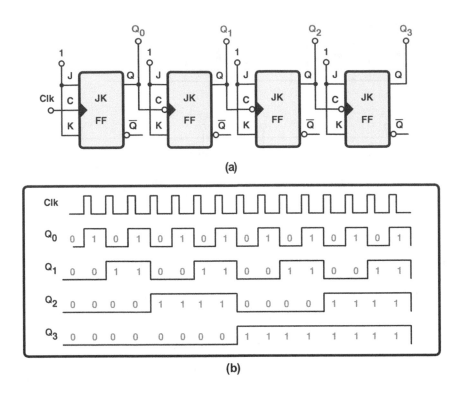

(a)

(b)

Figure 8.26 (a) Four-bit counter realized with edge-triggered JK-flip-flops; (b) modulo-N counter made from an M-stage register.

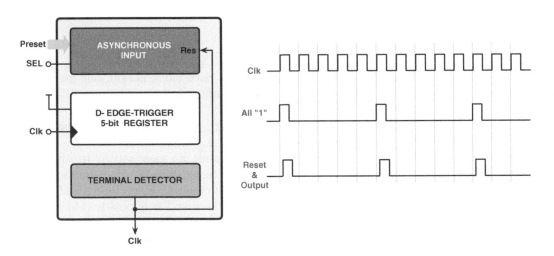

Figure 8.27 Architectural configuration of modulo-5 counter and its relevant waveforms. The SEL pin decides between a predefined starting value or zero.

This counter uses combinatorial logic that detects the status with all "1"s, denoting the terminal condition. The signal activates asynchronous inputs that set the status of the register

Table 8.3 Sequences produced by modulo-3 and modulo-5 counters

Clock	Modulo-3	Modulo-5
1	000	00111
2	001	01111
3	011	$11111 \rightarrow 00000$
4	$111 \rightarrow 000$	00001
5	001	00011
6	011	00111
7	$111 \rightarrow 000$	01111
8	001	$11111 \rightarrow 00000$
9	011	00001
10	$111 \rightarrow 000$	00011

to all "0"s or to a predefined word, depending on a selecting command. In this case the terminal condition occurs before. Figure 8.27 illustrates the block diagram of a modulo-5 counter. The scheme uses a D-edge-triggered register whose input is constantly "1". The register, initially empty or equal to the initial value, depending on the selection bit *SEL*, starts to fill up; when the combinatorial logic detects the terminal condition, a logic command activates the asynchronous reset. There is some delay after all the bits in the register are "1". Moreover, there is a delay in resetting the latch contents. However, all is completed before the next rising of the clock.

8.6 CIRCUIT DESIGN STYLES

Every design, in addition to optimizing processing functions, pursues multiple goals. The most important are cost, reliability, speed, and power consumption. These objectives, valid for any design, are more critical for digital systems, because the huge number of elements determines a large degree of design freedom and many trade-offs.

The cost of an integrated circuit depends on the so-called Non-Recurrent Engineering (NRE) cost given by the cost of the design and that of the masks used for fabrication. Moreover, there are recurrent costs, proportional to volume and silicon area plus the cost of processing, packaging, and testing. Reliability measures the circuit's ability to perform its intended function. Customer satisfaction is important for the company's reputation. Speed and power compete with each other because higher speed requires power. Both depend on the technology used.

All the above elements influence the design style of digital circuits, because they can generate demands for special attention to detail, reuse of methodologies or knowledge, and fast design cycles. We normally distinguish between the following design styles.

- **Full custom design** This is a method that leads to minimum silicon area. A significant part of such a scheme is specifically designed without reusing already designed cells. The circuit is normally designed all the way from specification to layout. The method requires much more design effort than other approaches; thus long design time is the key limitation. This category includes the so-called ASIC (Application-Specific Integrated Circuit).

■ **Standard cell design** The availability of pre-designed basic functions organized in cell libraries speeds up the design process. The cells are prefabricated, tested, and verified experimentally. The experimental results provide design parameters useful for estimating performance and architectural limits. The cell layout is rectangular with a standard height. This allows cells to be put from one side to the other on lines separated by so-called routing channels.

■ **Gate array** The integrated circuits of this category are fabricated as an array of basic cells performing standard functions. Their fabrication is the same until the final steps. The unfinished hardware is the basis for implementing digital functions. The "customization" is done by specific interconnections made with metal layers. The result is that the cost of the basic array is shared by many circuits while the specific design costs cover just the last fabrication steps and the metal masks. The use of silicon area is not optimal because any design uses only a fraction of the available gates.

Since the customization is done with metal connections, this type of circuit is also called a Metal-Programmable Gate Array (MPGA).

■ **Field programmable devices** This design style does not need extra fabrication steps because it just requires the "on field" interconnections of already defined cells in an integrated circuit to be defined. The interconnection scheme is stored in a memory used to operate electrically programmable switches. There are many field programmable devices, distinguished by various acronyms. We have PAL (Programmable Array Logic), PLD (Programmable Logic Devices), CPLD (Complex Programmable Logic Devices), and FPGA (Field Programmable Gate Arrays). They differ in their basic cell complexity, how the device programming is accomplished, capacity, type of memory (Flash, SRAM, EEPROM, etc.), and establishing connectivity.

Even if each of the above design styles matches some specific design needs, the trend is in two directions. The first virtually merges custom design and standard cells, because of reusing already designed cells or micro-functions, the so-called IP (Intellectual Property). The second is the use of field programmable solutions, because the cost of memories is negligible and the cost of metal masks very high.

Custom design style (including the use of IP) is for high performances, optimal power, and low cost, at very high volumes. The advantages of field programmable solutions are low development costs, off-the-shelf availability, short time-to-market, part reusability, and reconfigurability. The cost, for medium–high production volume, is competitive.

8.6.1   Complex Programmable Logic Devices (CPLDs) and FPGAs

It is worth knowing a bit more about CPLDs and FPGAs because they are increasingly used by system designers for implementing digital processing functions. The logic is programmed with dedicated units, even if in some cases the chips can be directly configured by the system that uses them. Substantially, the design uses a configuration of switches that interconnect the array of digital functions in a custom manner. The configuration of that specific design is then stored in memory. There are various technologies enabling us to store programs, as will be studied shortly. Nevertheless, it is good to know at this stage that using fuses or anti-fuses defines the program in a permanent manner. An alternative solution is to use the so-called EPROMs, which can be reprogrammed out-of-circuit. EEPROM and Static RAM permit on-circuit reprogrammability.

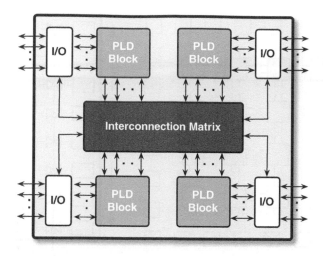

Figure 8.28   Possible architectural arrangement of four sections of a CPLD.

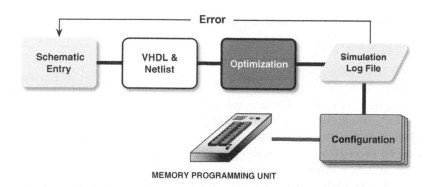

Figure 8.29   CAD logic flow for PLD design. The final result gives rise to the configuration to be written in the device with the memory program unit.

The CPLD, rather than enabling programmable interconnection of simple logic functions, as other programmable devices do, uses macro-blocks that implement complex combinational or sequential functions. There are different types of CPLD on the market. Their architectures depend on the vendor. To give you an idea of possible organization, Figure 8.28 illustrates a style based on sets of four programmable logic devices. There is a single interconnection controller for the four logic devices. Each PLD has its own programmable input/output interfaces that communicate with the external world. Replicating the Figure 8.28 module may form a large array of PLDs. In practice this means incorporating memories on board and making available special function cells placed on the periphery.

The PLD block performs combinational or sequential functions by the use of hardware-defined networks, or, frequently, a Look-Up Table (LUT) realizes these functions. A LUT, as mentioned above, is a small memory with the address (or the input code) at input and

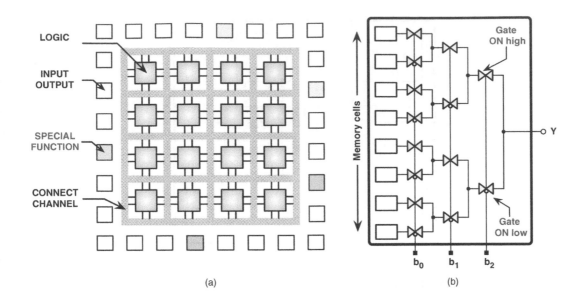

Figure 8.30   (a) Possible FPGA architecture with horizontal and vertical interconnecting channels; (b) look-up table made of transmission gates.

combinational results at output. The LUT is able to realize any logic function flexibly, because it is enough to program its truth table into the memory.

Most types of CPLDs can be programmed multiple times. This feature permits the designer to develop a processing function at preliminary level, and later, after testing the hardware, to make corrections by reprogramming the device. By reprogramming the designer may correct malfunctioning or add new functions not contemplated in the original design.

The design of circuits realized with CPLDs needs special Computer-Aided Design (CAD) programs. The first step in their use creates a schematic diagram using a graphical tool. Then it is necessary to describe the design in a suitable hardware description language, such as VHDL (VHSIC Hardware Description Language). Since the initial logic entry is not optimal, the tool uses algorithms to make the description more effective, using techniques that optimize the architecture while fitting the logic equations thus obtained into the CPLD. The resulting architectures must be checked by simulation to verify correct operation. Possibly, the simulator returns detailed information to fix errors. Figure 8.29 shows the logic flow of the design. After defining the configuration, the system controls a hardware unit that serves to write the memory.

Like CPLDs, the FPGA is an assembly of uncommitted logic functions and interconnected resources. They integrate different building blocks, including combinational and sequential logic, dedicated arithmetic units, input/output (I/O) interfaces, and other service blocks such as clock generators and RAM (Random-Access Memory). The interconnection channels can be parallel rows or go in both horizontal and vertical directions as shown in Figure 8.30(a). The configurable logic blocks (only an array of 4×4 of them is depicted) contain logic and arithmetic circuitry such as LUTs, adders, multipliers, registers, and others. Around the array there are input/output cells and other cells providing special functions, such as data converters, embedded microprocessors or micro-controllers, and parallel or high-speed serial transceivers.

It is possible to have various circuit schemes. Figure 8.30(b) shows an example of a LUT with three-bit input, realized with transmission gates. The bit stored on the fourth memory cell from the bottom, for example, flows to output when the three bits are 011.

Depending on the processing function required, the designer must choose the right FPGA. The decision depends on various factors such as, among others, the number of logic cells, which can range from a few tens of thousands to several hundreds of thousands, the RAM, the number of I/O cells, which can go from 64 to a few hundred, and power consumption.

8.7 MEMORY CIRCUITS

With digital memories it is possible to store and retrieve binary data. The process of storing data in a memory is called *writing*; the action of retrieving data is *reading*. There are devices that permit both reading and writing; others that do not allow for writing are "pre-written" by the manufacturer. These are typically referred to as Read-Only Memory, or ROM. Other types of memories can be rewritten, or furnished blank and written fresh by the user. This type is called *Read-Write* memory. Another feature of memories is their volatility, or data permanence without power. Many memory circuits operate as a latch that stays in the "high" or "low" status as long as the supply power is on. In this case we have the volatile feature. Other media or devices are non-volatile, because they do not need the power supply to preserve stored data.

Some of the parameters used to classify memories are as follows.

- **Capacity** is the number of individual memory cells. It is normally measured using the byte, i.e., a group of eight bits. For large capacities the multiples kbyte (as is well known, kilo = 10^3), Mbyte (mega = 10^6), Gbyte (giga = 10^9), or Tbyte (tera = 10^{12}) are often used. In addition to capacity the compactness of stored information is important. Modern memories are able to store huge amounts of data in very small volumes.

- **Data transfer speed** is the rate at which data can be transmitted in and out of the memory. This is often expressed in kbit per second or Mbit per second, abbreviated as kbps and Mbps respectively. In some cases the data transfer speed is expressed in kbytes or Mbytes per second, or kB/s and MB/s. Notice that with the symbol "b" we indicate a bit, while the capital letter "B" denotes the byte.

- **Data storage mechanism** As already mentioned there are volatile and non-volatile storage methods. Volatile memories often store data as charge on a capacitor. The information can be destroyed when reading or deteriorates because of leakage, thus requiring a periodic refreshing. The non-volatile memories, also known as ROM, can be: mask programmed (which are written during fabrication); programmable (written by burning out internal connections called fuses); or erasable, where data is stored as charge on an isolated gate (the so-called *floating gate*). The cell can be erased using ultraviolet light. With the electrically erasable, or flash, type, it is possible to reprogram the content by using a high voltage.

- **Type of access** There are two forms of access. Sequential access features a group of elements, such as data in a memory array or a disk file, that are accessed in a predefined, sequential order. The other kind is random access, which denotes that information from any storage location is equally accessible in random order at a fixed rate, independent

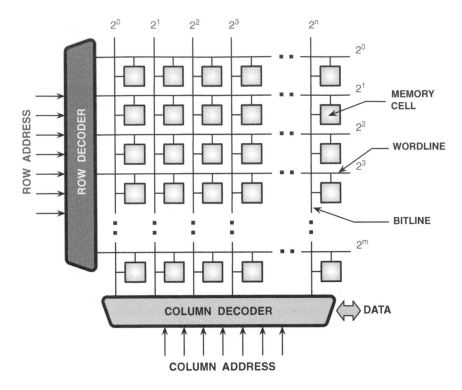

Figure 8.31   Typical organization of memory with random access (RAM).

of physical location, for reading or writing. This kind of memory is known by the acronym RAM.

8.7.1   Random-access Memory Organization and Speed

The organization of memory shown in Figure 8.31 is called random access. The storage array, or core, is made up of a block of single memory cells arranged on a rectangular matrix. They share connections in horizontal rows and vertical columns. The horizontal lines, called word-lines, are driven by a set of signals that identify the horizontal address, while through vertical lines, called bit-lines, flows the data of the line cells. The row address of the block, deciphered by decoding the row binary address, simultaneously selects the n columns. The column decoder selects one of the memory cells connected to it via the n bit-lines. In total, the core array stores $n \times m$ cells. For example, the architecture of a 64-kbit random-access memory with $n = m = 256$ has a total of $65\,536$ cells. It needs a 16-bit address to identify the position and access to a single bit output. The number of rows and columns can differ.

The circuit used to implement a memory cell can be any bi-stable scheme; however, since complexity and silicon area consumed critically depend on the memory cell, it is crucial to use very simple circuits to give rise to compact and area-effective solutions. Proper peripheral circuits surround the memory core. In this category there are decoders, sense amplifiers, column pre-charge, data buffers, and others. They perform any necessary amplification, buffering,

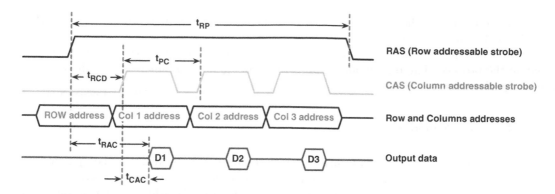

Figure 8.32   Three memory accesses in the same row. This access uses page mode.

and translation of voltage levels. The peripheral circuits use more sophisticated schemes, because many memory cells share them. In the case of Read/Write (R/W) memories there are peripheral circuits that determine whether data is being retrieved or stored.

For volatile memories the read operation can be destructive. In these cases it is required to have the peripheral circuit able to restore the data after reading, and, because of the volatility, to refresh the information periodically.

Memory speed is obviously crucial for the operation of digital circuits. Speed depends on the single memory cell and peripherals. Memories are made up of multiple banks, each of which is like the scheme of Figure 8.31. To access a specific memory cell, the system must first activate the bank containing the desired row (Row Address Strobe, RAS). Then the column selection is activated. This is done by the CAS (Column Access Strobe) followed by a command that identifies the desired row. Once the row is selected, it is possible to issue read or write commands to specific columns in the row. When reading or writing is completed, a further command closes the row. It is necessary to allocate suitable time for each step as determined by various operational time intervals. We have, for example:

- t_{RP}, the time used to switch on the internal memory banks (or RAS precharge);

- t_{RCD}, the delay occurring between the time of row activation and the read or write operation;

- t_{CAC}, the column access time;

- t_{PC}, the page mode cycle time. This is the access mode that enables fast access to the bit of a selected row;

- t_{RAC}, the random access time, given by the addition of t_{RCD} and t_{CAC}.

In order to speed up access to memory, memory cells of the same line share the row address. The first access time is the same as for normal access but the following ones are available more quickly, because it is not necessary to switch off row and memory bank access. The group of line bits is called a *page*, and the type of access is called *page mode*. Figure 8.32 shows an example of timing for three memory accesses on the same line in page mode. When receiving the row address the RAS signal starts the process. After the delay t_{RCD}, the row is activated. Then the column addresses give rise to a sequence of three output data items that occur with a t_{PC} time separation. The column address strobe delimits each column acquisition. The RAS terminates the line acquisition.

The above is just a specific example of memory access, used to introduce methods for speeding the process. The page mode, applied in a more general manner, is a method used to improve performance.

8.7.2   Types of Memories

Memory, other than being an essential part of computers, is used in any equipment that performs signal processing. Indeed, digital computation has become popular because digital processors are properly supported by semiconductor memories. In addition, new applications such as digital cameras, PDAs, cell phones, and many other devices use memories extensively.

The different roles of memory demand a careful choice between the various available types, such as PROM, EPROM, EEPROM, Flash, DRAM, SRAM, SDRAM, and MRAM. Expanding what we have already stated, the following gives more details on the types and their features.

- **PROM (Programmable Read-Only Memory)** This is programmable, but data can be written only once. Therefore, the data is written in permanent form. This type of memory is sold in a blank format to be programmed with a special PROM programmer. Often the memory consists of an array of fuses "blown" to define the data pattern. This type of memory is often used to store fixed instructions and set-up information that must not change or be changed by the user.

- **EPROM (Erasable Programmable Read-Only Memory)** This is a memory that enables programmability but also erasability, by exposure to an ultraviolet (UV) light for many minutes through a quartz window on the memory package. Each memory cell is made of a special type of transistor that has an isolated electrode, called a floating gate, such that it can store information as a trapped charge. Ultraviolet exposure favors release of the trapped charge. The operation totally erases the memory, which must be fully rewritten even to change a single bit. For programming, high voltages applied to the input of the cell cause large pulse currents and strong attraction of electrons toward the floating gate. Because of the very low oxide thickness the electrons penetrate and reach the floating gate. In use, the quartz window must be covered to preserve the data for extended periods.

- **EEPROM (Electrically Erasable Programmable Read-Only Memory)** This is a memory whose writing and erasing is done with electrical voltage. The memory cell also uses a floating gate, but its erase operation is not with UV but with an electrical pulse. Unlike EPROM, the EEPROM may be selectively erased. This type of memory is non-volatile, as it retains stored data when power turns off. The read and write cycles are relatively slow. Thus this type of memory is suitable for operations that do not affect the overall speed, typically when data must be downloaded at start-up. Another important feature is that the write and erase operations are on a byte-per-byte basis. The number of rewrite cycles is very large (10 million or more); this feature is called *endurance*. The *retention time* of data is limited because the floating gate insulation is not perfect, and charge can be lost, especially at high temperatures. The typical data retention time is 10 years.

- **Flash memory** Flash memory is an evolution of EEPROM technology. The key feature of non-volatile flash memory is that blocks of memory may be erased simultaneously. Data is read on an individual cell basis. There are two different technologies for flash memory, distinguished by the logic functions performed: NOR and NAND. The NAND type is preferred for memory cards because of the lower cost and higher storage capacity. The memory can store a single bit or multiple bits per cell. Having many bits per cell augments the capacity but slows the transfer speed, increases power consumption, and reduces cell endurance. Often the assembly of flash memories uses chip stacking technology.

- **DRAM (Dynamic Random-Access Memory)** This is a volatile form of random-access memory. It uses a capacitor to store each bit of data, with the charge on it used to determine the logical value "1" or "0". The cell is very simple, being made of a capacitor and a switch. However, the capacitor does not hold its charge indefinitely, because of leakage. To overcome this problem the data is periodically refreshed. The sensing of data allows reinstatement of the value on the same cell. Typically refreshing occurs every 64 ms. This activity is accomplished over single rows or portions of the memory by relatively simple refresh circuitry, and causes minimal interference with system operation. DRAM of various types is used for most system memories because it is cheap and small. The access time is relatively long, typically not less than 60 ns.

- **SRAM (Static Random-Access Memory)** Static memory cells do not need their contents refreshing. The access time is fast (10 ns, as against 60 ns for DRAM). Also, the cycle time is short because pauses between accesses are not necessary. The power consumption can be competitive with that of DRAM or even give rise to better performance. However, SRAM is less dense and more expensive than DRAM. In computers, SRAMs are normally used for caches.

- **SDRAM (Synchronous DRAM)** This is a special DRAM architecture that provides fast speeds. The key to its architecture is the synchronization of operation with the clock of the signal processor that uses it. The control of the memory keeps two sets of memory addresses open at the same time and transfers data alternately from the two sets. Overlapped operation avoids the delay caused by needing to close one address bank before opening the next one.

- **MRAM (Magneto-resistive RAM, or Magnetic RAM)** This is a type of memory that uses magnetic features instead of charges to store data. It uses materials with magneto-resistive properties. MRAM can provide high-density non-volatile memories that are extremely fast (access time less than 10 ns). Its commercial use depends on the maturity level of the technology employed.

The use of memory in fast, powerful CPUs demands quick access to enormous quantities of data. The CPU must get the data it requires, avoiding halts and waits. Modern CPUs, running at speeds measured in GHz, need massive quantities of data. This need is satisfied by the so-called cache memory. It is a very fast memory built into the CPU (level 1), or located next to it on a separate chip (level 2). In the cache the CPU stores instructions that are repeatedly required to run programs, avoiding the need to rely on the system bus for data transfer.

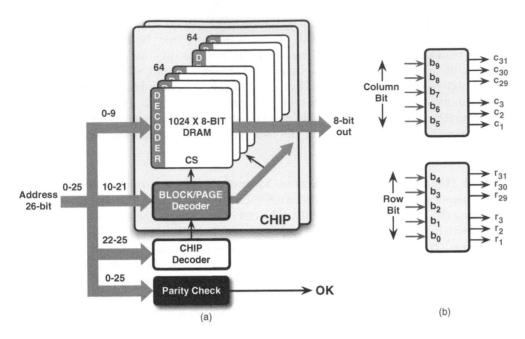

Figure 8.33 (a) Decoding of hypothetical memory made from four chips each with 64 blocks and 64 pages of 1-kB DRAM; (b) row and column decoder with five bits at input.

Other than using fast memories there are many applications with diverse needs, requiring, for example, low- or medium-speed memories when storing massive amounts of data. The information given in this subsection is an initial basis for understanding the proper choices for signal processor architectures.

8.7.3 Circuits for Memories

Reading and writing of memory, and refreshing data, require dedicated circuits surrounding the memory core. What they are depends on how the memory cells are organized. There are various high-level architectural choices, not studied here. However, the memory is usually divided in a hierarchical manner. The functions that are always needed are the address decoder, the sense amplifier, and on-chip voltage generators. The following paragraphs briefly discuss these functions.

8.7.3.1 Address Decoder

The address length needed for reading or writing memory cells depends on total memory capacity, which can possibly be realized with multiple memory chips. Each chip may be hierarchically divided into pages and blocks. Address decoding is the process that selects the right chip and generates signals to select rows and columns on that chip.

Suppose, for example, you have a 26-bit address to select one of the bytes of a (1024×8)-bit (or 1024-byte) DRAM block. The DRAM is one of the 64 pages of the 64 blocks of a chip. The first 10 bits select the memory cells, the next six identify the page, then another six are for the block and a further three select one of the four chips with which the entire

Table 8.4 Look-up table of a five-bit row or column decoder

b_9 b_4	b_8 b_3	b_7 b_2	b_6 b_1	b_5 b_0	c_0 r_0	c_1 r_1	c_2 r_2	c_3 r_3	c_4 r_4	c_5 r_5	c_6 r_6	c_7 r_7	.	.	c_{29} r_{29}	c_{30} r_{30}	c_{31} r_{31}
0	0	0	0	0	1	0	0	0	0	0	0	0	.	.	0	0	0
0	0	0	0	1	0	1	0	0	0	0	0	0	.	.	0	0	0
0	0	0	1	0	0	0	1	0	0	0	0	0	.	.	0	0	0
0	0	0	1	1	0	0	0	1	0	0	0	0	.	.	0	0	0
0	0	1	0	0	0	0	0	0	1	0	0	0	.	.	0	0	0
0	0	1	0	1	0	0	0	0	0	1	0	0	.	.	0	0	0
0	0	1	1	0	0	0	0	0	0	0	1	0	.	.	0	0	0
.	.	.	.	.	.	.	.	.	.	.	.	.	.	.	.	.	.
1	1	1	1	0	0	0	0	0	0	0	0	0	.	.	0	1	0
1	1	1	1	1	0	0	0	0	0	0	0	0	.	.	0	0	1

memory is assembled. The last bit is used to check possible errors in the overall address. The decoder, as shown in Figure 8.33(a), splits the input address into two four parts: bits 1–9 are the input from the row and column decoder; the others identify block, page, and chip. The eight outputs of the DRAM make the addressed byte, which is validated by the address check circuit, normally based on so-called *parity check verification* (i.e., the total number of bits set to "1" is even).

Each line and column decoder has five bits at input and generates 32 signals at output, only one of which is "1", to identify the eight memory cells (Figure 8.33(b)). Table 8.4 provides the look-up table of the row or column decoder. It is supposed to be the same for row and column decoding, because it is assumed that the memory uses a square matrix of cells. Thus the table indicates equal results for the same row and column address and identifies a row or column with equal ordering number. The addresses increase using a binary number and the result is just one "1" out of 32, as indicated by the gray line joining the "1"s.

The above, which refers to a hypothetical example, indicates a possible use of the address decoder. Single unique addressing is when each address corresponds to a single addressable memory location. There is a second possibility: some address lines are not used, or a component or page responds to multiple addresses. That option can be used to simplify the address decoder in complex situations.

The circuit that implements the address decoder can be combinational logic made by suitable interconnection of boolean cells (AND, NAND, OR, NOR), a matrix of switches, or a programmable logic device with volatile or non-volatile definitions of addresses.

8.7.3.2 Sense Amplifier

A sense amplifier, as the name implies, is used to sense or detect stored data from a selected memory cell. It is a critical peripheral block, because the signals that travel over long bit-lines can be very small. The bit-line behaves like a resistive delay line, causing attenuation and delay. Therefore, other than accounting for the delay in column access time, it is necessary to amplify the small signal received to attain a full logic level.

COMPUTER EXPERIMENT 8.4

Understanding Look-up Table

The logic function of a Look-Up Table (LUT) is a relationship between multiple digital inputs and multiple digital outputs. This Computer Experiment uses a random decoder that receives 4-bit signals at input and generates 8-bit at output. Pushing a button on the ElvisLab board you change the unknown look-up mapping scheme. A simple data generator alternates two 4-bit words at a variable rate. Data change at the rising edge of a clock (shown on Channel 3). Instead of combining inputs with logic circuits, this experimental board uses an input decoder that activates a single column of the LUT. The signal stored in the cells of that column goes to the outputs. Channel 1 shows the signal of one line of the input decoder. Channel 2 displays a serialized representation of the output. In this virtual experiment you can:

- generate the 4-bit words and set the clock, used to control serialization and triggers of the oscilloscope. The oscilloscope time scale is four times the clock period;
- select one of the input decoder columns in order to observe its value on the scope;
- change the clock frequency to measure delays and observe jitter.

MEASUREMENT SETUP

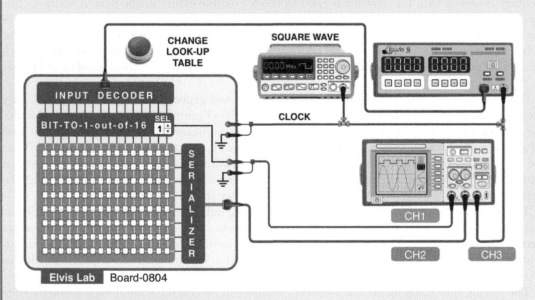

TO DO

- Set the 4-bit pair equal and verify that the output is a 1-out-of-16 code: i.e. each 4-bit is expected to have a "1" logic value and 15 logic "0".
- Find the sequence of 4-bit words that select the 8 outputs sequentially.
- Increase the clock frequency for measuring propagation delay and observing the jitter.
- The LUT can be used as an electronic key. Sketch a conceptual diagram of the state machine that carries out the key function.

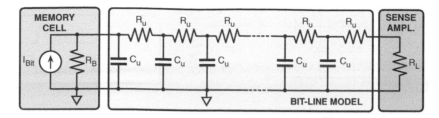

Figure 8.34   Model of sensing operation with distributed resistive delay line for modeling the bit-line.

Increase in memory capacity, technology scaling, and reduction in supply voltage require the designer to augment the number of cells per bit-line. Scaling of metal width reduces the distributed capacitance but increases the distributed resistance of bit-lines. As a result, the delay remains almost unchanged but the attenuation increases. Scaling of supply voltage also gives rise to further, smaller signals. Moreover, noise does not diminish. Thus the sense amplifier must be able to deal with the resulting challenging design requirements.

There are two types, the voltage sense amplifier and the current sense amplifier. The former focuses on measured voltage, and the latter assumes that the signal from the bit-line is mainly caused by current. The choice depends, obviously, on the type of memory cell. In order to reduce noise sensitivity memory cells often provide differential signals so that the voltage amplifier senses differential voltages. The challenge is to find the best trade-off between gain and speed. A frequent solution is to use a fast amplifier with medium gain followed by a latch, a positive feedback circuit to be discussed shortly. Indeed, it is not necessary to amplify a signal linearly but to rebuild a logic signal from its attenuated differential version.

Some voltage sense amplifiers are just a positive feedback loop. (Again you are hearing about feedback. Be patient! You will learn about it in a few chapters). A positive feedback loop detects the imbalance of the differential bit-line voltages and enhances the difference until full swing is reached. In this case the sense amplifier also restores the information on the memory cell, compensating for the disturbance caused by the load of the bit-line during the read operation.

Using current-sensing amplifiers does not suffer the limitation of low voltage signals. What matters is to have current in the bit-line to indicate a measured high/low status or high/low voltage. This method leads to a possible reduction in the swing of the bit-line voltage and reduced sensing delay. Current sensing is studied using equivalent circuits to the one in Figure 8.34. A Norton equivalent circuit models the memory cell. The bit-line is like a distributed resistive delay line and a load resistance is a feature of the sense amplifier.

For voltage sensing, the load resistance established by the sense amplifier can be assumed to be infinite; for current sensing it must be zero. The sensing after a preset takes place with a delay given by the time of bit-line crossing plus the time required for the minimum detectable voltage or, with current sensing the delay is what causes minimal current difference. We won't study the problem in detail; however, be informed that when a signal reaches the sense amplifier this quickly determines current imbalance but needs more time to give rise to the crossing of voltages. Therefore the delay in voltage sensing is longer.

An important design parameter is the power consumed by sensing. A dominant term is the dynamic power associated with charging and discharging the parasitic capacitance of the bit-lines. Assuming that the voltage swing of the bit lines is V_{BL}, the power is

$$P_{read} = C_{BL} V_{BL}^2 f_{read}, \qquad (8.10)$$

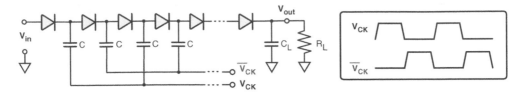

Figure 8.35   Circuit diagram of a Dickson charge pump.

where C_{BL} is the total parasitic capacitance of the bit-line and f_{read} is the read frequency. Since the bit-line voltage swing of the current sense input (close to zero) is much lower than the voltage sense case, the power consumed is much less.

Circuit details on how to realize sense amplifiers are found in specialized literature. However, the following study of analog cells will provide enough information on general design techniques.

8.7.3.3   Voltage Booster

Electrical programmability of memories requires high voltages to inject or remove charges in floating electrodes. Since the operational voltage is lower than that required for memory programming, it would be necessary to use two supply voltages. This is a limitation that on-chip high-voltage generators avoid. Fortunately, programming memories requires high voltage but limited current. Thus the so-called voltage multiplier, or charge pump, based on a scheme proposed by Dickson, is the solution that is mostly used. The circuit uses capacitors and diodes with complementary clock signals. How to boost the supply voltage to obtain high voltages is studied in a later chapter. However, it is worth anticipating the circuit diagram and briefly describing the charge pump.

Figure 8.35 shows the circuit diagram of a Dickson charge pump. The scheme increases the voltage across capacitors by pumping charge onto them. The output voltage can be double, triple, or an even higher multiple of the supply. As is well known, diodes permit the current to flow in only one direction. Thus the phase biasing the bottom plate of the capacitors charges them to the voltage of the previous stage when the phase is low. When the phase goes high, the voltage of the top plate of the capacitor rises. The pushed-up voltage turns on the diode of the next cell to charge its capacitor to a higher voltage. The phases used are generated by a clock and its inverse. Notice that, as the figure shows, the phases are non-overlapped; i.e., one phase returns to zero before the other goes high. This avoids a useless state with all of the scheme pushed up.

Notice that charging and discharging of capacitors causes power consumption. It is dissipated in the equivalent on-resistance of the diodes used to the current in a single direction. Thus the generation of high voltages involves, as it is normal to expect, power consumption and efficiency lower than one.

The voltage at output depends on the values of capacitors, the switching frequency, and the resistive load. The use of various stages gives rise to a boost in the supply voltage to a level that is enough for the required programming operation. Obviously the voltage generated must not exceed the safety limit for the technology used. This is particularly relevant for modern technologies for which oxide thickness, transistor dimensions, and voltages continually scale.

In most semiconductor technologies used for logic design, diodes are not available; the charge pump chain is implemented using a replacement for diodes: a MOS transistor with a configuration that operates in a non-linear manner that mimics the diode. Details of the diodes and MOS transistors mentioned above are given in the next chapter.

PROBLEMS

8.1   Estimate the frequency response of the equivalent circuit of Figure 8.7(b). Consider four cells with equal resistors and capacitors. To facilitate the study, determine the input starting from the output. Suppose that the load of the output is a resistance equal to $10R$. What changes in the result if you add an extra cell? Is it possible to find a recursive equation?

8.2   A soft drink vending machine accepts 5-, 10-, and 50-cent coins. It has a coin-input unit, a change-dispensing unit and a drink-dispensing unit. The coin-input unit generates three types of signals: "coin deposited" of 5, 10 or 50 cents. The change-dispenser unit must receive two signals: "coin dispense" by 5 or 10 cents. The drink-dispenser must receive only one signal: "dispense a can." Both change- and drink-dispenser must generate an "executed" signal. Sketch the block diagram of the combinational logic and specify the sequence of operations. (You know the cost of a soft drink, I assume.)

8.3   Expand the carry-saver adder scheme of Figure 8.17 to give rise to the addition of three numbers of eight bits. You can use half-adder or full-adder and enter the bits as regular input or carry. Minimize the hardware.

8.4   Compare the architecture of a matrix multiplier 4×6 and a matrix multiplier 6×4. Which is, in your opinion, the more effective solution?

8.5   Use the building blocks of Figure 8.22 to realize a delay line by six clock periods. You can use the clock and its inverse to control the circuit or to activate positive and negative edge-triggered flip-flops.

8.6   We analyzed in Chapter 1 a digital clock based on quartz, and briefly discussed the high-level architecture. Use the building functions studied in this chapter to go inside the high-level functions of Figure 1.4. What is the combinational logic that add the alarm clock function? What kind of logic realizes the time setting?

8.7   A decimal numbering system generates 10 symbols depicted by a seven-segment output device. Design a look-up table that performs the coding of the ten outputs into seven segments. Use transmission gates.

8.8   Draft a flow diagram that describes the functional steps of the read function of a DRAM memory. The organization of the memory is the one shown in Figure 8.33.

CHAPTER 9

BASIC ELECTRONIC DEVICES

Electronic circuits use solid-state devices to give non-linear operation and voltage- or current-controlled functions. We shall study the diode and various types of transistors: the Metal–Oxide–Semiconductor (MOS), either p-type or n-type, the Junction Field-Effect Transistor (JFET), and the bipolar one (either NPN or PNP). We shall learn about the fundamental physics as well as the features of these devices, and we shall outline the models that are used to study the operation of electronic devices in circuits.

9.1 INTRODUCTION

Not long ago electronic circuits were bulky and power-hungry. They were based on vacuum tubes, which use a hot filament to emit electrons in a vacuum. The bias of one or more metal grids interposed between two main electrodes, anode and cathode, controls the flow of electrons and causes current modulation between the emitting filament and the collecting plate. The electrical control of current is the function required of any active component, as it is the basis of amplification. This is what all electronic active devices do, including the modern active solid-state devices: bipolar or CMOS transistors.

After the pioneering vacuum-tube era, which enabled the invention of radio, radar, and television, solid-state devices boosted the electronics discipline to the relevance it has today. The credit for this goes to the transistor, invented in 1947 at the Bell Laboratories in Murray Hill,

Understanding Microelectronics: A Top-Down Approach, First Edition. Franco Maloberti.
© 2012 John Wiley & Sons, Ltd. Published 2012 by John Wiley & Sons, Ltd.

Figure 9.1   A picture of the first transistor. Reproduced by permission of © Alcatel-Lucent.

N.J., by three researchers: John Bardeen, Walter Brattain, and William Shockley. The three scientists were awarded the Nobel Prize in physics in 1957. The device they invented was, in fact, not very miniature. Figure 9.1 shows a picture of it. It was called a *point contact transistor* because it uses pointed metal contacts pressed onto the surface of a semiconductor material. The contacts were made of gold, and the semiconductor was germanium.

After that, many other inventions drove the progress onwards. Another key milestone was the invention of the integrated circuit, conceived by Jack Kilby in 1958 (Nobel Laureate in 2000). It uses a body of semiconductor material to completely integrate all the components of the electronic circuit in a single chip. The device produced by Kilby was still bulky and primitive; however, those initial ideas were the seeds of a tumultuous progress with continuous and important improvements, all aiming at the miniaturization of circuits and systems and the improvement of their performance. The integrated circuits roadmap, shown in Figure 9.2, foresees an increase of speed to many hundred of GHz or even THz and the shrinking of minimum features according to the sequence 65 nm → 45 nm → 32 nm → 28 nm → 16 nm (and below) for the most advanced technologies such as the ones used for memories. The complexity is also astonishing, with numbers of transistors in a single chip that exceed many billions of devices.

What is relevant for the electronics and microelectronics designer is that the evolution of technology determines important changes in the design approach. While in the past active devices together with passive components were the building blocks used by the designer, now electronics engineers work with high-level functional blocks such as op-amps, comparators, glue logic, or, at a higher level, with data converters, filters, ALUs, microprocessors, DSPs, memory, and so forth. This could, erroneously, lead to a reduced attention being given to functional blocks and, below them, to basic devices. However, even if not directly involved in

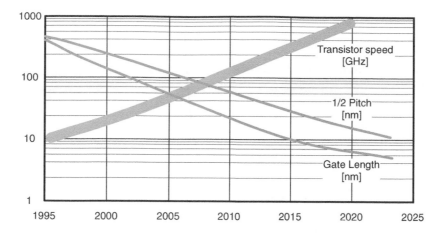

Figure 9.2   Technology roadmap for the physical dimensions of the transistor and its speed.

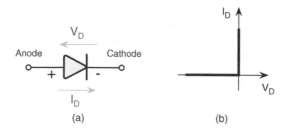

Figure 9.3   (a) Symbol for the diode; (b) *I–V* ideal characteristic.

the integrated circuit design activity, an electronics specialist should go down the complexity ladder and become aware of the features, properties, and limitations of the basic components. Nevertheless, the study must be at the right level, without going into too many fine details of physics and technology.

Obviously, for electronics engineers working directly in the integrated circuit design area, an in-depth knowledge of the physics and the fabrication methods of electronic devices is vital. The information and study provided here is only the first brick in building this kind of device knowledge, but perhaps it will be enough for system designers.

9.2 THE DIODE

The diode is the simplest two-terminal solid-state electronic device. The reader probably knows that a diode is the junction of two materials: a *p*-doped and an *n*-doped semiconductor (often silicon). "Doping" and "junction" are new concepts; however, before clarifying what they are, let us look at the symbol and the key feature of a diode: its non-linear voltage–current characteristic. Figure 9.3(a) shows the symbol. It is an arrow with a bar on the tip. The arrow points in the main direction of the current flow. The bar indicates that the diode resists a flow in the opposite direction. The two terminals are called the anode (or positive terminal) and the cathode (or negative terminal). Accordingly, the current prefers to flow from anode

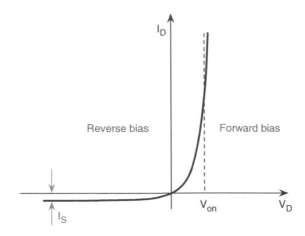

Figure 9.4   Shockley diode current–voltage characteristic.

to cathode. Ideally, the diode allows current to flow only in one direction. The corresponding ideal voltage-to-current characteristic is the one shown in Figure 9.3(b). A real diode only approximates the ideal response. It needs some voltage in the direct region to enable the flow of current and leaks some current in the reverse bias region. This non-ideality is modeled in various ways, as will be discussed shortly.

What is a diode?

A diode is a non-linear element with two terminals, which permits current flow with a given direction of bias (forward bias) and opposes the flow of current with voltages in the opposite direction (reverse bias).

An accurate theoretical understanding of the behavior of a p–n junction involves complex equations of solid-state physics in two or three dimensions. The analysis leads to elaborate final results not suitable for convenient study. Since what we can intuitively understand are models with relatively low complexity, simplified descriptions lead to more tractable results. One of these is the Shockley diode equation

$$I_D = A_D I_{SS}(e^{V_D/nV_T} - 1), \tag{9.1}$$

where V_D is the voltage at the diode terminals, A_D is the area of the diode junction, and n is the so-called emission coefficient. The value of n varies from 1 to 2 depending on the fabrication process and the material properties. In many cases n is set equal to 1. I_{SS}, called saturation current per unity area, depends on the p-doped and n-doped materials. V_T, the thermal voltage, is calculated by

$$V_T = \frac{kT}{q}, \tag{9.2}$$

where k is the Boltzmann constant $(1.38 \cdot 10^{-19} \text{ J/K})$, T is the absolute temperature in degrees Kelvin, and q is the magnitude of the electronic charge $(1.6 \cdot 10^{-19} \text{ Coul})$. At room temperature (300 K) V_T is approximately equal to 26 mV.

Figure 9.4 graphically represents the Shockley diode equation. It distinguishes two regions of operation: the forward bias for which the voltage across the diode is positive and causes

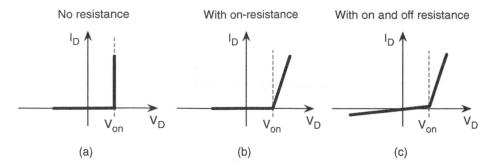

Figure 9.5 (a) I–V characteristic of a real diode with a given on-voltage; (b) I–V characteristic that accounts for an on-resistance; (c) I–V characteristic of a diode with a non-infinite reverse bias resistance.

a significant flow of current, and the reverse bias with negative voltage across the diode and negligible current. Actually, in reverse conditions the current is not zero, as indicated by Figure 9.3(b), but small and negative. It reaches the value $I_S = A_S \cdot I_{SS}$ for a reverse bias of few V_T. Since at room temperature $V_T = 26$ mV, the reverse current saturates with just about -200 mV across the diode. Because of the exponential, the current quickly increases with forward voltage. We can simply assume that when the forward voltage exceeds a particular level, called the on-voltage, V_{on}, the current increases indefinitely. Therefore, the current–voltage (I–V) characteristic becomes the one shown in Figure 9.5(a).

Example 9.1

Two diodes with equal junction structures are biased with a forward and a reverse bias voltage. The currents are 10 mA and -10 nA respectively; the area of the reverse-biased diode is 100 times larger than that of the other. Determine the forward bias, assuming the reverse to be much less than -50 mV.

Solution

If the reverse bias is much higher than the V_T value, the current equals $-A_D I_{SS}$. The use of it in the Shockley diode equation gives

$$I_{D,F} = A_{D,F} \frac{-I_{D,R}}{A_{D,R}} (e^{V_F/V_T} - 1),$$

which, neglecting the 1 in the parentheses, produces

$$V_F = V_T \ln\left(\frac{I_{D,F}}{-I_{D,R}} \frac{A_{D,R}}{A_{D,F}} \right) = V_T \cdot 18.42 \cong 479 \text{ mV}.$$

The subscripts F and R in the above equations mean forward- and reverse-biased devices. Moreover, the diodes' areas are quite different. Notice that a forward bias of less than 600 mV gives rise to a very large ratio between forward and reverse currents.

An effect to account for is the resistive contribution of the material making the diode and the series resistance of connecting wires. These resistances change the I–V characteristic as shown in Figure 9.5(b). Another effect to account for in reverse bias is an almost constant saturated

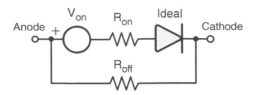

Figure 9.6   Equivalent circuit of the diode with the approximate I–V response of Figure 9.5.

current, as predicted by the Shockley equation. In addition, a mechanism called generation and recombination causes currents, much higher than $A_S \cdot I_{SS}$, almost linearly proportional to the reverse voltage. A non-infinite resistance in reverse conditions describes that effect. Figure 9.5(c) depicts the corresponding I–V characteristic by using quite different slopes in the *on* and the *off* state. The current *generation–recombination current* that determines the off slope strongly depends on temperature. An increase of temperature by 10°C almost doubles the generation–recombination current. Therefore, the reverse slope of the approximate plot of Figure 9.5(c) changes significantly with temperature.

9.2.1   Equivalent Circuit

As discussed earlier, the designer works with symbolic representations of circuits. Using the symbol for the diode, for instance, implies the non-linear operation of the device within given approximations. In some cases the symbol represents an ideal element, i.e., a device with the I–V characteristic of Figure 9.3(b). Possibly, a note near the symbol specifies that the component is ideal. In other cases the designer wants to account for some non-ideal features like the ones depicted in the I–V response of Figure 9.5(c). In that case a suitable equivalent circuit must include, in addition to an ideal element, other components, namely voltage sources and resistors. The result is the scheme of Figure 9.6 with $R_{on} \ll R_{off}$. The ideal diode activates the top branch when the forward bias exceeds V_{on}. The voltage across the circuit compensates for the source generator V_{on} and turns the ideal diode on. Then resistance R_{on} determines a current from anode to cathode: $(V_D - V_{on})/R_{on}$. A bias voltage lower than V_{on}, or negative, opens the upper branch, which is the ideal diode with reverse bias; the high value R_{off} establishes the reverse current. To be precise, R_{off} also contributes to the direct current. Its effect, albeit negligible, can be compensated for with a slight change of R_{on}.

The equivalent circuit of Figure 9.6 also models the I–V responses depicted in Figure 9.5(a) and (b). It is just necessary to set $R_{off} = \infty$ to have zero reverse current and to use $R_{on} = 0$ to obtain a steep increase of current for $V_D > V_{on}$.

When the diode voltage and current change slightly around a given point, instead of the large-signal model of Figure 9.6, it is more convenient to focus on the variations of electrical variables with respect to the quiescent values, which means using the small-signal equivalent circuit. Equation (9.1) leads to

$$I_D - \bar{I}_D = i_d = \left.\frac{\partial I_D}{\partial V_D}\right|_{\bar{V}_D} (V_D - \bar{V}_D) \simeq \frac{\bar{I}_D}{V_T}v_d, \tag{9.3}$$

giving the small-signal behavior equivalent to a conductance, g_d, equal to the quiescent current divided by $V_T = kT/q$, $g_d = \bar{I}_D/V_T$. Figure 9.7 summarizes the result. Figure 9.7(a) shows an expanded view of the response. The small-signal equivalent circuit is the one of Figure 9.7(b).

COMPUTER EXPERIMENT 9.1

Input–Output Diode Response

The diode is a non-linear two terminal device used to rectify waveforms. This simple computer experiment obtains the feature with a diode and variable resistance toward ground (20 kΩ is the default value). There are two equal schemes. The first one uses a sine wave input signal and the second uses a square waveform. A dual DC voltage generator produces two DC voltages used for shifting the input signals. The board enables you to observe the two outputs and one of the two inputs. In reverse bias conditions the diode is approximated by a 1 pF capacitor. The input capacitance of the oscilloscope is 20 pF. You can:

- set the amplitude and frequency of the two input signals. The sine wave generator can be set from 10 kHz to 100 MHz and the square wave generator from 1 kHz to 10 MHz;
- set the DC voltages added to sine wave and square wave;
- change the equal value of the two grounded resistances;
- use a normal p–n diode or a Schottky diode.

MEASUREMENT SET–UP

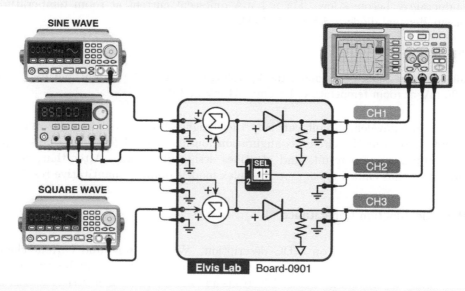

Elvis Lab   Board-0901

TO DO

- Set the sine wave and triangular wave with zero offset. Observe the waveform in the forward and reverse regions of operation.
- Change the value of the resistance and explain the small changes of the waveform that you observe.
- Increase the sine wave frequency at very high values and explain the response you observe with reverse biases.
- Use the square wave and offset to determine the I–V response of the diode with forward and reverse bias.

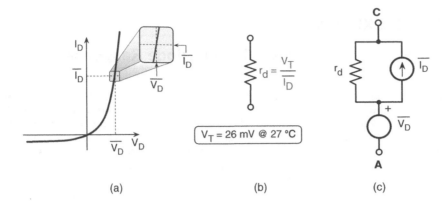

(a)   (b)   (c)

Figure 9.7   Equivalent circuit of the diode with small signals; (a) expanded view of the I–V response; (b) equivalent small-signal resistance; (c) equivalent circuit for small variations of voltage and current.

Figure 9.7(c) describes the diode model, including the quiescent values. Since in the forward region the current increases exponentially with voltage, the equivalent resistance (the inverse of the conductance) becomes low. For a 1 mA quiescent current at room temperature the equivalent small-signal diode resistance is 40 Ω.

A variable feature

The small-signal resistance of a diode is the V_T voltage divided by the current, but $V_T = KT/q$, equal to 26 mV at room temperature, becomes 34 mV at 120°C.

The use of an equivalent circuit can help the designer because extra nodes and jumped elements, not present in the circuit configuration, outline special features. This approach obtains first approximation results and identifies design directions. After that, the use of circuit simulators that embed much more complex models obtains quantitative results.

9.2.2   Parasitic Junction Capacitance

The previous study focused on a DC description. When voltage and current change in time it is necessary to account for capacitive effects that, in diodes, occur because of charge variations at the p–n junction interface. Electrons and holes create an electrical barrier at the junction, whose extent facilitates or opposes the current flow. A detailed study would need to describe the mechanism by using equations from solid-state physics. For our purposes, it is enough to be aware of the presence of bipolar charges at the junction. The charge value depends on the voltage across the junction. Thus if a variation of voltage changes the stored charge we have a junction capacitance. Since the voltage/charge relationship is non-linear, depending on the specific use, the designer employs a large-signal capacitance, C_J, or a small-signal capacitance, c_j. The large-signal element estimates the overall charge exchange. The small-signal component accounts for the small changes (normally assumed to be signals).

The two capacitances are given by

$$C_J = \frac{Q_D}{V_D}, \quad c_j = \frac{dQ_D}{dV_D}. \tag{9.4}$$

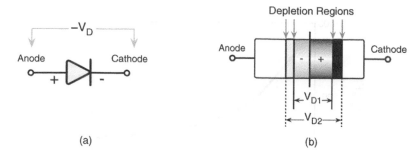

(a)   (b)

Figure 9.8   (a) Reverse-biased diode; (b) pictorial representation of the depletion region for two different reverse voltages $|V_{D1}| < |V_{D2}|$.

The junction capacitance is non-linear: it is small in the forward region and becomes large for reverse biasing. Since in the forward region the voltage across the diode is essentially constant, the junction capacitance is more noticeable in reversed biasing because, in addition to the larger capacitive value, the voltage across the diode can be large and the DC reverse current is very small.

We do not go into deep details of the theory; however, to understand why the diode capacitance is non-linear, we should know that the reverse bias augments the so-called depletion region. This is a region around the junction that is increasingly depleted by mobile charges, as outlined in the pictorial diode representation of Figure 9.8. A given V_{D1} causes depletion. Higher reverse biasing, V_{D2}, augments the depleted region. Thus the new charges are at increasing distances from the junction. Remembering the expression for parallel-plate capacitors,

$$C = \epsilon_0 \epsilon_r \frac{A}{d}, \tag{9.5}$$

where A is the plate area and d is the plates' separation distance, we can conclude that an increasing distance of charges diminishes the capacitance. An approximate expression for c_j is

$$c_j(V_R) = \frac{A_J c_{j0}}{(V_R/V_J + 1)^m}, \tag{9.6}$$

where A_J is the area of the junction, c_{j0} is the junction capacitance at zero bias, and m is a coefficient that depends on the type of junction. For sharp doping transitions $m = 1/2$.

The large signal junction capacitance, C_J, is important in power electronic circuits where large diodes turn on and off. The power consumed by charging and discharging junctions at large reverse voltages often limits the power efficiency. On the other hand, the small signal junction capacitance c_j and, in particular, its dependance on reverse biasing is valuable for electronic applications that need electrically tunable capacitors. When used for that purpose the diodes, called *varicap* or *varactor*, are the tuning elements in Voltage-Controlled Oscillators (VCOs), phase shifters, and frequency multipliers. A voltage-controlled oscillator generates an output periodic signal with the frequency controlled by voltage. This voltage typically DC biases the varactor, whose controlled capacitance tunes the frequency of operation of the oscillator circuit.

Figure 9.9 shows the voltage dependence of a hypothetical varactor diode. Notice that the values on the axis refer to a discrete part. Integrated components have a much smaller capacitance and operate with much lower reverse voltages.

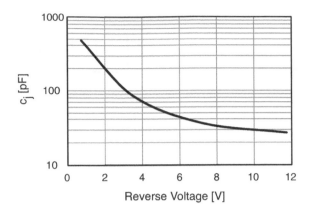

Figure 9.9   Typical voltage dependence of c_j of an hypothetical discrete varactor diode.

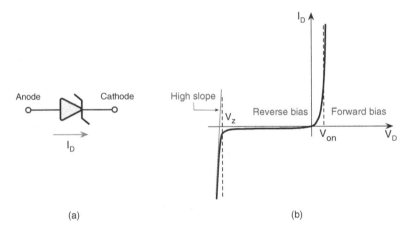

Figure 9.10   (a) Symbol for the zener diode; (b) typical I–V characteristic of a zener diode.

9.2.3   Zener and Avalanche Breakdown

A real diode, like all electronic devices, is unable to sustain high voltages across its terminals. An ideal diode in reverse conditions drains very low current, and the power is negligible. However, above a given reverse voltage, the current sudden augments because of physical mechanisms called *zener* and *breakdown*. The power eventually goes up and the raised temperature can reinforce the growth of current, leading to a destructive positive feedback. This may happen in uncontrolled situations, but zener and breakdown are not disruptive by themselves. They can satisfy electronic necessities if a control circuit keeps the current below the maximum allowed.

Figure 9.10(a) shows the symbol for a diode featuring the rapid change of current above a critical reverse voltage. The I–V response is illustrated in Figure 9.10(b). The symbol is a slight modification of the common diode: a stylized z replaces the bar at the tip of the arrow. The I–V response looks like that of a normal diode up to V_z. For higher reverse voltages the slope of the curve becomes high, denoting a small-signal resistance.

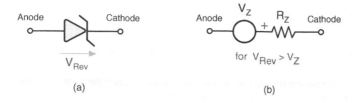

Figure 9.11   (a) Conventions used for the zener diode; (b) its small-signal equivalent circuit.

Two mechanisms give rise to the response of Figure 9.10(b). The zener effect, named after Carl Zener, the physicist who discovered it, relies on special doping profiles to obtain high electric fields at the p–n junction. A high electric field breaks the covalent bonds of electrons and generates electron–hole pairs. The electrons flow toward the cathode and the holes to the anode, giving rise to the zener current. The value of the zener voltage, well controlled by the technology, can be as low as 3 V. Commercial zener diodes obtain zener characteristics at 3.3 V, 3.6 V, 3.9 V, and higher voltages.

The zener mechanism serves to produce a steep current change at low voltage. Breakdown operates well for more than 5 V. Breakdown is also caused by a high electric field, but one whose value does not break covalent bonds. It accelerates trigger electrons that acquire large energy during the mean free path. The energy, released when the electrons collide with atoms, makes the probability of extracting an electron high. Then the electron–hole pair, separated and accelerated by the electric field, acquires energy for another collision. A reverse voltage higher than a certain value makes the process self sustained, causing the so-called avalanche current. The reason for this name is that the increase of current is similar to what happens with snow: an avalanche grows and sustains itself after being started by a trigger.

As shown by Figure 9.10, the slope of the I–V characteristic in the zener region is almost vertical. The equivalent resistance is

$$r_Z = \left.\frac{dV_D}{dI_D}\right|_{V_D<V_Z}. \tag{9.7}$$

Thus, the small equivalent circuit of Figure 9.11 models the zener (or avalanche) diode. The scheme holds for reverse bias beyond the zener voltage. The voltage generator V_z requires the application of a higher reverse voltage; the part that exceeds V_z and the low value of R_z give rise to the output current.

9.2.4   Doping and p–n Junction

A p–n junction produces the diode. Assuming we start from an n-type material, the technology realizes the p-type section by reversing the substrate doping. The selective addition of p-type dopants by a dose that exceeds the n concentration gives the result. The fabrication steps are similar to what are used in integrated circuit fabrication. They are the selective protection of surfaces, ion implantation, and the diffusion and deposition of metal lines for interconnections. Let us recall them.

- **Selective protection** This is used to deposit a suitable shielding layer that protects part of the silicon surface. When silicon must be exposed to high temperatures, resistant materials such as silicon dioxide or silicon nitride form the shield. For mechanical protection

COMPUTER EXPERIMENT 9.2

The Zener Diode

With this computer experiment you can verify the use of a zener diode for voltage regulation. The zener is a special diode that, when reverse biased, starts conducting if the voltage across it exceeds the zener value. The circuit on the board includes a series resistance that gives rise to a voltage drop approximately equal to the difference between input and zener voltage. The regulator feeds a load resistance. The input voltage is the addition of a DC quantity and a sine wave. The zener value is 1.8 V and the maximum power the zener can dissipate is 1 W. Above that value the part stops working properly, and after a short time the diode breaks down and becomes an open circuit. In this virtual experiment you can:

- change the DC bias and the sine wave amplitude. Choose the frequency of the sine wave equal to 1 kHz (default value);
- change the value of series and load resistance;
- chose the right value of the load resistance that can dissipate maximum 2 W. If that value is exceeded you see a red warning light blinking. If breakdown occurs you can change the part after reducing voltages by pushing the "Replace Part" button.

MEASUREMENT SET–UP

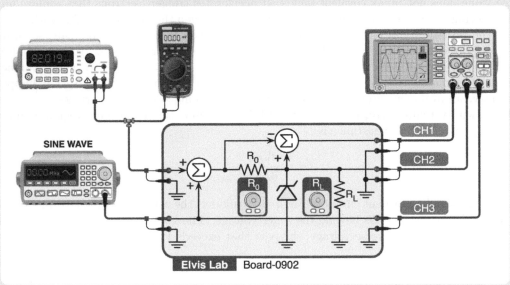

SINE WAVE

CH1

CH2

CH3

R_0

R_0

R_L

R_L

Elvis Lab   Board-0902

TO DO

- Set the sine wave amplitude to zero and vary the DC level. Derive the I–V response of the zener. Change the value of resistance, to limit the current in the zener to the maximum allowed by power consumption. If the zener blows up, you have to change the part.
- Set the amplitude of the sine wave and observe the effect on the output voltage.
- Use an input voltage slightly higher than the zener voltage with high load resistance. Observe the output at different amplitudes of the sine wave. Change the value of the load resistance and observe the effect.
- Set the value of the load to 700 Ω and find the optimal value of series resistance that admits an input voltage ranging from 2.4 V to 3 V.

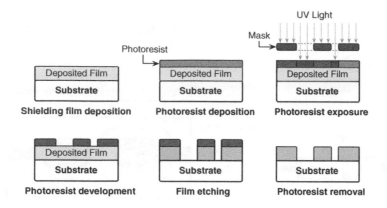

Figure 9.12   Process steps of the photoengraving of a deposited film using a positive photoresist.

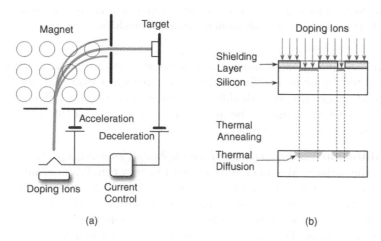

Figure 9.13   (a) Schematic structure of an ion implanter; (b) effect of the ion implantation and the successive thermal annealing.

a soft material, such as a polymeric layer, is enough. Photolithography or photoengraving defines the pattern of the surface coating. These processes employ photoresist, a material that, when exposed to light, becomes soluble or insoluble to a developer. Positive photoresists become soluble as a result of light exposure; negative resists are made insoluble by light. The light is Ultraviolet (UV), deep UV, or electromagnetic waves at very short wavelengths. Illumination through a mask defines the pattern. Figure 9.12 shows details of the fabrication steps. The first of them is the deposition of the protective film over the entire surface. Then the photoresist (positive in the case illustrated) is deposited. The UV light that passes through the mask makes, after development, part of the resist soluble and removable. A suitable chemical etches the protective film that is not covered by photoresist. Finally, eliminating the photoresist leaves the substrate selectively protected.

- **Ion implantation** This step employs ions of the doping material that are accelerated by a very high voltage to hit the silicon surface. The ions enter (or implant) the first atomic layers of the unprotected surface with a dose that depends on ion density and

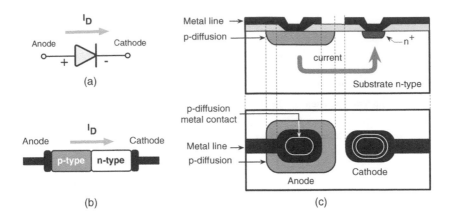

Figure 9.14   (a) Symbol for a diode; (b) its simplified physical structure; (c) cross section and top view of a typical diode produced by a semiconductor planar technology.

implant time. Figure 9.13(a) shows the schematic structure of an ion implanter. The ions, accelerated by several million volts, pass through a grid and are deflected by a magnetic field. Since atoms with different weights deflect in a different manner, only the desired ions hit the target. The arrangement also includes the deceleration voltage generator and the current controller.

- **Thermal diffusion** The high energy of the implanted ions damages the crystallographic structure of the surface. Moreover, ions penetrate a very thin layer of material. A thermal cycle at high temperature for a suitable period of time, called annealing, heals the damage in such a way that implanted atoms replace silicon atoms in the crystal structure. The thermal cycle, as shown in Figure 9.13(b), diffuses the doping atoms under the implanted region in all directions. Doping atoms go inside the volume including a limited region below the shielding protection.

- **Metals interconnection** The diode, obviously, needs interconnection with the rest of the circuit. A pattern of deposited metal, aluminum or sometimes copper, performs the task. Figure 9.14(c) shows the result. The metal lines run over an oxide layer opened in the contact regions. The openings are on the top of the p-type diffused region and the substrate, which is n-type. The metal lines also permit connection to other components fabricated on the same piece of silicon or lead the signal outside the chip.

Self training

Do a search on the Web and find information on metal deposition for integrated circuits. What kind of material makes the low-level interconnections? Which layers are used on the top? Sketch the thickness and separation between interconnections for a typical modem technology.

Figure 9.14(c) also shows the cross section and the top view of a p–n junction. The figure portrays the diode structure in a more precise manner than does Figure 9.14(b), which is often used. The substrate forming the cathode is n-doped silicon. The p-type diffused region makes the anode. Further, the n^+ diffusion on the top of the n-substrate ensures a good contact between the n-type material and its metal connection. Finally, as Figure 9.14(c) shows, a

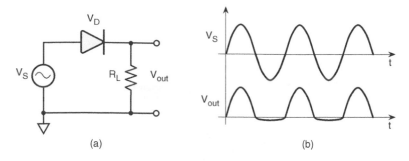

Figure 9.15   (a) Simple half-wave rectifier; (b) input and output waveforms.

major part of the junction is parallel to the surface, and thus the diode current flows in a direction orthogonal to the junction. The opening on the oxide protecting the surface permits the metal to connect the p and n regions to the external circuitry.

9.2.5   Diode in Simple Circuits

Many circuits exploit the non-linear characteristic of diodes. Figure 9.15(a) shows a sine wave generator that feeds the series resistance of a diode and a resistor. The output voltage is across the resistor. When the polarity of the sine wave is positive, the directly biased diode favors the flow of current. On the other hand, when the voltage is negative the expected flow is zero or very low, as it is opposed by the reverse-biased diode, resulting in an asymmetrical operation.

For a more detailed study, assume that the directly biased diode corresponds to the small on-resistance, R_{on}, while in reverse bias conditions the large resistance, R_{off}, describes the component. The voltage across the load, R_L, is the result of two resistive dividers, one for the forward and the other for the reverse operation:

$$V_{out} = V_S \frac{R_L}{R_L + R_{on}}, \quad \text{for } V_S > 0, \tag{9.8}$$

$$V_{out} = V_S \frac{R_L}{R_L + R_{off}}, \quad \text{for } V_S < 0. \tag{9.9}$$

Figure 9.15(b) displays the input and the output voltages. If $R_{off} \gg R_L \gg R_{on}$, the *on* attenuation is very low; the positive peak voltage at the output almost equals that at the input. For reverse bias the large off-resistance makes the output a small fraction of the input. Thus, the current flows mainly in the forward direction, and the average output voltage is positive, as indicated by the name of the circuit: the half-wave rectifier.

The result of the previous analysis is enough for a qualitative comprehension of the operation. If needed, the use of a complex diode model running on a circuit simulator obtains quantitative results. The procedure accounts for the I–V response and solves the non-linear problem defined by the Kirchhoff loop equation

$$\overline{V}_S = V_D + R_L I_D, \tag{9.10}$$

together with the non-linear characteristic of the diode. The unknowns are V_D and I_D.

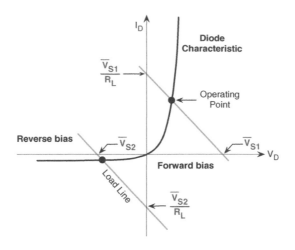

Figure 9.16   Graphical solution of the scheme of Figure 9.15.

Alternatively, we can use equation (9.10) to find the diode operating point in a graphic manner. We have

$$I_D = \frac{\overline{V}_S - V_D}{R_L}, \tag{9.11}$$

which, in the (I_D, V_D) plane, is a straight line crossing the current axis at $\overline{V}_S/R_L$ and the voltage axis at $\overline{V}_S$. Since R_L often represents a load, the line is called a *load line*.

Figure 9.16 plots the I_D–V_D characteristic and two load lines, one for direct bias, $\overline{V}_{S1}$, and the other with reverse bias, $\overline{V}_{S2}$. The load lines have equal slope, determined by R_L.

The crossings of the diode curve with the load lines give the operating points, i.e., the voltage and current of the diode for the two supply voltages considered. The results of the figure correspond to a value of R_L that is relatively low, not much higher than the on-resistance of the diode. Moreover, the direct supply voltage is low with respect to the on-voltage.

Example 9.2

Determine the output voltage of the circuit in the figure. The diode is a zero resistance when $V_D > 0.6$ V and becomes an infinite resistance when $V_D < 0.5$ V. The value of the source generator varies from 0 to $4V$. The resistances are $R_1 = 1$ kΩ, $R_2 = 2$ kΩ, and $R_3 = 1$ kΩ.

Solution

The circuit has a non-linear response consisting of two linear regions. The first is when the diode is off and $I_2 = 0$; the second is when the diode is a zero resistance. The transition point is when $V_{out} = 0.6$ V. Above that level the diode turns on and current flows through R_2. When the diode is off the output voltage is

$$V_{out} = E\frac{R_3}{R_1 + R_3} = \frac{E}{2};$$

the transition point is when $E = 1.2$ V. When the diode is on the current through R_2 is

$$I_2 = \frac{V_{out} - V_{D,ON}}{R_2}.$$

Moreover,

$$E - V_{out} = R_1 I_1$$

$$I_1 = I_2 + I_3 = \frac{V_{out} - V_{D,ON}}{R_2} + \frac{V_{out}}{R_3},$$

which produce the output voltage

$$V_{out} = \frac{E + V_{D,ON}\, R_1/R_2}{1 + (R_1/R_2) + (R_1/R_3)} = \frac{E + 0.3}{2.5};$$

the slope of the input–output relationship is $1/2$ for $E < 1.2$ V and becomes $4/10$ for higher source voltages.

A second simple circuit with a diode is the one shown in Figure 9.17. A battery, E_0, and a series resistance bias a zener diode in the reverse region. Supposing that the voltage across the diode exceeds the breakdown limit, V_Z, we can conclude that the output voltage is approximately equal to V_Z. For a more accurate estimation it is necessary to use the loop equation

$$E_0 = R_0 I_0 - V_Z(I_D), \tag{9.12}$$

which employs $V_Z(I)$, the voltage–current non-linear relationship of the zener diode. Since the forward current in the diode flows in the opposite direction to the current I_0 and the output voltage is the opposite of the one across the diode, we have

$$I_0 = \frac{E_0 + V_Z(-I_0)}{R_0}, \quad V_0 = -V_D, \tag{9.13}$$

which assumes $I_{out} = 0$. In the $(I_0, -V_{out})$ plane (or the $(-I_D, -V_Z)$), equation (9.13) is the load line of Figure 9.18. The crossing of the load line with diode characteristics gives the operating point $\overline{V}_{out}, \overline{I}_0$.

Notice that changing R_0 alters the slope of the load line, but the voltage value of the crossing point almost remains unchanged because the I–V slope in the zener region is very steep. For applications with zero current at output, we need the current $\overline{I}_0$ just to obtain V_Z at output. The total power that the circuit consumes is $E_0 \overline{I}_{out}$. That is, obviously, a cost that should be minimized by choosing a minimum $\overline{I}_0$. Looking at Figure 9.18 it is evident that a suitable choice is to have $\overline{I}_0 = I_{min}$ just above the knee of the zener region. The corresponding load line is the dotted one of Figure 9.18. The series resistance is

$$R_{0,max} \cong \frac{E_0 + V_Z}{I_{min}}. \tag{9.14}$$

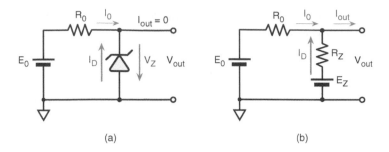

Figure 9.17   (a) Simple voltage regulator with a zener diode. (b) Simplified equivalent circuit.

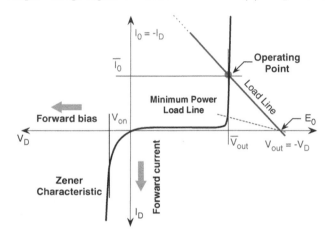

Figure 9.18   Graphical solution of the scheme of Figure 9.17.

However, a finite output current must satisfy

$$I_{out} = I_0 + I_D,$$ (9.15)

where I_D is negative (reverse current). Therefore the design must use a larger value of I_0 to provide both output current and current in the zener diode. Depending on the expected value of the output current, the resistance $R_{0,max}$ becomes

$$R_0' \cong \frac{E_0 + V_Z}{I_{out,max} + I_{min}}.$$ (9.16)

If the output current becomes larger than $I_{out,max}$, the current in the zener diode becomes lower than I_{min} and the output voltage is no longer controlled.

The output voltage is not constant but changes slightly with the output current. The simplified equivalent circuit of Figure 9.17(b), which models the zener region with a voltage generator V_Z and a series resistor R_Z, helps in the analysis. By inspection we have

$$I_{out} = \frac{E_0 + V_{out}}{R_0} + \frac{V_Z + V_{out}}{R_Z},$$ (9.17)

which yields

$$V_{out} = \frac{E_0 R_Z + V_Z R_0}{R_0 + R_Z} - \frac{R_0 R_Z}{R_0 + R_Z} I_{out},$$ (9.18)

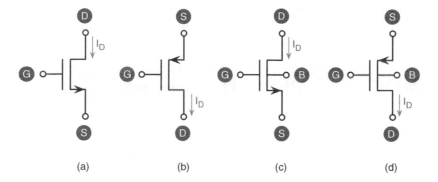

Figure 9.19   Symbols for the MOS transistor with and without the bulk (or substrate) node outlined.

which corresponds to a voltage source, $(E_0 R_Z + V_Z R_0)/(R_0 + R_Z)$, with, in series, the parallel connection of R_0 and R_Z. In order to have a small change in the output voltage it is necessary to have $R_0 > R_Z$, which, in turn, requires the use of a suitably large source voltage.

Use of the zener diode

The zener diode serves for the regulation of voltage with low accuracy. The scheme is simple but the power efficiency is low, especially at low output currents.

9.3 THE MOS TRANSISTOR

The MOS transistor is very important for modern microelectronics. Almost all electronic equipment uses MOS transistors to implement analog or digital functions. The designer uses two types of MOS transistors, the N-channel and the P-channel. The so-called CMOS (Complementary MOS) technology makes available both types of transistors on the same piece of silicon. MOS devices are voltage-controlled active elements. They enable amplification, because with them it is possible to use a voltage with almost zero current to control a relatively large current.

Figure 9.19 shows the symbols for the two types of transistors. The terminals are source (S), drain (D) and gate (G). The region connecting source and drain is called the *channel*.

The N-channel symbol has an arrow exiting the source; on the other hand the arrow of the P-channel symbol enters the source. Figures 9.19(a) and (b) outline just the three main terminals: source, drain, and gate; to be precise it would be necessary to outline a fourth terminal, the body(B) (or bulk or substrate) which is the material inside which the transistor is fabricated. The bulk connection can be the same for all the devices of a given type. In this case it is not necessary to indicate the terminal in the symbol because it is obvious where the body is connected: V_{DD}, the higher voltage of the circuit, for p-type devices, and V_{SS}, the lower voltage of the circuit, for the complementary N-channel transistors. When the bulks are at different voltages, made possible by electrically insulated regions that house the transistor, it is necessary to use the symbols of Figure 9.19(c) and (d).

Current flows from drain to source for N-channel devices and from source to drain for P-channel elements, provided that the voltage of the N-channel drain is higher (or lower for P-channel) than the source. Indeed, as discussed shortly, the MOS transistor is a symmetrical

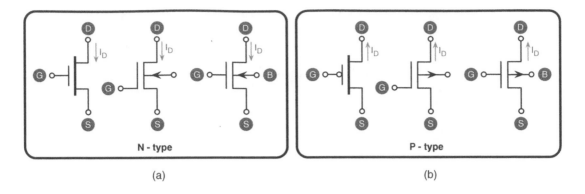

Figure 9.20 Other symbols used to represent the N-MOS and the P-MOS transistors. Arrows show the direction of current flow with direct bias.

device with source and drain interchangeable. The source is not a defined physical connection but is the terminal whose biasing favors flow of current in the direction indicated by the arrow; the drain is the other terminal.

Some circuit schematics use different MOS symbols. Figure 9.20 shows three of them. They distinguish between N-channel and P-channel by adding a small circle on the gate connection or with an arrow in the body connection. Notice that two of them do not distinguish between source and drain. This outlines once more the fact that the MOS transistor is a symmetrical structure.

9.3.1 MOS Physical Structure

The fabrication of CMOS transistors requires a number of complicated processing steps, especially for modern nanometer devices. The structure of modern MOS transistors is three dimensional. For simplicity, we consider a two-dimensional device whose cross section is shown in Figure 9.21. Key to the architecture are two diffusions, source and drain, both n-type or p-type, placed at very short distance from one another and separated by the channel. The distance between diffusions is the channel length, L. This is the most relevant feature of the technology. Years ago the channel length was a few microns; now, the continuous scaling of the technology has reduced the distance between source and drain by two orders of magnitude, down to tens of nm. Figure 9.21 depicts two transistors, one p-type and the other n-type, fabricated on a p-doped substrate. Two n^+ diffusions in the substrate implement the source and drain of the N-channel device. The sides of the source and drain facing each other are below the gate, because the technology uses the gate as a defining element. Moreover, as shown by the figure, the channel is shorter than the length of the gate because of lateral diffusion of the doping that realizes source and drain. The result is some overlap between the electrode that realizes the gate and the source and drain. Thus, the channel obtained, as shown by the figure, is shorter than the gate on top. The figure also shows the metal connections of the two diffused regions, and there is a third connection for bulk biasing. In order to ensure good contacts, the structure uses a p^+ diffusion to reduce the conductance of the p substrate material. A thin oxide separates the silicon surface from the gate. The gate oxide of modern technologies is very thin. This is critical because oxide layers of a few nm limit the bias voltage.

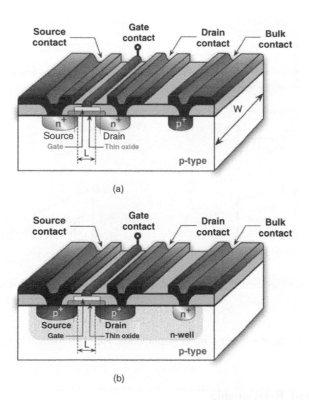

Figure 9.21   Cross sections of transistors: (a) N-MOS; (b) P-MOS.

Old technologies permitted supply voltages of 5 V or more. Nanometer technologies admit supply voltages no higher than 1 V or so.

The P-channel transistor configuration is similar. The difference is that the doping of body, source, and drain are complementary with respect to the N-channel device. Since the substrate used in the figure is n-type a so-called n-well diffused in the substrate creates the bulk of the N-channel transistor. The n^+ diffusion biases the body.

The path from source to drain is through the channel made by a material with opposite doping with respect to source and drain. Thus from source to drain there are two back-to-back p–n junctions. Having back-to-back diodes would prevent flow of current. However, the third electrode, the gate, changes the conductive properties at the silicon–oxide interface and creates a conductive channel.

Even source–bulk and drain–bulk are p–n junctions. The bulk does not participate to the conduction because it is just the material used to contain the transistor and to establish electrical insulation. Since a p–n junction with zero or reverse bias drains zero or a negligible current, two junctions are zero biased or reverse biased. This is done via the metal connections that contact the bulk.

It is worth outlining again that the MOS cross sections of Figure 9.21 are just a simple indication of real MOS structures. Also, the figure shows just one metal layer for connections. Modern technologies use several metal layers, some made of aluminum, and others of copper.

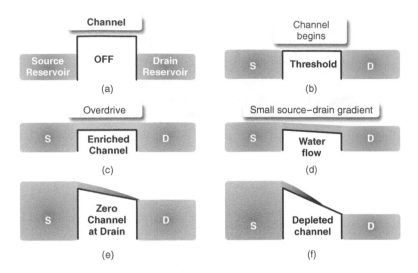

Figure 9.22   Hydraulic equivalent of the MOS behavior with two reservoirs at different levels separated by a channel.

Having many conductive layers favors interconnections, like the multilevel junctions in modern cities and highways.

9.3.2   Voltage–current Relationship

The channel of an N-MOS transistor is p-type. Electrons, the majority carriers in source and drain, would not flow through the intermediate region except as the result of a special action. That is a positive voltage applied to the gate that pushes back the holes from the oxide semiconductor interface, thus creating the conductive channel. The voltage between source and gate that is capable of creating the channel is called the threshold voltage, V_{th}. In reality, the threshold level does not establish a sharp off/on transition: even below the threshold (in the sub-threshold region) there is some conduction. The sub-threshold region of operation is used in micro-power analog circuits.

Typically the threshold voltage ranges between a few hundred mV to less than 1 V. Obviously the value of the threshold must be lower than the supply voltage. Transistors made from nanometer technologies that need low supply voltages have low thresholds.

When V_{GS} exceeds V_{th} the voltage controlling the gate gives an overdrive, V_{ov}, given by

$$V_{ov} = V_{GS} - V_{th}. \tag{9.19}$$

The overdrive voltage adds to the threshold that has already created the channel. It attracts electrons, which can now flow from source and drain to enrich the conduction.

Keep in mind

The threshold voltage is the value of V_{GS} that conventionally produces a conductive channel at the silicon–oxide interface. A V_{GS} voltage lower than V_{th} does not mean that the conduction is zero. The threshold is not an on/off transition point.

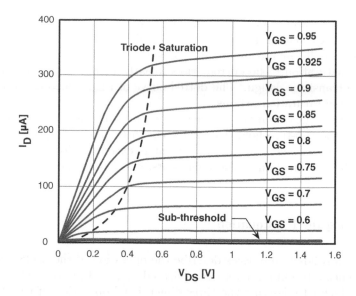

Figure 9.23   Voltage–current relationship of a MOS transistor with V_{GS} as parameter.

A voltage between source and drain with the drain positive with respect to the source moves the charges on the channel from source to drain, producing current. If the value of V_{DS} is small, the current is almost proportional to it, but when it becomes large its action starts depleting the channel, and the voltage–current relationship bends until it becomes almost constant.

The operation can be better understood by looking at the hydraulic equivalent shown in Figure 9.22. Two reservoirs, the source and the drain, are separated by a channel. A mechanical system controlled by V_{GS} moves the bottom of the channel. In some conditions the bottom is higher than the level of the water in the two reservoirs, and, independently of their levels, no water flows (Figure 9.22(a)). By changing the control it is possible to reach the condition of Figure 9.22(b). The bottom of the channel is at the same level as the water in the two reservoirs, assumed to have equal levels. A further increase of the control variable enriches the channel that connects the two reservoirs. If the level of the drain diminishes there is then a flow of water from the source to the drain reservoir (Figure 9.22(d)), but when the difference of level is too high the channel depletes near the drain, as shown in Figure 9.22(e) and (f). The first of these figures indicates the condition in which the channel depletes at the end, just before the drain. Two other physical parameters, μ, the carrier mobility, and C_{ox}, the oxide capacitance, have an equivalent in the hydraulic example. Mobility means liquid fluidity, and the effectiveness of the control corresponds to the closeness of the gate to the channel or to lower oxide capacitance.

The voltage–current relationship of MOS transistors with different values of V_{GS} looks like the diagram of Figure 9.23. With low drain voltages the current increases in a parabolic manner, but when $V_{DS} = V_{ov}$ it increases linearly with a limited slope.

Figure 9.23 describes the transistor response in three different regions of operation. They are conventionally distinguished as:

- sub-threshold;

- triode;

- saturation.

Sub-threshold is is for $V_{GS} < V_{th}$. The current is very small, almost independent of V_{DS}, with an approximate exponential increase with V_{DS}. The other two regions of operation are for $V_{GS} > V_{th}$. With small V_{DS} the current increases in a parabolic manner. When $V_{DS} > V_{ov}$ the transistor is in saturation. The current rises linearly with a slope that depends on the technology and the transistor design. The dotted curve of Figure 9.23 defines the transition between the triode and saturation regions.

Depending on the specific circuit needs, the designer uses transistors in one of these regions. Notice that, even if the sub-threshold does not seem to be interesting, micro-power circuits largely use that region because, even if the current is extremely low, its value depends on the gate voltage: the key function of the transistors, controlling the current in an electrical manner, is obtained anyway.

9.3.3   Approximating the I–V Equation

Suitable studies in solid-state physics describe the operation of the MOS transistor. They account for many effects to obtain a very complicated model. Indeed, with deep sub-micron technologies it is difficult to describe the three-dimensional operation of the device, and this makes it necessary to use fitting parameters that do not have any physical meaning but are just used to trim the response of a model to obtain the fitting with experimental measurements.

Obviously, complicated models serve for computer simulations that must obtain precise estimations of the circuit response. The models are almost incomprehensible, so the designer must rely on what given by the simulation. For simplified study, which is useful to acquire design experience, it is better to use simple models that do not use a unique description but distinguish the I–V relationship in the three regions of operation. The details of the simple model are not given here, since they are the topic of more specific courses. For us it is enough to know the relationship

$$I_D = I_{D0} e^{V_{GS}/nV_T} e^{V_{BS}/nV_T} (1 - e^{V_{DS}/V_T}), \tag{9.20}$$

which is valid for MOS transistors in the sub-threshold region. The equation shows an exponential dependence on the ratios V_{GS}/V_T and V_{BS}/V_T. Both are attenuated by the factor n, used to account for the fraction of the gate voltage that sustains the drop voltage across the oxide. The value of I_{D0} is extremely small as the current in the sub-threshold is very low.

In the triode region we have

$$I_D = \mu C_{ox} \frac{W}{L} \left[(V_{GS} - V_{th}) V_{DS} - \frac{1}{2} V_{DS}^2 \right] (1 + \lambda V_{DS}), \tag{9.21}$$

which is a parabola in the (I_D, V_{DS}) plane as manifested in the input–output curves. The equation holds until the limit of the triode region, defined by $(V_{GS} - V_{th}) = V_{DS}$, is reached.

The response in saturation approximately follows the relationship

$$I_D = \frac{1}{2} \mu C_{ox} \frac{W}{L} (V_{GS} - V_{th})^2 (1 + \lambda V_{DS}), \tag{9.22}$$

an equation that displays a quadratic increase of drain current with overdrive, $(V_{GS} - V_{th})$. Moreover, the current depends on V_{DS} through the parameter λ. It has a small value because the slope of the responses in saturation is low.

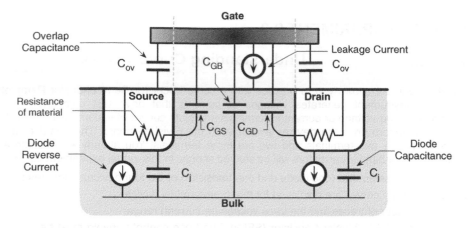

Figure 9.24   Stylized cross section of MOS with parasitic components.

9.3.4   Parasitic Effects

To study the dynamic operation of circuits with MOS transistors it is necessary to know the dynamic parameters of the device. Since they are undesired because they limit the speed performance, we include them in the parasitic category. Figure 9.24 illustrates the components discussed here above a stylized cross section of the MOS transistor.

The presence of conductive materials and insulators gives rise to parasitic terms. One is due to the reverse-biased source–bulk and drain–bulk diodes. As we know, the parasitic capacitance of p–n junctions is proportional to the junction area and is non-linear:

$$C_j = A_j C_j(V_D). \tag{9.23}$$

Accordingly, to enhance the speed it is necessary to design source and drain regions with minimal area. However, the transverse extent of source and drain must allow the space for metal contact.

Another source of parasitic capacitance is the gate electrode and the closer conductive materials: source, drain, and bulk. Since part of the gate directly overlaps source and drain, we have an overlap term, C_{ov}. The rest of the gate gives rise to three other capacitances, C_{GS}, C_{GD}, and C_{GB}. The value of these capacitances depends on the gate area and changes in the three mentioned regions of operation. In saturation C_{GS} dominates the others. In the triode region C_{GS} and C_{GD} are almost equal, while in the sub-threshold C_{GB}, which is almost zero in the other regions, becomes substantial. There are other capacitive contributions like those due to the crossing of a metal connection with other metals on top. These terms are normally small but can become important if the overlap occurs over long distances.

Important limitations, especially with nanometer technologies, come from leakage currents. Very thin oxides are not able to secure solid electrical insulation. An effect (called a tunnel), explained by quantum mechanics, produces current through the gate oxide. The tunnel current increases exponentially with voltage and temperature and in some situations can become significant for the transistor operation and affect the power consumption. The problem of the gate current mainly concerns digital circuits but is also a limitation to analog schemes.

The parasitic diodes towards the bulk also cause leakage. As already mentioned, the reverse current, dominated by the generation–recombination term, is proportional to junction area and

COMPUTER EXPERIMENT 9.3

Understanding Curve Traces

This virtual experiment allows you to practice with a curve tracer (or Semiconductor Parameter Analyzer). The instrument generates voltages or currents that drive test components. The instrument detects the voltage or current at terminals and plots curves of various responses. The input/output connectors drive the devices available on the ElvisLab board. The board contains four devices: two are two–terminals and two are three–terminals. You already know diode and zener. The MOS and bipolar transistors will be studied shortly. In this virtual experiment you can:

- use switches to select two (one diode and one transistor) of the four devices under test;
- decide how many curves are displayed for the three-terminal devices;
- select the number of traces and the trace voltage (or current) range;
- change from MOS to bipolar transistor (SEL2). The trace control variable must be changed from voltage to current;
- change the temperature of the experiment with a numerical stepper.

MEASUREMENT SET–UP

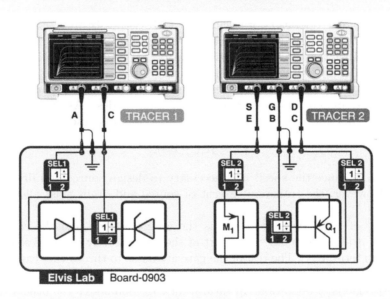

Elvis Lab   Board-0903

TO DO

- Select one of the devices and choose, for transistors, five curves. Observe that when the parameter is below a given level the output current is almost zero. You will learn why shortly.
- Observe the I–V characteristics of the diode and the zener. Compare the response of both devices for low reverse bias.
- Notice that it is not permitted to drive the gate of the MOS with a current. If you do that, you will see a red warning light.
- Observe the I–V curve of the diodes at different temperatures. With equal flowing current, determine the voltage change across the diode when the temperature goes from 300°K to 360°K by 10°K steps. What is the law that describes the result?

increases exponentially with reverse voltage and junction temperature. Since reverse current doubles every 10°C of increase of the junction temperature, a device at a temperature that is 100°C more than the ambient leaks a 1000× reverse current.

A final parasitic term worth mentioning is the resistance of the semiconductor material making source and drain. The actual source and drain are at the beginning of the channel, but the external connection is via the metal lines. The piece of material from those two points determines a series parasitic resistance.

The parasitic elements of Figure 9.24 are used together with the main MOS equivalent circuit (studied below) in accurate circuit simulations.

Example 9.3

The oxide thickness of an MOS transistor is 4 nm. The relative dielectric constant is four. Determine the gate total capacitance of a device with length and width 65 nm and 350 nm respectively.

Solution

We suppose the capacitor to be a parallel plate structure whose capacitance is given by

$$C_p = \epsilon_0 \epsilon_r \frac{A}{n}.$$

ϵ_0, it is known, is $8.86 \cdot 10^{-12}$ F/m; therefore

$$C_p = 8.86 \cdot 10^{-12} \cdot 4 \cdot \frac{22.75 \cdot 10^{-15}}{4 \cdot 10^{-9}} = 0.2 \cdot 10^{-15} \text{ F},$$

an extremely low value.

9.3.5   Equivalent Circuit

The transfer curves of Figure 9.23 provide useful information for drafting the main equivalent circuit of the MOS. The plots show that the drain current changes with voltages V_{DS} and V_{GS}, but, in reality, current also slightly depends on the bulk voltage; therefore

$$I_D = f(V_{DS}, V_{GS}, V_{GB}). \tag{9.24}$$

That is all we need to derive the large-signal equivalent circuit: a current generator controlled by the three voltages V_{DS}, V_{GS}, and V_{GB}. Since one of them (V_{DS}) is across the generator itself, possibly the dependence on V_{DS} can be distinguished from the others. Equation (9.24) is approximated as

$$I_D = f'(V_{GS}, V_{GB}) + G_m(V_{GS}, V_{GB})V_{DS}, \tag{9.25}$$

which describes the parallel connection of a voltage-controlled current source and a non-linear conductance across it. Figure 9.25(a) and (b) show the two large-signal equivalent circuits derived from equations (9.24) and (9.25).

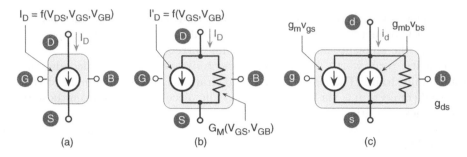

Figure 9.25   Large-signal equivalent circuit of the MOS transistor with different complexity.

Small changes in the electrical variables around the quiescent values $\bar{V}_{DS}$, $\bar{V}_{GS}$ and $\bar{V}_{GB}$ enable a linearization of equation (9.24). The result is

$$I_D = f(\bar{V}_{DS},\ \bar{V}_{GS},\ \bar{V}_{GB}) + \left.\frac{\partial I_D}{\partial V_{DS}}\right|_{op} \Delta V_{DS} + \left.\frac{\partial I_D}{\partial V_{GS}}\right|_{op} \Delta V_{GS} + \left.\frac{\partial I_D}{\partial V_{GB}}\right|_{op} \Delta V_{GB}, \qquad (9.26)$$

where ΔV_{DS}, ΔV_{GS}, and ΔV_{GB} are small changes of V_{DS}, V_{GS}, and V_{GB} around quiescent values. The subscript op means operating point. We use small letters to denote small variations:

$$v_{ds} = \Delta V_{DS} - \bar{V}_{DS}, \quad v_{gs} = \Delta V_{GS} - \bar{V}_{GS}, \quad v_{gb} = \Delta V_{GB} - \bar{V}_{GB}. \qquad (9.27)$$

The derivatives of equation (9.26) define the small-signal parameters

$$g_m = \left.\frac{\partial I_D}{\partial V_{GS}}\right|_{op} \quad \text{transconductance gain} \qquad (9.28)$$

$$g_{m,b} = \left.\frac{\partial I_D}{\partial V_{GB}}\right|_{op} \quad \text{bulk transconductance gain} \qquad (9.29)$$

$$g_0 = \left.\frac{\partial I_D}{\partial V_{GD}}\right|_{op} \quad \text{output conductance,} \qquad (9.30)$$

which, together with small-signal current $i_d = \Delta I_D - \bar{I}_D$, produce the relationship

$$i_d = g_m \cdot v_{gs} + g_{m,b} \cdot v_{gb} + g_0 \cdot v_{gd}. \qquad (9.31)$$

Two voltage-controlled current sources and a conductance across the drain and source terminals make the small-signal drain current, as pictorially represented by the small-signal equivalent circuit of Figure 9.25(c).

The use of the simple relationship given in subsection 9.3.3 derives approximate expressions for the small-signal parameters. We obtain

$$g_m = \frac{I_D}{nV_T} \quad \text{in sub-threshold region;} \qquad (9.32)$$

$$g_m = \mu C_{ox} \frac{W}{L} V_{DS} \quad \text{in triode region;} \qquad (9.33)$$

$$g_m = \frac{2 I_D}{V_{GS} - V_{th}} = \mu C_{ox} \frac{W}{L}(V_{GS} - V_{th}) = \sqrt{\mu C_{ox} \frac{W}{L} I_D} \quad \text{in saturation region.} \qquad (9.34)$$

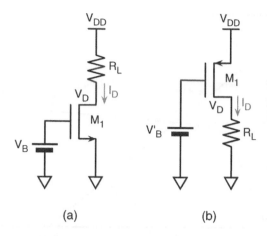

Figure 9.26   Simple transistor stage with resistive load: (a) with an N-channel device; (b) with a P-channel device.

The (9.34) relationships are important because, often, analog circuits use the MOS in saturation. One of them, for example, indicates that the transconductance is the ratio of twice the drain current and the overdrive voltage. We should remember this result for a comparison with the transconductance of a bipolar transistor.

The output conductance is the slope of the I–V curves on the operating point. It is worth noting that in the triode region the output conductance is

$$g_0 = \mu C_{ox} \frac{W}{L}(V_{GS} - V_{th} - V_{DS}),\tag{9.35}$$

showing that for low values of V_{DS} the output conductance is proportional to the overdrive voltage. Thus, in those conditions we can use the transistor as a voltage-controlled resistance.

The small-signal output conductance in saturation is

$$g_0 = I_D\lambda,\tag{9.36}$$

a result that, combined with one of the transconductance equations, states that the ratio g_m/g_0 decreases as the square root of the current. Since that ratio is the intrinsic gain of the transistor, to increase the achievable small-signal gain we have to diminish the drain current.

9.4 MOS TRANSISTOR IN SIMPLE CIRCUITS

Even for the MOS transistor it is useful to analyze simple circuits first. Figure 9.26 is a scheme with a MOS transistor and a resistance. Figure 9.26(a) employs an N-channel device; Figure 9.26(b) is the complementary scheme with a P-channel component. We only discuss the scheme with the N-channel transistor, because the study of the other is just symmetrical.

The battery V_B establishes V_{GS}. The drop voltage across the resistance reduces the V_{DD} voltage to obtain the value of V_{DS}, according to

$$V_{DS} = V_{DD} - R_L I_D.\tag{9.37}$$

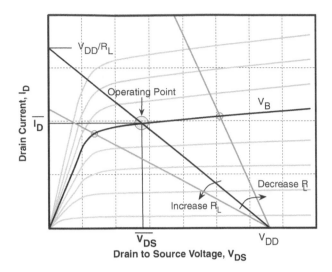

Figure 9.27   Graphic solution of the simple circuit of Figure 9.26.

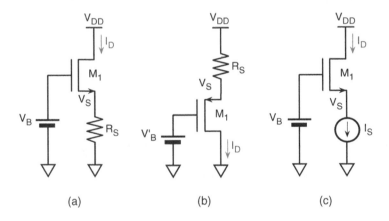

Figure 9.28   Simple transistor stage: (a) with resistance on the source and an N-channel device; (b) with resistance on the source and a P-channel device; (c) with current source.

A straight line on the plane (V_{DS}, I_D) with slope $1/R_L$ depicts equation (9.37). The line crosses the voltage and current axes at V_{DD} and V_{DD}/R_L respectively. The name of the line, as in the case of the circuit with the diode, is a load line.

To obtain the operating point we need a second relationship linking V_{DS} and I_D: the transistor response with $V_{GS} = V_B$. Its plot and load line give, in a graphic manner, the operating point, as depicted in Figure 9.27. The figure outlines the current–voltage curve with $V_{GS} = V_B$ but also shows, using gray lines, the curves with different gate-to-source voltages. The crossing point of that black curve and the load line produces the operating voltage and current, $\overline{V}_{DS}$ and $\overline{I}_D$. It is in the saturation region. If we increase the load resistance, the slope of the load line diminishes, and, at a given point, the transistor enters the triode region, as shown in Figure 9.27 by the gray load line. By contrast, a lower R_L increases the current, and the operating point goes further into the saturation region.

If V_B changes, a different curve to represent the transistor response must be used. In particular, if V_B decreases, the current of the operating point becomes lower, going in the sub-threshold direction. If V_B increases, the circuit goes towards the triode region. Thus, there is a range of values of V_B that, for a given V_{DD} and R_L, keeps the transistor in saturation.

Notice that if V_{GS} increases, the operating voltage decreases and vice versa. This behavior implies inversion. Moreover, with a relatively large value of the load resistance, small changes in V_{GS} can produce large variations of the operating point. For the above reasons the circuit of Figure 9.26 is called an inverting amplifier with resistive load.

A second interesting scheme is the circuit of Figure 9.28. It looks similar to the one already studied. The difference is that the resistance, whose name is changed to R_S, is at the source side. For the N-channel version (Figure 9.28(a)), the drain is at V_{DD}. The P-channel version connects the drain at the ground terminal. The previous scheme has a drain voltage that depends on the operating conditions. This circuit has a source voltage that depends on the drop voltage on R_S. Again, we have a load line. Its equation is

$$V_{DS} = V_{DD} - R_S I_D. \tag{9.38}$$

V_{GS} is given by

$$V_{GS} = V_G - V_S = V_B - R_S I_D. \tag{9.39}$$

The operating point is on the load line, but the current–voltage curve we have to use depends on the operating point current. A possible method to obtain the quiescent values of source voltage and flowing current is to do a search using successive attempts. Assume we have a given V_{GS} whose transistor curve determines the operating point. The current that establishes the drop voltage $R_S I_D$, using equation (9.39), should equal the guessed V_{GS}. If the result is lower, the assumed value of V_{GS} is increased for the next attempt. The process continues until it reaches the desired accuracy. The procedure needs I–V responses for a continuous number of V_{GS} values. The data sheet gives the I–V curves for discrete V_{GS} values. The search may use interpolated curves to obtain a detailed description around the result, as shown in Figure 9.29.

The scheme can use, instead of R_S, a current source (or, better, a circuit that operates like a current generator) as a biasing element, as shown in Figure 9.28(c). The load line becomes parallel to the current axis and the crossing point of the load line with $I_D = I_S$ is the operating point. The I–V curve that passes on that point is the V_{GS} that sustains that source current.

A second way to find the quiescent point is to use the V_{DS} voltages that cross the load line for given V_{GS}. These values determine V_B

$$V_B = V_{DD} - V_{DS} + V_{GS}, \tag{9.40}$$

which results in a diagram like the one of Figure 9.30. The values of V_B that produce various points of the load line decrease almost linearly with V_{DS}: when V_{DS} is near V_{DD}, reducing V_B is almost ineffective. The crossing of the curve in Figure 9.30 and the value of the gate bias used in the circuit produce the operating voltage and, consequently, the current.

9.5 THE BIPOLAR JUNCTION TRANSISTOR (BJT)

Now we shall study the Bipolar Junction Transistor (BJT). It was invented and used many years before the MOS transistor, but the possibility of realizing digital functions with low power (almost zero in static conditions) meant that CMOS was increasingly used. Bipolar

COMPUTER EXPERIMENT 9.4

Finding the Operating Point

The solution of non-linear equations defines the operating point of non-linear electronic circuits. For simple circuits it is possible to resolve the problem graphically. Two equations describing the circuit are represented on the same voltage–current plane; the crossing point of the two curves provides the operating values. This computer experiment allows you to study a MOS transistor with constant gate voltage and a resistive load from drain to supply voltage. The board splits the scheme into two parts for measuring voltage and current across transistor and resistor. The current to voltage converter (I-to-V) measures the current across it with zero voltage drop. Four signals drive one of the oscilloscopes on the X–Y mode. The results are two curves, whose intersection gives the operating point. In this virtual experiment you can:

- set the upper and lower limits of the sawtooth waveform to explore small parts of the voltage–current plane;
- change the value of V_{DD} and R_L;
- change the voltage at the gate of the MOS transistor.

MEASUREMENT SET–UP

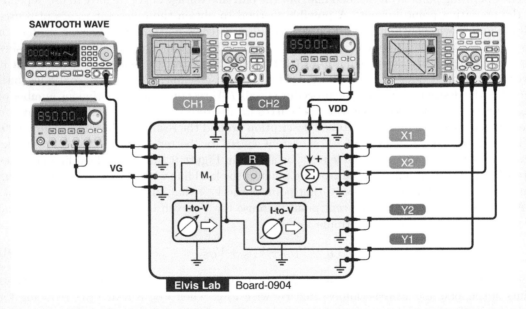

TO DO

- Use R_L=10 kΩ and set V_{DD}=1.8 V. Change the voltage at the gate terminal to ensure that the MOS transistor moves from triode to saturation region. Change R_L and observe what happens.
- Set R_L=20 kΩ (or other values), V_{DD}=1.8 V and determine the value of V_{GS} that gives 0.9 V at output. What is the variation of V_{GS} that changes the position of the operating point by ±50 mV? Repeat the measurement with V_{DD} = 3.3 V. Explain the difference.

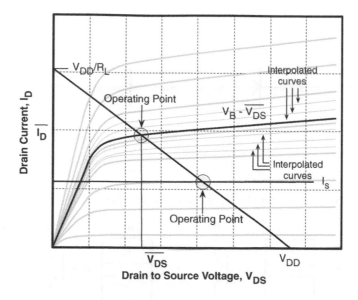

Figure 9.29   Graphic solution of the simple circuit of Figure 9.28.

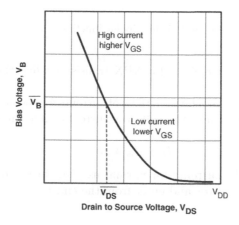

Figure 9.30   Alternative graphic solution to find the quiescent point of the circuit of Figure 9.28.

transistors are now available as discrete parts or employed in integrated circuits for power or high-speed applications. Bipolar transistors are also integrated together with CMOS devices to combine the benefits of the two technologies. The result is the so-called BiCMOS circuits.

The bipolar transistor, unlike a MOS transistor, operates under the control of current. Because of this difference, the MOS is described as a voltage-controlled active device while the BJT is a current-controlled active component. Actually, input current changes modify the input voltage, but the variation is minor and often neglected.

The bipolar transistor has three terminals: collector (C), base (B) and emitter (E). There are two different types, the NPN and the PNP. Figure 9.31 shows the symbols for both. The leg with the arrow is the emitter. The other slanted leg is the collector, and the third terminal

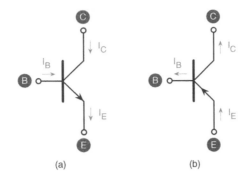

Figure 9.31   Symbols for the bipolar junction transistors: (a) NPN; (b) PNP.

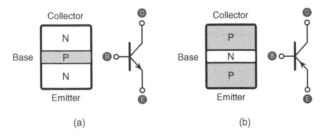

Figure 9.32   Physical architecture of the bipolar junction transistor: (a) NPN; (b) PNP.

is the base. The arrow exits the emitter for NPN transistor while the inward-pointing arrow denotes PNP device.

The symbols of Figure 9.31 show the directions of currents for the normal region of operation. For an NPN transistor the base and the collector currents enter the device; their addition is what exits from the emitter,

$$I_E = I_B + I_C. \tag{9.41}$$

For the PNP transistor the currents for normal use flow in the opposite directions: they exit from the base and the collector and enter from the emitter.

9.5.1   The BJT Physical Structure

Three layers with alternate doping make the bipolar transistor. We have an n-type followed by p-type and ending with another n-type for the NPN, and vice versa, starting from the p-type, for the PNP. The first terminal of the sequence is the emitter, the intermediate is the base, and the ending is the collector. Thus the NPN structure, shown in Figure 9.32, is the series of two back-to-back diodes. They are special diodes because the key feature of bipolar transistors is that the base is very thin, so that the operations of the diodes interfere with each other. The current of the base–emitter junction strongly influences the current of the other junction and determines a control, as will be discussed in some detail shortly.

The scheme of Figure 9.32 just indicates a sequence of doped materials. The physical scheme obtained by real fabrication is not that but a much more complicated architecture. Figure 9.33 shows the cross section of a possible simplified realization. The substrate of the entire structure is a p-doped silicon. On the top of it there is a thin layer of n-type material

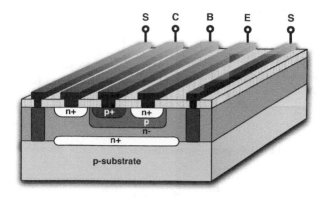

Figure 9.33   Cross section of an NPN transistor built on a p-type epitaxial layer.

grown by a technique called epitaxy. This layer forms the collector. Two diffusions, one p-doped and the other n-doped, one inside the other, make the emitter and the base. The result is a vertical NPN sequence with the thickness of the base controlled by the extent of the two diffusions.

The resulted doping sequence is not symmetrical. The doping of the emitter is higher than the opposite sign featuring the base, which in turn is higher than the n-doping of the collector. Therefore, as shown in the figure, the emitter is n+, the base is p and the collector is n−. Having asymmetry is not a negative feature but, actually, is good because it obtains high effectiveness in the transistor operation.

Observe that the device uses a p+ diffusion inside the p-type base and an n+ diffusion onto the n− collector. Heavily doped regions below metal connections and diffused inside the low doped layers obtain good contacts. Moreover, the structure facilitates vertical flow of current and directs it toward the collector contact via a heavily n-doped region, below the NPN sequence. This is produced before the epitaxy. A suitable insulation around the collector and the substrate bias ensures electrical insulation of the transistor.

What is illustrated in Figure 9.33 requires many steps for fabrication. Modern technologies make it even more complicated, with many masks for fabrication and several processing steps, especially when the BJT is fabricated together with CMOS devices (the so-called BiCMOS technology).

9.5.2   BJT Voltage–current Relationships

The static electrical characteristic of any three-terminal device employs one terminal as reference and measures the voltage of the other two.

The current entering or exiting the reference terminal is the addition of the currents through the others. For the bipolar transistor the emitter is often used as the reference, the base and emitter make the input port, and the collector–emitter realizes the output of a two-port representation (see Figure 9.34). Base and collector currents can be written as

$$I_B = I_B(V_{BE}, V_{CE}) \tag{9.42}$$
$$I_C = I_C(V_{BE}, V_{CE}), \tag{9.43}$$

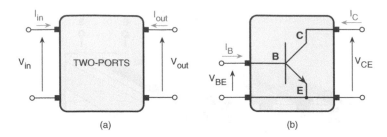

Figure 9.34   Two-port representation of an NPN transistor.

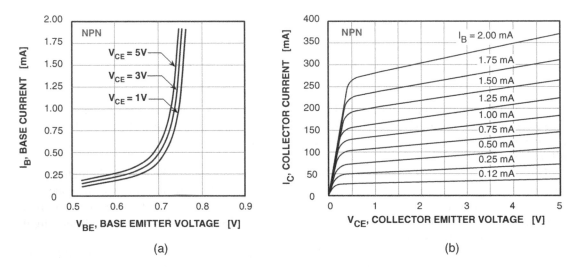

Figure 9.35   Input and output I–V responses of a typical bipolar transistor.

which are non-linear functions of the voltages V_{BE} and V_{CE}. Equations (9.42) and (9.43) give rise to a family of curves of current versus one of the voltages, with the other as the parameter. Figure 9.35 shows typical plots for a bipolar transistor. Notice that the base current increases rapidly when V_{GS} exceeds 0.7 V. Moreover, the control established by V_{CE} is almost irrelevant. Because of this it is convenient to use in the second set of curves the base current, instead of V_{BE}, as the control quantity. This is done in Figure 9.35(b), which gives the collector current versus V_{CE} with I_B as the parameter. Both of the diagrams are for an NPN transistor.

Observe that the collector to base voltage is

$$V_{CB} = V_{CE} - V_{BE}, \qquad (9.44)$$

and moreover

$$I_E = I_C - I_B. \qquad (9.45)$$

Since V_{BE} is almost constant, Figure 9.35(b) obtains the $I_C - V_{CB}$ plots by just a shift of the voltage axis. Another important observation concerns the value and the sign of V_{CB}. Notice that V_{CE} in Figure 9.35 is positive and larger than V_{BE} in a wide region of operation. There V_{CB} is positive; i.e., the collector that is n-type is at a higher voltage than the p-type

COMPUTER EXPERIMENT 9.5

Bipolar Transistor as a Two-Port

The bipolar transistor can be viewed as a two-port device with the base–emitter pair as input port and the collector–emitter pair as output port. This computer experiment enables you to derive the low–frequency two-port relationships. The scheme determines the voltage–current plot for each port, with constant voltage across the other port. The board uses two equal NPN transistors, one with constant collector voltage the other with constant base voltage. Sawtooth generators vary the port voltage while a current-to-voltage converter (I-to-V) measures the current with zero voltage drop across it. A supply generator determines the constant bias across the other port. In this virtual experiment you can:

- set the range of the sawtooth voltages;
- set the DC level of the other port bias;
- select the emitter current (SEL = 2) of Q_1 to verify the operating conditions of the output port when deriving the I–V relationship of the input port.

MEASUREMENT SET–UP

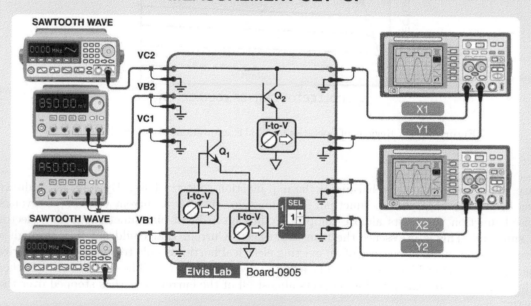

TO DO

- Draft the two–port representation of the bipolar transistor on paper, and identify the electrical variables of the two ports. Set the base voltage to 0.7 V. Derive the current–voltage plot of the output port with positive and negative collector biasing.
- Change the base voltage and measure the collector voltage at which the transistor goes into saturation.
- Derive the input port current–voltage relationship with positive and negative collector voltages. Estimate the ratio between collector current and base current.

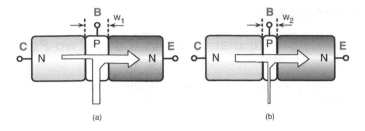

Figure 9.36 Currents in an NPN transistor: (a) with thick base; (b) with thin base.

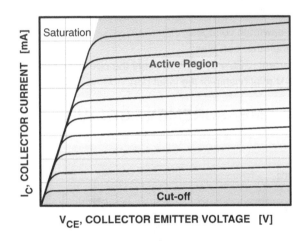

Figure 9.37 Regions of operation of a BJT: saturation, active, and cut-off.

base. This means a reverse biasing of the n–p junction collector base. By contrast, a direct bias drives the base–emitter junction. Summing up, there is a current entering a reverse-biased junction that exits almost completely through a direct-bias junction (from collector to emitter). This is in essence the transistor effect. Current that could not flow through a reverse-biased junction because of an existing electrical barrier is able to step over the barrier thanks to the direct-biased junction placed at a very short distance. The emitter, on the other side of the forward bias junction, collects almost all of the current that has stepped over the barrier.

This operation is pictorially described by Figure 9.36. The amount of current from collector to emitter depends on the base width. A thin base produces more emitter current than does a thicker base. If the voltages are the same, a thick base gives rise to lower currents, with, proportionally, a higher base current fraction.

The I–V output responses of Figure 9.35(b) distinguish three regions of operation. They are the counterparts of the regions defined for the MOS transistor. Illustrated in Figure 9.37, they are as follows.

- **Cut-off** This features an almost-zero collector current. It results from a reverse biasing of both collector–emitter and base–emitter junctions.

- **Active region** This is the normal region of operation that produces amplification. The collector current, controlled by the base current, slightly increases with the collector–emitter voltage. The MOS counterpart of this region is saturation.

- **Saturation** Unlike the MOS case, with a BJT this is where the collector current steeply increases. This is for low values of the V_{CE} voltages, making the base–collector junction direct-biased.

There are other regions of operation corresponding to voltages reversed with respect to the normal cases discussed above. For example, we have the active reverse, which exchanges the roles of collector and emitter. In this condition the transistor works in a similar manner to the normal active region, but the efficiency of the device (measured by the β of the transistor, to be discussed shortly) is much lower.

9.5.3 Bipolar Transistor Model and Parameters

Studies in solid-state physics derive the voltage–current relationships of the bipolar transistor. They account for the voltage–current non-linear response of a p–n junction and include the interfering action of the two junctions. Since the V–I relationship of a diode is exponential, for the bipolar transistor the approximate model also gives exponential relationships. The well-known equation of the Ebers–Moll model prescribes for the collector and emitter currents

$$I_C = -I_{BC} + \alpha_F I_{BE} \tag{9.46}$$

$$I_E = I_{BE} - \alpha_R I_{BC}, \tag{9.47}$$

where the transistor is assumed to be NPN. I_{BC} and I_{BE} are the currents of the reverse- and forward-biased diodes BE and BC.

$$I_{BC} = I_{CS}(e^{V_{BC}/V_T} - 1) \tag{9.48}$$

$$I_{BE} = I_{ES}(e^{V_{BE}/V_T} - 1), \tag{9.49}$$

where I_{CS} and I_{ES} are the saturation currents of the two diodes, both proportional to the junction areas:

$$I_{CS} = A_C I_{SS_C}, \quad I_{ES} = A_E I_{SS_E}. \tag{9.50}$$

Equations (9.47) and (9.46) state that the emitter and collector currents are those of one diode corrected by an "interfering" action due to the other. The parameters α_R and α_F quantify the strength of diode current interdependencies. When the base thickness becomes very large, the two αs go to zero and the emitter and collector currents are those of simple diodes.

The two αs are linked by the equation, not verified here,

$$\alpha_F I_{SB} = \alpha_R I_{SC} = I_S, \tag{9.51}$$

and therefore

$$I_E = \frac{I_S}{\alpha_F}(e^{V_{BE}/V_T} - 1) - I_S(e^{V_{BC}/V_T} - 1) \tag{9.52}$$

$$I_C = I_S(e^{V_{BE}/V_T} - 1) - \frac{I_S}{\alpha_R}(e^{V_{BC}/V_T} - 1). \tag{9.53}$$

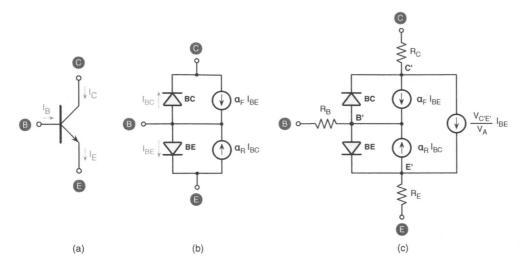

(a)   (b)   (c)

Figure 9.38   (a) Symbol for an NPN transistor; (b) its large-signal Ebers–Moll equivalent circuit; (c) Ebers–Moll equivalent circuit with included parasitic components.

Recalling that $I_B = I_E - I_C$, this results in

$$I_B = \frac{I_S}{\beta_F}(e^{V_{BE}/V_T} - 1) + \frac{I_S}{\beta_R}(e^{V_{BC}/V_T} - 1), \tag{9.54}$$

where

$$\beta_F = \frac{\alpha_F}{1 - \alpha_F} \quad \text{and} \quad \beta_R = \frac{\alpha_R}{1 - \alpha_R}. \tag{9.55}$$

Obviously, the values of α_F and α_R are lower than 1 because the interfering term cannot exceed the origin. Properly fabricated transistors have α_F close to 1 and α_R smaller than 1. The resulting values of β_F and β_R are a high and a low value respectively.

Notice that if $V_{BC} \ll 1$, the second terms of equations (9.53) and (9.54) are negligible, being equal to a saturation current. Therefore

$$\frac{I_C}{I_B} = \beta_F \quad \text{and} \quad \frac{I_C}{I_E} = \alpha_F, \tag{9.56}$$

showing that a small base current controls a large collector current if β_F is large. Moreover, being α_F close to 1, the emitter and collector currents are almost equal.

The above equations describing the NPN transistor of Figure 9.38(a) give rise to the large-signal equivalent circuit of Figure 9.38(b), which outlines the two contributions to emitter current and collector current.

The large-signal equivalent circuit does not consider parasitic terms. They are caused by the series resistances of the three terminals and parasitic capacitances. The addition of the parasitic components to the scheme of Figure 9.38(b) gives the equivalent circuit of Figure 9.38(c). The effective base, emitter, and collector are nodes B', E' and C' respectively. The equivalent circuit of Figure 9.38(b) does not account for a finite slope of the current curves in the active region. A slight reduction of the effective base width caused by the reverse biasing $V_{B'C'}$ explains this behavior. The limit, called the Early effect, is accounted for by the additional

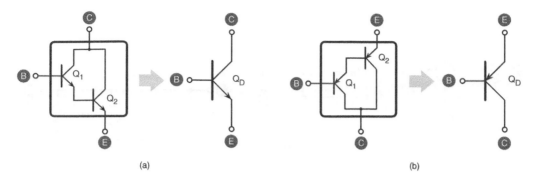

(a)   (b)

Figure 9.39   Darlington configuration for increasing the transconductance gain.

current source equal to $[V_{C'E'}/V_A] \cdot I_{BE}$. It is a resistance across the $C'E'$ terminals inversely proportional to the diode direct current.

The large-signal description with equations and the equivalent circuit of this subsection permit you to understand the bipolar transistor behavior. The use of the equation helps in identifying design strategies and directions. This is what you have to expect, because quantitative studies of circuits that include bipolar transistors require much more complicated equations and models. CAD tools incorporate complicated models with many parameters justified by complex physical descriptions or useful to fit the response of the model with experimental verifications. The results obtained are, obviously, much more precise than the ones given by hand estimation. Nevertheless, by having simple equations in mind it is possible to better assess the complex results obtained by CAD tools.

9.5.4   Darlington Configuration

The β_F of transistors is typically high, and is enough for many applications. There are situations that require very high gain, and instead of using multiple stages the requirement can be satisfied by technology providing devices with an extremely thin base. The result is a very high value of β. These components are called super-β transistors.

A second way to enhance the β is to use the so-called Darlington configuration, shown in Figure 9.39. Two transistors of the same type make up the scheme. The emitter current of the input device is the base current of the output device. Therefore

$$I_{E1} = I_{B2} = (\beta_{F1} + 1)I_{B1} \tag{9.57}$$

$$I_{C2} = \beta_{F2}I_{B2} = (\beta_{F1} + 1)\beta_{F2}I_{B1}, \tag{9.58}$$

showing an overall β equal to the product of the two βs.

The current through the combined collector is, indeed, the addition of the two collector currents

$$I_{C,D} = I_{C1} + I_{C2} = (\beta_{F1}\beta_{F2} + \beta_{F1} + \beta_{F2})I_B, \tag{9.59}$$

which is almost equal to the result of equation (9.58).

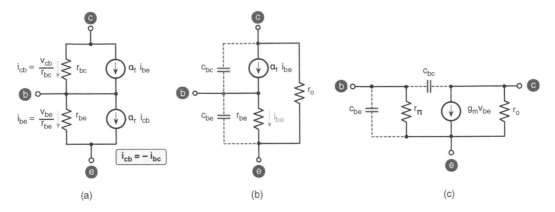

(a)   (b)   (c)

Figure 9.40   Small-signal equivalent circuits of the bipolar transistor: (a) small-signal Ebers–Moll model; (b) T-small signal model; (c) hybrid-π small-signal model.

9.5.5 Small-signal Equivalent Circuit of the Bipolar Transistor

The small-signal equivalent circuit describes the bipolar transistors when the amplitude of signals is small enough to justify a linearization of the set of equations describing the device. The low-frequency Ebers–Moll model gives rise to the scheme of Figure 9.40(a). The equivalent resistances, r_{be} and r_{bc}, replace the non-linear diode BE and BC. They, according to the model describing the diode, are

$$r_{be} = \frac{V_T}{I_{BE}} \quad \text{and} \quad r_{bc} = \frac{V_T}{I_{BC}}. \tag{9.60}$$

Since I_{BE} is a forward current and I_{BC} is a reverse current, r_{be} is small while r_{bc} is rather large. The parameters α_r and α_f are roughly the same as those in the large-signal model, α_R and α_F.

The linearization of parasitic components of Figure 9.38(b) adds to the scheme of Figure 9.40(a) linearized parasitic capacitances and resistances in series with the terminals. The current source accounting for the Early effect becomes a resistance between collector and emitter. It is called the output resistance, r_o. The contributions of the BC junction (i_{cb} and r_{bc}) are negligible. Without them, we obtain the so-called T-small signal model of Figure 9.40(b). The small-signal current entering the collector is the α_f fraction of the emitter current. The base current is $(1 - \alpha_f)i_e$.

Another circuit configuration that describes the small-signal behavior is that of Figure 9.40(c). It is the so-called small-signal hybrid-π model. The name comes from the π-scheme shape of the circuit. It results from the two-port description of the bipolar transistor. Base and emitter make the input port, and the collector and emitter are the output port. The circuit theory describes a two-port network as a "black box" whose equivalent circuit is specified by four parameters. The scheme of Figure 9.40(c) uses the hybrid (or h) parameters to describe the input network as a real voltage source and the output variable as a real current source.

Figure 9.34 recalls the two-port description that estimates the input and the output currents using voltages. The following equations describe the relationships outlining a small signal voltage and a current estimated using the other two electrical variables:

$$v_1 = h_{ie}i_1 + h_{re}v_2 \tag{9.61}$$

$$i_2 = h_{fe}v_1 + h_{oe}v_2. \tag{9.62}$$

They show that the output voltage may affect the input voltage. The output current may depend on the input voltage. The scheme of Figure 9.40(c) does not use the term controlled by the output voltage because its contribution is negligible. Including the parasitic terms, the input network is just a resistance in parallel with the small-signal capacitance c_{be}. The input controls the output current source though the parameter g_m, transconductance gain.

Suitable relationships link the small-signal parameters of the T-model and the hybrid-π model. By inspection of Figure 9.40(b) we obtain

$$i_e = \frac{v_{be}}{r_{be}} + \frac{v_{ce}}{r_o}, \quad i_b = (1 - \alpha_f)\frac{v_{be}}{r_{be}}. \tag{9.63}$$

The scheme of Figure 9.40(c) gives

$$i_e = v_{be}\left(g_m + \frac{1}{r_\pi}\right) + \frac{v_{ce}}{r_o}, \quad i_b = \frac{v_{be}}{r_\pi}; \tag{9.64}$$

equations (9.63) and (9.64) yield

$$r_\pi = \frac{r_{be}}{1 - \alpha_f}, \quad g_m = \frac{\alpha_f}{r_{be}}, \tag{9.65}$$

showing that the resistance at the input port r_π is that of the forward diode BE augmented by $1/(1 - \alpha_f)$. The transconductance gain is approximately the inverse of the forward diode equivalent resistance. Moreover, the small-signal current gain established by the output current source is

$$\frac{i_c}{i_b} = g_m r_\pi = \frac{\alpha_f}{1 - \alpha_f} = \beta_f, \tag{9.66}$$

which is almost independent of the emitter and collector currents.

9.6   BIPOLAR TRANSISTOR IN SIMPLE CIRCUITS

Let us consider now a BJT in simple circuits. The goal is to understand the analysis methods and perform preliminary study of circuits.

Consider the scheme of Figure 9.41(a). It is a general configuration with a single bipolar transistor. The circuit gives two possible schemes. If $R_C = 0$ the collector is at the fixed V_{DD} voltage. With $R_E = 0$ the source voltage is at ground.

The use of the Thévenin theorem transforms the resistive divider R_1–R_2 into the series of a voltage generator, V_{Th}, and a resistance, R_{Th}:

$$V_{Th} = \frac{V_{DD}R_1}{R_1 + R_2}, \quad R_{Th} = \frac{R_1 R_2}{R_1 + R_2}, \tag{9.67}$$

as shown in Figure 9.41(b).

Our task is to find voltages and currents. The nodal equations of the circuit are

$$V_{Th} - R_{Th}I_B - V_{BE} - R_E I_E = 0 \tag{9.68}$$

$$V_{DD} - R_C I_C - V_{CE} - R_E I_E = 0 \tag{9.69}$$

$$I_C + I_C - I_E = 0. \tag{9.70}$$

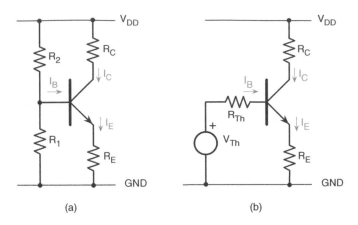

Figure 9.41 (a) Simple generic bias scheme with a single bipolar transistor; (b) scheme modified by the use of the Thévenin theorem.

In addition we should account for the transistor characteristics. The results are $\hat{I}_B$, $\hat{I}_E$, $\hat{I}_C$, and $\hat{V}_{BE}$ and $\hat{V}_{BC}$ (the "hat" superscript denotes DC or quiescent values).

The problem requires us to solve a non-linear system of equations, possibly by graphic methods. However, we can obtain quick approximate results. Let us suppose that

$$\hat{I}_C \simeq \hat{I}_E, \quad \hat{V}_{BE} \approx const = V_{BE,0}. \tag{9.71}$$

A bipolar transistor in the active region has $V_{BE,0} \approx 0.7$ V. Moreover, $I_E = \beta_F I_B$. Therefore, equation (9.68) becomes

$$V_{Th} - R_{Th}I_B - V_{BE,0} - \beta R_E I_B = 0, \tag{9.72}$$

which gives rise to the base current without knowledge of the details of the transistor characteristics. Knowing β is enough to estimate the current, which becomes

$$\hat{I}_B = \frac{V_{Th} - V_{BE,0}}{R_{Th} + \beta R_E}. \tag{9.73}$$

Then, $\hat{I}_B$ gives the voltages

$$\hat{V}_E = \beta R_E \hat{I}_B \tag{9.74}$$

$$\hat{V}_C = V_{DD} - \beta R_L \hat{I}_B \tag{9.75}$$

$$\hat{V}_B = \hat{V}_E + V_{BE,0}. \tag{9.76}$$

The currents through R_1 and R_2 are, obviously, $I_1 = \hat{V}_B/R_1$, $I_2 = (V_{DD} - \hat{V}_B)/R_2$.

The simplified solution does not require the solustion of non-linear equations or the use of graphic methods. It is an example of how a skilled designer analyzes a circuit at a "first glance." Knowing what the acceptable approximations are is a matter of experience that you can acquire with time. What is important is to consider problems with an engineer's attitude rather than a mathematician's. Certainly the results are not accurate, but often the errors are acceptable, especially in the preliminary phases of the design.

COMPUTER EXPERIMENT 9.6

Working with Simple Circuits

This computer experiment studies two simple circuits with transistors. One uses an MOS device, the other employs a bipolar transistor. The experimental setup enables you to set the DC and AC parts of the input signal and to establish the supply voltage. The resistance R_B in series with the base of the bipolar transistor makes the input signal generator real. An ideal current-to-voltage converter measures the base current. The bipolar scheme employs a resistance from emitter to ground to augment the emitter voltage when current increases. The MOS scheme is a simple inverting amplifier. The bipolar circuit is an inverting amplifier with emitter degeneration.
In this experiment you can:

- set the DC levels of supply voltages and input bias;
- set the value of load resistors, degeneration resistor and base resistor;
- define the amplitude of the (small) sine wave added to the input bias.

MEASUREMENT SET–UP

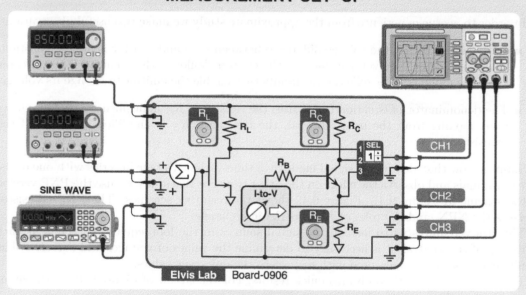

TO DO

- Set the load resistors and send the degeneration resistance R_E to zero. The supply voltage is 1.8 V. Zero the sine wave amplitude and measure the quiescent voltages in the MOS and the bipolar circuit. Explain why corresponding voltages differ.
- Choose a proper input voltage that brings the drain voltage to 0.9 V. Find the values of R_B and R_E that give rise to a 0.9 V voltage at the collector. Apply a 10 mV sine wave and observe the outputs. Change the load resistances and see the outcomes.
- Measure the DC drop voltages across R_C and R_E and justify the result. Apply the signal and measure the signal swings at various points of the schemes.
- Estimate the ratio between the signal currents across the base and the emitter.

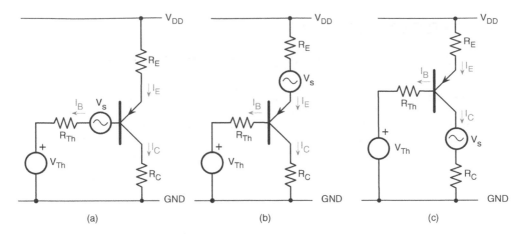

Figure 9.42   PNP general scheme with resistors and signal generator: (a) in series with the base; (b) in series with the emitter; (c) in series with the collector.

In order to accrue experience from the approximate study we make two useful observations.

- In the active region the voltage difference between base and emitter is almost constant. Therefore, if the base voltage changes, the emitter "follows" that change. This outcome gives the name *emitter follower* to circuits that enable the emitter to change its voltage.

- The denominator of equation (9.73) uses the resistance R_E multiplied by β. This shows that, looking from the base terminal, the resistance R_E is amplified by the β of the transistor.

Suppose now that we apply a signal made by a time-variant voltage in series with one of the three terminals and observe the effect on the other two. For this study we use the PNP version of the scheme in Figure 9.41. Figure 9.42 shows the resulting schemes. The PNP transistor replaces the NPN, and the roles of R_E and R_C interchange.

The study requires us to include the signal sources in the loop equations. Then at each time step of the analysis it is necessary to determine the values of voltages and currents. The results can possibly be measured with respect to the quiescent values.

The approximate method obtains quick results. For the circuit of Figure 9.42(a) the input loop equation obtains a new base current I'_B

$$V_{DD} - V_{Th} - V_S - R_{Th}I'_B - V_{BE,0} - \beta R_E I_B = 0. \tag{9.77}$$

The same equation gives the quiescent base current by setting $V_S = 0$. Therefore

$$\Delta I_B = I'_B - \hat{I}_B = \frac{V_S}{R_{Th} + \beta R_E}. \tag{9.78}$$

Knowing the base current or its change with respect to the quiescent value determines the emitter and collector currents and voltages at terminals or, if needed, the variations with respect to the quiescent values. The sign of the signal voltage is not specified in Figure 9.42(a) because it is assumed that it varies in time around zero. Equation (9.77) assumes that a positive V_S has the same polarity as V_{Th}.

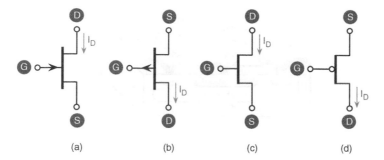

Figure 9.43   (a) and (c) Symbols for the N-channel JFET; (b) and (d) symbols for the P-channel JFET.

For the Figure 9.42(b) scheme it is easy to verify that the input loop equation is exactly the same as that of the previous case. The results coincide but the difference is that the currents through the signal sources are significantly different. That of Figure 9.42(a) must provide I_B. The signal generator of Figure 9.42(b) is required to supply I_E.

The scheme of Figure 9.42(c) places the signal generator out of the input loop. The base and emitter currents are not influenced, and the effect of the signal generator is just to change the collector voltage. This is almost inessential, and actually the configuration is never used in processing schemes.

The results with the generic configuration apply to the case of one of the bipolar terminals connected to a fixed voltage. They occur with $R_E = 0$, $R_C = 0$ or with a second battery biasing the base. These configurations are called *common emitter*, *common collector*, or *common base*, because the terminal at fixed voltage can serve as the reference of input and output voltages.

9.7   THE JUNCTION FIELD-EFFECT TRANSISTOR (JFET)

The Junction Field-effect Transistor (JFET) is an active device that almost operates as a MOS transistor but has a different physical structure. The device has three terminals, whose names are the same as those of the MOS transistor: source, drain, and gate. The JFET symbols, shown in Figure 9.43 refer to two different types of transistors, N-channel and P-channel. They look like the MOS symbol, but the bar used in the MOS to separate the source and the drain terminals from the gate is missing. This indicates that there is no galvanic insulation (that occurs when there is an oxide between terminals) between conductive channel and gate. The current flows from drain to source in the N-channel transistor and, vice versa, from source to drain in the P-channel component.

Figure 9.44 shows a physical three-dimensional sketch of the JFET for both types of transistor. Two p-doped regions squeeze a thin n-doped region, which makes the N-channel. An opposite scheme defines the P-channel JFET. At the two sides of the channel, n^+ regions connect the source and the drain to the metal terminals. The gate is the top p-doped material. To ensure insulation between gate, substrate, and channel all of the junctions must be reverse-biased.

Notice that the transistor channel is an actual doped material and not, as it is for the MOS, a region whose doping is reversed by the action of the gate. Suitable technological steps of

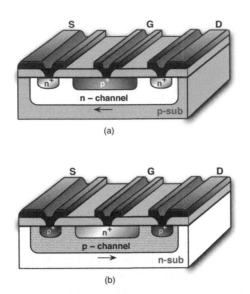

(a)

(b)

Figure 9.44   (a) Cross section of N-channel JFET; (b) cross section of P-channel JFET.

doping and diffusion on the low-doped substrate produce the device. The absolute value of the channel doping is higher than for the substrate and lower than for the gate.

The current flowing through the channel depends on doping, channel length, and channel cross section. Figure 9.45 outlines that the actual JFET cross section depends on bias conditions. It shows that the effective cross section of the channel is diminished because of the depleted regions on the two sides caused by reverse-biased diodes. If the voltage across the channel is low, the gate-channel and the bias of the diodes' channel–substrate biases are almost constant along the gate; thus the depletion regions are uniform all over the channel. Increasing the gate voltage augments the depletion on the top of the channel, and, consequently, the source-to-drain current diminishes. With a large voltage drop across the channel the depletion increases moving towards the drain. The result is that increasing V_{DS} does not augment the I_S or I_D much, because of a counterbalancing action of the channel resistance. For a given V_{DS} the channel vanishes near the drain: electrons ballistically cross the depleted region, as in a vacuum, attracted by the drain voltage. In that region of operation, called saturation, the increase of current is negligible, as happens for the MOS. The voltage that makes the channel vanish at the drain location is called the pinch-off, V_{GD}, voltage.

The input–output characteristics of the JFET show an equivalent behavior to that of the MOS: the gate current is almost zero and the curves with corresponding transistor sizes are analogous. The extraction of equivalent circuits and the study of simple circuits using the JFET are not done here, since they are similar to what has already been done for the MOS.

The functional difference between a MOS and a JFET is that the channel conduction in the former takes place at the interface between the semiconductor and the oxide that separates the channel from the gate. The current in the latter flows in the body of the channel. Therefore, since at the semiconductor–oxide interface there are more localized traps and crystal defects than in the body of the semiconductor, the JFET has better $1/f$ noise performances than does the MOS. This feature is valuable and is used in many low-noise realizations. However, the technology is less frequently used and, when combined with CMOS, requires additional masks.

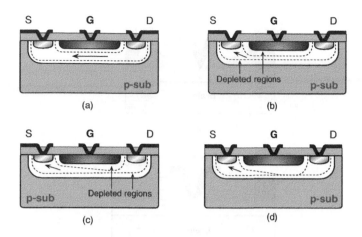

Figure 9.45   Depletion regions in N-channel JFET: (a) with low V_{DS} and low V_{GS}; (b) with low V_{DS} and large V_{GS}; (c) with high V_{DS} and V_{GS}. (d) Pinched-off transistor.

9.8   TRANSISTORS FOR POWER MANAGEMENT

The transistors studied in the previous sections are implicitly intended for a use with low or very low currents. For handling large currents, in the ampere or tens of amperes range, as required by power-intensive applications, it is necessary to ensure control with high power effectiveness. The power consumed by electronic circuits produces heat that must be carefully dissipated to avoid the high temperatures causing deterioration in performance and affecting long-term reliability. You have to be aware that hot-spot temperatures higher than $100°C$ quickly degrade semiconductor devices. Temperatures higher than what is prescribed reduce the time-to-failure exponentially. If a component's expected lifetime is t_L at the maximum of the temperature range T_{max}, the lifetime for a continuous exposure to an over-temperature T_{ov} (or, with a relative over-temperature $\Delta_{ov} = (T_{ov} - T_{max})/T_{ov}$) is estimated by

$$t_{L,ov} = t_L \cdot e^{-\Delta_{ov} \cdot E_a/(kT_{max})}, \tag{9.79}$$

where E_a is an activation energy parameter ranging from 0.7 eV to 1 eV for bipolar processes and 0.5 eV to 0.7 eV for CMOS processes. Therefore, the lifetime diminishes by a factor of 10 at steady operation with $\Delta_{ov} = 15\%$ for bipolar and $\Delta_{ov} = 10\%$ for CMOS circuits.

Heat sinks or fans for air cooling improve heat dissipation. They are useful in bulky systems but not in small-volume apparatus. Anyway, with tens of W or more at output, the power efficiency, η_P – the ratio between power at output and total power – must be very high.

Power depends on static and dynamic contributions. The static power is proportional to resistances; the switching on and off of capacitances causes dynamic consumption. A transistor with series on-resistance R_{on} and C_p parasitic capacitance that switches from ground to V_{DD} at the switching frequency f_{ck} consumes

$$P_{term} = R_{on}I_{av}^2 + C_p V_{DD}^2 f_{ck}, \tag{9.80}$$

where I_{av} is the total average current flowing between the supply generator V_{DD} and ground.

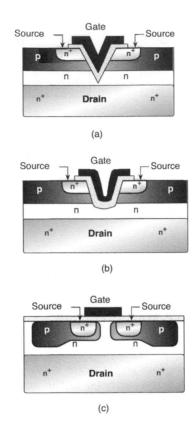

Figure 9.46   (a) Cross section of VMOS transistor; (b) variant with U-shaped groove; (c) cross section of DMOS device.

In order to reduce the on-resistance the device area must be large, but the increase augments the parasitic capacitances. Thus the optimal choice of the power transistor depends on both average current and switching frequency.

Requirements for very low series resistances and low parasitic capacitances have encouraged the development of special technologies and new devices. The goal is to maximize the cross section of the current flow with minimum highly resistive material along the current path. Obviously, with large device areas and a lot of power to dissipate it is possible to include a single or a few power transistors in an integrated circuit. Moreover, in addition to electrical studies the designer must carefully analyze the thermal power flow and verify temperature profiles. However, for the purposes of this study it is not necessary to go into the details of special technologies for power transistors and to study equations and methods for thermal flow analysis. It is enough to be aware of the problem and to know the features and configuration of the transistors mostly used in power schemes.

The Vertical Metal–Oxide–Silicon (VMOS) is one of the MOS structures used for power applications. It has a V-shaped gate for obtaining a vertical flow of the current. The VMOS is also used for medium power RF (Radio Frequency) applications because of its ability to switch very quickly from *on* to *off*. Figure 9.46(a) depicts an N-channel VMOS. There is a sequence of doped layers made by epitaxial growth. The substrate of the N-channel device is n$^+$ drain; on the top are two layers of n and p, and inside the p an n$^+$ diffusion making the source.

A V-groove shape vertically cuts the p layer to make room for the gate. The series resistance is very small thanks to the large cross section. Moreover, the V shape of the gate allows a larger amount of current from source to drain with a relatively low parasitic capacitance.

Figure 9.46(b) depicts a variant of the VMOS: the U version. The shape of the cut is U-like instead of V-like. The operation is almost the same. Notice that in both schemes the drain is in the substrate. This needs a back contact that collects the drain current. Some devices bring the current from the substrate to the surface to enable a top drain contact.

A second type of power transistor is the DMOS, shown in Figure 9.46(c). It looks like a VMOS with the n-type materials making a source and drain that are not separated by a V-shaped cut but by p-diffusions surrounding the source. The gate is flat, but the electric field and the physical structure favor the current flow downwards.

PROBLEMS

9.1 A fixed voltage equal to 0.65 V biases the series connection of a diode and a 100 Ω resistance to determine a DC current. The $A_D I_{SS}$ product of the diode is 10^{-11} A. Calculate the current and voltage at 300 K and 200 K. The parameters used in the Shockley equation are $K = 1.38 \cdot 10^{-23}$ J/K and $q = 1.6 \cdot 10^{-19}$ coulomb.

9.2 Use the numerical values of the Boltzmann constant, K, and electron charge, q of the previous problem to determine the thermal voltage at temperatures of $-40°C$, $0°C$, $25°C$, and $100°C$. Assume the use of a temperature dependent current source $I_s = 12(1 - T/260)$ mA to directly bias a diode whose $A_D I_{SS}$ product is $3 \cdot 10^{-12}$ A. Estimate the voltage across the diode at the above temperatures.

9.3 A sine wave generator with 1 V amplitude supplies the series of a diode and a 1 kΩ resistance. Use the equivalent circuits of Figure 9.6 to model the diode and $V_{on} = 0.6$ V, $R_{on} = 40\,\Omega$, $R_{off} = 56$ kΩ. Determine the waveform of the voltage across the resistance.

9.4 The voltage across a reverse-biased diode changes from -0.4 V to -1 V. Estimate the change of charge across the diode, assumed to be made by a sharp silicon junction. Use $A_J C_{J0} = 0.1$ pF and $V_J = 0.6$ V. What is the further change if the reverse bias increases by -0.6 V?

9.5 Determine the output voltage of the circuit below. The diode is modeled by a 200 kΩ resistor when the voltage across is less than 0.6 V, with zero current at 0.6 V. Above 0.6 V the current increases with slope 0.025 mA/V. The resistances are $R_1 = 2.3$ kΩ, $R_2 = 1.8$ kΩ, and $R_3 = 5$ kΩ. The input voltages are -0.8 V and 1.2 V.

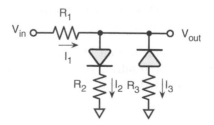

9.6 A zener diode drains a reverse current of 0.2 A with $V_R = -2$ V and 0.3 A with $V_R = -2.02$ V. For $V_R > -2$ V the device is like a normal diode. Derive a simplified equivalent circuit that describes the operation in the normal diode and zener regions.

9.7   The voltage regulator of Figure 9.17 uses a zener diode with $E_z = 1.8$ V and $R_z = 30\ \Omega$. Use $R_0 = 1$ kΩ and determine the voltage–current characteristic of the output port.

9.8   Suppose that the load of the zener regulator scheme of the previous problem (Figure 9.17) is just a 10 pF capacitor. The source voltage E_0 suddenly jumps from 1 V to 3 V. Determine the output transient response.

9.9   An N-channel transistor with threshold $V_{Th} = 0.45$ V discharges a capacitor from 1.2 V to zero. The gate of the transistor is at constant voltage 1.2 V. Write an approximate expression of the discharge transient, assuming the small-signal output conductance of the transistor in saturation is zero. Define a suitable normalization factor for the time axis.

9.10   Using the diagram of Figure 9.27 draw a diagram that plots the operating voltage as a function of the load resistance. Assume that, in saturation, the drain current extrapolated to $V_{DS} = 0$ V is $I_D = k(V_{GS} - V_{Th})^2$ and $V_{Th} = 0.6$ V. Suppose that the tick spacings on the horizontal and vertical axes are 0.2 V and 100 µA.

9.11   The circuit of the diagrams below controls the base current of the BJT transistor, whose β is 95. Estimate for scheme (b) the transistor currents with $V_{BE} = 0.6$ V, $V_{DD} = 3.3V$, $V_B = 1.8V$, and $R_C = 8R_E = 16$ kΩ. Calculate the value of R_B that gives rise to the same collector current. Determine for scheme (a) the collector voltage as a function of the supply voltage.

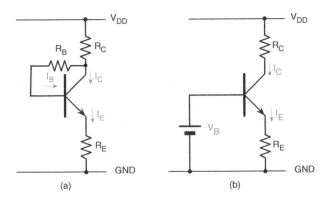

9.12   The hybrid π model of the bipolar transistor of Figure 9.38(b) models a device whose emitter is connected to ground. The voltages of base and collector are 0.6 V and 2.3 V respectively. The base current is 0.06 mA, and the collector current is 4.8 mA. Calculate the model parameters, assuming the BC diode current is zero.

9.13   Expand the Darlington configuration of Figure 9.39 to a scheme with three bipolar transistors, and estimate the resulting equivalent β. What happens if the collectors of extra transistors are connected to one of the supply voltages?

CHAPTER 10

ANALOG BUILDING CELLS

This chapter goes down the hierarchical ladder of the microelectronic building and looks at the internal structures of basic analog building blocks like operational amplifiers and comparators. We shall learn that a few basic cells construct the more complex analog functions. Among these we have the simple gain stage, the differential pair and the output stage. A suitable interconnection of those cells enables the designer to build functions, to optimize their features, and to compensate for limitations of which you will become aware while studying this chapter.

10.1 INTRODUCTION

The analog cells analyzed in this chapter are essential parts of the building blocks studied in Chapter 6. Figure 10.1 shows the typical architecture of an operational amplifier and that of a clocked comparator. Both use at their inputs a differential to single-ended stage, because they handle differential signals. After the input stage one or more gain stages amplify the signal. Finally, special cells like an output stage for op-amps or a latch for clocked comparators might complete the architecture. The differential input stage also provides gain: the overall gain is the product of the differential gain and the amplification of the following stages.

Cascading multiple stages can give rise to stability issues when the op-amp operates in certain configurations, as established by external components. To avoid problems the scheme uses active or passive networks to comply with stability conditions.

Understanding Microelectronics: A Top-Down Approach, First Edition. Franco Maloberti.
© 2012 John Wiley & Sons, Ltd. Published 2012 by John Wiley & Sons, Ltd.

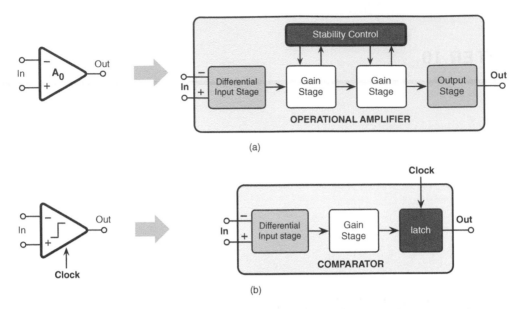

Figure 10.1   (a) Block diagram of a typical operational amplifier; (b) block diagram of a latched comparator.

Since every analog cell influences the overall operation, it is necessary to comprehend the operation of cells, to know their key features, and to be aware of limitations. In addition, the estimation of the effect of noise and spurs is required.

This chapter studies building cells as outlined in Figure 10.1. They are parts of complex architectures but also can implement a basic function when used alone. Indeed, there are situations that need not a complete op-amp but simpler and cheaper schemes.

10.2   USE OF SMALL-SIGNAL EQUIVALENT CIRCUITS

The previous chapter derived small-signal equivalent circuits for electronic devices. Before using these for circuit analysis it is worth considering some useful points. These circuits describe the response of non-linear elements with linear components because the underlying assumption is that with small variations of voltages or currents around a fixed operating point we can linearize the "large-signal" equations. The models result from the mathematical function "partial derivative," but, certainly, the result obtained cannot be viewed as a mathematical concept that holds just for infinitesimal changes. The outcome is an "engineering" way to describe the operation of circuits: variations in electrical quantities are small but not infinitesimal. Therefore, the equivalent circuit always gives approximate results with accuracies that get better and better with smaller and smaller signals.

Small-signal equivalent circuits replace non-linear devices in electrical networks. For their use it is important to remember that small-signal circuits deal with variations of voltages or currents and not with their quiescent values. Therefore, since nodes with constant voltage experience zero variations they are the same as ground for small-signal studies. Also, since a constant current undergoes zero variations, wires carrying a constant current operate as

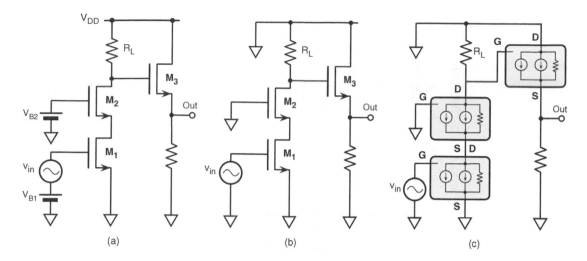

Figure 10.2 (a) Generic scheme with transistors; (b) its small-signal scheme; (c) replacement of transistors with their small-signal equivalent circuits.

an open circuit. Accordingly, we draft the small-signal equivalent circuit of a large circuit schematic by replacing devices with their linear models, connecting to ground nodes that are at fixed voltages, and opening connections carrying constant currents.

Figure 10.2 illustrates an exemplar schematic with non-linear components (MOS transistors), which is then translated into its small-signal equivalent circuit. Figure 10.2(a) is a generic configuration whose function and benefits are not relevant at this level. The scheme includes three N-channel transistors, two resistors and two bias generators. The voltages V_{B1}, V_{B2} and V_{DD} are constant. Therefore, the nodes connected to them are ground for small signals. No current generators are used in the circuit. Removing fixed voltage and current sources transforms Figure 10.2(a) into the scheme of Figure 10.2(b). Finally, Figure 10.2(c) replaces the three transistors with their small-signal equivalent circuits. The scheme is for small signals: i.e., variations. Therefore, the actual electrical quantities are the addition (or superposition) of quiescent values and small signal changes. Observe that, conventionally, the schemes indicate the names of large signals with capital letters and the names of small signals with lower-case ones.

What is illustrated by Figure 10.2 can be applied to any non-linear electronic scheme. The two steps needed are: first, short all the constant voltage sources, and, if any, open connections with constant current; second, replace all devices with their small-signal equivalent circuits.

10.3   INVERTING VOLTAGE AMPLIFIER

The first scheme studied is the inverting voltage amplifier. This, as the name implies, generates a replica of its input with augmented amplitude and opposite polarity. Without signal applied, all nodes, including input and output, measured using a ground level, have suitable quiescent voltages. Those levels have "zero small signals." Thus, quiescent values are disregarded when performing the small-signal analysis. However, they are important for

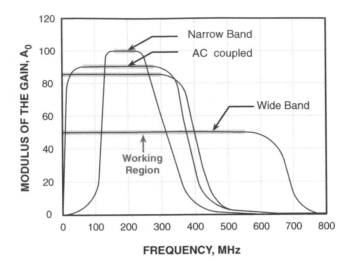

Figure 10.3 Gain versus frequency of possible inverting amplifier analog cells.

circuit operation and must be carefully chosen. For example, with a 0–1.8 V supply interval, having 0.9 V as quiescent output can be a good choice because it is likely to allow a large output swing in both positive and negative directions. However, the quiescent input is not any value but with single ended input is the constant voltage that brings the output to the desired 0.9 V. An output quiescent value close to a supply limit is problematic for symmetrical swing but can be the optimal choice for unipolar output signals. However, when the quiescent value goes too close to rails the saturation degrades the small-signal results because small-signal study relies on quiescent values.

Another relevant point concerns the use of small-signal results. If the gain of an amplifier is, for instance, −60, a tiny input equal to 2 mV causes an output change of −120 mV. If the quiescent output is 0.9 V we expect that the output voltage goes to 0.78 V, and that is what is likely to happen. However, if the quiescent output is at 0.15 V we cannot expect that a real circuit is able to go so near ground. Moreover, a larger input, say 20 mV, would cause a swing of −1.2 V, a large value that saturates output if the quiescent point is 0.9 V. Therefore, small-signal gain of amplifiers is an "engineering" concept that needs care in its use.

Remember

Small signal gain implicitly refers to linear operation. Large signal gain denotes increase of signal amplitude without care about distortion.

Frequency affects any small-signal gain. That can happen at low frequencies because of particular circuit configurations or the type of input coupling and at high frequencies where all circuits start lagging behind. Thus expected gain can be large starting from DC or can reach its value at an intermediate frequency. Inevitably the gain drops at very high frequency. Figure 10.3 plots typical situations. All the diagrams refer to simple amplifiers like the ones studied in this section, with gain that rarely exceeds 100 (40 dB). The thick gray lines indicate working regions where gain is almost constant. The plots refer to either narrow-band or wide-band gain stages. One of the responses denotes AC coupling.

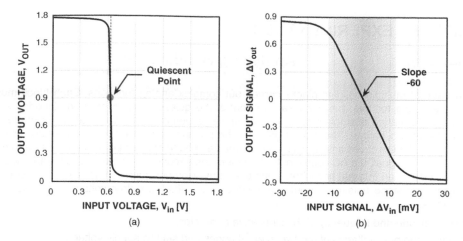

Figure 10.4 (a) Input–output characteristic of an inverting amplifier with absolute values on the axis; (b) expanded input–output characteristic.

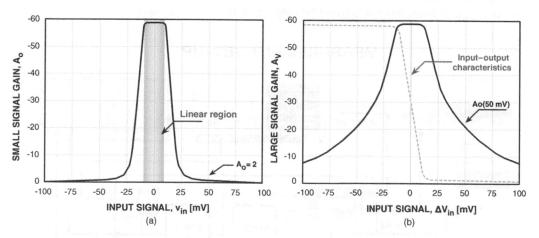

Figure 10.5 (a) Small-signal gain of an non-inverting amplifier; (b) large-signal gain of an non-inverting amplifier.

The slope of the static input–output transfer characteristics provides the DC gain at various quiescent points. Figure 10.4(a) shows a typical plot with actual values of input and output, both spanning the supply interval (0–1.8 V). Figure 10.4(b) expands the horizontal axis around the marked quiescent point. At small inputs the output is near V_{DD}; when V_{in} becomes large the output goes to zero. Figure 10.4(b) better quantifies the input–output relationship and delineates the region of almost constant slope. Despite a relatively low gain (equal to 60, 35.6 dB) the region with constant slope is small. Outside that tiny region output saturates.

For gain we use two definitions: small-signal gain, A_v, and large-signal gain, A_V, given by

$$A_v = \frac{dV_{OUT}}{dV_{IN}} = \frac{v_{out}}{v_{in}}, \quad A_V = \frac{\Delta V_{OUT}}{\Delta V_{IN}}. \tag{10.1}$$

A_v, the slope of the curves in Figure 10.4, is displayed in Figure 10.5(a). It is negative and almost constant in the ± 10 mV input range. Outside that, it quickly drops to low values.

COMPUTER EXPERIMENT 10.1

Analyzing Building Blocks

The cascade of multiple building blocks carries out complex analog functions. Each block must work properly and for this it is necessary to control the quiescent level of internal nodes and check the performance of single blocks. This computer experiment studies an amplifier made by the cascade of two stages. The input signal can be provided by a sine wave generator or the sweep output of a spectrum analyzer. A selector enables you to choose between one of the other and to send that signal, via AC coupling, to the first or the second stage input. Each stage needs the proper bias of an internal node that shifts the input or output quiescent levels. The first bias controls the first stage quiescent input, the second bias shifts the output voltage of the second stage. You can:

- set amplitude and frequency of the sine wave generator;
- regulate the bias voltages of the two internal nodes and set the supply voltage;
- set the parameters controlling the spectrum analyzer, including the amplitude of the swept signal provided on the back panel of the instrument;
- select the two-way input switch and the output switch.

MEASUREMENT SET–UP

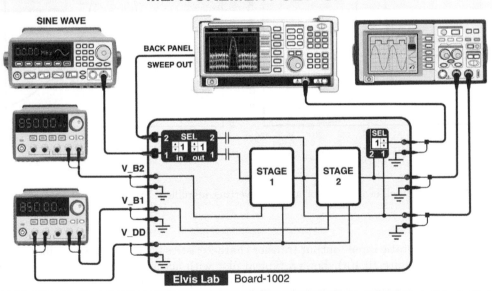

TO DO

- Use a sine wave with zero amplitude. Regulate the bias of the first stage until its output voltage is in the supply range. Observe the output voltage of the second stage and trim the first bias until the second stage output is "approximately" within the supply range. With the second bias bring the output voltage at half of the supply interval.
- Set a small value of the sine wave amplitude and measure the voltage gain of first and second stage in a wide frequency interval. Increase the amplitude and look at the spectra.
- Use the spectrum analyzer with its sweep output at the input of the first and second stages. Determine the frequency interval in which the gains are flat.

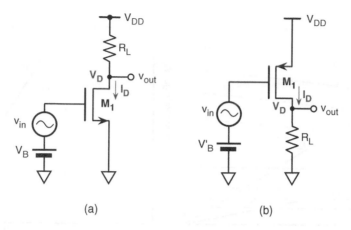

Figure 10.6   (a) N-MOS inverter with resistive load; (b) P-MOS inverter with resistive load.

With just 50 mV the gain becomes 2, diminishing by a factor of 30. The gray background area of Figure 10.5(a) highlights the working region that gives rise to low distortion.

Figure 10.5(b) displays the large-signal gain versus the shift with respect to the quiescent value. For low inputs large- and small-signal gains are equal. For wider inputs the large signal diminishes smoothly. That is a useful comparison. However, notice that the results hold for that particular quiescent value, the symmetric point of the input–output response.

10.4   MOS INVERTER WITH RESISTIVE LOAD

One of the simplest circuits that give rise to inverting amplification is the MOS inverter with resistive load. Figure 10.6 displays its complementary simple configurations. One uses an n-type the other a p-type transistor. These schemes were studied in Chapter 9. Here, in addition, they use a signal generator in series with the gate bias. The signal can be any waveform even if conventionally the diagram uses the sine wave symbol. Since the n-type and p-type circuits are complementary with one another, we can study the response of only one type because for the other the same procedure leads to the same expressions.

The input is the superposition of DC bias and signal. The DC study determines the quiescent point before; then, the small-signal analysis describes the variations brought about by a signal.

For the quiescent point estimation, already studied, we can use the load line

$$V_{DD} = V_{DS} + R_L I_D \tag{10.2}$$

and the current–voltage relationship of the transistor with $V_{GS} = V_B$

$$I_D = f_I(V_{DS}, V_B); \tag{10.3}$$

both establish the crossing of the two curves to obtain the quiescent point.

Notice that, in addition, the crossings of the load line with other I–V curves give rise to the input–output transfer curve (Figure 10.7). A gate voltage as low as V_{G1} brings the output near to V_{DD}. A gate voltage as high as V_{G6} takes the transistor into the triode region. Changing

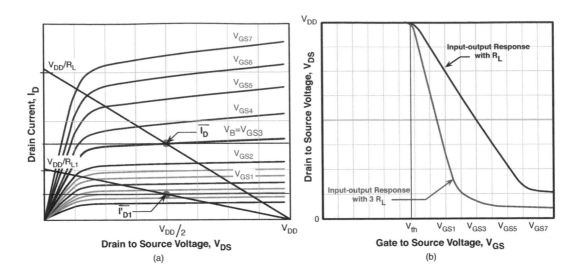

Figure 10.7   (a) Load lines and I–V responses of MOS inverter; (b) input–output transfer characteristics.

the value of load resistance, both quiescent and crossing points change. If, for example, the load resistance increases to $R_{L1} = 3R_L$, as depicted in Figure 10.7(a), the slope of the load line becomes three times lower. In order to keep a quiescent output at $V_{DD}/2$, the gate bias must diminish giving rise to a smaller quiescent current. Moreover, the lower slope of load line gives rise to crossing points that swing over the output range with small variations of gate voltage. Therefore, the input–output transfer curve becomes steeper, as shown in Figure 10.7(b).

10.4.1   Small-signal Analysis of the CMOS Inverter

To derive the small-signal equivalent circuit Figure 10.8(a) shorts V_B and connects to ground the supply terminal, since it is at constant voltage. The use of the equivalent circuit of Figure 9.25(c) leads to Figure 10.8(b). The scheme does not include the body transconductance generator because the source is connected to ground. Since the MOS body is also connected to ground, the small-signal generator $g_{mB}v_{GB}$ is equal to 0.

A simple analysis leads to

$$v_{out} = -g_m v_{in} \frac{r_{ds} R_L}{r_{ds} + R_L} = v_{in} A_v, \tag{10.4}$$

where $r_{ds} = 1/g_{ds}$. Notice that the sign of the gain is, as expected, negative because the transconductance generator points toward ground.

Equation (10.4) shows that the small signal gain increases with R_L, a feature already noticed with large-signal considerations. If $R_L \gg r_{ds}$ the modulus of the gain equals $g_m r_{ds}$, the transistor property called *intrinsic gain*, or A_{MOS}.

Using the intrinsic gain, the small-signal gain becomes

$$A_v = \frac{v_0}{v_{in}} = -\frac{g_m r_{ds}}{1 + (r_{ds}/R_L)} = -\frac{A_{MOS}}{1 + (r_{ds}/R_L)}, \tag{10.5}$$

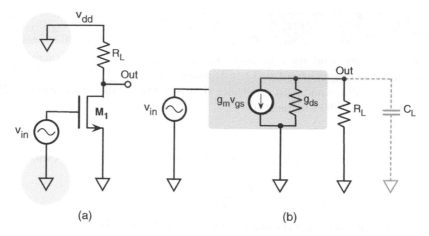

Figure 10.8 (a) Small-signal equivalent of the scheme of Figure 10.6; (b) use of the small-signal scheme of the MOS transistor.

highlighting that the maximum possible amplification is just the intrinsic gain. To obtain the result the circuit should use a load resistance much higher than r_{ds}, but since r_{ds} is high, very large loads and medium–low supply voltages would give rise to extremely low bias currents.

Notice that ...

the intrinsic gain g_m/g_{ds} of the input transistor is the maximum gain that a simple gain stage can generate. For higher gain a cascade of gain stages is necessary.

The small-signal gain obviously depends on frequency. The limit comes for parasitic and real capacitances possibly included in the scheme. The equivalent circuit is very simple; it is only necessary to account for capacitors connected across three nodes: input, output, and ground. Any element between input and ground does not produce consequences in the scheme because the input source is supposed to be an ideal signal generator. A capacitance between input and output is normally small and its effect is assumed to be negligible. The only frequency limit comes from the capacitance loading the output. This is in parallel with resistances $1/g_{ds}$ and R_L to receive the transconductance current $g_m v_{in}$. The output voltage becomes

$$v_{out}(s) = -g_m v_{in} \frac{R_{out}}{1 + sR_{out}C_L},\qquad(10.6)$$

where the output resistance R_{out} is given by

$$R_{out} = \frac{r_{ds}R_L}{r_{ds} + R_L}.\qquad(10.7)$$

Equation (10.6) shows that a single pole establishes the frequency response of the inverter. The low-frequency gain is $g_m R_{out}$. At the angular frequency of the pole, $\omega_p = 1/(R_{out}C_L)$, gain starts dropping with a 20 dB/dec slope. The Bode diagram crosses the zero dB axis at the angular frequency g_m/C_L. This is the unity gain of the inverter with resistive load. Notice that the result does not depend on the value of the resistive load. The reason is that if R_L increases the low-frequency gain increases but the time constant of the pole also increases by an equal extent. Thus gain goes up and pole frequency goes down.

Example 10.1

An inverter with resistive load uses an MOS transistor in saturation. Operating with an overdrive of 140 mV the drain current is 0.1 mA; moreover, $V_{DS} = 0.75$ V and λ is $0.107\ V^{-1}$. Determine the supply voltage and load resistance that realize a small-signal gain of -25. What is the intrinsic gain of the MOS transistor?

Solution

The values of the current and overdrive voltage determine the transconductance of the MOS transistor

$$g_m = \frac{2I_D}{V_{ov}} = \frac{0.2 \cdot 10^{-3}}{0.14} = 1.4 \text{ mA/V}.$$

The equivalent resistance that determines the expected gain must be

$$r_{eq} = \frac{A_v}{g_m} = \frac{25}{1.4 \cdot 10^{-3}} = 17.86 \text{ k}\Omega.$$

The output conductance, calculated by using the relationship $g_o = \lambda I_D$, is $10.66 \cdot 10^{-6}/\Omega$. The resulting intrinsic gain of the MOS is equal to $g_m/g_o = 131$. Therefore, the resistance to be placed in parallel with $1/g_o = 93.5$ kΩ that obtains 17.5 kΩ is $R_L = 21.51$ kΩ. The drain current 0.1 mA flowing through the estimated value of R_L determines a drop voltage of 2.21 V. The supply voltage V_{DD} becomes 2.96 V. The result is a bit surprising; a relatively low gain (25) requires a resistance value that leads to a large supply voltage, about four times more than the voltage across the transistor.

10.5 CMOS INVERTER WITH ACTIVE LOAD

The current–voltage curve of a MOS transistor in the saturation region is almost straight with a very low slope. That indicates high equivalent resistance (or low g_{ds}). Moreover, the line does not pass through zero, but its extrapolation crosses a current level that depends on overdrive. That is an ideal behavior to perform the task of the load line, because it is desirable to have a large load resistance. Moreover, at the same time, the current should have a suitable value so as to meet the required operational speed.

The above gives rise to the scheme shown in Figure 10.9, called an inverter with active load. The circuit resembles Figure 10.6 but instead of a resistance a complementary transistor makes the load. The figure illustrates the configuration with N-channel input and a complementary version with a P-channel input transistor. We study the scheme with the N-channel input. The values of bias voltages V_{B1} and V_{B2} should give rise to a "good" operating point. Suppose that the voltage V_{B1} drives the input transistor to establish the I–V curve of Figure 10.10. The I–V curves of the P-channel load can be drawn on the same I–V plane because of the relationships

$$-V_{DS,P} = V_{DD} - V_{DS,N} \tag{10.8}$$

$$-I_{D,P} = I_{D,N}, \tag{10.9}$$

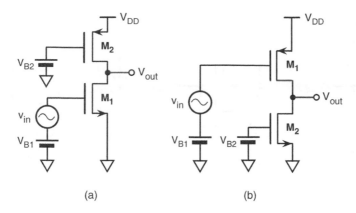

Figure 10.9   (a) CMOS inverter with active load and n-MOS as input device; (b) the same circuit with p-MOS as input device.

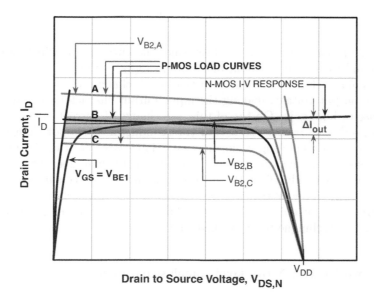

Figure 10.10   I–V responses of a CMOS inverter with active load for three different P-load biases.

which assert that the current is the same (with negative sign because of the complementarity of transistors) and the drain-to-source voltage of the active load is the difference between V_{DD} and $V_{DS,N}$. The results are flipped and shifted I–V curves of the P-transistor on the I–V input plane, as shown in Figure 10.10. The crossing of the input transfer characteristic and the P-channel I–V plot established by $V_{B2,B}$ gives a good quiescent point.

$V_{B2,A}$ gives an overdrive higher than what is necessary; the P-channel current in saturation is larger than that of the input device, making the crossing point close to V_{DD}. A reduced overdrive ($V_{B2,C}$) gives rise to the plot C with crossing point at too low a voltage. Both curves give rise to inadequate operating points, being the active load or the input transistor in the

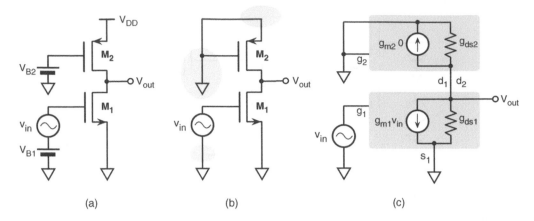

Figure 10.11   (a) Inverter with active load; (b) intermediate step for obtaining the small-signal equivalent circuit; (c) small-signal equivalent circuit – notice that the control of the active load transconductance generator is zero.

triode region. The optimal plot is, as has been said, the curve B for which both transistors are in saturation and the output is about at the mid-supply voltage.

The quiescent point of Figure 10.10 is such that small changes of input voltage determine large variations in output voltage, making the input–output characteristics steep and, consequently, giving rise to high gain. On top of this, the change of current in the linearity region is small. That is good news because transistor transconductances do not change significantly. Figure 10.10 outlines the operating region in gray. That is where the circuit should work, because both transistors are in saturation. Outside it, one of them goes into triode.

Since V_{B1} and V_{B2} critically affect the operating point, it is necessary to design the bias generators carefully. Often one of the supplies determines the quiescent current and the other supply ensures the value of quiescent voltage. The control of one of the two biases often establishes feedback, a technique to be studied shortly.

10.5.1   CMOS Inverter with Active Load: Small-signal Analysis

The small-signal analysis of the inverter with active load follows the steps needed to transform a large-signal scheme into its small-signal equivalent circuit. Figure 10.11 shows the details. The first diagram reports the large-signal scheme. Figure 10.11(b) replaces the constant voltage generators with ground connections, while Figure 10.11(c) replaces transistors with their equivalent circuits. The constant bias of the P-transistor nulls its transconductance generator. The bodies of both transistors, albeit not specified in the scheme, are at fixed voltages, thus making the body term $g_{mb}v_{gb}$ zero. The resulting simple scheme of Figure 10.11(c) that does not include capacitances is for low-frequency analysis.

The current of the transconductance generator that flows into the parallel connection of g_{ds1} and g_{ds2} determines the small-signal output voltage

$$v_{out} = -v_{in} \frac{g_{m1}}{g_{ds1} + g_{ds2}}. \tag{10.10}$$

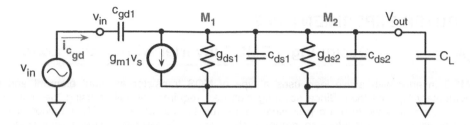

Figure 10.12   Equivalent circuit of the inverter with active load and parasitic capacitances.

The intrinsic gain of the input transistor, $A_{MOS} = g_{m1}/g_{ds1}$, used in equation (10.10), gives rise to the inverter gain, expressed by

$$A_v = -\frac{g_m}{g_{ds1} + g_{ds2}} = -A_{MOS}\frac{g_{ds1}}{g_{ds1} + g_{ds2}}, \tag{10.11}$$

showing that if g_{ds1} and g_{ds2} are equal the gain is just half of the intrinsic gain, a significant improvement with respect to the inverter with active load counterpart.

The above small-signal study is the starting point of the frequency analysis. That is done again by adding capacitors and parasitic terms to the small-signal equivalent circuit. Figure 10.12 shows the result with, in addition, the capacitive load C_L. The scheme includes capacitances loading the output and parasitic between input and output. The capacitance between input node and ground is not included because it would be in parallel with an ideal signal generator.

The effect of capacitance c_{gd}, neglected in the study of the inverter with resistive load, is actually minor because its value is much smaller than the capacitance on the output node. However, it is worth studying possible consequence of its connection between input and output. The analysis examines the current flowing through c_{gd} from input to output and the one from output to input. They are

$$i_{c_{gd}} = v_{in}(1 - A_v)sc_{gd}, \quad i_{c_{dg}} = -i_{c_{gd}} = v_{out}\left[1 - \frac{1}{A_v}\right]sc_{gd}, \tag{10.12}$$

that is, $(1 - A_v)$ times higher current through the same capacitor connected between gate and ground and almost the same current flowing through an equal capacitor connected between output and ground. This is what the Miller theorem states: a capacitance across two nodes is equivalent to two capacitors, one between input and ground, and the other between output and ground, with values as obtained from above equations.

The result is that large negative gains increase the input capacitance but give rise to a load almost equal to c_{gd} at output.

Miller effect

A capacitance across the input and output of inverting amplifiers causes an input load equal to the across capacitance multiplied by the gain. Output load almost equals the across capacitance.

Let us go back to the schemes of Figure 10.12. The capacitance at input is inessential. The one at output establishes a pole whose time constant is the output resistance multiplied by the output capacitance:

$$\tau_{out} = \frac{c_{gd1} + c_{ds1} + c_{ds2} + C_L}{g_{ds1} + g_{ds2}}. \tag{10.13}$$

COMPUTER EXPERIMENT 10.2

Designing a CMOS Inverter with Active Load

The MOS inverter with active load uses a type of MOS transistor as input element and a complementary type as load. Since securing high gain requires carefully biasing the gates of both transistors, it is worth doing preliminary experimental studies of the DC behavior of the selected devices. This computer experiment enables you to choose bias voltages and measure how the operating point changes with those biases. There are three different types of N-channel and P-channel devices to build the inverter. The two curve tracers permit you to display the static characteristics of the two devices selected. The Monitor Driver combines them and plots the two families of curves with the P-channel ones mirrored and shifted with respect to a supposed supply voltage. The result gives a set of possible quiescent points. In this virtual experiment you can:

- choose two transistors to test, one N-channel, the other P-channel;
- select the supposed supply voltage used to construct the combination of curves;
- set the parameters of the two curve tracers.

MEASUREMENT SET–UP

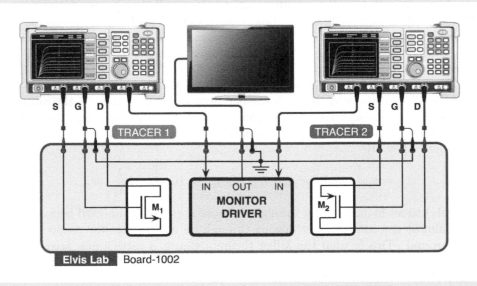

Elvis Lab   Board-1002

TO DO

- Choose a pair of transistors and examine their static curves. Suppose that the supply voltage is in the range 1.2 V and 1.8 V. Observe the combination of the two families of curves.

- Set the parameters of the two curve tracers and obtain a good number of crossing points. Derive the pairs of gate voltages that determine an operating point at $V_{DD}/2$ ($V_{DD} = 1.8$ V).

- Repeat the study with other pairs of transistors, and select the one that ensures the widest region of operation ($V_{GS,N} = 0.7$ V).

- Consider the spectrum analyzer which controls the N-channel device and set the gate voltages around 0.7 V with small steps. Draw the input-output curve supposing that the small steps mimic an input signal. What is the DC gain with a possible active load?

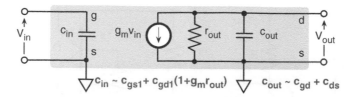

Figure 10.13   Two-port representation of MOS inverter.

The gain becomes

$$A_v(s) = \frac{A_v(0)}{1 + s\tau_{out}}, \tag{10.14}$$

where $A_v(0)$ is the already estimated DC small-signal gain of the inverting amplifier.

The low-frequency Bode diagram is $20\log_{10}|A_v(0)|$ until the frequency $1/(2\pi\tau_{out})$ is reached. Then it rolls down by 20 dB/dec to reach zero dB at $f_T = |A_v(0)|/(2\pi\tau_{out})$. The use of equations (10.11) and (10.13) yields

$$f_T = \frac{g_{m1}}{c_{gd1} + c_{ds1} + c_{ds2} + C_L}, \tag{10.15}$$

which, again, depends on input transconductance and is independent of output resistance.

In some cases it can be useful to handle the amplifier with a two-port equivalent. That, for the schemes of Figure 10.12, leads to the circuit of Figure 10.13. The input port is a capacitive load

$$c_{in} = c_{gs1} + A_v c_{gd} \tag{10.16}$$

made by input capacitance plus the capacitance across input and output multiplied by the Miller coefficient, $(1 + A_v)$, approximated by A_v.

A transconductance generator controlled by the input voltage connected in parallel to the output impedance and output capacitance realizes the output port. The output resistance, r_{out}, is $1/(g_{gd1} + g_{gd2})$. The output capacitance is

$$c_{out} = c_{gd1} + c_{ds1} + c_{ds2} + C_L. \tag{10.17}$$

The output port of the scheme of Figure 10.13 is a parallel connection. It can be easily transformed into a series connection by using the Thévenin equivalent. The result gives rise to a voltage-controlled voltage source $g_m r_{out}$ in series with output resistance loaded by c_{out}.

10.6   INVERTING AMPLIFIER WITH BIPOLAR TRANSISTORS

The bipolar counterpart of the CMOS inverter is similar to the scheme already studied, but the quantity relevant for input control is current and not voltage. The input port is the base-to-emitter pair while the output port is the collector-to-emitter pair. Obviously there are two schemes, one with an NPN and the other with a PNP at input, but the following considers only the NPN input configuration because extending the schematic and results to the complementary circuit with PNP at input is straightforward.

Figure 10.14 shows inverting amplifiers with resistive and active loads. There are various methods to deliver the quiescent base current. The circuits in Figure 10.14 simply use a

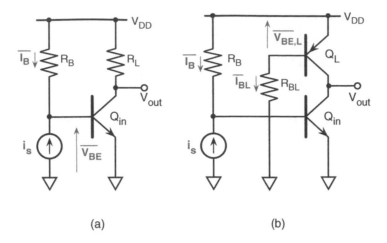

(a)   (b)

Figure 10.14 Inverting amplifier with NPN input bipolar transistor: (a) with resistive load; (b) with active PNP transistor load.

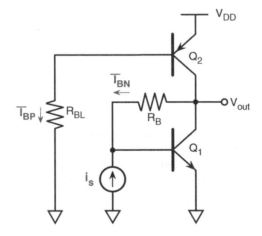

Figure 10.15 Auto-polarization of an inverting amplifier with NPN input bipolar transistor.

resistance R_B connected to the supply voltage. In addition, there is a current generator, i_s, that provides the small input signal. The quiescent bias current of the input transistor and the active load are, respectively,

$$\overline{I}_B = \frac{V_{DD} - \overline{V}_{BE}}{R_B}, \quad \overline{I}_{BL} = \frac{V_{DD} - \overline{V}_{BE,L}}{R_{BL}}. \tag{10.18}$$

A simple procedure analyzes the circuit. The load line and the I–V transfer characteristics of the BJTs determine the quiescent point. The designer places it in the best position that gives rise to the widest swing at the output while ensuring good linearity.

As happens for the MOS scheme, the circuits yield high gain when transistors are in the active region. For the configuration with active load, the crossing of the NPN characteristic and the PNP load curve is difficult to obtain. It is necessary to match the nominal collector currents of transistors. As a first approximation, it is required to verify the condition

$$\beta_{B,N}\overline{I}_B = \beta_{F,P}\overline{I}_{BL}. \tag{10.19}$$

Obviously the above condition is not enough because the matching of currents depends on the accuracy of the fabrication steps that determine the two β_F and the matching accuracy of resistors. Possible errors and drifts, even if small, make the operating point control problematic.

A possible method that keeps the quiescent value stable even with process variations and drifts is the self-bias scheme of Figure 10.15. The resistance R_{BL} sets up the PNP base current as before. Instead, the base current of Q_1 is

$$\overline{I}_{BN} = \frac{V_{DD} - \overline{V}_{CE,2} - \overline{V}_{BE,1}}{R_B}, \tag{10.20}$$

showing that the output voltage, which determines the voltage across R_B, gives rise to some extent to the bias current of Q_1 and, consequently, favors the current matching.

If output voltage tries to increase, I_{BN} also increases, contrasting with the change of output voltage. The method, as verified with a small-signal analysis, costs a gain reduction.

Example 10.2

An inverting amplifier like the one of Figure 10.15 has β_F of NPN and PNP transistors equal to 140 and 107 respectively. The supply voltage is $V_{DD} = 3.3$ V. Design the bias network that gives rise to $I_{C,1} = 2.53$ mA. Suppose that the V_{BE} voltages are equal to 0.7 V. Estimate the change of output voltage caused by an increase by 5% of R_{BL}.

Solution

The quiescent current, $I_{C,1}$, and the two β factors determine the base currents. These, neglecting the base currents with respect to the collector currents, are approximated by

$$I_{BN} \approx \frac{I_{C,1}}{\beta_N} = 18.1 \text{ μA}, \quad I_{BP} = \frac{I_{C,1}}{\beta_P} = 23.6 \text{ μA}.$$

Supposing that a good value of the quiescent output is the mid-supply voltage, 1.65 V,

$$R_B = \frac{V_{out} - V_D}{I_{BN}} = \frac{V_{DD}/2 - V_D}{I_{BN}} = \frac{1.65 - 0.7}{18.1 \cdot 10^{-6}} = 52.5 \text{ k}\Omega$$

$$R_{BL} = \frac{V_{DD} - V_D}{I_{BP}} = \frac{3.3 - 0.7}{23.6 \cdot 10^{-6}} = 110.2 \text{ k}\Omega;$$

where V_D is a diode voltage, supposed equal to 0.7 V.

These are resistors with affordable values for discrete realizations. For integrated circuits they are a bit high but still possible. If R_{BL} becomes 115.7 kΩ the collector current of the active load, supposing that V_{BE} in not affected, becomes

$$I_{CP} = \frac{V_{DD} - V_D}{R_{BL}}\beta_P = \frac{3.3 - 0.7}{115.7 \cdot 10^3}107 = 2.40 \text{ mA}.$$

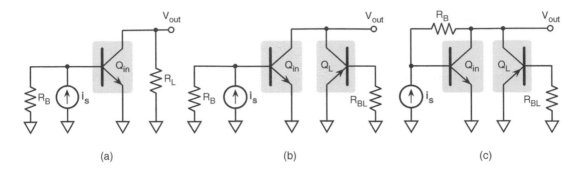

Figure 10.16   (a) Small-signal equivalent circuit; (b) the scheme of Figure 10.14; (c) the auto-polarized circuit shown in Figure 10.15.

The base current of the input transistor must become $I'_{BN} \approx I_{CP}/\beta_n = 17.14~\mu$A. The output voltage that gives rise to that base current is

$$V_{out} = V_D + I'_{BN} R_B = 0.7 + 57.5 \cdot 10^3 \cdot 17.14 \cdot 10^{-6} = 1.6~\text{V},$$

a relatively small change $(-50~\text{mV})$ that corresponds, approximately, to an increase by 5% of the voltage across R_{BL}.

10.6.1   Small-signal Analysis of BJT Inverters

For the small-signal analysis of the two schemes of Figure 10.14 and the one of Figure 10.15 we follow the already defined procedure. The first step, recalled again in Figure 10.16, consists of shorting the supply and bias generators. The current signal generator i_s, which is a signal, remains in the circuit. Notice that the symbol for a transistor inside the gray rectangles does not refer to the physical devices but, customarily, represents the small-signal equivalent circuits.

In Figure 10.16(a) and (b), the resistance R_B is in parallel with the input source. It drains part of the signal current to ground because the resistance of the input transistor is not zero. The consequent signal leakage is the cost that we have to pay for the simple method used for biasing Q_{in}. In addition we should account for the parallel resistance of the current source, which, in real cases, is not infinite. The above and this leak of signal current require a resistance at the input of Q_{in} much lower than the ones responsible for leakage. Otherwise the circuit loses part of the signal current directly at input before amplification. However, the requirement to ensure low input resistance is not difficult to fulfill for the schemes of Figure 10.16(a) and (b) because the small-signal input resistance of a BJT is r_π, whose value is of the order of a couple of $k\Omega$ or less. In any case we have to remember that the issue can be problematic for other circuit configurations that, for instance, have a relatively high resistive input.

Be aware that ...

the network used to bias a BJT stage withdraws part of the input signal current. Make sure that the lost current is only a small fraction of the total.

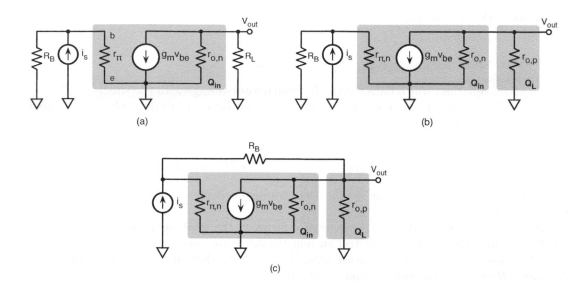

(a)   (b)

(c)

Figure 10.17   (a) Small-signal equivalent circuit of Figure 10.14(a) with the hybrid π equivalent circuit of Figure 9.40(c); (b) the schemes of Figure 10.14(b); (c) the auto-polarized circuit shown in Figure 10.15.

The resistance R_B in the scheme of Figure 10.16(c) is across input and output. Since the Miller effect reduces the value of R_B by the inverting gain, we can expect a limit to the gain. That is what actually happens. However, since R_B establishes an active part of the circuit operation its effect is studied at global level together with the active contribution of transistors.

The above preliminary analysis permits us to better understand the subsequent small-signal study and to be aware of the significance or irrelevance of terms that possibly give rise to second-order corrections. Now we can move on with the study of small-signal circuits.

Figure 10.17replaces the symbols for transistors used in Figure 10.16 with the π-model of bipolar equivalent circuit shown in Figure 9.40(c). The diagrams do not include parasitic capacitances because we analyze the low-frequency behavior first. For studying high-frequency operation we shall add parasitics afterwards. The equivalent circuit of Figure 10.17(a) refers to the resistive load scheme with R_L. The circuit of Figure 10.17(b), which describe the configuration with active load, is a small alteration of the previous one; it just replaces R_L with $r_{o,p}$ as the transconductance generator of Q_2 is zero. Notice also the difference between Figures 10.17(b) and (c). The former connects the resistance R_B to ground, and the latter has the same resistance across input and output, creating a feedback that influences performance.

For the schemes of Figure 10.17(a) and (b) R_B is in parallel with r_π, and therefore

$$v_{be} = i_s \frac{R_B r_\pi}{R_B + r_\pi}. \tag{10.21}$$

The use of the equivalent circuit of Figure 10.17(a) easily gives rise to the small-signal output voltage

$$v_{out} = -g_m v_{be} \frac{R_L r_{o,n}}{R_L + r_{o,n}}, \tag{10.22}$$

which, using the relationship (10.21), yields

$$v_{out} = -i_s \left[g_m \frac{R_B r_\pi}{R_B + r_\pi} \frac{R_L r_{o,n}}{R_L + r_{o,n}} \right]. \tag{10.23}$$

The above equation, which links the small-signal output voltage with the small-signal input current, defines a transresistance gain. This is the ratio between output voltage and input current, expressed by

$$G_r = \frac{v_{out}}{i_s} = -g_m \frac{R_B r_\pi}{R_B + r_\pi} \frac{R_L r_{o,n}}{R_L + r_{o,n}} = -G_{BJT} \frac{R_B R_L}{(R_B + r_\pi)(R_L + r_{o,n})}, \tag{10.24}$$

which, similarly to what was done for the MOS transistor, defines a BJT intrinsic transresistance gain, $G_{BJT} = g_m r_o r_\pi$.

The above supposes that a current generator furnishes the small signal at input. It can be a voltage source that yields the control current thanks to the relationship $v_s = v_{be} = i_s / r_\pi$. In such a case we can define the voltage gain, $A_v = v_{out}/v_s$, that equation (10.23) quantifies. Assuming $R_{BN} \gg r_\pi$, the voltage gain results as

$$A_v = \frac{v_{out}}{v_s} = -g_m \frac{R_L r_{o,n}}{R_L + r_{o,n}} = -A_{BJT} \frac{R_L}{R_L + r_{o,n}}, \tag{10.25}$$

which defines the intrinsic voltage gain of the bipolar transistor, $A_{BJT} = g_m r_{o,n}$.

The equivalent circuit of Figure 10.17(b) gives rise to the same results as the scheme of Figure 10.17(a) since the only difference is that $r_{o,p}$ replaces R_L. However, since the output resistance of bipolar transistors is likely to be much higher than the load resistances typically used, the values of small signal gains are higher. Supposing $R_B \gg r_\pi$ and $r_{o,n} = r_{o,p} = r_o$, the transresistance and voltage gain, disregarding the sign, are

$$G_r = \frac{1}{2} \frac{g_m r_o}{r_\pi} = \frac{1}{2} G_{BJT} \tag{10.26}$$

$$A_v = \frac{1}{2} g_m r_o = \frac{1}{2} A_{BJT}, \tag{10.27}$$

half of the respective intrinsic gains – the maximum achievable with a single-stage scheme.

The scheme of Figure 10.17(c) is a bit more complex than others because of the action of R_B. The equations describing the circuit are

$$v_{be} = r_{\pi,n} \left[i_s + \frac{v_{out} - v_{be}}{R_B} \right] \tag{10.28}$$

$$v_{out} \simeq - \frac{g_m v_{be}}{1/r_{o,n} + 1/r_{o,p} + 1/R_B} = -g_m v_{be} r_{eq}, \tag{10.29}$$

whose solution gives to the output voltage

$$v_{out} = \frac{-i_s g_m r_{\pi,n} r_{eq}}{1 + r_{\pi,n} g_m r_{eq}/R_B}. \tag{10.30}$$

The result can be simplified supposing valid the conditions

$$r_{\pi,n} g_m r_{eq} \gg R_B \quad \text{and} \quad r_{o,n} = r_{o,p} = r_o \tag{10.31}$$

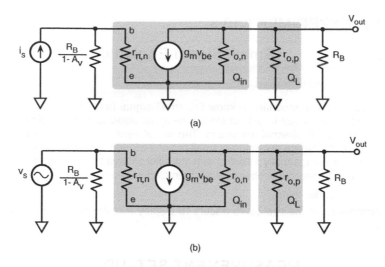

Figure 10.18   Small-signal equivalent circuit of the scheme in Figure 10.17 with the Miller equivalent of R_{BN}: (a) current source at input; (b) voltage source at input.

to obtain a more comprehensible equation:

$$v_{out} = \frac{1}{2} i_s g_m r_\pi r_o \frac{R_B}{R_B + 2r_o}. \tag{10.32}$$

Therefore,

$$G_r = \frac{1}{2} G_{BJT} \frac{R_B}{R_B + 2r_o}, \quad A_v = \frac{1}{2} A_{BJT} \frac{R_B}{R_B + 2r_o}. \tag{10.33}$$

The transresistance and voltage gain are lower by $(R_B/(R_B + 2r_o))$ because of the limiting action of R_B, connected between input and output. This is the price we pay for stabilizing the quiescent point against voltage fluctuations, aging, or temperature changes.

We further understand the cost of self biasing with the help of the Miller theorem. This splits R_B into a resistance at input and another at output, as shown in Figures 10.18(a) and (b). Since at input R_B is divided by $(1 - A_v)$, the leakage of current with a current generator is high. However, with a voltage source, as shown in Figure 10.18(b), the resistance across the input does not reduce the gain. That is not surprising because the use of a voltage source v_{in} across a finite input resistance r_π gives rise to a current $i_{in} = v_{in}/r_\pi$ independently of the rest of the circuit. Therefore, limitations caused at input by R_{BN} are eliminated.

The small-signal circuit schemes can give rise to two-port equivalent schemes. This is done as shown in Figure 10.19 for the small-signal circuits of Figures 10.17(b) and (c). The transconductance generator remains unchanged. The input and output resistances are the parallel connections of resistances at input and output. The two-port equivalent quickly shows the effect on the input generator or the consequences of possible impedances loading the output.

After the low-frequency study we analyze the frequency dependence. This is done by adding to small-signal schemes the parasitic capacitances and other capacitances that are possibly used by the circuit. Figures 10.20(a) and (b) show the small-signal circuits of the inverter with active load and the auto-polarized inverter already studied. The schemes symbolize transistors and outline parasitic capacitances. They also include at output a load capacitance C_L.

COMPUTER EXPERIMENT 10.3

Small–Signal Equivalent Circuit

For analog applications it is important to know the small–signal features of active devices. This computer experiment allows you to extract the small–signal equivalent circuit (in DC conditions) of two N-channel and two P-channel transistors. The experiment uses two curve tracers suitably regulated to measure the device operation in small regions around a defined operating point. The terminals of all devices are manually selected with the different available selectors. The input port of the transistors (gate-to-source) is obviously an open circuit. You can:

- select the device under test;
- set the parameters of the two curve tracers to display transfer responses at various bias conditions.

MEASUREMENT SET–UP

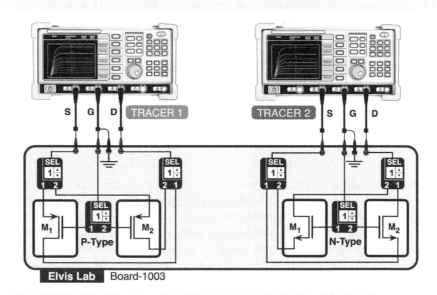

TO DO

- Select a pair of transistors. The ones with the same number are from the same technology. Compare the features of the N-channel and P-channel transistors assumed to have the same *W/L* ratio. Estimate the threshold voltages.
- Choose a convenient operating point and expand the region of study. Measure the slope of the curves, when they are almost straight lines and estimate the transconductance gain.
- Repeat the experiment for the other devices and different regions of operation (go close to the triode region and near sub-threshold).

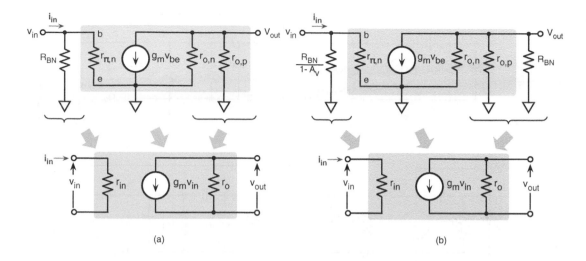

Figure 10.19 Two-port equivalents of small-signal circuits: (a) of Figure 10.17(b); (b) of Figure 10.17(c).

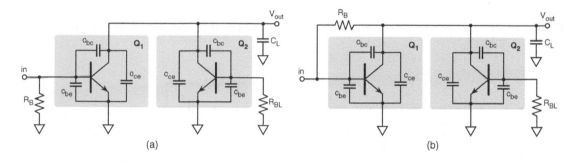

Figure 10.20 Small-signal equivalent circuits used to study frequency dependence of BJT inverters.

The diagrams of Figure 10.21(a) and (b) expand the symbolic equivalent circuits of Figure 10.20 by using for transistors the π schemes. The result is a little complex. However, we can identify in the diagrams three nodes: input, output, and ground. From output to ground there is an intricate passive network that can be simplified by supposing $r_{\pi,p}$ small. A small $r_{\pi,p}$ shorts, in practice, to ground the base of Q_2, thus removing the effect of c_{be2} and R_{BL}. With this approximation at output there is the parallel of resistances $r_{on,n}$ and $r_{on,p}$ and three capacitances c_{ce1}, c_{ce2}, and c_{bc2}. Figure 10.22 depicts the results of approximation performed on the schemes in Figures 10.21(a) and (b). The circuit uses

$$r_{in} = \frac{r_{\pi,n} R_B}{r_{\pi,n} + R_B}, \quad r'_{in} = r_{\pi,n}, \quad c_{in} = c_{be1} \tag{10.34}$$

for the input port, and

$$r_{out} = \frac{r_{o,n} r_{o,p}}{r_{o,n} + r_{o,p}}, \quad c_{out} = c_{ce1} + c_{ce2} + c_{bc2} + C_L \tag{10.35}$$

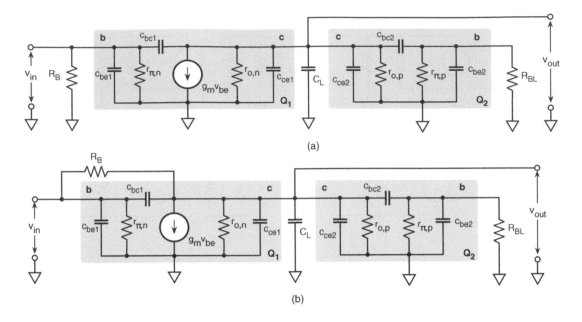

(a)

(b)

Figure 10.21 (a) Expansion of the scheme of Figure 10.20(a); (b) expansion of the scheme of Figure 10.20(b).

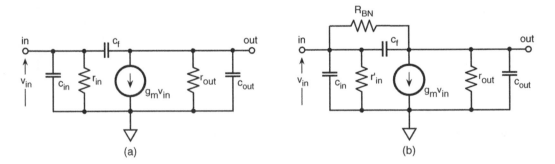

(a)　　　　　　　　　　　　(b)

Figure 10.22 Simplified small-signal equivalent circuits of the schemes of Figures 10.21(a) and (b).

for the output port. Moreover, $v_{in} = v_{be1}$ and $c_f = c_{bc1}$.

The circuits of Figure 10.22 hold for any input source. With a voltage source the parallel of resistance r_{in} and capacitance c_{in} is, obviously, ineffective, and components can be neglected. The nodal equation becomes

$$g_m v_{in} + v_{out}/r_{out} + v_{out} s c_{out} + (v_{out} - v_{in}) s c_f = 0, \tag{10.36}$$

which obtains

$$A_v = \frac{v_{out}}{v_{in}} = -g_m r_{out} \frac{1 - s c_f/g_m}{1 + s r_{out}(c_{out} + c_f)}. \tag{10.37}$$

The result shows a frequency response with one zero and one pole. The time constant of the zero is $\tau_z = c_f/g_m$. It is much smaller than the time constant of the pole, $\tau_p = r_{out}(c_{out} + c_f)$, because $r_{out} \gg 1/g_m$ and c_{out} is large as it includes the load capacitance C_L.

   The Bode diagram of frequency response is constant at low frequency, starts dropping by 20 dB/dec at the pole frequency and returns flat at very high frequency. Notice that the pole is in the right-hand half s-plane while, on the other hand, the zero is in the right-hand half s-plane, because of the minus sign in the numerator. This is bad news because the phase shift of zeros in the right-hand half s-plane causes phase shift as the pole. We do not know at the moment why that feature is not good. We shall understand that point when studying feedback. However, be assured that having the zero at a very high frequency circumvents the possible problem.

   If the input is a current source it is necessary to consider, in addition to the output nodal equation (10.36), the input nodal relationship

$$i_{in} = v_{in}sc_{in} + v_{in}/r_{in} + (v_{in} - v_{out})sc_f, \tag{10.38}$$

which determines a second pole in the output voltage. The result is

$$v_{out} = \frac{-i_{in}g_m r_{in} r_{out}(1 - sc_f/g_m)}{1 + s(\tau_1 + \tau_2) + s^2\tau_1\tau_2}, \tag{10.39}$$

where

$$\tau_1 + \tau_2 = r_{out}[c_{out} + c_f(1 + r_{in}g_m)] + r_{in}(c_{in} + c_f) \tag{10.40}$$

$$\tau_1\tau_2 = r_{in}r_{out}[c_{in}c_{out} + c_f(c_{in} + c_{out})]. \tag{10.41}$$

   The zero featuring in equation (10.39) is the same as in the voltage gain of equation (10.37). The poles, both in the left-hand half s-plane, are the solutions of the denominator of equation (10.39). The result is not difficult to obtain. However, we can make some approximations because in real schemes $r_{out}c_{out} \gg r_{in}(c_{in} + c_f)$. The calculation with the numerical values of typical cases shows that the two poles are both real and far away from each other. The one at low frequency dominates the response; the one at high frequency typically occurs after the zero.

   Figure 10.23 shows the modulus Bode diagrams of voltage gain and transresistance gain. The transresistance gain $G_{BJT} = g_m r_o r_\pi$ of the input transistor normalizes the latter diagram. As shown, the single pole dominates the voltage gain at low and medium frequency. Then the zero makes the diagram flat when the gain is already well below the 0 dB axis. However, in real circuits the gain drops at high frequencies because of additional parasitic poles, not revealed by the approximate model utilized.

   The Bode diagram of the normalized transresistance gain places the first pole approximately in the same position of the gain response. The zero occurs before the second pole. That is the more likely outcome.

Rules of thumb

The output time constant $\tau_{out} = r_{out}c_{out}$ gives the angular frequency of the dominant pole. The time constant $\tau_f = c_f/g_m$ mainly controls the angular frequency of the right-plane zero.

   The analysis of the scheme with the resistance R_B across input and output modifies the nodal equations: the $(v_{in} - v_{out})sc_f$ term becomes $(v_{in} - v_{out})(1/R_B + sc_f)$. The solutions lead to complex equations with, again, one zero and one or two poles. We do not go into details: with a simple study, if you desire, you can obtain the positions of the zero and the poles.

COMPUTER EXPERIMENT 10.4

Frequency Response

The small–signal response of analog cells always depends on frequency. At medium frequency the capacitors used by the circuit give rise to some frequency dependence. This computer experiment considers two inverters with active load: one with MOS transistors, the other with bipolar devices. The active loads of both circuits have their own supply generators to produce bias current. An analog adder adds bias voltage to a small sinusoidal signal; the result drives the gate of the input MOS and, through a resistance, furnishes the input current to the bipolar transistor. At source and emitter there are two RC networks. Notice that the outputs are both high impedance nodes. Therefore, the quiescent output voltage must be set by careful control of the DC bias conditions. This computer experiment enables you to:

- set all the supply voltages of the circuits generated by two accurate double voltage sources;
- select one of the currents or one of the output voltages for display on the scope;
- set the (equal) value of the capacitors and the base resistance, R_B. The resistors in parallel to the capacitors are both equal to 10 kΩ;
- set frequency of the sine wave; when the generator is turned on, the sine wave amplitude is 0.01 mV, a value small enough for performing a small signal study.

MEASUREMENT SET–UP

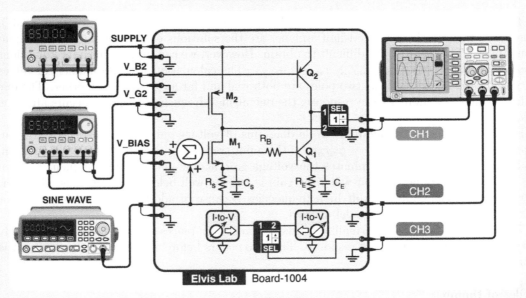

Elvis Lab   Board-1004

TO DO

- Set the supply and bias voltages. Leave off the sine wave generator and regulate the bias voltages V_{B2} and V_{G2}, so that the quiescent output nodes are about 0.7 the supply.
- Apply the sine wave signal and measure the low frequency voltage gain. Increase the frequency until the gain drops by 3 dB. Verify that the phase shift is close to 45°.
- Change the value of capacitors and base resistance to estimate the effect on frequency response. Extend the study ahead of the unity gain frequency.

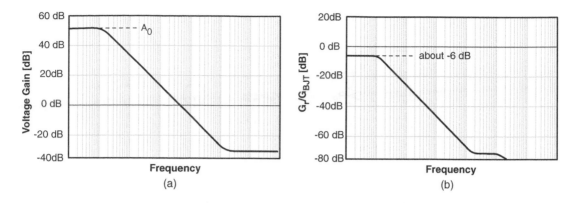

Figure 10.23 Amplitude Bode diagrams: (a) of equation (10.37); (b) of equation (10.39).

What this subsection has taught us is that with the small-signal equivalent circuit we can derive analytical equations that describe relevant features like the gain or the parameters of a two-port description of our circuit. If the schematic includes few elements the equations are simple but approximate. If we include extra elements (such as parasitic capacitances) the description becomes complex, and we obtain equations with many terms that are difficult to interpret. These must be approximated with reasonable assumptions to derive rules of thumb that predict operation and what happens when changing components' values. Obviously making approximations gives rise to inaccuracies, but that is the only way to get usable results. With computer simulations we can obtain quantitative performances.

10.7   SOURCE AND EMITTER FOLLOWER

Inverting amplifiers give rise to relatively large gains because of the transconductance gain and resistance of the stage. The output resistance must be high because, actually, it directly enhances the gain. However, there are situations where gain is not an important feature but, instead, the design interest is in ensuring high input and low output resistances. These features are what the follower stage provides. These schemes are so-called because the output voltage is almost a replica of the input – or the output follows the input.

Figure 10.24 shows the circuit configuration of simple followers made with n-MOS and NPN bipolar transistors. Obviously, we also have complementary configurations made by p-MOS and PNP devices. As usual, we consider just one type because the study of the others is straightforward. What we have in the figure are two different versions of a follower. One simply uses a resistance between output and ground, and the other employs a current generator. We shall see shortly how a current generator is realized with transistors. The supply rail (V_{DD}) directly biases the collector or the drain. The input signal is applied to the gate or the base, superposed with a suitable bias, and the output is taken at the source or the emitter.

The methods already discussed for circuits studied previously give rise to the quiescent point. They need to solve non-linear equations or obtain the result in a graphic manner. Here, we analyze the scheme in a different way, just by making simple considerations. If the circuit is in normal conditions of operation with current through source or emitter, the bias voltage

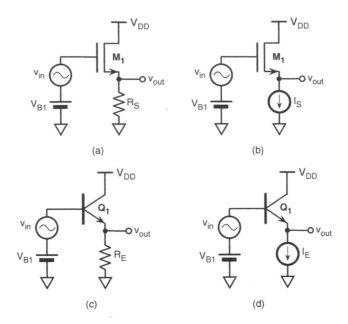

Figure 10.24 Circuit schematic of source follower and emitter follower with resistive and current source bias.

V_{B1} and the signal will have a level that drives the device suitably. The $\overline{V}_{GS}$ or $\overline{V}_{BE}$ DC voltage that gives rise to the flowing current establishes a voltage drop between input and output. Therefore

$$V_S = V_G - \overline{V}_{GS} \quad \text{for the MOS circuit} \tag{10.42}$$

$$V_E = V_B - \overline{V}_{BE} \quad \text{for the BJT circuit,} \tag{10.43}$$

giving, to a first approximation, the output voltage as the input shifted down by about the MOS threshold or an on-diode voltage.

A more precise study shows that the shift can depend on input voltage and current through the active devise. For the bipolar transistor, V_{BE} does not change with the input voltage but varies with the logarithm of current. If we use a reference base-to-emitter voltage $\overline{V}_{BE}$ corresponding to a given emitter current $\overline{I}_E$, it results in

$$\Delta V_{BE} = V_T \ln\left(\frac{I_E}{\overline{I}_E}\right); \tag{10.44}$$

changing current by a factor e modifies the input–output shift by $V_T \simeq 26$ mV.

For the MOS scheme, assumed in saturation, V_{GS} changes because the threshold depends on source-to-substrate voltage (the so-called body effect) and because the overdrive voltage, V_{ov}, depends on current:

$$V_{ov} = \sqrt{\frac{I_D}{\mu C_{ox}(W/L)}}, \tag{10.45}$$

if current doubles the overdrive becomes $\sqrt{2}$ larger.

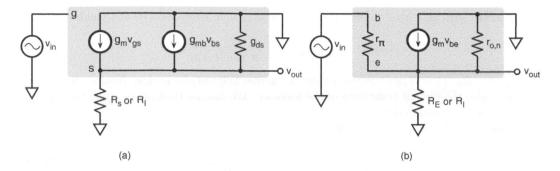

Figure 10.25 Small-signal equivalent circuit of source and emitter follower.

The above does not give the exact output voltage for the given bias and signal. It just indicates what the designer can expect and gives the possible level of confidence on guessed results. If the BJT follower uses a "normal" current the expected shift in quiescent conditions is about 0.7 V. If the aspect ratio of the MOS is suitable for the quiescent current the overdrive is likely to be a couple of hundreds of mV, giving rise to a quiescent shift equal to the threshold plus overdrive. That is, often, enough for hand calculations. Precise results are obviously important, but in many circumstances it is necessary to perform rough estimations before moving to more accurate studies. Moreover, simple considerations give rise to design recommendations such as, for the follower, the ones given below.

■ A constant input–output shift requires constant current in the active device. This is difficult to realize with a resistive load or a very high frequency. If the input changes output follows, but because of the resistive load current also changes. Remember that an output capacitive load drains current proportional to frequency. At 1 GHz a capacitance as small as 1 pF becomes an impedance of 160 Ω.

■ The so-called body effect in a MOS transistor changes the threshold because of the dependence of V_{Th} on V_{SB}. We avoid the limit with a transistor inside a well connected to the source. However, the follower must drive the parasitic capacitance of the well substrate diode.

10.7.1 Small-signal Equivalent Circuit of Source and Emitter Follower

Figure 10.25 shows the small-signal equivalent circuits of source follower and emitter follower. The schemes are valid with either a resistance or a current generator between output and ground. In the latter case the resistance of the diagram depicts the finite output resistance of the current source, R_I. With an ideal current source that resistance is infinite. The equivalent circuit of the MOS version includes the transconductance generator controlled by the body voltage because the body-to-source voltage can vary. As usual, the circuits do not include parasitic capacitances because we describe the low-frequency behavior first.

The circuit analysis of the network of Figure 10.25(a) gives rise to

$$v_{out} = [g_m(v_{in} - v_{out}) - g_{mb}(v_b - v_{out})]\frac{R_S}{1 + R_S g_{ds}}, \tag{10.46}$$

where the term with g_{mb} goes to zero if $v_b = v_s$. For $v_b = 0$, equation (10.46) yields

$$A_{v,M} = \frac{v_{out}}{v_{in}} = \frac{g_m r_{eq,M}}{1 + (g_m + g_{mb}) r_{eq,M}}, \tag{10.47}$$

where $r_{eq,M} = R_S/(1 + R_S g_{ds})$ is the resistance across the output nodes. Notice that such a resistance is also the output resistance of the follower. We discuss that property of the follower in detail in the next subsection.

The equivalent circuit of 10.25(b) leads to

$$v_{out} = \left(g_m + \frac{1}{r_\pi} \right) (v_{in} - v_{out}) \frac{R_E r_{o,n}}{R_E + r_{o,n}}, \tag{10.48}$$

which gives the voltage gain

$$A_{v,B} = \frac{v_{out}}{v_{in}} = \frac{(g_m + 1/r_\pi) r_{eq,B}}{1 + (g_m + 1/r_\pi) r_{eq,B}}, \tag{10.49}$$

where $r_{eq,B} = R_E r_{o,n}/(R_E + r_{o,n})$ is, again, the resistance across the output nodes.

The above equations show that the gain of source follower and emitter follower are positive but not unity. To give rise to a unity value it is necessary that the denominators of equations (10.47) and (10.49) are very close to the numerators. For the MOS stage it is required that $v_b = v_s$ and $g_m r_{eq,M} \gg 1$. Similarly, for the bipolar scheme, $g_m r_{eq,B}$ must be large. Since the transconductance of a bipolar transistor is larger than for MOS with the same bias current, the emitter follower performs better than the source follower.

10.7.2 Small-signal Input and Output Resistance

We have seen that it is often convenient to characterize a one-input one-output linear block with its two-port equivalent. The two-port example of Figure 10.26(a) uses the input resistance and, in parallel, a current generator controlled by output current (it is a Current-Controlled Current Source, or CCCS). The output port is a Norton equivalent made by the parallel of a Voltage-Controlled Current Source (VCCS) and an output resistance. There are other two-port equivalents with a CCCS at output or with voltage sources at input or output controlled by voltage or current (VCVS and CCVS); however, all of these schemes have input and output resistances. The two ports have been recalled here because through them it is possible to outline methods that determine the input and output resistances.

The estimation of the input resistance of the two-port scheme of Figure 10.26(a) requires us to null the input CCCS. The circuit of Figure 10.26(b) does that because it uses a test voltage at input and opens the output port to make the output current zero. A test current i_x injected at the input terminal and the determined voltage give rise to the input resistance. Figure 10.26(c) shows how to measure the output resistance. It uses a test voltage generator and shorts the input terminals to make the output VCCS zero. After estimating i_x, the ratio between test voltage and test current gives the output resistance.

Let us go back to the followers to observe that for the scheme of Figure 10.25(a) we have, obviously, infinite input resistance. For the output resistance we short the input terminals to make v_g zero. The result is that only the source voltage controls the two transconductance generators. Both inject their signal current, $-v_x g_m$ and $-v_x g_{mb}$, into the output node. The test generator results in

$$i_x = v_x g_m + v_x g_{mb} + v_x/r_{eq,M}. \tag{10.50}$$

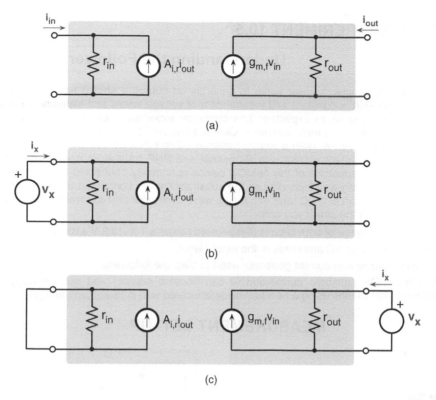

Figure 10.26   (a) Small-signal dual-port scheme; (b) configuration for the input resistance estimation; (c) configuration for measuring the output resistance.

Thus output resistance is estimated as

$$r_{out} = \frac{r_{eq,M}}{1 + (g_m + g_{mb})r_{eq,M}}. \tag{10.51}$$

If $g_m r_{eq,M} \gg 1$ we obtain the approximated expression

$$r_{out} = \frac{1}{g_m + g_{mb}}, \tag{10.52}$$

a result to keep in mind: output resistance of the source follower is the inverse of the addition of the two transconductances, g_m and g_{mb}, provided that the body is not connected to the source. With body connected to source, we simply have $r_{out} = 1/g_m$.

In order to quantify the value of output resistance, we should remember that for MOS transistors in saturation it satifies

$$g_m = \frac{2I_D}{V_{ov}}, \tag{10.53}$$

showing that if the drain current is 0.1 mA and the overdrive voltage is 200 mV, the output resistance is 1000 Ω, a relatively low value that can be achieved with a bias current whose value is affordable for medium-power applications.

The calculation of the output resistance of the BJT scheme shown in Figure 10.25(b) follows the same steps previously studied. Instead of the transconductance body generator the circuit

COMPUTER EXPERIMENT 10.5

Understanding the Follower

The source and the emitter follower serve to generate, at output, a shifted replica of the input voltage. Ideally the shift is constant and independent of voltage levels and frequency. However, real circuits do not operate as expected. This computer experiment allows you to understand features and limits with two simple schemes. One is a source follower (with an MOS transistor) the other is an emitter follower (with a bipolar transistor). The schemes use an N-channel and an NPN device. The equivalent circuits with P-channel and PNP transistors operate in a similar manner. However, the substrate of this N-MOS device is normally connected to ground. For P-MOS devices or with twin-well technologies the substrate can be connected to the source. The input signal is the addition of a DC term and a sine wave. A resistor and a current source bias the circuit. For the study of the circuit you can:

- set the supply voltage of both circuits (the allowed range is 1.2 – 3.3 V ±10%);
- set the DC level and AC amplitude of the input signal;
- choose resistance and current generator used to bias the followers;
- define the MOS substrate connection (it can become source) and set the body effect coefficient (that means using a new transistor fabricated with different technology).

MEASUREMENT SET–UP

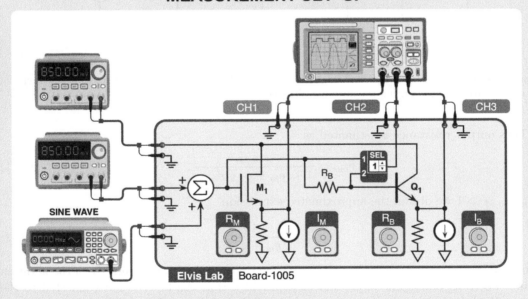

Elvis Lab   Board-1005

TO DO

- Set the supply voltage at 1.5 V and zero the sine wave amplitude. Determine the input–output response with ideal bias current (R_M and R_B infinite); use 0.05, 0.1, 0.5 and 1 mA.
- Repeat the measures with finite values of R_M and R_B and zero current sources. The resistances must establish the same set of DC currents.
- Apply the sine wave and observe the outputs with various sine wave amplitudes and DC input levels. Measure the response at different frequencies with a small input amplitude.
- Connect source and substrate or use MOS transistors with different body effect coefficient.

has r_π, which gives rise to a similar limit. It drains current from the output node toward ground equal to v_x/r_π. The current of the test generator becomes

$$i_x = v_x g_m + v_x/r_\pi + v_x/r_{eq,B}, \tag{10.54}$$

which yields

$$r_{out} = \frac{r_{eq,B}}{1 + (g_m + 1/r_\pi)r_{eq,B}}. \tag{10.55}$$

In order to get an approximate expression of the result we use a condition stating that $g_m r_\pi = \beta_f$ and we suppose that $g_m r_{eq,M} \gg 1$. The second and third terms of equation (10.54) become negligible, yielding

$$r_{out} = \frac{1}{g_m}, \tag{10.56}$$

whose value is V_T/I_C. For a bias current of 0.1 mA the output resistance is 260 Ω, lower by a factor of four than the estimated MOS counterpart.

The calculation of the input impedance uses the setup of Figure 10.26(b) implemented in the scheme of Figure 10.25(b). The nodal equation of the input is

$$v_x = i_x r_\pi + i_x(1 + g_m r_\pi)r_{eq,B}, \tag{10.57}$$

where, remember, $r_{eq,B} = R_E r_{o,n}/(R_E + r_{o,n})$. Therefore, as $g_m r_\pi = \beta_f$, it results in

$$r_{in} = r_\pi + (1 + \beta_f)r_{eq,B}. \tag{10.58}$$

The result is that the input resistance is given by r_π plus the resistance at the output node multiplied by $\beta_f + 1$. Since the multiplication factor can be as large as 100 or even more and output resistance is large, the input resistance, dominated by the second term, is very large.

In summary, both MOS and bipolar followers obtain a very high input resistance, a low output resistance, and a gain slightly lower than 1. Therefore, both followers are suitable solutions for implementing voltage interfaces able to "read" input in voltmetric manner and to generate output in an ideal voltage fashion.

A significant feature of followers is the DC voltage shift between input and output. The shift is downward for n-type input circuits and upward for the complementary schemes. The feature can be a limitation because it constrains the possible swing at input or output, but in some schemes the feature is positive because it gives rise to a voltage shifting.

10.8 CASCODE WITH ACTIVE LOAD

The inverting amplifiers with active load give rise to good gain but not enough for demanding applications. Moreover, the Miller amplification of the parasitic capacitance between input and output can be significant in integrated implementations. The latter issue finds a solution with the *cascode* scheme. The former is satisfied by the cascode with cascode load, to be studied shortly.

Figure 10.27 shows the circuit schematic of the inverter cascodes with active load. The figure refers to n-MOS input and NPN-input schemes. Complementary versions and features are, as usual, simply derived from what will be studied in the following.

The scheme is the same as the inverter with active load, with an additional active device interposed between input transistor and active load. The new element has its gate or base at

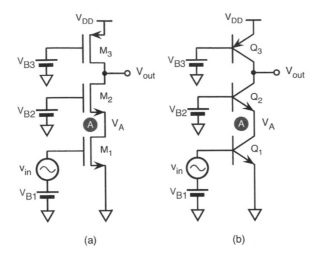

Figure 10.27   Cascode version of the inverter with active load: (a) MOS scheme; (b) BJT scheme.

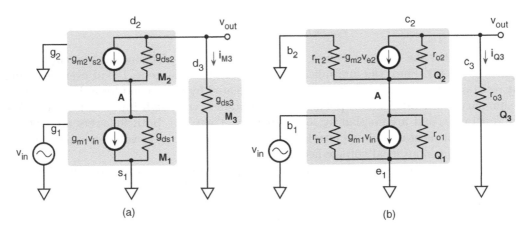

Figure 10.28   Small-signal equivalent circuit of the cascode configurations of Figure 10.27: (a) MOS scheme; (b) BJT scheme.

constant voltage, established by the bias V_{B2}. Since to operate a device it is necessary to have some voltage across it, the output dynamic range diminishes.

The circuit highlights a node, A, whose voltage mainly depends on the bias V_{B2} because the extra device operates as a follower. Thus from V_{B2}, V_A almost follows. For a small signal the action of the input signal influences the voltage of node A but the follower control limits the variation, giving rise to a limited gain from input to node A. As a result, the Miller amplification of the capacitor between input and output becomes affordable.

What is discussed above in a descriptive manner can be verified using the small-signal equivalent circuits shown in Figure 10.28. For the sake of simplicity this does not account for the body effect of M_2. The equations describing the circuit of Figure 10.28(a) are

$$v_{out} = \frac{i_{M3}}{g_{ds3}} \tag{10.59}$$

$$v_{s2} = \frac{-i_{M3} - g_{m1}v_{in}}{g_{ds1}} \tag{10.60}$$

$$v_{out} = v_{s2} + \frac{-i_{M3} + g_{m2}v_{s2}}{g_{ds2}}, \tag{10.61}$$

yielding the solutions

$$v_{out} = -v_{in}g_{m1}\frac{g_{ds2} + g_{m2}}{g_{ds3}(g_{ds1} + g_{ds2} + g_{m2}) + g_{ds1}g_{ds2}} \tag{10.62}$$

$$v_{s2} = -v_{in}g_{m1}\frac{g_{ds2} + g_{ds3}}{g_{ds3}(g_{ds1} + g_{ds2} + g_{m2}) + g_{ds1}g_{ds2}}. \tag{10.63}$$

The above equations are difficult to interpret despite the approximation that neglects the body term of M_2. Since it is advisable to derive equations that can be easily memorized as rules of thumb, we simplify the results, knowing that $g_m \gg g_{ds}$. Therefore,

$$v_{out} = -v_{in}\frac{g_{m1}}{g_{ds3}} \tag{10.64}$$

$$v_{s2} = -v_{in}\left(\frac{g_{m1}}{g_{m2}}\right)\frac{g_{ds3} + g_{ds2}}{g_{ds3}}, \tag{10.65}$$

showing that the gain equals the intrinsic gain of M_1 multiplied by a ratio between output conductances g_{ds1}/g_{ds3}. The voltage gain from input to the node A is low, being the ratio of transconductances multiplied by the ratio of output conductances. That low gain gives rise to a limited Miller amplification of the parasitic capacitance from input to node A, as desired. Therefore, the transistor M_2 interposed between input transistor and output adds a decoupling node whose effect is also to moderately increase the voltage gain.

The small-signal equivalent circuit of the BJT cascode of Figure 10.28(b) is described by

$$v_{out} = i_{Q3}r_{o3} \tag{10.66}$$

$$v_{e2} = (-i_{Q3} - g_{m1}v_{in})\frac{r_{\pi,2}r_{o1}}{r_{\pi,2} + r_{o1}} \tag{10.67}$$

$$v_{out} = v_{e2} + (-i_{Q3} + g_{m2}v_{e2})r_{o2}, \tag{10.68}$$

which, using r_p for the parallel value $r_{\pi,2}r_{o1}/(r_{\pi,2} + r_{o1})$, yields

$$v_{out} = -v_{in}g_{m1}r_{o3}\frac{r_p(1 + g_{m2}r_{o2})}{r_p + r_{o2} + r_{o3} + g_{m2}r_pr_{o2}} \tag{10.69}$$

$$v_{e2} = -v_{in}g_{m1}r_p\frac{r_{o2} + r_{o3}}{r_p + r_{o2} + r_{o3} + g_{m2}r_pr_{o2}}. \tag{10.70}$$

Again, it is worth approximating the expressions. This is done using the relationship $g_m \gg 1/r_o$. The results are

$$v_{out} = -v_{in}g_{m1}r_{o3} \tag{10.71}$$

$$v_{s2} = -v_{in}g_{m1}\frac{r_{o2} + r_{o3}}{g_{m2}r_{o2}}. \tag{10.72}$$

The results conform to what was found for the MOS scheme. The gain equals the intrinsic gain of Q_1 multiplied by r_{o3}/r_{o1}. The gain from input to node A is low, as requested.

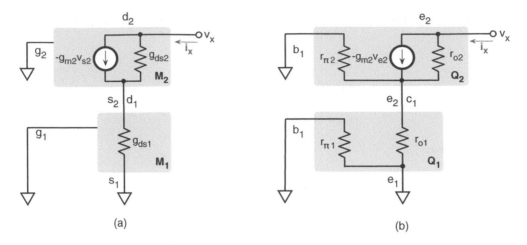

Figure 10.29   Small-signal equivalent circuit for measuring the output resistances of the cascodes of Figure 10.27: (a) MOS scheme; (b) BJT scheme.

10.8.1   Equivalent Resistances

Equations (10.64) and (10.71) show that the cascode gain is higher than that of the inverter with active load. The benefit is small, but the result indicates a remarkable feature. Since the gain is the product of transconductance and output resistance, the cascode configuration is such as to augment the output resistance, making it almost equal to the one of the active load. In addition, the signal current of the input transconductance generator almost completely reaches the output node. Therefore, the pathway through M_2 is greatly preferred with respect to the "shorter" one across the input transistor towards ground.

These qualitative observations can be confirmed with small-signal studies. For verifications we need to perform two checks: estimating the resistance of the path from the output node to ground established by the cascode and the input device, and measuring the resistance from node A to ground passing through the cascode and the active load.

The first estimation uses Figure 10.29. The schemes utilize the small-signal equivalent circuits of transistors and a test current i_x that gives rise to the voltage v_x. The resistance is the ratio v_x/i_x. For this operation we set the input generator to zero. The zero transconductance generator of the input device is therefore not present in the equivalent circuit. By inspection, this results in

$$v_x = i_x/g_{ds1} + (i_x + g_{m2}v_{s2})/g_{ds2}, \quad v_{s2} = i_x/g_{ds1} \quad \text{for the MOS circuit and} \quad (10.73)$$

$$v_x = i_x r_p + (i_x + g_{m2}v_{e2})r_{o2}, \quad v_{e2} = i_x r_p \quad \text{for the BJT circuit,} \quad (10.74)$$

where r_p is the parallel of r_{o1} and $r_{\pi2}$. If there is an equivalent resistance used to bias the base of Q_2, it must be accounted for in series with $r_{\pi2}$.

Therefore, the equivalent resistance of the scheme of Figure 10.29(a) is

$$r_{casc,MOS} = \frac{1}{g_{ds1}} + \left(1 + \frac{g_{m2}}{g_{ds1}}\right)\frac{1}{g_{ds2}} \simeq \frac{1}{g_{ds1}}\frac{g_{m2}}{g_{ds2}} = r_{ds1}(g_{m2}r_{ds2}), \quad (10.75)$$

which approximately equals the output resistance of the input transistor multiplied by the intrinsic gain of the cascode transistor.

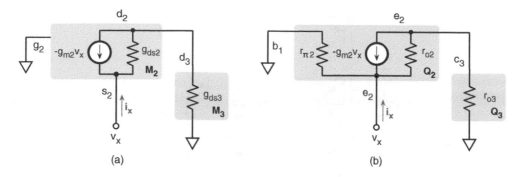

Figure 10.30 Small-signal equivalent circuit for measuring the resistance from the input of the cascoding transistors in Figure 10.27: (a) MOS scheme; (b) BJT scheme.

Similarly, for the BJT scheme, we have

$$r_{casc,BJT} = r_p + (1 + g_{m2}r_{o2})r_p \simeq r_p(g_{m2}r_{o2}), \tag{10.76}$$

which, in similar fashion to the MOS case, is r_p multiplied by the intrinsic gain of the cascode transistor.

The equivalent resistances $r_{casc,MOS}$ and $r_{casc,BJT}$ are both higher than a simple transistor by a factor approximately equal to an intrinsic gain. Therefore, such high resistances do not alter the output resistance of the active loads when they are parallel to them.

The second check makes use of the equivalent circuits of Figure 10.30. Test currents i_x and voltage v_x in the MOS and BJT circuits are related by

$$v_x = \frac{i_x - g_{m2}v_x}{g_{ds2}} + \frac{i_x}{g_{ds3}} \quad \text{for the MOS circuit and} \tag{10.77}$$

$$v_x = \left(i_x - g_{m2}v_x - \frac{v_x}{r_{\pi 2}}\right)r_{o2} + i_x r_{o3} \quad \text{for the BJT circuit,} \tag{10.78}$$

which give rise to the equivalent resistances

$$r_{s,MOS} = \frac{g_{ds2} + g_{ds3}}{g_{ds2}g_{ds3}} \cdot \frac{1}{1 + g_{m2}/g_{ds2}} = \frac{r_{ds2} + r_{ds3}}{1 + g_{m2}r_{ds2}}, \tag{10.79}$$

$$r_{s,BJT} = \frac{r_{o2} + r_{o3}}{(1 + g_{m2}r_{o2}) + r_{o2}/r_{\pi 2}} = \frac{r_{o2} + r_{o3}}{(1 + g_{m2}r_{o2})}//r_{\pi 2}, \tag{10.80}$$

where the two slanted lines '//' means parallel connection.

If the output resistances of devices 2 and 3 are equal, the above equations become

$$r_{s,MOS} = \frac{2}{g_{m2}} \quad \text{with } r_{ds2} = r_{ds3} \tag{10.81}$$

$$r_{s,BJT} = \frac{2}{g_{m2}}//r_{\pi 2} \quad \text{with } r_{o2} = r_{o3}. \tag{10.82}$$

The result shows that the equivalent resistances are twice $1/g_m$, much smaller than the output resistance of a transistor. For the bipolar case a more complex calculation may be necessary when $r_{\pi 2}$ is possibly augmented by the resistance of the fixed bias that provides the gate voltage to Q_2. In any case the equivalent resistance is well below the output resistance of a transistor. Therefore, the current of the transconductance generator almost completely flows through the cascode device.

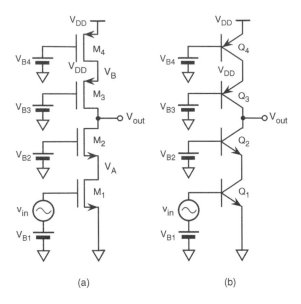

Figure 10.31 Cascode with cascode load: (a) MOS scheme; (b) BJT scheme.

10.8.2 Cascode with Cascode Load

The previous subsection shows that the output resistance of a cascode from the drain or the collector of the cascoding transistor is high and is given by an output resistance multiplied by an intrinsic gain. This feature is used to augment the voltage gain, because it is made by the product of transconductance and output resistance. For a given transconductance, augmenting the output resistance increases the gain. The circuit that implements that concept is the cascode with cascode load shown in Figure 10.31.

The circuits use stacks of four transistors with four bias voltages that define quiescent operating conditions. The biases of the cascoding elements 2 and 3, V_{B2} and V_{B3}, are not critical because they determine the drain or collector voltage of the transistors ending the stacks. On the other hand, biases V_{B1} and V_{B4} determine the gate or base voltage of the same transistors, and that is much more critical. For proper operation it is necessary to have MOS transistors in saturation (or BJT in the active region) and this mainly depends on the quiescent bias of the ending transistors. Matching of V_{B1} and V_{B4} to keep the quiescent output at the desired voltage is difficult or even, because of limitations such as fabrication inaccuracies and fluctuations of voltages, impossible to obtain in practice. The designer often sets the two intermediate biases and one of the others, and uses some sort of low-frequency control for regulating the remaining one. We shall not go into further details; it is enough to be aware of the issue.

Instead of a detailed small-signal circuit analysis it is convenient to use some of the previously obtained outcomes to derive approximate results. Suppose that the quiescent points are well controlled; the small-signal input develops a proper transconductance gain whose produced current mainly flows through the second device of the stack. This conforms with what we have obtained for a cascode with active load. Having a cascode load changes the configuration, but, within limits, we can suppose the approximation that signal current reaches the output node to be valid. The developed small-signal voltage equals that current multiplied by the output resistance. In turn, the resistance is the parallel connection of two cascode configurations. Therefore, the following relationship approximates the small-signal voltage gain:

$$A_v = g_{m1} \frac{r_{casc,U} r_{casc,D}}{r_{casc,U} + r_{casc,D}}, \tag{10.83}$$

where the resistances $r_{casc,U}$ and $r_{casc,D}$ of the up and down cascode are estimated by equations such as (10.75) and (10.76). The resulting voltage gain is large, being approximately the square of the voltage gain of simple inverters with active load. Thus the use of a cascode with cascode load gives rise to the same gain as the cascade of two stages.

The frequency study of the circuit requires an analysis that includes parasitic and actual capacitors. The schemes have three relevant nodes: the output and the source or emitter of the intermediate transistors. The resistance of the output is, as already discussed, very large: the parallel of two cascodes. The resistance of the other two nodes is relatively low because on those nodes there is a source (or an emitter). Because of the difference between the resistances we can expect that the pole caused by the output node is at a much lower frequency than those determined by the other nodes. That is actually what happens. The details of the result require a computer simulator or the analysis of the complete equivalent circuit.

10.9   DIFFERENTIAL PAIR

Analog processing frequently needs to handle differential signals. The circuits that do that are symmetrical schemes performing a symmetrical processing of inputs. The simplest differential circuit, often used as the first block of a differential analog processor, is the differential pair. It is made up of two matched transistors with the source or the emitter connected together, as shown in Figure 10.32 for n-type and NPN devices. A resistor or a current source provides the bias current. The schemes with resistance between the common source (or common emitter) and ground give rise to a current that depends on the quiescent input voltage. The other schemes use an ideal current source, a conceptual approach because real current generators have a finite resistance in parallel. How to realize real current sources will be studied shortly.

If the transistors are identical there is a symmetry with respect to the median vertical line. Equal input voltages ($V_{in+} = V_{in-} = V_B$) split the bias current, I_B, into equal parts. A differential input that changes symmetrically around the quiescent bias V_B gives rise to balanced changes if

$$V_{in+} = V_B + \frac{V_d}{2} \tag{10.84}$$

$$V_{in-} = V_B - \frac{V_d}{2} \tag{10.85}$$

and V_d is small enough to not alter the symmetric operation of transistors across the quiescent point. The currents can be approximated by

$$I_{M1} = \frac{I_B}{2} + I_d, \quad I_{M2} = \frac{I_B}{2} - I_d, \tag{10.86}$$

for the MOS scheme and, similarly, for the bipolar circuit

$$I_{Q1} = \frac{I_B}{2} + I_d, \quad I_{Q2} = \frac{I_B}{2} - I_d. \tag{10.87}$$

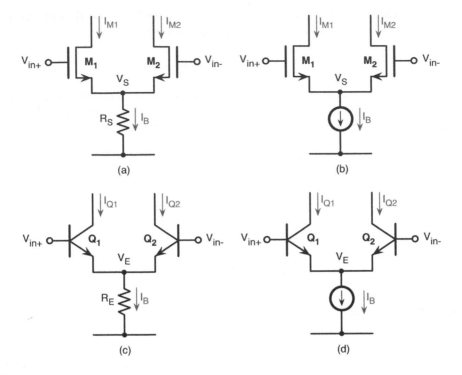

Figure 10.32 Differential pairs: (a) and (b) with MOS; (c) and (d) with bipolar transistors – a resistor or a current generator provides the bias current.

The use of the large-signal equations introduced to model the large-signal operation of transistors permits us to study the circuit in more detail. For the MOS scheme we suppose that the devices are in saturation, and, for the sake of simplicity, we neglect $(1 + \lambda V_{DS})$. This results in

$$I_{M1} = \frac{1}{2}\mu_n C_{ox} \left[\frac{W}{L}\right]_1 \left(V_B - V_S + \frac{V_d}{2} - V_{th}\right)^2 \tag{10.88}$$

$$I_{M2} = \frac{1}{2}\mu_n C_{ox} \left[\frac{W}{L}\right]_2 \left(V_B - V_S - \frac{V_d}{2} - V_{th}\right)^2, \tag{10.89}$$

where V_S is the source voltage. Since $(W/L)_1 = (W/L)_2$, the above equation gives rise to

$$I_{M1} = \frac{I_B}{2} + G_m \frac{V_d}{2} \tag{10.90}$$

$$I_{M2} = \frac{I_B}{2} - G_m \frac{V_d}{2}, \tag{10.91}$$

where

$$I_B = \mu_n C_{ox} \left[\frac{W}{L}\right]_1 \left\{(V_B - V_S - V_{th})^2 + \frac{V_d^2}{4}\right\}, \quad G_m = \mu C_{ox} \left[\frac{W}{L}\right]_1 (V_B - V_S - V_{th}). \tag{10.92}$$

The result, actually, shows a balanced linear behavior, which holds even for large signal within the validity of the large-signal model used.

COMPUTER EXPERIMENT 10.6

Understanding Differential Stages

The differential stage transforms a differential voltage into differential current. Input and output are made with a differential part and a common–mode term. Ideally the differential processing just accounts for the difference but in real circuits the common mode term influences the result. This virtual experiment studies two types of differential stages; one with MOS, the other with bipolar transistors. A sine wave generator provides the input term. Its voltage is added to and subtracted from a signal generated by a second sine wave generator with controllable offset to give rise to the common mode term. The output of the differential pair is a differential current. The ElvisLab board measures various currents with electronic ammeters (I-to-V) which generate a pair of voltages. Their differences give rise to output signals in the voltage domain. With the parallel of a current source and a resistance, we describe the non-ideal behavior of the current bias generator. This experiment allows you to:

- set supply voltage and offset level of the second sine wave generator. The default value of the supply voltage is 1.8 V, the offset is 0.9 V;
- change the amplitudes and the frequencies of the two sine waves;
- choose resistance and current of the current bias generators (the values are equal).

MEASUREMENT SET–UP

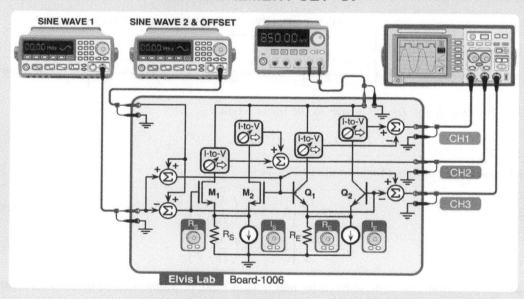

TO DO

- Leave the amplitude of the second sine wave at zero and increase the amplitude of the differential signal. Observe the differential output currents at different input amplitudes. Plot output current versus input voltage and observe the difference between MOS and bipolar schemes. Use various bias currents and offset levels.
- Bring the differential signal to zero and increase the amplitude of the common–mode sine wave. Measure the transconductance gain with different offset levels. Estimate, using small signals, the transconductance differential and common mode gain at various frequencies.

Indeed, the quantity that really matters for differential circuits is the output current difference. This, according to equations (10.90) and (10.91), is proportional to the differential signal V_d: $\Delta I_M = G_m V_d$. That good result is a direct consequence of the quadratic relationship between current and overdrive voltage for a MOS transistor in saturation.

However, G_m is almost constant only for small differential amplitudes; moreover the left-hand equation in (10.92) envisages a dependence of the common source voltage on the differential signal that, in turn, changes G_m. In addition, with a resistance between the common source or emitter and ground or a real current generator the transistors total current is not constant. It is not difficult to estimate the value of that bias current. The use of the Ohm's law relates bias current and V_S. Accounting for all the above points, it can be verified that a large $|V_d|$ diminishes G_m.

When current in one of the MOS transistors becomes very low the device goes into sub-threshold. The current difference saturates because the bias current flows entirely through one transistor and the other conducts almost zero current.

The analysis of the bipolar scheme makes use of the following equations:

$$I_{C1} = I_S \cdot e^{(V_{BE} + V_d/2)/V_T} \tag{10.93}$$

$$I_{C2} = I_S \cdot e^{(V_{BE} - V_d/2)/V_T}. \tag{10.94}$$

A simple mathematical transformation gives rise to the subsequent relationships

$$I_{C1} = I_S e^{V_{BE}} \left\{ \frac{e^{V_d/(2V_T)} + e^{-V_d/(2V_T)}}{2} + \frac{e^{V_d/(2V_T)} - e^{-V_d/(2V_T)}}{2} \right\} \tag{10.95}$$

$$I_{C2} = I_S e^{V_{BE}} \left\{ \frac{e^{V_d/(2V_T)} + e^{-V_d/(2V_T)}}{2} - \frac{e^{V_d/(2V_T)} - e^{-V_d/(2V_T)}}{2} \right\}. \tag{10.96}$$

The addition $I_{C1} + I_{C2}$ gives rise to the bias current I_B,

$$I_B = I_S e^{V_{BE}} [e^{V_d/(2V_T)} + e^{-V_d/(2V_T)}], \tag{10.97}$$

which, used in equations (10.95) and (10.96), yields

$$I_{C1} = \frac{I_B}{2} \left\{ 1 + \tanh\left(\frac{V_d}{2V_T} \right) \right\} \tag{10.98}$$

$$I_{C2} = \frac{I_B}{2} \left\{ 1 - \tanh\left(\frac{V_d}{2V_T} \right) \right\}. \tag{10.99}$$

The results show that the differential pair with a bipolar transistor responds to an input differential voltage with a differential output current equal to

$$\Delta I_C = I_B \tanh\left[\frac{V_d}{2V_T} \right], \tag{10.100}$$

since for small values of the argument $\tanh(x) \simeq x$, small differential inputs give rise to a linear differential response. However, just a differential input equal to V_T (26 mV at room temperature) reduces the transconductance gain by 7.5%. Twice that voltage brings the attenuation factor to 76.2% of the maximum. Therefore, it is necessary to make sure that the differential input is a low fraction of V_T.

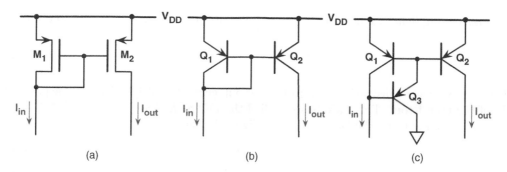

Figure 10.33   (a) MOS current mirror; (b) bipolar current mirror; (c) bipolar current mirror with base current compensation.

10.10   CURRENT MIRROR

The operation of the building blocks of this chapter and others in general analog schemes, including data converters, often depends on the bias current generators that they use. Well-defined and controlled quiescent currents must flow in circuits, possibly together with signal currents. Since we measure performances by having quiescent points as reference, it is important to ensure that the values of those quantities are what is expected, and this often depends on the accuracy and quality of DC current generators. Moreover, in addition to the absolute value, it is repeatedly necessary to ensure well-controlled ratios between currents: electronic circuits often rely more on relative values and less on absolute amplitudes.

A convenient way to generate bias currents is to start from a main source and to generate replicas, perhaps attenuated or amplified if necessary. Since we reproduce an image or a copy of the original, the circuit that performs this function is called a *current mirror*. Figure 10.33 shows its simplest implementation with p-type MOS or PNP devices. The current to be reproduced flows through transistors connected using the so-called diode configuration, with gate connected to drain or base to collector. The diode connection gives rise, actually, to a transformation of current to voltage. The resulting voltage drives another transistor engaged in generating the output current. The transistors' type of the pairs is the same, but their aspect ratios or areas can differ.

The circuit supposes that the input current comes from a good source generator and that voltages across transistors are such as to allow a functioning that properly mimics the required current source function. Also, fabricating the transistors together, one close to another, with a careful layout design, gives rise to the same or well-matched technological and physical parameters.

Both PMOS devices, assumed in saturation, give rise to

$$I_{in} = \frac{1}{2}\mu C_{ox}\left(\frac{W}{L}\right)_1 (V_{GS1} - |V_{Th,p}|)^2(1 + \lambda V_{DS1}) \qquad (10.101)$$

$$I_{out} = \frac{1}{2}\mu C_{ox}\left(\frac{W}{L}\right)_2 (V_{GS1} - |V_{Th,p}|)^2(1 + \lambda V_{DS2}); \qquad (10.102)$$

for the sake of simplicity, we neglect the $(1 + \lambda V_{DS1})$ term in equation (10.101) to obtain

$$I_{out} = I_{in} \frac{(W/L)_2}{(W/L)_1} (1 + \lambda V_{DS2}), \tag{10.103}$$

which is a replica (or mirrored version) of the input current multiplied by the aspect ratio (W/L) of the transistors. This ratio is also called the *mirror factor*.

The equations featuring the BJT scheme of Figure 10.33(b) are

$$I_{in} = I_{C1} + I_{B1} + I_{B2} \tag{10.104}$$

$$I_{C1} \simeq \alpha_{F1} A_1 I_S e^{V_{BE1}/V_T} \tag{10.105}$$

$$I_{C2} = I_{out} \simeq \alpha_{F2} A_2 I_S e^{V_{BE1}/V_T} \tag{10.106}$$

$$I_{C1} = \beta_1 I_{B1}, \quad I_{C2} = \beta_2 I_{B2}, \tag{10.107}$$

where A_1 and A_2 are the emitter junction areas and I_S is the saturation current of the base-to-emitter junction with unity area.

Since the transistors are matched it is reasonable to assume that they have the same α and β. Thus the base current I_{C1} is equal to I_{C2} if $A_1 = A_2$. Otherwise, the current differs just because of the area ratio. To be more precise, after some simple mathematics we determine the condition

$$I_{out} = I_{in} \left(\frac{A_2}{A_1} \right) \frac{\beta}{\beta + 1 + (A_2/A_1)}, \tag{10.108}$$

which, actually, gives for output a replica of the input multiplied by the ratio of the base–emitter junction areas, but denotes a correcting factor accounting for the fraction of input current used by the base of transistors. The correcting factor can be notable with large area ratios. It also decreases if the same main source serves many mirror branches. In those cases it is worth using the base-current compensation shown in Figure 10.33(c). The input current just feeds the base of Q_3, β times smaller than the overall base currents.

10.10.1   Equivalent Circuit

The input–output response of the transistor used to generate the output current determines the large-signal response of the circuit. The transistor needs a minimum voltage across it to give rise to a current with a small dependence on output voltage. Then, as shown in Figure 10.34, the current increases, hopefully with a low slope. The curve in this region of operation is a straight line whose value is $I_{out,0}$ for zero voltage across the transistor. The large-signal equivalent circuit depicting the straight line is a constant-current source with, in parallel, a resistance, R_{out}, as shown in Figure 10.34(b). The figure distinguishes between the current source generating current from the positive rail (the source) and the case of current entering ground or the negative rail (the sink). Figure 10.34(c) shows the small-signal equivalent circuit, just made by the output resistance $r_{out} \simeq R_{out}$ because the current source has constant value. In parallel there is the output capacitance depicting the parasitic element and a possible capacitance actually used in circuit.

The value of the equivalent resistance is, for MOS and bipolar current source, approximated respectively by

$$r_{out,MOS} = \frac{1}{\lambda I_D} \quad \text{and} \quad r_{out,BJT} = \frac{V_A}{I_C}, \tag{10.109}$$

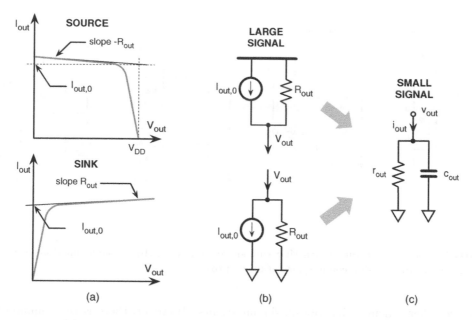

Figure 10.34   (a) Voltage–current response of real current source and current sink; (b) large-signal equivalent circuits; (c) small-signal circuit.

both large values but not enough for demanding applications. However, the circuit maintains the value of output resistance for a wide range of large-signal output voltages.

10.10.2   Current Mirror with High Output Resistance

Some applications need to increase the output resistance because that of the simple current mirror of Figure 10.33 is not enough. More complex schemes must be used. These exploit the same methods that increase the output resistance of active loads. As is already known, a cascode scheme secures high resistance by interposing an extra transistor between the input device and output node. The same is done by the cascode current sources (with exiting current) shown in Figures 10.35(a) and (b). They use MOS or BJT transistors. Equivalent diagrams made by N-channel or NPN transistors make the current sink counterparts (with entering current).

The generated current comes from a stack of two transistors. The one connected to the positive rail or ground determines the current. The second transistor follows the voltage at its gate or base to give rise to a "protective" voltage at the drain or collector of the main transistor. The reduced influence of the output voltage increases output resistance.

We can estimate the output resistance with a small-signal analysis. The study exactly follows what was done for the cascode load that gave rise to the results expressed by equations (10.75) and (10.76). These are approximated by

$$r_{casc,MOS} \simeq r_{ds2}(g_{m4}r_{ds4}) \tag{10.110}$$

$$r_{casc,BJT} \simeq r_{o2}(g_{m4}r_{o4}), \tag{10.111}$$

showing that the output resistance increases by the gain of the cascoding element.

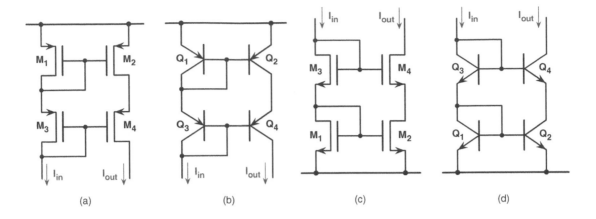

Figure 10.35 (a) Current source with MOS cascode configuration; (b) the same function with bipolar transistors; (c) and (d) the sink counterparts of (a) and (b).

There are other schemes that obtain similar results. However, there is the common limit established by the requirement of bias voltage across the extra transistor. The drop voltage needed by the current source to operate properly limits the usable output swing of the current source or sink. This issue is the topic of further studies in advanced courses, because, indeed, low supply voltages imposed by modern technologies require a careful management of voltages. Large drops for current generators cannot be allowed.

10.10.3 Differential to Single-ended Converter

The current mirror is the basis for converting differential current signals into a single-ended current signal. As is already known, the currents at the output of a differential stage are a common-mode term plus a differential quantity:

$$I_+ = I_{cm} + \frac{i_s}{2} \tag{10.112}$$

$$I_- = I_{cm} - \frac{i_s}{2}. \tag{10.113}$$

It is often desirable to separate the common-mode part from the differential part. This is done by subtracting the two currents: $(I_+ - I_-)$. The circuit performing that operation is one of the schemes of Figures 10.36(a) and (b). The differential currents, represented by two real current generators, are connected to the two terminals of the current mirror. The positive current of the differential pair is the input current of the mirror. The output is the junction between the output of the current mirror and the negative current of the pair. The operation of the circuit is evident by inspection, assuming that the output resistance of the differential current generators is infinite and the mirror operates in an ideal manner. The mirror generates an exact replica of the current of the left-hand-side generator. Thus the current exiting the output node is the difference between $(I_{cm} + i_s/2)$ and $(I_{cm} - i_s/2)$.

Obviously, if the differential current is zero the output should be zero. In reality there is a mismatch that produces a constant current at output. It depends on the voltage of the

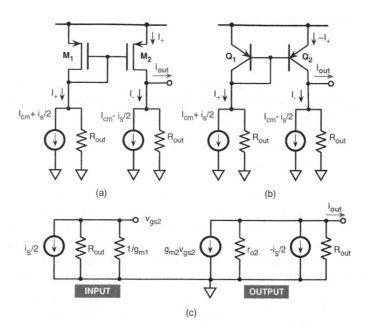

Figure 10.36   Differential to single-ended converter: (a) with MOS transistors; (b) with BJT transistors; (c) small-signal equivalent circuit of scheme (a).

output node, established by the circuit that receives the signal from the output terminal. If that voltage equals the drain of M_1 and the components match, the scheme of Figure 10.36(a) is in a symmetrical condition and the output current is zero. For the circuit of Figure 10.36(b) a voltage at output equal to the common base would give rise to a constant output current equal to the addition of the two base currents, entering the circuit.

The analysis of the small-signal operations makes use of the small-signal circuits. That of the MOS configuration is shown in Figure 10.36(c). The transistor M_1 in the diode configuration is equivalent to a resistance $1/g_{m1}$ that loads the small-signal current source $i_s/2$. The current flows into the parallel of R_{out} and $1/g_{m1}$, yielding

$$v_{gs2} = -\frac{i_s}{2}\frac{R_{out}}{1 + g_{m1}R_{out}}. \tag{10.114}$$

The voltage $v_{gs1} = v_{gs2}$ controls the transconductance generator of transistor M_2 to give rise to the output signal current $g_{m2}v_{gs2}$. That current and the second small-signal current produce

$$i_{out} = \frac{i_s}{2}\left[1 + \frac{g_{m2}R_{out}}{1 + g_{m1}R_{out}}\right] = \frac{i_s}{2}\frac{1 + 2g_{m1}R_{out}}{1 + g_{m1}R_{out}}, \tag{10.115}$$

where the last expression supposes $g_{m1} = g_{m2}$.

The signal current given by equation (10.115) is equivalent to a signal generator i_s if $g_{m1}R_{out} \gg 1$. As shown by the equivalent circuit, it has in parallel an output resistance made up of R_{out} and r_{o2} in parallel. The current flows out almost completely when the external resistance is much smaller than the output resistance.

10.11   REFERENCE GENERATORS

Voltage and current reference generators are important analog cells. We have seen that many analog blocks use voltages to bias the gate or base of transistors. Moreover, circuits need current that, possibly, is delivered by a current mirror starting from a main source.

Voltage and current references are also necessary for higher-level processing, mainly for defining the full scale or the quantization step of data converters. The accuracy either for biasing analog cells or for other uses determines the precision of results. Therefore, the availability of a "good" reference generator is essential for many analog designs.

There are situations that are not very exacting; such cases use resistances, capacitors and Ohm's law. For example, a resistive or a capacitive divider made from two resistors, R_1 and R_2, or two capacitors, C_1 and C_2, which are initially discharged, determines a fraction of the supply voltage. A drop voltage across a resistance gives rise to a current. That is what the simple schemes of Figure 10.37 do. The scheme with capacitors uses switches to discharge capacitors C_1 and C_2 periodically, to make sure that possible charges trapped in the output node do not alter the generated reference voltage. The resulting voltages and current are

$$V_{ref} = V_{DD} \frac{R_1}{R_1 + R_2}, \quad V_{ref} = V_{DD} \frac{C_2}{C_1 + C_2}, \quad I_{ref} = \frac{V_{DD} - V_{GS1}}{R_1}, \quad (10.116)$$

where the output of the capacitive divider is available during phase Φ_2.

For generating a reference current the simplest way is to exploit Ohm's law, which "transforms" a voltage across a resistance into a current flowing though it (Figure 10.37(c)).

Obviously, demanding schemes need more complex solutions. Here we do not go into deep details of the circuit architectures that generate reference quantities. However, let us observe that the limits of the simple circuits of Figure 10.37 concern the dependence on spur signals corrupting supply lines and temperature sensitivity. Accuracy in values is also a concern.

Precise reference generators use some basic "ingredients" to generate the output. These are:

- **The diode-on voltage, V_D (or base-to-emitter voltage, V_{BE})** Its dependance on current is quite limited (it changes by V_T if current changes by $e = 2.72$) but changes with temperature by -2.2 mV per °C.

- **The difference between two diode-on voltages, ΔV_D** Since the diode voltage changes with the exponent of V_T, the difference is proportional to $V_T = kT/q$, equal to about 26 mV at room temperature.

- **The threshold voltage of a MOS transistor** This ingredient gives rise to the V_{GS} voltage whose value hardly changes with current with very low biases. The threshold, however, is temperature dependent.

- **The difference between threshold voltages, ΔV_{Th}** This is an ingredient that is usable if the technology makes transistors of the same type and different threshold voltages available.

A suitable combination of the above "ingredients" gives rise to current references with very low sensitivity to changes in the supply voltage or to voltage references with low temperature sensitivity. This is what is done by the so-called band-gap reference generators that combine the diode voltage "ingredients" that have a negative temperature coefficient with the difference

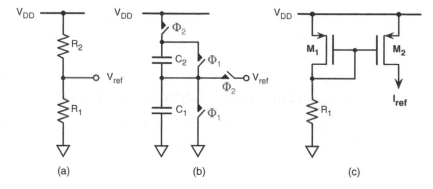

Figure 10.37   (a) Simple reference generator made by a resistive divider; (b) capacitive divider with periodically reset capacitors; (c) simple current reference.

between diode voltages, which has a positive temperature coefficient. The following weighting of two "ingredients,"

$$V_{BG} = V_D + 22.8\frac{kT}{q}, \qquad (10.117)$$

gives rise to zero temperature dependence at around room temperature.

PROBLEMS

10.1   Draft the frequency response of an amplifier featuring a 15 dB gain from DC to 7 MHz. Then the gain ramps up with 20 dB per decade slope until 28 MHz. After that frequency, the gain remains constant for another 20 MHz before dropping down at 20 dB per decade. What is the frequency at which the gain crosses the 0 dB axis?

10.2   Use the voltage–current characteristic curves of Figure 10.7(a) and suppose that the gate voltages are $V_{GS1} = 0.4$ V, $\Delta V_{GS} = 0.1$. The full scales of the diagram are 3 V and 0.2 mA. What is the load resistance that, with $V_{GS} = 0.5$ V and supply voltage 1.8 V, determines a current of 60 μA?

10.3   The voltage–current response of the N-channel MOS transistor of Figure 10.6 is approximated by the relationship $I_D = (2 + 0.013\ V_{DS})$ mA. What is the operating point with 1.8 V supply voltage and load resistance 4 kΩ? Determine the load value that brings the transistor to the limit of the triode region, $V_{DS} = 0.28$ V. If the threshold voltage of the transistor is 0.42 V what is its V_{GS}?

10.4   The MOS transistor of an inverter with a 10 kΩ resistive load is in the saturation region. The small-signal gain is -45 and the overdrive voltage is 230 mV. The supply voltage is 1.8 V and the quiescent output 0.9 V. What is the bias current of the stage? What bias current gives rise to a gain equal to -30? Estimate the new quiescent output.

10.5   The voltage current curves of two transistors in saturation, one N-channel and the other P-channel, connected as an inverter with active load (Figure 10.9), are approximated by

$$I_{D,N} = 1.510^{-3}(V_{GS,N} - 0.5)^2(1 + 10^{-4}V_{DSN}) \tag{10.118}$$

$$I_{D,P} = 1.210^{-3}(V_{GS,P} - 0.4)^2(1 + 2 \cdot 10^{-4}V_{DS,P}). \tag{10.119}$$

Determine the gate voltage of the P-channel transistor that gives rise to a quiescent-point output voltage of 0.7 V. The supply voltage is 1.5 V and the V_{GS} of the N-channel device is 0.75 V.

10.6   An inverter with active load uses a bipolar transistor with $\beta = 97$. The resistive load is 2 kΩ and the base current is 3.2 μA. What is the supply voltage if the collector voltage is 0.9 V? Determine the base current that brings the collector voltage to 0.63 V. What happens below that value of collector voltage?

10.7   Suppose that the signal generator of the equivalent circuit of an inverter with active load shown in Figure 10.12 is real. Its series resistance is $r_{in} = 0.1/g_{ds1}$. Estimate the frequency response. Assume a low-frequency gain of 68 and $g_m = 0.2$ mA/V. The pole of the amplifier with zero r_{in} is at 98 kHz, $C_{gd1} = 0.06$ pF. Calculate the input current at 100 MHz and 300 MHz.

10.8   An inverting amplifier with resistive load equal to 10 kΩ uses a bipolar transistor with $\beta = 100$. The supply voltage is 3.3 V and the collector voltage is 1.2 V. What is the base current?

10.9   The amplifier of Figure 10.15 uses $R_B = 10$ kΩ. The current generated by the PNP transistor is 1 mA and the β of the NPN is 112. What is the quiescent output voltage? The supply voltage is 2.5 V and the β of the PNP is 87. What is, approximately, the value of R_{BL}?

10.10   A source follower made by a P-channel transistor with threshold 0.5 V has its gate connected to a sawtooth generator with amplitude 1.7 V. The minimum level of the sawtooth is 0 V. The supply voltage of the circuit is 1.8 V. Draft the expected output voltage. What assumption have you made about the threshold voltage and the bias current generator? What happens if the sawtooth changes its zero level from the initial 0 V to -0.7 V?

10.11   The cascode scheme of Figure 10.27(a) uses suitable voltages to keep the output at a proper quiescent point. The bias generator V_{B2} is affected by noise. Use the small-signal equivalent circuit to estimate the noise at output and its input referred equivalent.

10.12   Draft the circuit schematic of a triple cascode with double cascode load (this means five MOS transistors, one on top of the others). The supply voltage is 1.8 V, the threshold of the N-channel device is 0.4 V, and that of the P-channel is 0.5 V. What are the possible bias voltages of the gates, assuming that the saturation voltages of all of the transistors are 0.2 V?

CHAPTER 11

DIGITAL BUILDING CELLS

This chapter helps you by explaining the digital cells hidden inside the macro-blocks of electronic systems. You will also learn about the logic circuitry needed to interface macro-functions. With those purposes in mind, we shall analyze simple Boolean cells, their functional implementations, and their circuit schematics at the transistor level. We shall study latches, half- and full-adders, and many other simple logic cells. We shall also discuss the limits of interconnections and how to describe them to give a reliable estimate of circuit performance.

11.1 INTRODUCTION

Complex logic functions are often implemented with a few impenetrable integrated circuits capable of performing diverse tasks. The operation of those integrated circuits is typically soft-defined; the designer's assignment is just to develop software with suitable computer tools using a given programming language. However, that is not all that is needed, because it is occasionally necessary to design special hardware or construct specific interfaces for complex integrated circuits. In any case it is important to be familiar with the basic "ingredients," to be aware of their transistor-level schemes, and to know their performance limits.

We have already studied, in Chapter 8, how to realize logical and mathematical functions for digital processing systems. We have seen that the circuit implementations are principally based on elementary logic gates that are able to express Boolean relationships. With the

Understanding Microelectronics: A Top-Down Approach, First Edition. Franco Maloberti.
© 2012 John Wiley & Sons, Ltd. Published 2012 by John Wiley & Sons, Ltd.

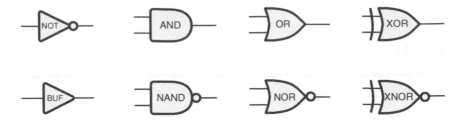

Figure 11.1   Symbols for logic gates, as defined by the ANSI/IEEE standard. The name inside denotes the type of gate, but it is often omitted.

combination of those logic gates, and with possible clock control, we realize the higher-level functions.

11.2 LOGIC GATES

The logic gate combines one or more logic signals at input to generate a logical output. It is represented by a symbol defined according to the standard established by ANSI/IEEE. These symbols are generally adopted for technical documentation and for published literature. The function of the logic gate is realized by integrated circuits or cell libraries. In both cases there is suitable documentation available to define both logic operation and circuit specifications. Possible types of logic gates are as follows.

- **Inverter** Also called a NOT gate, this has only one input and gives rise at output to the complement. A "true" (or "1") at input causes a "false" (or "0") at output, or vice versa.

- **Buffer/driver** This is a one-input gate that replicates at output what is applied at input. The use of the buffer can be for impedance matching or for isolating input from output. The main function is to boost a weak input and give rise to a signal able to drive low capacitive or resistive loads. A frequent scheme for a driver is a cascade of an even number of inverter gates with increasing driving capability.

- **AND gate** This is the logic that implements the AND Boolean function. The gate has multiple inputs and gives rise to a "true" (or "high") signal only if all of the inputs are "1".

- **NAND gate** This is the logic circuit that generates the inverse of the AND gate. The output is "0" if all of the inputs are at the "1" level. Otherwise the output is "high."

- **OR gate** With this cell we implement the OR Boolean function. The OR gate also has multiple inputs (normally between two and four). It generates a "1" if just one input is "high."

- **NOR gate** This is the inverse of the OR. The output is "0" when one of the inputs is "1". Otherwise the output is "1". The gate is equivalent to an OR followed by an inverter.

■ **XOR gate** This gate, also called *exclusive-OR*, has two inputs. It gives rise to "1" if either, but not both, of the inputs is "1". The output is "0" if both inputs are "1", or if both inputs are "0".

■ **XNOR gate** This is the inverse of the XOR, and is also called *exclusive-NOR*. The output goes to "0" if either, but not both, of the inputs is "1". On the other hand the output is "1" if both inputs are "1", or if both inputs are "0".

Figure 11.1 shows symbols for the elementary gates. There is, obviously, one input for the inverter or the buffer and just two inputs for the other gates. It is possible to have multiple inputs for AND, NAND, OR and NOR. The symbol with multiple inputs is the same but with more wires at the input side. For many inputs, instead of expanding the symbol to accommodate the input wires, some diagrams just expand the line at the input side.

11.2.1 Gate Specifications

The gate specifications give general information such as the nominal supply voltage, a possible indication of a three-state operation (this means that the output terminals have three possible states, low, high, and high impedance – which is like having the output terminal unconnected), complementary output (this means that there are two outputs for each logic output, one conforming to the required logic function and the other not), and the number of logic gates included in the part (in the case of discrete integrated circuits).

Just as for analog circuits, the specifications of logic gates provide absolute maximum ratings: the values beyond which damage to the device may occur. Those specifications should be met, without exception, to ensure that the circuit designed is reliable over its power supply, temperature, and output/input loading variables. There is also a section giving recommended operating conditions. These concern supply voltage, high and low input levels, operating temperature, and, in some cases, minimum input edge rate $\Delta V_{in}/\Delta t$. This last condition avoids having the logic cell staying for too long in the transition region.

A logic gate dissipates power. We have two different contributions: the power consumed continuously, with possible differences between the low and high states, and the power dissipated by the circuit only during switching between states. The former contribution is the static power; the latter is the dynamic power. Indeed, the static power can be almost zero, but we always have dynamic power because of the capacitances that must be charged or discharged when the state changes. If the output is switched low-to-high and high-to-low with average frequency f_{sw}, the dynamic power consumed by the gate is

$$P_{av} = (C_{p,l} + C_{p,h})V_{DD}^2 f_{sw}, \tag{11.1}$$

where $C_{p,l}$ is the parasitic capacitance of the nodes that are charged when the output goes high and $C_{p,h}$ is the parasitic capacitance of the nodes that are charged when the output becomes low.

For logic gates realized with modern technologies, the parasitic capacitances involved are very small – of the order of a few fF or less (remember that "f," femto, means 10^{-15}); therefore the consumed power of a single logic gate is very small even if the switching frequency is many hundreds of MHz. However, for complex circuits that contain many hundreds of millions or billions of transistors power becomes a serious design concern. Other important information involves the timing of operation. The specifications provide low-to-high (t_{PLH}) and high-to-low (t_{PHL}) propagation delays, the times at which the output is a valid "1" or "0" after a quick

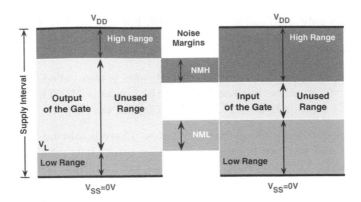

Figure 11.2   Permitted and unused voltage ranges for "low" and "high" logic signals at output and input of digital cells; noise margins (NMH, Noise Margin High, and NML, Noise Margin Low).

input transition. Other data featuring logic cells available in integrated circuits pertains to the capacitive loads at input and output, the type of package, and the pin configuration.

Chapter 8 defined voltage intervals within which logic signals are supposed to be valid. That was, actually, for generated output(s). In reality we have to account for both input and output, because what a cell generates is better delimited than what is wanted at input. For that reason the specifications provide the minimum "high" levels at input and output and the maximum "low" levels at input and output (see Figure 11.2). The difference between pairs of "high" and "low" boundaries indicates unused ranges. What matters is having large "high" and "low" regions and fairly wide unused ranges.

When the output of one cell feeds another cell, noise and interference that typically corrupt signals along connecting paths can affect the signal at input of the receiving cell. Also there is random fluctuation of the supply voltage. In order to avoid negative effects, the circuit must feature so-called *noise immunity*. Its measure is the *noise margin* estimated in Figure 11.2. Actually, there are two noise margins, for the low and the high range (Noise Margin High, NMH, and Noise Margin Low, NML). They give a voltage space for noise, which ensures noise immunity.

Remember that ...

the specifications of discrete logic cells or cells from a library give you valuable information for the design of real circuits. You need to know, for example, the delays so as to define the maximum speed of operation, capacitive load so as to interface cells properly, and power so as to comply with consumption requirements.

Since a gate realizes a Boolean relationship between a number of logic quantities, there are a number of input terminals equal to the number of these logic variables. The number is called the *fan-in* of the cell, a parameter useful for anticipating speed. In fact, a gate with a large fan-in is typically slower than gates with small fan-in because the input complexity causes a larger input capacitance. There is another parameter, called *fan-out*, used to indicate the driving capability of a gate. It provides the number of logic gates of the same type that the output can drive. Since the architectures of complex logic functions require use of the same logic signal at the inputs of multiple cells, the fan-out indicates the possibility of wiring the output of the cell directly to multiple inputs without the need to use buffers.

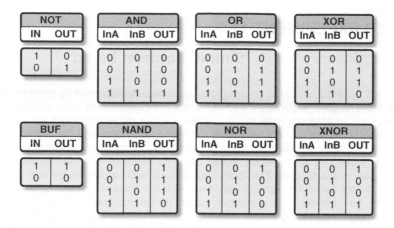

Figure 11.3   Truth tables of basic logic gates.

The truth tables laid out in Figure 11.3 give the logic relationships between the inputs and the output of basic cells. The tables shown allow for only one or two inputs; for multiple inputs (except in the case of XOR and XNOR) the extension is straightforward. As already discussed, this kind of table has a number of columns equal to the total number of inputs and outputs. The rows correspond to all the possible combinations of inputs. For a single input there are two possibilities: "low" and "high." For two inputs there are $2^2 = 4$ lines. Three inputs require $2^3 = 8$ lines and so forth. The truth table of the NOT (or inverter) is the simple negation of input. For the buffer, input and output are the same. The other tables provide the proper coupling between a set of inputs and the output. For example, equal inputs give a 0 for the XOR and a 1 for the XNOR.

11.3   BOOLEAN ALGEBRA AND LOGIC COMBINATIONS

The study of logic circuits uses Boolean algebra: a set of laws, rules and theorems that help in analyzing the combination of logic signals. The topic is studied in detail in specific courses. However, since the knowledge of Boolean expressions is essential for analyzing logic circuits, we give here a few basic elements of Boolean algebra.

The previous section introduced basic logic gates. Their functions correspond to Boolean operators. The inverter or NOT gate is the *complementary* function of Boolean arithmetic. It is represented with a bar superscripting the name of the variable. The OR and AND are respectively the Boolean addition and multiplication: the Boolean addition of two numbers equals one if one of the adders is one; multiplication gives zero if one of the multipliers is zero. With three Boolean variables $A = 0$, $B = 1$, and $C = 1$, for example, we have the equalities

$$\bar{A} = 1$$
$$A + B + C = 1$$
$$A \cdot B = 0 \tag{11.2}$$
$$B \cdot C = 1$$
$$\bar{A} \cdot B \cdot C = 1.$$

COMPUTER EXPERIMENT 11.1

Simple Boolean Logic

Digital processing uses Boolean logic to link digital quantities represented by voltages or, rarely, currents. With this computer experiment you can verify the operation of logic schemes focusing on functions and not on transistor level circuits. There are four mini-boards that you can plug into this ElvisLab board. Use one of them and verify the logic combinations of five digital codes. The mini-boards use a clock for controlling a five-bit counter. In this manner the outputs a, b, c, d, e, are all possible sequences of a five-bit digital set. The cascade of three levels of Boolean functions gives rise to the outputs. The integrated circuits used in the board are CMOS static parts. However, do not forget that special needs are satisfied by other technologies and that the logic circuit can be clocked. The expected speed is limited because, in discrete device implementation, large parasitic capacitances slow the operation down. However, observing waveforms gives you an idea of propagation delays with different input signals and parasitic loads. The oscilloscope does not affect measurement. You can:

- choose the experimental mini-board you want to study;
- set the frequency of the square wave generator or put it in a single shoot mode for manual generation of pulses; reset the counters.

MEASUREMENT SET–UP

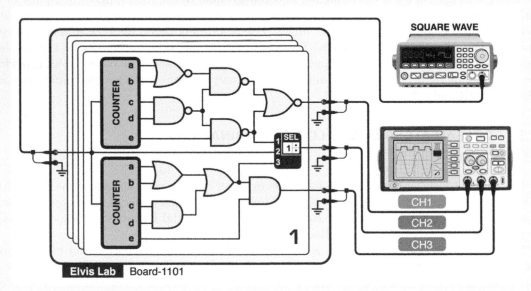

Elvis Lab Board-1101

TO DO

- Choose the first mini-boards and analyze its logic function on paper. Set the square wave generator in the single shoot mode and reset the counters. Push the manual button to increase the counter content one by one, every time you push the button.
- Set the frequency of the clock and increase its value until you see jitter in the output patterns. Measure the delay between input and output signals.
- Change the mini-board and repeat the study.

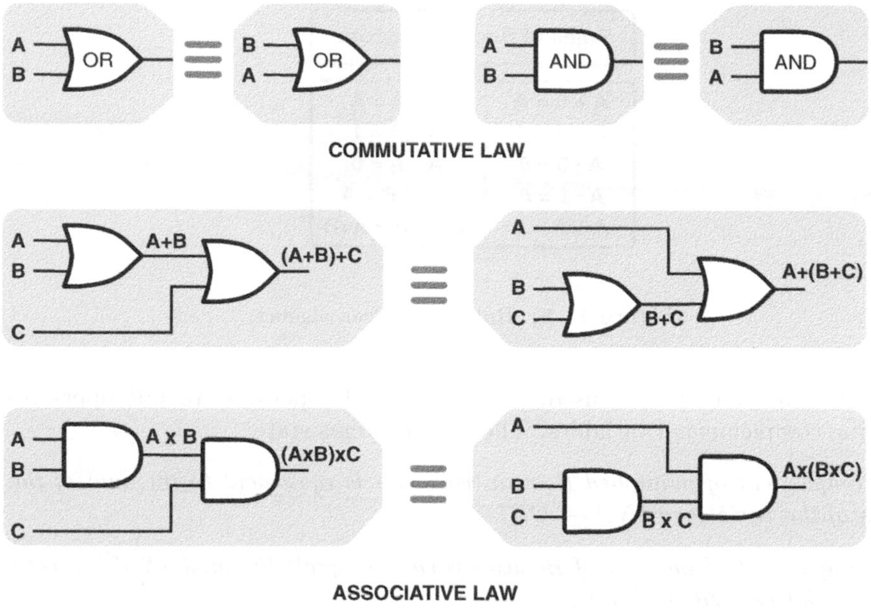

COMMUTATIVE LAW

ASSOCIATIVE LAW

Figure 11.4   Equalities of Boolean implementations as defined by the commutative and associative laws.

Boolean algebra defines a set of laws analogous to those of ordinary algebra. We have the *commutative law*, which holds for additions and multiplications,

$$A + B = B + A$$
$$A \cdot B = B \cdot A, \tag{11.3}$$

saying that it is possible to swap the terms of addition and multiplication and obtain the same result. The law, from a circuit point of view, states that logic OR or AND perform the same Boolean addition independently of how A and B are connected to the inputs, as Figure 11.4 shows.

The associative law states that the AND or the OR of more than two variables can be realized by associating any pair of them. For example,

$$(A + B) + C = A + (B + C)$$
$$(A \cdot B) \cdot C = A \cdot (B \cdot C). \tag{11.4}$$

This means that circuits can use two ORs or two ANDs indifferently connected, as shown in Figure 11.4, where a two-input logic gate combines two of the inputs first, and a second gate receives the output of the first gate and the third input to generate the overall output.

The distributive law is another relevant Boolean algebra tool. It applies to the AND of a variable with the OR of two (or more) variables. The distributive law, similar to the equivalent distributive law in mathematics, affirms that

$$A \cdot (B + C) = A \cdot B + A \cdot C. \tag{11.5}$$

In addition to the Boolean algebra laws, we also use the rules of Figure 11.5 for handling logic expressions. These are obvious relationships which, anyway, you can easily verify.

Rules of Boolean Algebra	
$A + 0 = A$	$A \cdot A = A$
$A + 1 = 1$	$A + \overline{A} = 1$
$A \cdot 0 = 0$	$A \cdot \overline{A} = 0$
$A \cdot 1 = A$	$A + AB = A$
$A + A = A$	$A + \overline{A} B = A + B$

Figure 11.5   Rules of Boolean algebra.

De Morgan's theorems permit us to move from AND operators to OR operators or vice versa by using complemented variables. The two theorems state:

The complement of multiplied Boolean variables is equivalent to the sum of the complements of the variables (AND → OR).

The complement of the sum of Boolean variables equals the product of the complements of the variables (OR → AND).

For two variables, A and B, de Morgan's theorems state:

$$\overline{A \cdot B} = \overline{A} + \overline{B}$$
$$\overline{A + B} = \overline{A} \cdot \overline{B}. \tag{11.6}$$

For three variables we have

$$\overline{A \cdot B \cdot B} = \overline{A} + \overline{B} + \overline{C}$$
$$\overline{A + B + C} = \overline{A} \cdot \overline{B} \cdot \overline{C}, \tag{11.7}$$

and so forth.

The above provides you with initial elements that, possibly, should be expanded in other courses or further developed by reading specific books. For now it is enough to add that Boolean expressions can be written in standard forms such as Sum-of-Products (SoP) or Product-of-Sums (PoS) and that the same logic function enables different implementations. There are techniques that help in finding the most convenient logic scheme.

Example 11.1

A seven-segment display indicates each of the 10 digits of the decimal numbering system by lighting the proper segments. The input is a four-bit digital word used for the first 10 decimal codes. The circuit includes seven drivers, each of which energizes a segment when the associated control generates a logic "1". Determine the SoP Boolean equations for each input and the logic needed to activate the segments. Figure 11.6 defines the seven-segment arrangement and recalls the truth table.

Solution

Each segment must switch on with a given set of decimal numbers that, in turn, correspond to the coincidence (AND) of a mixture of four address bits or their inverses. For example,

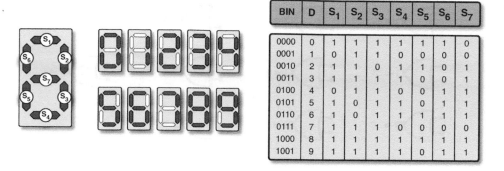

BIN	D	S_1	S_2	S_3	S_4	S_5	S_6	S_7
0000	0	1	1	1	1	1	1	0
0001	1	0	1	1	0	0	0	0
0010	2	1	1	0	1	1	0	1
0011	3	1	1	1	1	0	0	1
0100	4	0	1	1	0	0	1	1
0101	5	1	0	1	1	0	1	1
0110	6	1	0	1	1	1	1	1
0111	7	1	1	1	0	0	0	1
1000	8	1	1	1	1	1	1	1
1001	9	1	1	1	1	0	1	1

Figure 11.6   Seven-segment display and truth table.

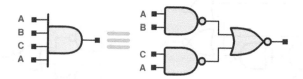

Figure 11.7   Equivalence of NAND and negate logic.

segment S_4 lights to show the decimal number "0", occurring when the AND of the four inverted bits is "1". The logic must perform various ANDs for generating the decimal codes, and then must take the OR of the decimal codes that require it to switch a segment on. The logic functions are

$$S_1 = \bar{b}_3\bar{b}_2\bar{b}_1\bar{b}_0 + \bar{b}_3\bar{b}_2b_1\bar{b}_0 + \bar{b}_3b_2\bar{b}_1\bar{b}_0 + \bar{b}_3b_2b_1b_0 + \bar{b}_3b_2b_1\bar{b}_0 + \bar{b}_3b_2b_1b_0 + b_3\bar{b}_2\bar{b}_1\bar{b}_0 + b_3\bar{b}_2b_1b_0$$
$$S_2 = \bar{b}_3\bar{b}_2\bar{b}_1\bar{b}_0 + \bar{b}_3\bar{b}_2b_1\bar{b}_0 + \bar{b}_3b_2\bar{b}_1\bar{b}_0 + \bar{b}_3b_2b_1b_0 + \bar{b}_3b_2b_1\bar{b}_0 + b_3\bar{b}_2\bar{b}_1\bar{b}_0 + b_3\bar{b}_2b_1b_0$$
$$S_3 = b_3 + b_2 + \bar{b}_1 + b_0$$
$$S_4 = \bar{b}_3\bar{b}_2\bar{b}_1\bar{b}_0 + \bar{b}_3\bar{b}_2b_1b_0 + \bar{b}_3b_2\bar{b}_1b_0 + \bar{b}_3b_2b_1\bar{b}_0 + \bar{b}_3b_2b_1\bar{b}_0 + b_3\bar{b}_2\bar{b}_1\bar{b}_0 + b_3\bar{b}_2b_1b_0$$
$$S_5 = \bar{b}_3\bar{b}_2\bar{b}_1\bar{b}_0 + \bar{b}_3b_2\bar{b}_1b_0 + \bar{b}_3b_2b_1b_0 + b_3\bar{b}_2\bar{b}_1\bar{b}_0$$
$$S_6 = \bar{b}_3\bar{b}_2\bar{b}_1\bar{b}_0 + \bar{b}_3b_2\bar{b}_1\bar{b}_0 + \bar{b}_3b_2b_1b_0 + \bar{b}_3b_2b_1\bar{b}_0 + b_3\bar{b}_2\bar{b}_1\bar{b}_0 + b_3\bar{b}_2b_1b_0$$
$$S_7 = \bar{b}_3b_2\bar{b}_1\bar{b}_0 + \bar{b}_3b_2\bar{b}_1b_0 + \bar{b}_3b_2b_1\bar{b}_0 + \bar{b}_3b_2b_1b_0 + b_3b_2\bar{b}_1\bar{b}_0 + b_3\bar{b}_2\bar{b}_1b_0.$$

Notice that even if it is possible to perform ANDs of four inputs, it may be better to accomplish the tasks in two steps by using a cascade of two elementary gates:

$$A \cdot B \cdot C \cdot D = \overline{\overline{A \cdot B \cdot C \cdot D}}$$
$$= \overline{\overline{A \cdot B} + \overline{C \cdot D}}.$$

Figure 11.7 shows the equivalence.

Notice that, since the four bits at input serve for 10 symbols only, the NAND of b_3 and b_2 is not necessary; therefore, the first-level logic generates

$$\overline{b_1 \cdot \bar{b}_0}, \quad \overline{\bar{b}_1 \cdot b_0}, \overline{b_1 \cdot \bar{b}_0}, \overline{\bar{b}_1 \cdot b_0}, \quad \overline{\bar{b}_3 \cdot \bar{b}_2}, \quad \overline{\bar{b}_3 \cdot b_2} \quad \text{and} \quad \overline{b_3 \cdot \bar{b}_2}.$$

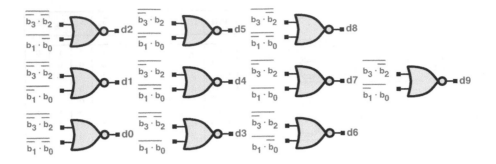

Figure 11.8   Generation of decimal codes.

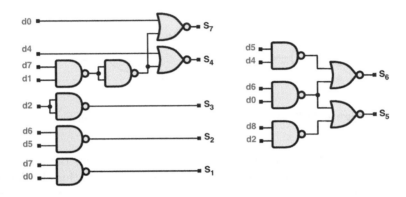

Figure 11.9   Seven-segment coding.

The NOR of pairs of the first-level outputs gives rise to the decimal codes as shown in Figure 11.8. The signal generated by the NANDs of the previous stage are used as input of the two NORs, except the one that generated $\overline{\overline{b_1} \cdot b_0}$, which is used three times. That, even if is a minor difference, can require the use of a cell with higher fan-out.

The last step is to perform the OR of the relevant decimal codes. This operation can be done with negating logic.

Custom logic optimizes the straightforward implementation of the above logic relationships and the given table. Observe that segment S_1 is not illuminated for codes d_0 and d_4. S_2 does the same with the digital codes d_5 and d_6; S_3 is off only when d_2 is "1". These three conditions require simple logic as shown in Figure 11.9. For the other segments it is possible to verify that the networks in the figure fulfill the related logic relationships.

11.4   COMBINATIONAL LOGIC CIRCUITS

The interconnection of simple logic cells, like AND, OR, NAND, and NOR, gives rise to the functions of Chapter 8. Since AND and OR need more transistors than NAND or NOR, logic

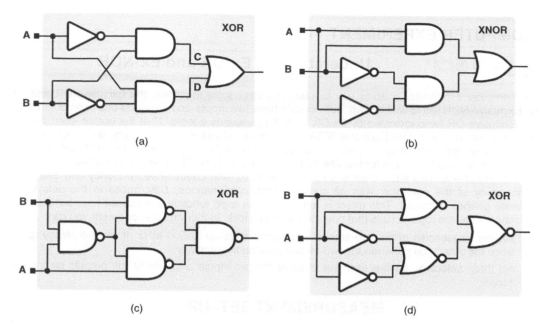

Figure 11.10   (a) SoP circuit scheme for XOR; (b) SoP circuit scheme for XNOR; (c) and (d) alternate schemes for XOR with negate logic, NANDs, and NORs.

schematics should mainly be based on negating logical cells. This section mainly describes functional diagrams of the relevant logic blocks.

11.4.1   Exclusive-OR and Exclusive-NOR

The *exclusive-OR* (XOR) generates a logic 1 when the two inputs are different. The XNOR, on the other hand, gives rise to a logic 1 when both inputs are equal. The Boolean expressions in the SoP form are

$$\text{XOR}(A, B) = A \cdot \bar{B} + \bar{A} \cdot B \tag{11.8}$$

$$\text{XNOR}(A, B) = A \cdot B + \bar{A} \cdot \bar{B}. \tag{11.9}$$

Figures 11.10(a) and (b) show the logic diagrams that directly translate the above equations into circuits. They employ NOT, AND, and OR, counting five logic cells of three different types. Since AND and OR require more transistors, the schemes are not optimal. Figures 11.10(c) and (d) realize the XOR function more effectively. In order to verify the operation of Figure 11.10(c) we can start from Figure 11.10(a) and transform it into two steps. The first observes that the AND outputs remain the same if one input changes from "0" to "1" provided that the other is "0". Thus using the input NAND of Figure 11.10(c) in Figure 11.10(a) preserves the logic operation. The second step applies the de Morgan theorem to the output OR:

$$\overline{C + D} = \overline{\overline{C} \cdot \overline{D}}. \tag{11.10}$$

This changes OR into NAND and, transferring the NOT to the input ANDs, makes them NANDs.

COMPUTER EXPERIMENT 11.2

Understanding Ex-OR and Ex-NOR

This computer experiment enables you to study two basic logic functions: the Exclusive-OR and the Exclusive-NOR. As is well known, both circuits have two inputs and generate one logic output. The Exclusive-OR (also indicated by Ex-OR or XOR) generates a logic "1" at the output when the two inputs are different. The Exclusive-NOR (Ex-NOR of XNOR), on the contrary, gives rise to a logic "1" when both inputs are equal. There are various implementations of the functions. This board implements one of them for the XNOR and two for the XOR. The first uses only NANDs, the second only NORs. Two square wave generators at input, with controllable frequency and delay, permit study of the operation with all possible input combinations. Differences in the delays generate glitches at output. This effect is not just for the logic function considered here but also happens with all the logic circuits that operate without a clock. In this virtual experiment you can:

- set the frequencies of the two square wave generators up to 500 MHz. If one frequency is twice the other, you generate the two-bit combination sequentially;
- set three selectors, so that you can observe the two inputs and one of the outputs on the scope.

MEASUREMENT SET–UP

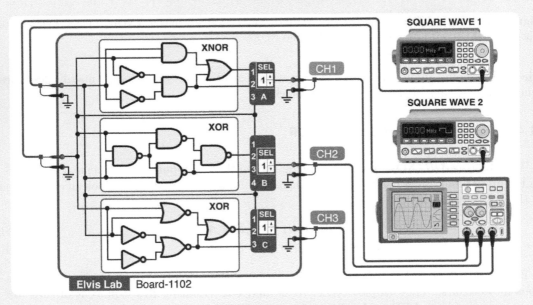

TO DO

- Set the frequency of one square generator to 10 MHz and the other to 20 MHz. Observe the outputs. Change the frequency of the first generator and observe the results.
- Suppose that you are not able to observe the inputs and you just know the function of the two processors. Based on that can you guess the value of the inputs?
- Increase the clock frequencies and observe glitches and their duration. Be aware that there is some jitter in the clock signals. Is one of the XORs better than the other? Why?

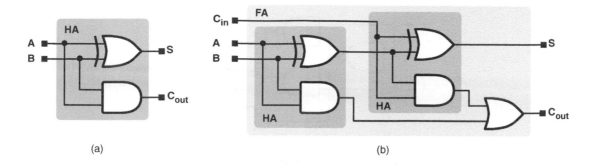

(a)   (b)

Figure 11.11   (a) Circuit implementation of a half-adder; (b) circuit implementation of a full-adder based on two half-adders.

The XOR of Figure 11.10(d) is based on the XNOR of Figure 11.10(b). The de Morgan theorem changes the input from ANDs to NORs because

$$A \cdot B = \overline{\overline{A \cdot B}} = \overline{\overline{A} + \overline{B}} \tag{11.11}$$

$$\overline{A} \cdot \overline{B} = \overline{\overline{\overline{A} \cdot \overline{B}}} = \overline{\overline{\overline{A}} + \overline{\overline{B}}} = \overline{A + B}. \tag{11.12}$$

This equation, used in Figure 11.10(d), gives rise to the scheme of Figure 11.10(d), which exchanges the last OR with a NOR to invert the logic from XNOR to XOR.

The scheme of Figure 11.10(c) uses only NANDs; since inversion can be realized with a NOR with the same signal at the inputs, the circuit of Figure 11.10(d) can be made using only NORs. Because of that, NAND and NOR are frequently called "universal gates:" with them it is possible to build any logical function.

11.4.2   Half-adder and Full-adder

Logic gates and XORs construct half-adders and full-adders. There are various possible architectural implementations. Here we discuss the ones that directly translate the truth table. The schemes obtained can become implementations using universal gates after possible transformations based on the rules of Boolean algebra and the de Morgan theorems.

If the inputs of a half-adder are A and B, the sum is "1" only if one of A and B is "high." The carry equals "1" when both A and B are "high." Therefore an XOR gives rise to the sum and an AND provides the carry, as shown in Figure 11.11(a). Notice that the possible use of the XOR scheme of Figure 11.10(d) would avoid the AND needed to generate the carry, because that signal is already generated by the NOR at the bottom of the diagram.

The inputs of a full-adder are, say, A and B plus the input carry, C_{in}. The outputs are the sum and the output carry, C_{out}. As is well known, the sum is "1" if only one of the three inputs is "high." The output carry is "1" when two or three inputs are "high." The following Boolean relationships describe the rules for generating sum and carry:

$$S = (A \oplus B) \oplus C_{in} \tag{11.13}$$

$$C_{out} = A \cdot B + (A \oplus B) \cdot C_{in}, \tag{11.14}$$

where the symbol $\oplus$ indicates XOR.

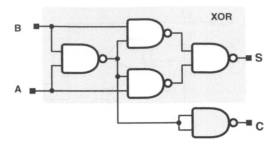

Figure 11.12   Half-adder made only of NAND gates.

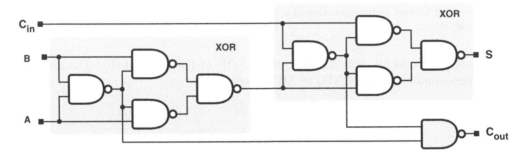

Figure 11.13   Full-adder made only of NAND gates.

The sum requires a cascade of two XORs, the first with A and B at input and the second with the output of the first XOR and the carry C_{in} at input. The output of the first XOR also serves for determining C_{out}, as shown by the scheme of Figure 11.11(b). The first XOR and the AND of the inputs construct a half-adder. The second XOR and AND make another half-adder. Therefore, as shown in Figure 11.11(b), two half-adders must be used, plus an OR, to construct a full-adder. The scheme is not at the lowest level, because it gives rise to the interconnection of logic cells. Direct use of transistor implementations of logic cells can lead to optimized solutions. There are, indeed, custom schemes determining the half-adder or the full-adder with a small number of transistors.

Example 11.2

Design a half-adder and a full-adder using only NAND gates.

Solution

The all-NAND scheme of Figure 11.10(c) is selected as the basis for both schemes. Notice that the first NAND combines two inputs. It is therefore just necessary to invert its output to produce the carry of a half-adder. The result is shown in Figure 11.12.

The use of the de Morgan theorem applied to the OR that gives rise to output carry in Figure 11.11(b) transforms that OR into a NAND and inverts the preceding ANDs. They are already available in the XOR of Figure 11.10(c). The resulting scheme is Figure 11.13.

For both schemes it is just necessary to assign an extra NAND to a single or a double XOR, of the type shown in Figure 11.10(c), made of four NANDs.

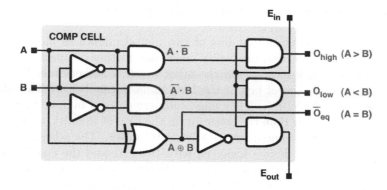

Figure 11.14   Comparator cell generating higher, lower, and not-equal logic signals. The cell operates if the input enable is "high" and generates an output enable signal.

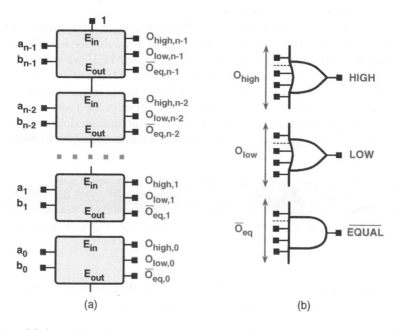

Figure 11.15   (a) Assembly of comparator cells for an n-bit comparison; (b) outputs of cells combined to obtain the overall outputs.

11.4.3   Logic Comparators

A logic comparator determines whether the values of two digital quantities are equal or whether one of the two inputs is bigger than the other. The equality of two n-bit numbers means having all of the pairs of bits with the same order equal. Their equality can be verified by XOR or XNOR with pairs of bits at input, depending whether we want to use "0" or "1" to indicate equality. Then the logic signals of the bit comparisons are combined together by suitable logic networks, mainly based on multiple input NOR or multiple input AND, to determine "higher," "lower," or "equal" results.

The inequality of the most significant bits of two digital quantities indicates that one is larger than the other. Moreover, that inequality makes any further comparison of pairs of bits with lower order useless and valueless. On the other hand, if higher-order bits are equal it is necessary to compare the pairs of bits at the level immediately below. This gives the rule for possible comparator architectures. They must compare pairs of bits, and give rise to logic signals stating that one is larger than the other, or, in the case of equality, they must generate an enable signal for activating comparison at a lower bit order. The scheme of Figure 11.14 realizes these functions simply. It is a logic cell with two bits and the enable from a higher order at input. There are four outputs, "higher," "lower," "equal," and the enable for the lower-order cell. If the input enable is "0", the output enable is also "0". The scheme of Figure 11.14 uses the obvious observation that if one bit, A, is greater than a second one, B, A is "1" and B is "0". Thus for verifying "higher" and "lower" the two Boolean operations $A \cdot \overline{B}$ and $B \cdot \overline{A}$ are required. This is done by the two left-hand sections of Figure 11.14. The XOR generates inequality signals. That, together with the input enable generated by the higher-order cell, determines the output enable.

A stack of n cells like the ones in Figure 11.14 builds the n-bit comparator. The enable of the highest order cell is "1". The outputs of all of the cells, suitably combined, give rise to three outputs: *HIGH*, *LOW*, and *EQUAL*, as shown in Figure 11.15(a) and (b).

The scheme is not optimal, being a mixture of diverse cells. Moreover, timing is disregarded. The reason is that when considering combinational logic the designer mainly focuses on the combination of inputs, and considers synchronization of results afterwards. It is clear that that results do not occur instantaneously because of the propagation delays. That effect may possibly be accounted for at the global level when the combinational cell becomes part of a sequential scheme.

Example 11.3

Redesign the logic of Figure 11.14 used in a digital comparator. Use only negate cells, NAND, and NOR.

Solution

The XOR of Figure 11.10(c), made using NANDs, is a good basis because it makes available opportune intermediate results. Analysis of the scheme yields

$$C = \overline{\overline{A \cdot B} \cdot A} = \overline{(\overline{A} + \overline{B}) \cdot A} = \overline{A \cdot \overline{B}}$$
$$C = \overline{\overline{A \cdot B} \cdot B} = \overline{(\overline{A} + \overline{B}) \cdot B} = \overline{\overline{A} \cdot B},$$

which are the inverse of the signals of Figure 11.14. The transformation of the AND output logic into a negate version turns the AND into NOR and asks for inverted inputs. This just requires inversion of the input enable and removal of the inverter at the XOR output. The scheme, shown in Figure 11.16, replaces the inverter needed with a NOR, whose inputs are both of the input enables.

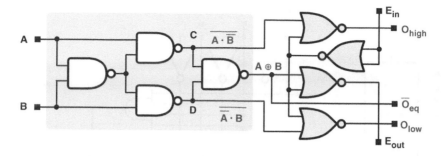

Figure 11.16   Logic for digital comparator made of NAND and NOR gates.

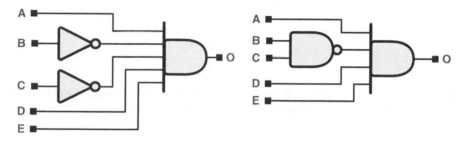

Figure 11.17   Two possible implementations of the simple decoding logic defined by equation (11.15).

11.4.4   Decoders

A decoder is a logic cell with multiple inputs and single or multiple outputs capable of detecting specified combinations of bits and acknowledging the result with an output code. Thus a decoder converts coded inputs into coded outputs. Suppose, for example, you want to control the automatic transfer of a liquid from one tank to another. The signal that acknowledges the valve opening (signal O) must verify that the liquid in the delivering tank is higher than a defined level (signal A) and that the receiving tank is below a safety threshold (signal B). Moreover, it is necessary to make sure that the flow towards a third tank is disabled (signal C) and that the cleaning valve and inspection door in the receiving tank are closed (signals D and E). The possible equation of the decoder is

$$O = A \cdot \bar{B} \cdot \bar{C} \cdot D \cdot E, \tag{11.15}$$

which requires two inverters and a five-input AND. The use of Boolean arithmetic gives rise to various implementations of the logic equation. Figure 11.17 shows the plain solution and another possible variant. The output is a single bit to acknowledge the inputs coincidence.

The circuit of Example 11.1 also includes a decoder with a four-bit input representing a decimal symbol in the output code used to control a seven-segment display. The truth table in Figure 11.6 gives the rules for designing the decoding network.

A binary n-to-2^n decoder is another kind of code converter. It uses n inputs and 2^n outputs. Only one output is active at any one time to give rise to input value decoding. The n-to-2^n decoder is frequently used to multiplex data. Suppose we have eight inputs identified by a three-bit address and want to select one of them. Figure 11.18(a) shows the functional scheme of the multiplexer (MUX) and Figure 11.18(b) provides the block diagrams for the selection

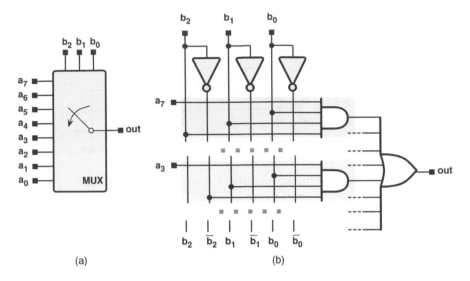

Figure 11.18 (a) Functional diagram of a multiplexer; (b) logic implementation of the eight-to-one multiplexer. It is the merging of a 3-to-8 selector and an 8-to-1 multiplexer.

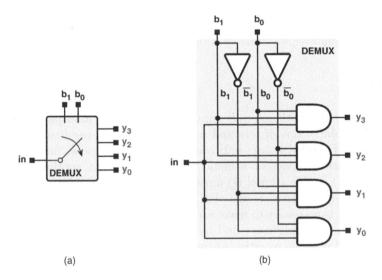

Figure 11.19 (a) Functional diagram of a demultiplexer; (b) logic implementation of the one-to-four demultiplexer.

of two of the inputs (a_3 and a_7). The decoder uses four-input ANDs. One of the inputs is the signal, and the remaining three are the address. Three inverters complement the address bits for the eight-to-one selection. Out of eight four-input ANDs there is only one with active output: the value of the selected input; all of the other ANDs' outputs are zero. The active signal gives rise to the output passing through an eight-input OR together with inactive signals.

The opposite of the multiplexer is the demultiplexer (DEMUX). It has a single input directed towards one of the multiple outputs under the control of a data-select input. Figure 11.19 shows the functional scheme and a possible logic implementation of a one-to-four demultiplexer.

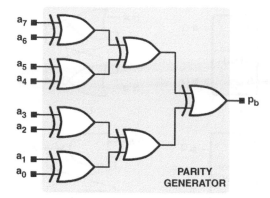

Figure 11.20 Parity even generator with eight bits at input.

The address code is two bits of data. It must be decoded four-to-one to decide where the input signal must go. For this function the logic employs three-input ANDs and two inverters necessary for the address decoder. It is easy to verify that the output of the selected address corresponds to the input and the others are zero.

11.4.5 Parity Generator and Parity Checker

The noise that affects digital signals can give rise to errors that must be detected and corrected. When transferring many bits it can be assumed that no more than one error occurs at the same time. Under this assumption, a tool useful for verifying correctness is an extra bit, called a parity bit. Its value makes the total number of "1" bits transferred either even or odd. In one case the extra bit determines even parity; in the other case it gives rise to odd parity. The parity bit is generated and added at transmission, and the parity is verified at reception by suitable combinational logic circuits.

A possible circuit that produces even parity is a cascade of XORs as shown in Figure 11.20. It combines pairs of bits to generate the parity bit. As we know, an XOR gives "0" at output if there are two "0"s or two "1"s at input. That corresponds to an even contribution to the sum. If the output of the XORs that combine the pair of bits of the next stage is again "0" that indicates an even sum, and so forth. The end of the tree of XORs gives "0" if the sum is even and "1" if it is odd.

Consider, for example, an eight-bit word whose correctness is ensured by an added extra bit that gives even parity. Depending on the signal value at input, we can have the following possible exemplary cases:

$$
\begin{array}{lcc}
\text{Data} & & \text{Parity} \\
00100101 & \rightarrow & 1 \\
10100101 & \rightarrow & 0 \\
11111001 & \rightarrow & 0 \\
11111101 & \rightarrow & 1.
\end{array}
\tag{11.16}
$$

The parity bit, added to the data, sums up into a nine-bit word. At reception a circuit similar to the parity generator examines the input data and the added extra bit to determine the sum of the "1" signals and, consequently, verifies the parity bit. Obviously, in the case of a correct parity check, the result must be affirmative. The wrong parity can give rise to

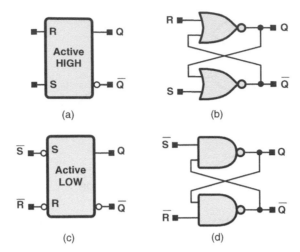

Figure 11.21 (a) Block diagram of a latch that is active when one of the inputs is high; (b) its NOR implementation; (c) block diagram of a latch that is active when one of the inputs is low; (d) its NAND implementation.

various decisions, such as, for example, a request to resend data. More sophisticated systems with more information transferred are able to correct the error.

11.5 SEQUENTIAL LOGIC CIRCUITS

This section shows how basic logic cells fulfill sequential functions. We already know that the operation of sequential circuits depends on past history. For that reason, a common feature of sequential circuits is the capability of storing data in a static or dynamic manner. The logic must have two stable states to represent the two symbols of binary arithmetic. There are many circuits that do that; we have the latch and various types of flip-flops used in shift registers, counters, and memories. It is obviously necessary to understand the basic operations and architectures. This is what we do in this section. However, this study is just an introductory step to more detailed examination.

11.5.1 Latch

The simplest kind of sequential logic is the latch. The circuit was briefly discussed in Chapter 8; it is a cell used to store logic signals temporarily, with two stable states sustained by positive feedback. The latch has two inputs with opposite effects on output, one conventionally called set (S) and the other called reset (R). There are two outputs (Q and $\bar{Q}$), one the inverse of the other, which can be utilized as a single output or as complementary signals, depending on circuit requirements. The latch is similar to the flip-flop, to be studied shortly. The difference is that the latch is not synchronous with clock transitions. On the other hand, the flip-flop switching is clocked.

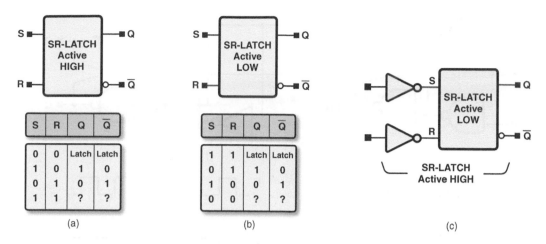

Figure 11.22 (a) Active-high S–R latch and truth table; (b) active-low S–R latch and truth table; (c) how to transform active-low into active-high.

Using universal gates as building blocks we construct two schemes of latch: the Set–Reset NOR (S–R NOR) and the S–R NAND. These, as shown in Figure 11.21, are a pair of cross-coupled two-input gates. For the S–R NOR, the so-called storage mode is when the S and R inputs are both low: under those conditions the feedback keeps the Q and $\bar{Q}$ outputs constant. If Q is "1" the control of the lower NOR keeps the value of $\bar{Q}$ at "0" independently of the value of the other input, S. In fact if S goes "high" nothing happens. By contrast, if R (reset) goes "high" Q becomes "0" because the inputs of the lower NOR change $\bar{Q}$ to "high". The set control behaves in a similar manner when Q is "low." Notice that this latch operates when one input is high. For this reason Figure 11.21(a) specifies in the block symbol that it is "active-high". However, as already indicated in Chapter 8, only one input can be "1" at the same time: the condition $R = S = 1$ is invalid (or illegal) because both NOR gates would force their output to become equal to "0". That breaks the assumption of having complementary outputs: the basis for sustaining the complementary outputs themselves. If both inputs are "1", the outputs are "0", and when both inputs return to a low value, the memorized outputs depend on the timing of the return to zero (i.e., the circuit experiences a race condition).

The latch scheme in Figure 11.21(b) uses NAND gates. Again, it has two cells in feedback configuration to memorize data. However, unlike the NOR-based cell, the circuit is in the storage mode when both inputs are "high," because using NAND cells makes the scheme active when one of the inputs becomes "low." The top input controls the data setting (i.e., it makes $Q = 1$). The bottom input determines the reset (leads to $Q = 0$). Moreover, the invalid mode is when the inputs are both "0", because both outputs go to "1". Since the signals that control the NAND-based latch are opposite to the ones driving an NOR-based scheme, the truth table of a latch is not really what is shown by Figure 8.22. For this reason Figure 11.22 updates the table and, in addition, shows how to transform an "Active-low" latch into an "Active-high" type by using two inverters.

A useful observation about this type of sequential circuit is that the output transition of both types of S–R latches occurs almost immediately after one of the inputs switches to the active condition. This feature is normally referred to by saying that the plain latch is a *transparent* cell, because the outputs reflect the input state directly without any blocking signal concealing the transitions that may occur at any time at input.

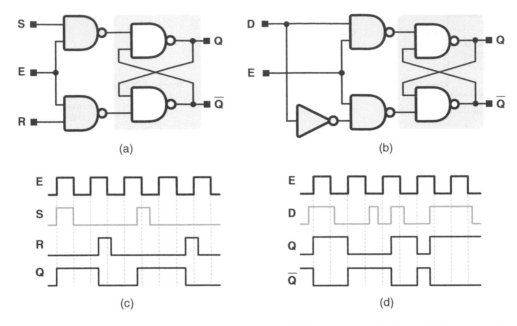

Figure 11.23   (a) Gated S–R latch; (b) gated D-latch ("D" means data); (c) and (d) input and output waveforms with periodic enable signal.

11.5.2   Gated Latch

There are situations where it is advisable to preclude the transition of the latch outputs because they can be badly influenced during the rough periods of input changes. The use of gates at inputs satisfies this requirement. This is done with additional logic gates that keep the inputs in storage mode when the enable signal is not active. Figure 11.23(a) shows a NAND-based latch with added control. The set and reset inputs do not operate the regenerative loop directly but pass though two NANDs. Only when the enable is "high" does a set or reset equal to "1" make one of the latch inputs "0", possibly changing the corresponding output. Notice that the NAND inverting action makes the latch "active-high" with the "set" control operating on the top input. Moreover, output transitions can occur at any time in the interval when signal E remains "high." It is also possible to gate the latch based on NOR. The principle of operation is the same; the inputs must be forced into the storage mode when the gating is not active. During the gating period it is still necessary to avoid the invalid mode, occurring when S and R are both equal to "0" or "1", depending on the latch type.

Take heed!

The latch may use, or not use, a clock, but not for defining transition times. It may just be used to gate the latch operation and avoid the signal bouncing caused by data uncertainty during transitions.

As was anticipated in Chapter 8, there is a useful variant of the gated S–R latch: the gated D-latch. This is a transparent cell that transfers data from input to output when the gate is closed and maintains data at output when the gate opens. As shown in Figure 11.23(b), it has the input data and its inverted replica at the set and reset inputs. When the enable signal is

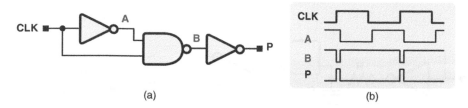

Figure 11.24   (a) Possible positive edge-transition detector; (b) waveforms at various points of the scheme.

"high" the data pass through. If the input remains "high" $Q = 1$. Conversely, when the input experiences a high-to-low transition during the enable period, Q can become "0", and remains at that level if no change happens until the end of the enable phase. Thus the gating signal determines when the latch is transparent to input data.

Figures 11.23(c) and (d) show the waveforms at the input and output of gated latches. Notice that for both schemes transitions occur only when the enable is "high." Any input status during the periods of gate-off does not modify the outputs. If one of the inputs changes at any time when the enable is "high," it possibly gives rise to output switching.

11.5.3 Edge-triggered Flip-flop

The change of state of edge-triggered schemes occurs only at the rising or falling edge of clock pulses. This is the key difference with respect to gated latches. The circuit that performs this function is the flip-flop. We have already briefly examined this type of sequential circuit in Chapter 8, where we discussed the features at a high level. There are four different types of flip-flops. The set–reset (S–R), the D-type, the JK and the toggle. Each type can be positive edge-triggered or negative edge-triggered, which means that the state changes at the low-to-high transitions of the clock or, the other way around, at the high-to-low transitions.

> **Notice that ...**
>
> the key benefit of edge-triggering techniques is that they limit the responsiveness of memory-type schemes for very short periods of time. This not only overcomes "spikes" at input but also defines the transition times precisely.

The key to the schemes is their ability to capture the rising or falling edges of the clock. This may be done by clock-edge detectors that generate very short pulses synchronous with the quick transitions of clock waveforms. Consider, for example, the circuit of Figure 11.24. The input clock passes through an inverter that generates the inversion of the clock within a very short delay. The propagation time depends on the type of transistors used, on the technology, and on the parasitic capacitances of the scheme. The clock and its delayed inverse are the inputs of a NAND that changes the normally "high" output to "0" for a very short time period: the inverter propagation delay. A cascaded inverter gives rise to a positive spike that is used for activating the latch gate. Since the pulse duration is very short, we can say that the latched output replicates the input status at the rising (or falling, by a slightly different logic) clock edge.

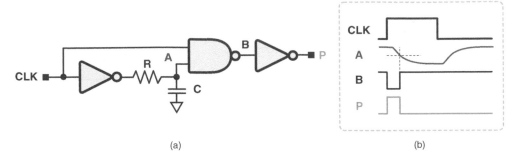

(a)   (b)

Figure 11.25   (a) Pulse generator with analog-controlled duration; (b) waveforms at various points of the scheme.

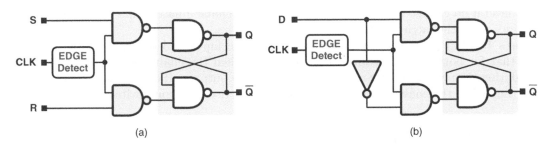

(a)   (b)

Figure 11.26   (a) Set–reset flip-flop; (b) D-type flip-flop.

The pulse duration must be short, to precisely indicate the edge transition, but should be long enough to make sure the latch operates. Modern technologies make inverter propagation delays and the speed of logic cells very fast, matching each other well. Since the edge-transition detector uses a simple circuit, the method is preferred to the traditionally used master–slave technique, described in the next subsection.

The method of generating a short pulse triggered by a clock-edge can be generalized to serve various needs. In some cases longer pulse generation may be required. The scheme in Figure 11.24 is slowed by using the analog delay network of Figure 11.25. The passive RC receives a steep square signal and gives rise to exponential-fall and exponential-rise waveforms after the delayed and inverted switchings of the clock. The NAND output remains zero until input A crosses the gate threshold. The pulse duration equals the inverter delay plus the extra time established by the analog time constant to reach the threshold.

The clock-edge detector replaces the enable input of a gated latch to construct a flip-flop. A possible flip-flop circuit scheme is shown in Figure 11.26(a). The operation is the same as that of a gated latch but the short pulse makes the transitions "clock-edge controlled." Also for this category of circuit there is an invalid input condition given by $S = R = 1$. However, invalid statuses that occur far from the clock edges do affect the operation because the latch is disabled. Thus the focus on the invalid condition must be just on the clock-edge times.

Figure 11.26(b) illustrates the second type of flip-flop, the D version. It derives from the D-type gated latch with an edge detector at input. Like the S–R flip-flop, the cell is transparent to input data only when clock edges occur. During the rest of the clock period, the cell is opaque to input and stores valid data that was applied at input at the time of clock transition.

The D-type flip-flop serves for storing data and for the parallel transfer of information in digital format. In addition, a cascade of D-flip-flops, with the output of one cell used as input

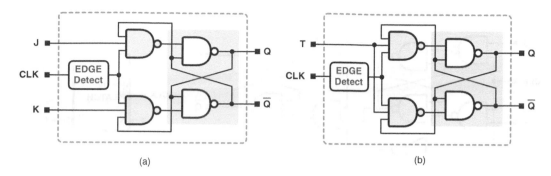

Figure 11.27   (a) JK flip-flop; (b) JK flip-flop used as toggle cell.

of the next one, shifts the stored data, each clock period, by one position. This feature is significant and is widely used for implementing many digital processing functions.

The third type of flip-flop, the JK flip-flop, essentially avoids the "invalid" or "illegal" state, a serious limitation of the S–R flip-flop. The JK version has two inputs, like the set and reset of the S–R scheme, but the input names change: they are the J and K inputs, for set and reset respectively. There is no explanation for this choice of new names; they just serve to distinguish between the two different types of flip-flop. The main difference with respect to the S–R flip-flop is that the JK replaces the two-input NANDs with three-input NANDs. The circuit feeds back $\overline{Q}$ and Q to the third inputs in such a way that the K input has an effect only when it must perform a reset and the J input has an effect just when it is necessary to execute a set. The rules, for an all-NAND scheme, bring $\overline{Q}$ to the input of the NAND controlled by K and employ Q together with J, as shown in Figure 11.27(a).

Observe that what the extra input of the NAND does is a sort of cross-inhibiting that avoids simultaneous activation of set and reset commands. If the output is already "set," the J input is inactive, because of the action of the "0" given by $\overline{Q}$. When the circuit is already "reset," the K input is neutralized by the value of Q. Other than avoiding the invalid state, there is an interesting consequence of the method used by the JK flip-flop that occurs when both J and K are "1". Since one of the outputs inhibits the input that tries to confirm that output itself, it is the other input that wins, switching the outputs to the complementary values. Therefore, with constant "high" inputs, the signals at output operate alternately on successive occasions. This manner of functioning is called a *toggle*.

The toggle property gives rise to the so-called toggle flip-flop, shown in Figure 11.27(b). There is only one input, applied to both terminals, J and K. How the scheme operates is straightforward, being a consequence of the toggle property. If the input is "0" the outputs freeze to a given status defined by the past output history. When the input becomes "high" the outputs toggle at each clock edge. The result is a waveform equal to dividing the clock by two, provided that the input value makes the circuit sensitive to the clock transitions. This feature is useful and is used in many logic circuits, including counters and dividers.

11.5.4   Master–slave Flip-flop

The traditional design of a flip-flop employs a cascade of two gated latches. The first serves as master, and the other operates as slave. The clock gates the master; its inverse activates the slave. Figure 11.28(a) shows the logic scheme. The two gated latches of the figure both use NAND gates. When the clock is "high" the master is in the transparent mode while the

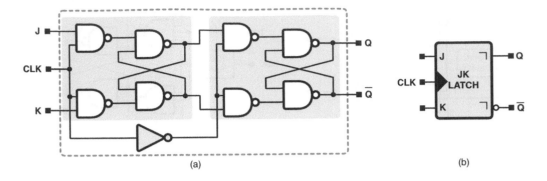

Figure 11.28   (a) Master–slave flip-flop; (b) logic symbol (ANSI/IEEE) indicating postponed operation.

slave, controlled by $\overline{CLK}$, is in the storage mode. When the clock switches to "low" the master stores the data, which passes to the slave, which switches into transparent mode.

The result is that the flip-flop receives input data when the clock signal is "high" and passes it to the output on the falling edge of the clock. The data is available at the slave outputs until the next falling edge. Moreover, since the master–slave flip-flop is sensitive to any input changes when the clock is "high," it can be necessary to set the inputs before the rising edge and keep them unchanged until the falling edge. Actually the memorized signal is what is at the input just before the transfer from master to slave. Therefore, the operation of the scheme corresponds to a clock-edge-triggered circuit, which catches the state exactly at each falling edge of the clock signal. A possible limitation is that data lasting for a clock period that changes at the leading edge of the period does not show up at output until the trailing edge. Thus the output postpones the input by half a clock period, as a small corner in the ANSI/IEEE symbol indicates (Figure 11.28(b)).

11.6   FLIP-FLOP SPECIFICATIONS

As for any electronic component, the performance, features, and limitations of flip-flops are reported in data sheets, when they are available as discrete parts, or in the library specifications when the cells are part of a design library. This kind of information is essential for their use in real circuits. Indeed, in addition to ensuring the correct implementation of Boolean functions, it is necessary so as to comply with timing responses, signal levels, and operating and power limit conditions. The specifications of flip-flops include the following.

- **Type** This defines the class of flip-flop. We have seen various categories and features; we have, for example, the D-type used to store and transfer data, the S–R, the JK for loading and temporarily retaining logic information and the T (toggling) flip-flop. In addition there are other types with, for example, extra features such as asynchronous preset and clear.

- **Triggering** This is the manner in which the clock controls the flip-flop. We can have positive-edge or negative-edge control. With a positive-edge-triggered option the output can change only at the clock's positive edges.

- **Output characteristics** The specification defines the output "low" and "high" levels for a given supply voltage. We can also have three-state output, for which a specific pin (Output Enable, OE) enables the output. It is also possible to have single output and complementary outputs.

- **Supply voltage** This depends on the technology. Obviously, when using multiple parts the same supply voltage determines a proper matching of logic signals.

- **Propagation delay** This is the time interval between application of the input signal and occurrence of the associated output. There are four different contributions: the low-to-high propagation delay, t_{PLH}, given by the time interval between the triggering edge and the low-to-high transition, the high-to-low propagation delay, t_{PHL}, defined in the same manner but considering the high-to-low transition, the preset time (t_{PH}), and the clear delay (t_{PL}). The last two concern the application of preset or clear commands.

- **Set-up time** This is the minimum time required before the clock triggering edge to recognize an input signal status.

- **Hold time** This is the minimum time required after the triggering edge to ensure that the effect of the input signal is held.

- **Maximum clocking frequency** This is the maximum clocking frequency at which the flip-flop operates in a reliable manner.

- **Power dissipation** This is the total power consumed by the device. For MOS technologies the specifications focus on dynamic dissipation, but modern technologies increasingly take the leakage terms into account.

- **Operating conditions** These specifications are similar to the ones given for generic integrated circuits. It is important to know them, because performance can change significantly when the part operates out of its specified operating ranges, even when working inside the maximum rating ranges.

Self training

Make a search on the Web for different types of logic integrated circuits. Consider the CMOS, low-power Schottky, TTL, and ECL implementations. Look at different features, namely speed, signal levels, and power consumption. Focus on one of these logic functions:

- AND/NAND, OR/NOR, XOR/XNOR;

- NOR- and NAND-gated latch;

- S–R and JK flip-flop;

- master–slave flip-flop.

Write a comparative report on the different technological implementations of one of the above logic functions. Pay attention to two key features: speed and power consumption.

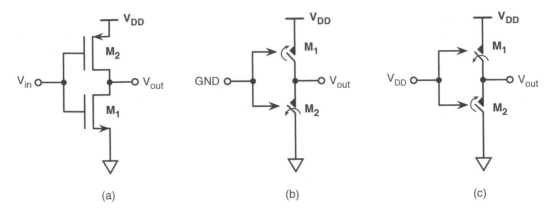

Figure 11.29 (a) CMOS digital inverter; (b) switch model of inverter with input voltage at ground; (c) switch model with input connected to supply voltage.

11.7 TRANSISTOR SCHEMES OF LOGIC CELLS

Until now we have described logic circuits as the interconnections of logic cells that realize from a functional point of view basic Boolean expressions. However, since the analysis assumes those cells to be ideal, the limitations and features defined in the specification list in the previous section are not noticeable, because they are evident only when circuits are examined at the transistor level. The previous self training activity presented various possible technologies that can be used for fabricating logic cells. Here we consider only the CMOS technology, as that is the most widely used. Since other technologies are for special needs, their study can be postponed to more advanced courses or done using specialized books or technical documents.

11.7.1 CMOS Inverter

The CMOS inverter implements logic negation. It uses very little power and operates at very high speed. With nanometer technologies an inverter can work at several tens of GHz. The CMOS inverter, other than accomplishing its logic function, often serves as the inverting or non-inverting buffer needed to drive large capacitive loads. A cascade of a number of inverters with tapered transistor sizes makes increasing driving capabilities secure.

Two transistors, one P-channel and the other N-channel, connected as shown in Figure 11.29(a), build a logic CMOS inverter. It looks like the inverter with active load studied in Chapter 10, but the difference is that the input signal drives the gates of both transistors. When input voltage is very low, at its extreme at ground level, $V_{GS,N}$ of the N-channel device is zero, while $V_{GS,P}$ equals the supply voltage V_{DD}. The result is that the N-channel device operates like a switch that is turned off, while the large driving voltage of the P-channel device (almost) makes it a closed switch. These simple observations justify an elementary switch model of the CMOS inverter's functioning, as shown in Figure 11.29(b). If the input signal is high, at the limit V_{DD}, we have complementary operative conditions. The N-channel device is like a closed switch while the P-channel device works as an opened switch, as shown in Figure 11.29(c).

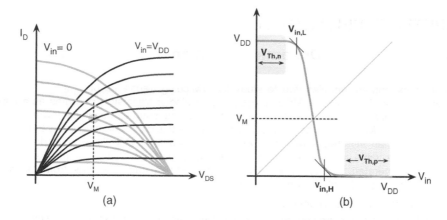

Figure 11.30 (a) Voltage–current plot of the input transistor and active load of a typical CMOS inverter; (b) input–output voltage characteristic of CMOS inverter.

The simple model made from switches verifies that the circuit performs the inverter function. When the input is "high" a closed switch connects the output to ground, giving rise to a logic "0". A "low" input closes the top switch, making the output equal to V_{DD}, a logic "1". A more accurate static study models the transistor in the "on" state as an equivalent resistance. For that analysis we observe that, assuming the load at output is just by capacitor, the quiescent current in the "on" transistor is zero. Since zero current brings the device into the triode region, we use equation (9.35), which defines the output conductance of a MOS transistor as a function of applied voltages. The result is

$$g_{out} = \mu C_{ox} \frac{W}{L}(V_{DD} - V_{th}),$$ (11.17)

showing that the output node is close to V_{DD} or to ground by a resistance that depends on technology, supply voltage, and transistor aspect ratio.

The study of the static behavior of a CMOS inverter involves the voltage–current curves of the N-channel and P-channel transistors. The study can be done in a graphic manner using, for example, the voltage–current plane of the N-channel transistor. The two sets of voltage–current curves give rise to the diagram of Figure 11.30(a). The procedure, like the one already discussed for analog circuits, identifies the intersection of pairs of curves for which $(V_{GS,n} + V_{GS,p}) = V_{DD}$ and determines the input–output operating points as a function of $V_{in} = V_{GS,n}$. The result is a curve like Figure 11.30(b). It drops from V_{DD} to ground at an intermediate voltage with a given slope. The slope and transition region depend on the technology and design used. The output voltage always equals V_{DD} until $V_{in} < V_{Th,n}$ and is zero for $V_{in} > V_{DD} - V_{Th,p}$.

A useful quantity is the crossing of the input–output curve with the $V_{out} = V_{in}$ line. The voltage at that point is called the *gate switching threshold*, V_M or the *voltage midpoint*. The value of V_M is normally estimated by assuming both transistors are in the saturation region. Neglecting the term accounting for output resistance and presuming the same C_{ox} for both transistors, V_M confirms the condition

$$I_{inv} = \frac{\mu_n C_{ox}}{2}\left[\frac{W}{L}\right]_n (V_M - V_{Th,n})^2 = \frac{\mu_p C_{ox}}{2}\left[\frac{W}{L}\right]_p (V_{DD} - V_M - V_{Th,p})^2,$$ (11.18)

COMPUTER EXPERIMENT 11.3

Dynamic Operation of Simple Cells

This computer experiment enables you to study the switching transients of simple CMOS logic cells. As explained in this book, an MOS transistor (or a bipolar transistor) operates as a current source, if in saturation. When the voltage, across drain and source, becomes lower than the overdrive voltage, the MOS works as a resistor. The board of this experiment contains three simple popular logic cells: an inverter, a two-input AND and a two-input OR. All circuits drive a load capacitance, whose value also accounts for the capacitive load established by the oscilloscope. The inputs are generated by square wave generators, whose output swings from the supply voltage to ground. In this virtual experiment you can:

- set the voltage of the supply generator;
- change the frequency of the square waves. The swing amplitude tracks the supply voltage;
- set the value of the load capacitances (equal for all the cells);
- change the *W/L* ratio of the P-channel and the N-channel transistors of all the circuits.

MEASUREMENT SET–UP

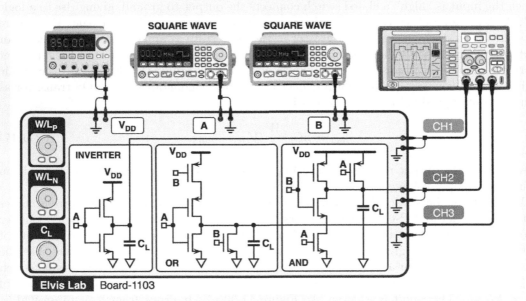

Elvis Lab   Board-1103

TO DO

- Set the frequency of square wave A to 10 MHz and the one of square wave B to 5 MHz. Set the supply voltage to 1 V. Observe charging and discharging transients of the output nodes that change status.
- Change the load capacitance and measure the propagation time as a function of the load.
- Repeat the above measurements with different *W/L* ratios of the two transistors.
- Increase the supply voltage to 1.8 V and repeat the above observations.
- Change the frequency of both input signals with one five times the other. Observe what happens at very high frequency.

which yields

$$V_M = \frac{V_{DD} - V_{Th,p} + V_{Th,n}\sqrt{r_{NP}}}{1 + \sqrt{r_{NP}}}, \quad r_{NP} = \frac{\mu_n W_n L_p}{\mu_p W_p L_n}. \tag{11.19}$$

This defines r_{NP} as the product of mobility and aspect ratio (W/L) of the N-channel transistor divided by the same product for the P-channel device.

Obviously, it is desirable to have V_M halfway to the interval supply voltage-ground. This would give symmetrical low and high ranges for logic "0" and "1". V_M equals $V_{DD}/2$ under the condition

$$\frac{W_n L_p}{W_p L_n} = \frac{\mu_p}{\mu_n} \left(\frac{V_{DD}/2 - V_{Th,p}}{V_{DD}/2 - V_{Th,n}} \right)^2, \tag{11.20}$$

giving a design rule for optimal transistor sizes with a defined supply voltage and technology.

For characterizing an inverter, there are two other useful quantities: $V_{in,L}$ and $V_{in,H}$ (input low voltage and input high voltage). They are the input voltages for which the slope of the input–output characteristics is -1. These two points correspond to input voltages that bring one transistor into the triode region and the other into saturation. How to estimate these quantities is discussed in the following example. The two voltages should be close to V_M so as to ensure there are wide voltage ranges permitted for logic "0" and logic "1".

Example 11.4

Determine the input low voltage, $V_{in,L}$, for a CMOS inverter operating with $V_{DD} = 1$ V. The threshold voltages are $V_{Th,n} = V_{Th,p} = V_{Th} = 0.3\, V_{DD}$. The transistor sizings are such that $r_{NP} = 1$. Observe what happens when the two thresholds become as low as $0.2\, V_{DD}$.

Solution

The use of equation (11.19) determines V_M. It gives

$$V_M = \frac{V_{DD} - V_{Th} + V_{Th}\sqrt{r_{NP}}}{1 + \sqrt{r_{NP}}} = 0.5 \text{ V}.$$

The input low voltage $V_{in,L}$ is when the N-channel is in saturation and the P-channel in triode. Since the current in the transistors is the same, the result is

$$
\begin{aligned}
I_{inv} &= \frac{\mu_n C_{ox}}{2} \left[\frac{W}{L} \right]_n \{ V_{in,L} - V_{Th} \}^2 \\
&= \mu_p C_{ox} \left[\frac{W}{L} \right]_p \left\{ (V_{DD} - V_{in,L} - V_{Th})(V_{DD} - V_{out}) - \frac{(V_{DD} - V_{out})^2}{2} \right\}.
\end{aligned}
$$

That yields

$$\{ V_{in,L} - V_{Th} \}^2 = \{ 2(V_{DD} - V_{in,L} - V_{Th})(V_{DD} - V_{out}) - (V_{DD} - V_{out})^2 \}.$$

The above equation determines V_{out} as a function of $V_{in,L}$ and the design parameters:

$$
\begin{aligned}
V_{out} &= (V_{in,L} + V_{Th}) + \sqrt{(V_{in,L} + V_{Th})^2 - (V_{in,L} - V_{Th})^2 + V_{DD}[V_{DD} - 2(V_{in,L} + V_{Th})]} \\
&= (V_{in,L} + V_{Th}) + \sqrt{(2V_{in,L} + V_{DD})(2V_{Th} - V_{DD})}.
\end{aligned}
$$

The use of the condition

$$\frac{dV_{out}}{dV_{in,L}} = -1$$

yields

$$1 + \frac{1}{2\sqrt{(2V_{in,L} + V_{DD})(2V_{Th} - V_{DD})}} = -1$$

and gives rise to the following relationship for input low voltage, $V_{in,L}$:

$$V_{in,L} = \frac{3V_{DD} + 2V_{Th}}{8} = 0.45 \text{ V}.$$

This is a good result, acceptably close to V_M. The above equation shows that if thresholds diminish $V_{in,L}$ also diminishes by an amount equal to $\Delta V_{Th}/4$. With $V_{Th} = 0.2$, $V_{in,L}$ becomes 0.425 V.

11.7.2 Dynamic Response of CMOS Inverters

A useful aspect of the CMOS inverter to study concerns the dynamic response when the output goes from low to high or discharges from high to low. Since the output load of CMOS inverters is a capacitor, it is the charge or discharge transient of that capacitance that controls the responses. For a first approximate study it is enough to include, in the switch model of Figure 11.29, the equivalent resistance of the transistor that goes on.

With RC transients the speed increases with low time constants: i.e., low resistance and low capacitance. Equation (11.17) provides, to a first approximation, the resistance; for the capacitance we have to take into account all of the elements involved in the process. These are the parasitics of the transistors and the loading circuitry. Consider, for example, the cascade of two inverters shown in Figure 11.31(a). The scheme includes all of the capacitances that limit the speed but omits the irrelevant ones. For example, source-to-ground and bulk-to-source parasitics are ignored because the voltage across them is zero. In addition, the circuit includes C_w, the parasitic capacitance of wire connections.

The low-to-high transition switches the transistor M_2 on. Figure 11.31(b) models M_2 with the resistance $R_{on,2}$ and comprises the capacitors that $R_{on,2}$ charges or discharges. Also, the figure indicates the status immediately after the switching on. Capacitors C_{db1}, C_w, and C_{gs3} are initially discharged because the previous output voltage was "0". The others are pre-charged to the supply voltage. The $+$ and $-$ signs denote the pre-charged condition. Capacitors C_{db1}, C_w, and C_{gs3} need to be charged while C_{db2} and C_{gd4} must be discharged. The case of the feedback capacitors, C_{f1} and C_{f2}, is different. They are already pre-charged to the supply voltage and must be discharged beforehand, and then recharged with the opposite polarity. Therefore they demand double charging, or must be considered twice in the transient estimation.

Analogous considerations describe the high-to-low transition, whose approximate equivalent circuit is shown in Figure 11.31(c). Here the capacitances are the same but the equivalent on-resistance is that of transistor M_1, $R_{on,1}$.

To conclude, the time constants become

$$\tau_{LH} = R_{on,2}(C_{db1} + C_{db2} + C_{gs3} + C_{gs4} + C_w + 2C_{f1} + 2C_{f2}) \tag{11.21}$$

$$\tau_{HL} = R_{on,1}(C_{db1} + C_{db2} + C_{gs3} + C_{gs4} + C_w + 2C_{f1} + 2C_{f2}), \tag{11.22}$$

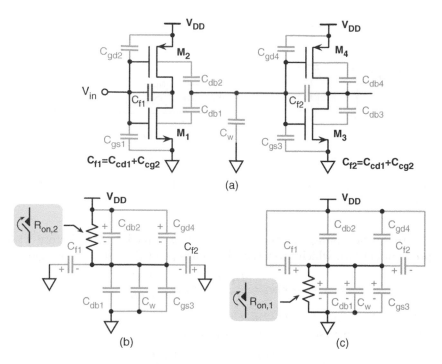

Figure 11.31   (a) Cascade of two CMOS inverters with highlighted parasitic capacitances; (b) equivalent circuit for the low-to-high transition; (c) the same for the high-to-low transition.

giving the delay for reaching a given output amplitude. With an exponential response, the output reaches 50% of the final value after 0.69τ and 80% of the final value after 1.61τ.

Reducing the time constant by increasing the on-conductance (equation (11.17)) for a given technology and fixed supply voltage would require augmenting the aspect ratio (W/L), or, better, increasing the width while the length is at its minimum. However, a larger area increases C_{gs}. Thus, augmenting the transistor width does not help much.

A more accurate study of the transient responses must take into account the fact that at the beginning of the transient the transistor is switched on and it remains in saturation until $V_{DS} < V_{sat} = V_{DD} - V_{Th}$. The current can be approximated by

$$I_D = \frac{\mu_n C_{ox}}{2}\left[\frac{W}{L}\right]_n (V_{DD} - V_{Th})^2. \tag{11.23}$$

That neglects the term accounting for the saturated output resistance, $(1 + \lambda V_{DS})$.

The (almost) constant current changes the output voltage linearly until the time at which the transistor enters the triode, say t_s. Then it is necessary to use the voltage–current relationship that is valid for transistors in the triode region.

We normally use the more accurate procedure for estimating a design parameter that is frequently used: the propagation delay, t_p. This measures the time required to go from 10% to 90% of the supply voltage for low-to-high transitions, and from 90% to 10% of V_{DD} for the opposite swings. The high-to-low estimation made here supposes that $V_{sat} < 0.9\,V_{DD}$ (i.e., the transistor goes into the triode region after the 90% crossing). Thus the current remains constant until the time $t_s > t_{0.9}$ (where $t_{0.9}$ is the time of the $0.9\,V_{DD}$ crossing).

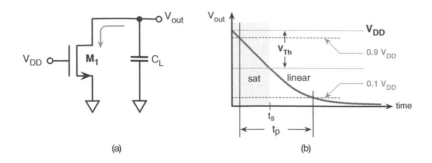

(a)   (b)

Figure 11.32   (a) Equivalent circuit for studying the high-to-low transition of a CMOS inverter; (b) transient response.

From $t = 0$ to t_s the voltage across the discharging capacitor C_L (Figure 11.32) drops linearly. Expression (11.23), which gives the MOS current in the saturation region, leads to

$$V_{out}(t) = V_{DD} - \frac{I_D \cdot t}{C_L} = V_{DD} - \frac{KV_{ov}^2 t}{C_L}, \qquad (11.24)$$

where $V_{ov} = (V_{DD} - V_{Th})$ and $K = (1/2)\mu_n C_{ox} W/L$.

With equation (11.24) we estimate the times $t_{0.9}$ and t_s:

$$t_{0.9} = \frac{0.1 V_{DD} C_L}{KV_{ov}^2}, \quad t_s = \frac{V_{Th} C_L}{KV_{ov}^2}. \qquad (11.25)$$

After t_s M_1 operates in the linear region, the successive transient follows the equations

$$\frac{dV_{out}}{dt} = -\frac{I_D(V_{out})}{C_L}, \quad I_D(V_{out}) = \mu_n C_{ox} \left[\frac{W}{L}\right]_n \left\{(V_{DD} - V_{Th})V_{out} - \frac{1}{2}V_{out}^2\right\}, \qquad (11.26)$$

which uses a simplified linear-region MOS current–voltage relationship.

The use of equation (11.26) and the initial condition $V_{out} = V_{DD} - V_{Th}$ yield

$$\int_{V_{ov}}^{V_{out}(t)} \frac{dV}{K\{2V_{ov}V - V^2\}} = -\int_{t_s}^{t} \frac{dt}{C_L}$$

$$\int_{V_{ov}}^{V_{out}(t)} \left\{\frac{1}{2V_{ov} - V} + \frac{1}{V}\right\} \frac{dV}{2KV_{ov}} = -\frac{t - t_s}{C_L}$$

$$\left\{\ln\left(\frac{V}{2V_{ov} - V}\right) \frac{C_L}{2KV_{ov}}\right\}\Bigg|_{V_{ov}}^{V_{out}(t)} = t_s - t, \qquad (11.27)$$

which implicitly determines V_{out}.

Equations (11.24) and (11.27) give rise to the required propagation delay, given by

$$t_p = t_{0.1} - t_{0.9} = \frac{C_L}{KV_{ov}^2}[V_{Th} + V_{ov}\ln(2\chi_{ov} - 1) - 0.1V_{DD}], \qquad (11.28)$$

where the symbol χ_{ov} indicates the ratio $(10\ V_{ov}/V_{DD})$: i.e., how many times the overdrive voltage V_{ov} exceeds 10% of V_{DD}.

Equation (11.28), determined after elaborate study, is just an approximation of the result, useful for first-order hand estimations. To obtain precise results, it is necessary to analyze more complicated models or, better, to use the computer to perform time-domain circuit simulations.

11.7.3 Power Consumption

There are two contributions to power consumption: static and dynamic. For MOS inverters made by traditional technologies, the static term is virtually zero, because there is no current flowing from V_{DD} to ground in static conditions. However, nanoscale technologies give rise to leakage currents through extremely thin gate oxides. The leakage term causes a significant increase in the static power, especially for very complex integrated circuits that incorporate hundreds of millions or billions of transistors.

The dynamic power comes from charging and discharging capacitors. As we know, the energy that a supply generator V_{DD} must provide to charge a capacitor C is

$$E_{ch} = V_{DD} \int_0^\infty I_C(t)\,dt = V_{DD} \int_0^\infty C \frac{dV_C}{dt}\,dt = V_{DD} \int_0^{V_{DD}} C\,dV_C = CV_{DD}^2. \qquad (11.29)$$

Half of that energy is stored on the capacitor, being

$$E_{st} = \int_0^\infty I_C(t)\,V_C(t)\,dt = \int_0^{V_{DD}} C\,V_C\,dV_C = \frac{1}{2}CV_{DD}^2; \qquad (11.30)$$

the other half becomes heat in the resistance that connects the supply generator and the capacitor.

Since the subsequent high-to-low transition dissipates the energy stored on the capacitor, a complete charge/discharge cycle consumes the entire CV_{DD}^2. If the inverter activity corresponds to f_{av} average charge/discharge cycles the dynamic power is

$$P_{inv,dyn} = C_{inv} V_{DD}^2 f_{av}. \qquad (11.31)$$

In addition to the power consumed in capacitors, there is a term caused by direct current flowing during inverter transitions. There are input voltages for which both N-channel and P-channel transistors are conductive. That situation gives rise to non-zero current flowing from V_{DD} to ground. With input voltages that switch very fast the current caused is a sharp spike.

The power consumed depends on the relationship between input voltage and output current. It shows zero current when the input is lower than the N-channel threshold and zero current again for $V_{in} > V_{DD} - V_{th,p}$. In between there is a peak, which, for an approximate study, is assumed to occur for $V_{in} = V_{DD}/2$. With that input voltage, both transistors are likely to be in saturation. Therefore

$$I_{peak} = \frac{\mu_n C_{ox}}{2} \left[\frac{W}{L} \right]_n \left(\frac{V_{DD}}{2} - V_{Th,n} \right)^2. \qquad (11.32)$$

The voltage current diagram looks like Figure 11.33(a).

The current that flows during input switching lasts for the rise or fall times, rising to I_{peak} and returning to zero with a waveform that resembles a triangle. Therefore, the switching consumed energy can be estimated by

$$E_{sw,rise} = V_{DD} \int_{t_{sw}-t_r/2}^{t_{sw}+t_r/2} I_{out}\,dt = \frac{V_{DD} I_{peak} t_r}{2} \qquad (11.33)$$

$$E_{sw,fall} = V_{DD} \int_{t_{sw}-t_f/2}^{t_{sw}+t_f/2} I_{out}\,dt = \frac{V_{DD} I_{peak} t_f}{2}. \qquad (11.34)$$

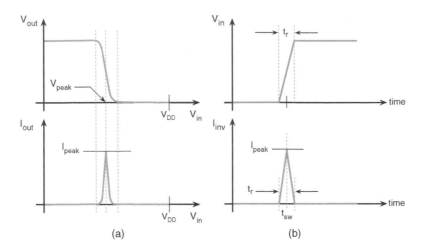

Figure 11.33   (a) Input–output static characteristics (voltage and current) of CMOS inverter; (b) possible current transient response with finite input rise time.

To sum up, the power consumed with an average activity f_{av} is

$$P_{sw,dyn} = V_{DD} I_{peak} \frac{t_r + t_f}{2} f_{av}. \tag{11.35}$$

Since the design of an inverter focuses on speed, I_{peak} cannot be diminished to limit the power consumption. The only such parameters are the rise and fall times. Therefore, critical wire connections that smooth the waveform must be as short as possible.

11.7.4   NOR and NAND

Schemes made by switches bring about circuits realizing NOR and NAND functions. The NOR cell with two inputs generates an output "0" when one of the inputs is "high." This requires the parallel connection of two switches that pull down the output under the control of a "high" input. To have a "1" at output it is necessary to have both inputs equal to "0". Thus pulling up the output node requires the concurrent action of two switches that, consequently, must be connected in series, as shown in Figure 11.34(a). Multiple inputs NORs determine similar schemes with multiple switches in parallel, for pulling down the output, and multiple switches in series to generate a "1". A scheme with switches also implements a two-input NAND. Having "0" at output needs both inputs at "1". This requires two switches connected in series pulling down the output node. For setting a "1" at output it is enough to have one of the inputs "low." Accordingly, two parallel pull-up switches made active by an input "0" produce the result. Figure 11.34(b) illustrates the switch diagram. The controls of the pull-down switches are the logic A and B; the controls of the pull-up are, on the other hand, the negate inputs.

The circuit transformation of the switch scheme is straightforward, because an N-channel transistor with source connected to ground operates as an on-switch when the gate is "high." A P-channel transistor with the source connected to the supply terminal is an on-switch when its gate is "low." The transistor schemes of NOR and NAND are given in Figure 11.35.

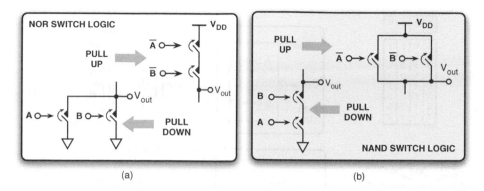

Figure 11.34   (a) NOR logic implemented with switches; (b) NAND logic realized with switches.

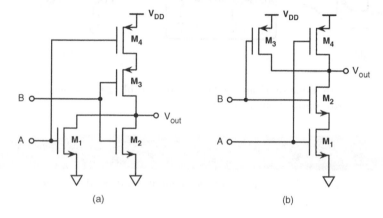

Figure 11.35   (a) Transistor implementation of two input NORs; (b) transistor implementation of two input NANDs.

The input signal A drives transistors M_1 and M_4: one is N-channel and the other is P-channel. The input B drives M_2 and M_3. Notice that the circuit configuration, for both NOR and NAND, is asymmetrical, because the series transistor driven by the input B is closer to the output node. This makes that transistor faster than the other one and, for this reason, is suitable for *critical paths*.

It is easy to extend the schemes of Figure 11.35 to a multiple-input logic. It is enough to add pull-up or pull-down transistors in parallel or in series, depending on the function. However, many transistors in series give rise to operational problems because the speed is lower. Often it is necessary to analyze the trade-off between a cascade and a multiple-input solution.

Be aware!

The multiple inputs of a logic gate are not equivalent. The ones closer to the output node give rise to better speed performance. Use them for critical paths.

Switch logic and associated circuits can also implement generic functions fulfilling a given truth table. They are often custom solutions used to minimize the transistor count. The next example is a simple illustration of the technique.

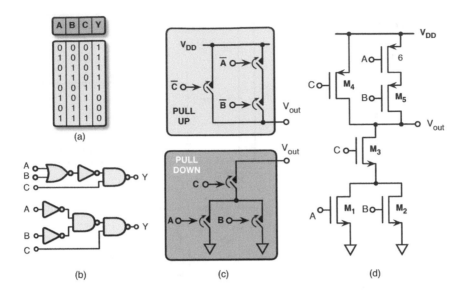

Figure 11.36 (a) Truth table of Example 11.5. (b) Possible gate implementation. (c) Switch configuration. (d) Transistor optimal implementation.

Example 11.5

Derive a switch network and its transistor equivalent that realizes the Boolean function $\overline{(A + B) \cdot C}$.

Solution

The truth table of Figure 11.36(a) gives the logic relationship for the switch configuration of the pull-up and pull-down sections. The corresponding gate interconnections for the two implementations are also given.

Note that "low" output corresponds to three input combinations, all of them with $C = 1$ and A or B equal to "1". The resulting pull-down network is the one in Figure 11.36(c). The "high" output is for five combinations. Four of them have $C = 0$ and one has $A = B = 0$. The pull-up network is straightforward.

The CMOS circuit directly derives from the switch logic as shown in Figure 11.36(d). It just uses six transistors. Three of them are N-channel and three are P-channel. The transistor circuits implementing the logic gate counterparts, illustrated by the functional circuit of Figure 11.36(b), would need 10 or 12 transistors respectively.

11.7.5 Pass-gate Logic

The logic circuits studied until now "pass" the supply or the ground voltage to output under the control of input signals. Instead of using one of the supply terminals, the pass-gate logic passes one of the logic inputs directly to the output via switches controlled by other logic signals.

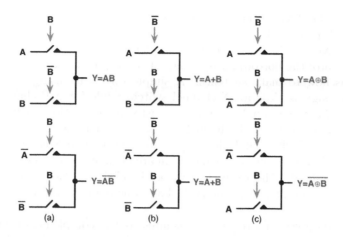

Figure 11.37   (a) Pass-gate logic implementing AND and NAND; (b) pass-gate logic implementing OR and NOR; (c) pass-gate logic implementing XOR and XNOR.

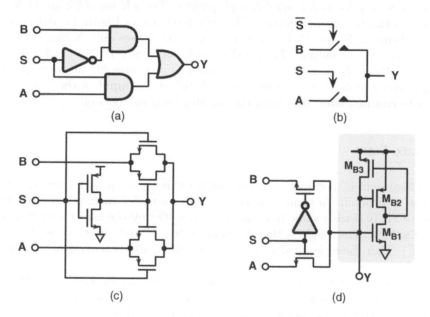

Figure 11.38   (a) Logic scheme of two-to-one multiplexer; (b) switch scheme; (c) gate-logic implementation; (d) pass-transistor implementation with buffer for weak "1" recovery.

In order to understand the method, let us consider the six schemes of Figure 11.37, which give rise to six simple two-input logic functions by using two switches and input signals or their complements. A simple analysis lets us understand the logic, provided that the wired connections after the switches are regarded as the superposition of signals able to pass through. Only one switch closes per controlling combination. Consider, for example, the NAND of Figure 11.37(a). The top switch transfers $\overline{A}$ when $B = 1$; the bottom switch transfers $\overline{B}$ when $B = 1$. Thus there is a "1" only for $A = B = 0$, as envisioned for NAND. When the input of the bottom switch of the AND is zero, output is zero: it just serves to provide "0" at output when the other switch is opened.

Now examine the XOR. The scheme directly implements the truth table. The top switch passes A when B is zero; the bottom switch sends $\overline{A}$ to output when B is "1".

The point now is how to make switches. A simple way is by using a simple transistor. Obviously, plain "1" signals close an N-channel switch, while for a P-channel switch a "0" control is necessary. Moreover, an N-channel transistor passes a "low" signal well when the gate is "1". The same happens for P-channel transistors when passing "1" with the gate at "0". On the other hand, "high" signals do not pass well through an N-channel switch, because the scheme is like a follower circuit. Gate and drain are both at V_{DD} and output is below V_{DD} by one N-channel threshold. The result is a weak "high" level. However, the result is still usable if it is within the admitted "1" interval. The same happens for a "0" passing through a P-channel switch.

The use of pass-gates gives safer results. Pass-gates are two complementary transistors with normal and inverted gate controls. The limit is the use of two devices, the body of one of which is in the well. Another solution uses single transistors followed by a digital buffer to recover from the voltage loss. This solution is preferred for medium pass-logic complexity.

Pass transistor logic is frequently used in multiplexers. The scheme of Figure 11.38(a) shows the gate logic of a two-to-one multiplexer. The switch scheme of Figure 11.38(b) gives rise to the pass-logic scheme of Figure 11.38(c). The circuit, which uses only N-channel transistors, can generate a weak "1" at output. The possible buffer of Figure 11.38(c) enforces that weak "1". The inverter M_{B1}–M_{B2} senses the "1" at input and generates "1" at output. This signal activates M_{B3}, which pulls up the output node. A "0" at the input of the inverter must be strong enough to contrast with M_{B3} until the inverter itself switches off.

11.7.6   Tri-state Gates

A common feature of CMOS logic gates is that there is always a path from V_{DD} to output or from output to ground. In some situations it can be useful to have no path or a high-impedance state. This is what some integrated circuits offer by the use of a control pin called OE (Output Enable). The result is that in addition to "1" and "0" there is the impedance output (High-Z). That, being equivalent to an open switch, allows sharing of the same output lines.

Tri-state logic is normal logic with a series switch incorporated. Figure 11.39(a) shows a simple tri-state inverter. This is a normal inverter followed by a pass-switch. The inverter output passes when the enable signal is "high." Figure 11.39(b) incorporates the enabling transistors in the inverter. They are in series with pull-up and pull-down transistors and are both "on" when enable is "high." The solution of Figure 11.39(b) is faster than Figure 11.39(a) because the input is closer to the output. The tri-state arrangement of Figure 11.39(b) is suitable for any combinational logic: it is enough to use that combinational logic in between two complementary enabling transistors connecting the logic to the supply terminal and ground.

Remark

The tri-state logic makes three possible logic signals available: the "1" or "high," the "0" or "low," and the "Z" or high impedance.

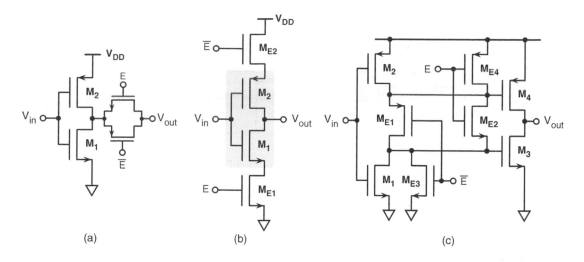

Figure 11.39   (a) Tri-state inverter made by normal inverter and output switch; (b) tri-state inverter with switch incorporated in the scheme; (c) tri-state buffer for capacitance driving.

The buffers used to drive large capacitive loads like the output pins of an integrated circuit can provide tri-state output. They are made up of a cascade of an even number of tapered inverters, with the last two of them having the tri-state option. Figure 11.39(c) shows a possible scheme with two enabling and two disabling transistors. M_{E1} opens the drain of the input inverter while M_{E2} separates the gates of the output inverter when the enable signal is "low." In addition, M_{E3} and M_{E4} zero the V_{GS3} and V_{GS3}, and short the source and drain of the input inverter. The condition makes sure that both input and output inverters are firmly turned off.

11.7.7 Dynamic Logic Circuits

Dynamic logic is a design method that relies on the infinite resistance of the gate of MOS transistors. This feature allows us to pre-charge a node, leave it alone, and possibly discharge it – only if necessary. The operation anticipates a pre-charge (or pre-discharge) phase, during which all relevant nodes are pre-charged (or pre-discharged), and an evaluation phase, during which the input data decides whether the node must remain at the pre-charged level or be brought to the complementary logic level. This method was popular many years ago and is still beneficial for high-speed digital circuits. Unlike other circuits, the dynamic logic one needs a clock, because it is necessary to distinguish between the pre-definition phase and the evaluation phase.

The two different types of dynamic logics are illustrated in Figure 11.40. One pre-charges the output node at V_{DD}, and the other sets the guessed output voltage to ground. The pre-charge phase is $\overline{\Phi}$ in the scheme of Figure 11.40(a) and Φ for the circuit in Figure 11.40(b).

If the output node is pre-charged at V_{DD}, and the expected output is "high," nothing happens; otherwise it is necessary to discharge the node by activating a pull-down path. The opposite operation holds for the scheme in Figure 11.40(b). Therefore, that scheme uses a pull-down network made only of N-channel devices with just one P-channel transistor, which

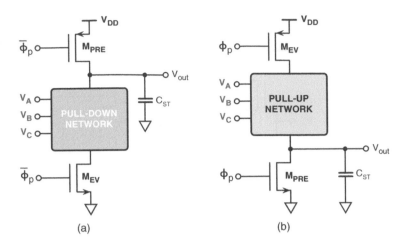

Figure 11.40   (a) Dynamic logic circuit with pull-down network; (b) dynamic logic circuit with pull-up network.

is necessary for pre-charging the output node. The opposite is the case for the other scheme. The figure indicates three inputs; that is a simplification. In reality it is possible to have several inputs planned for in the logic function.

The advantage of dynamic logic is that it avoids the transistor tree defining the path towards V_{DD} or ground. Therefore the circuit density is high, and, because of the smaller number of transistors, the power consumed is typically low. However, there are limits to the technique. They are as follows.

- **Charge leakage** The output node is not exactly an insulated node because of reverse-biased junctions that leak current. In addition, with nanometer technology the gate leaks current. The leakage paths charge or discharge the output nodes, causing limited reliability.

- **Charge sharing** The voltage of predefined nodes changes during the evaluation phase, because even if the complementary path is disabled there is the capacitive load of disconnected nodes possibly pre-charged to the complementary voltage. The result is a sharing of charge between storing capacitance C_{ST} and those parasitics.

- **Clock feedthrough** Opening the pre-charge transistor injects charge on the output node. It is a fraction of the charge channel, which partially flows into the storing capacitance. In addition, the capacitive coupling of clock and output injects charge.

There are various techniques for resolving these limitations. Among them we have the pre-charge of output at a voltage higher than V_{DD} (dynamic bootstrapping) or the pre-charge of some nodes of the pull-down network.

PROBLEMS

11.1   Design the logic diagram that realizes the seven-segment coding with only ANDs, and draft the scheme that uses only NORs.

11.2 Translate the schematic diagram of Figure 11.13 into a logic architecture made only with NOR gates.

11.3 Draft the block diagram of a decoder that realizes the following logic functions:

$$y_i = x_{a,i} \quad \text{for } i = 1, 2, 3, 4$$
$$y_{i+4} = x_{a,i} \quad \text{for } i = 1, 2$$
$$y_{i+4} = x_{4-i} \quad \text{for } i = 3, 4.$$

Sketch the circuit solution with two inputs and basic logic gates.

11.4 An eight-bit parallel word realizes the key of an electronic lock. The code is 10110101. Draft the combinational logic that enables unlocking. Two adjacent codes are a sign of sequential searching. That event and three wrong attempts both activate a temporary disabling of the input, lasting five minutes. Draft the logic.

11.5 Design an edge-transition scheme (or pulse generation) made by a high-pass network including a simple CR circuit. Use logic gates that detect the crossing of a threshold that is half the supply voltage. Calculate the pulse duration as a function of the component used. Use only two input NANDs.

11.6 Design a toggle flip-flop using only NOR gates. The edge detector is made by a XOR and delay cells are realized with NOT gates. Translate the logic function into a CMOS scheme, and, if possible, find an alternate solution that minimizes the number of transistors used.

11.7 Repeat the study done in section 11.7.2 to estimate the propagation delay of the transitions low-to-high and high-to-low of a two-input AND. What is a good design strategy that balances the propagation time of the two transitions?

11.8 Use the pass transistor logic to design a JK flip-flop. Use only N-channel transistors and, where necessary, use buffers to recover from weak "1"s.

CHAPTER 12

FEEDBACK

Feedback is a process mentioned many times in this book. It is a relevant tool used for the control of any type of system. The algorithms developed for the theoretical description of feedback are implemented with analog or digital electronic circuits. We shall study the fundamentals of the process and discuss the limitations and benefits of real circuit implementations.

12.1 INTRODUCTION

Feedback is what we use for the control of any action. When we drive a car, for example, we always utilize feedback. We measure the error between the real and expected car trajectories. Then we estimate the existing difference with respect to the foreseen trajectory and feed the error back for suitable processing to produce actions that control the steering. Also, when we turn right or left on a winding road we estimate the curvature of the path, and, after some processing, we feed back the information that controls the brake pedal, so as to slow down before bends. The correcting signals on the steering and the brake pedal must be suitable in amplitude and time to avoid ineffective or dangerous outcomes and obtain smooth results.

The same thing is done in electronic systems to control given electrical quantities. The measurement of output, its processing and the error obtained by comparison with an expected value serves to improve results. The output signal, after passing through suitable processing

Understanding Microelectronics: A Top-Down Approach, First Edition. Franco Maloberti.
© 2012 John Wiley & Sons, Ltd. Published 2012 by John Wiley & Sons, Ltd.

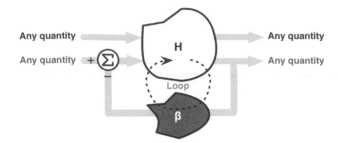

Figure 12.1   General configuration of an electronic system with feedback.

functions, returns back to the input to give rise to the output again. For this reason it is said that the system establishes a feedback loop, or, more simply, that it uses feedback.

Feedback can be positive or negative. It is positive when the feedback signal enhances the error; it is negative when it opposes the error. Obviously, for driving a car you have to use negative feedback. However, it may happen that in some situations feedback, which is intended to be negative, turns out to be positive. Suppose you are driving along a winding road. If the steering control is delayed, the car tends to turn right when it is necessary to turn left. Your reaction might be a sudden rotation to the right, so as to regain control and avoid a dangerous situation (and I hope you succeed). It may be that you turn too far to the right and you have to turn left again, causing the car to swing to and fro (in electronic circuits, a similar behaviour gives rise to oscillating waveforms). When the road is straight your steering keeps the car on the right path, but in reality the control is not perfect, and minimal corrections make the car imperceptibly fluctuate around the ideal straight trajectory.

In electronics, feedback does the same, so as to secure the stability control of electronic systems. The response can exhibit swinging or it can be smooth, depending on the effectiveness of the control loop. In addition to the control action, negative feedback gives rise to other effects. It improves linearity (at the expense of gain), varies the input and output impedance, and modifies the frequency response. These features, verified in some detail shortly, are beneficial side effects of the stabilizing action of negative feedback.

Electronic circuits also make use of positive feedback. It serves to sustain oscillation, at a controlled frequency, that otherwise would fade because of damping losses. It is also useful to enforce a logic level, making it firm against perturbations that tend to change its value.

12.2   GENERAL CONFIGURATION

Feedback is a process that applies to any quantity at output, to become any quantity at the input of a generic processing block, as shown in Figure 12.1. Some of the output signals are the input of a block β that, after processing, generates signals to be added to the input. The indicated sign is minus, because the scheme is assumed to have negative feedback. On the other hand, using a positive sign would anticipate positive feedback. The block H is the forward processor and the block β is the feedback processor. Moreover, the signals involved, voltages or currents, can be continuous-time or sampled-data, with analog or digital representations.

The key to the operation is the processing loop. When going around this loop, there are changes in amplitudes, and delays or phase shifts for periodic signals. The two blocks of the

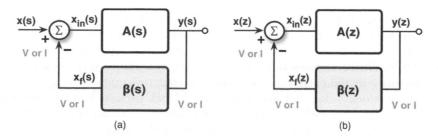

Figure 12.2   Linear feedback system with single signal: (a) in the continuous-time (s) domain; (b) in the sampled-data (z) domain.

figure are, in general, non-linear. However, as already mentioned, when the system is non-linear the outcome is linearized, provided that the feedback is negative. On the other hand, a circuit with positive feedback enhances the non-linearity.

Remember

Feedback is a process widely used in disparate fields, mainly for control. Electronic circuits are indispensable instruments to implement the process.

Whether feedback is positive or negative is assessed by going around the loop. Consider, for example, a single-signal linear system with feedback assumed to be negative and a sine wave applied at input. On opening the loop, what returns back is a sine wave with the same frequency but changed amplitude and phase. If the phase shift is less than 90°, the feedback is negative, because what feeds back and is subtracted from the input has an in-phase component opposed to the input itself. This is a sign of stimulus to stabilization. If the in-phase component has the opposite sign, the subtraction operation enhances the input and the feedback is positive. Observe that the condition of positive feedback always occurs because the time needed to travel around the loop is not zero. At high frequency, when the period of a sine wave becomes smaller and smaller, the feedback, expected to be negative, turns out to be positive.

What matters for avoiding instability is the amplitude of the returned signal. The next subsection studies the issue for the special case of feedback schemes made with linear blocks.

12.2.1   Linear Feedback Systems

A linear feedback system with single input and single output has the configuration of Figure 12.2. The input is a voltage or a current. The feedback signal, assumed to be of the same type (voltage or current) as the input, is subtracted from the input, giving rise to a result that passes through the block $A(s)$ to determine the output. The feedback transfer function $\beta(s)$ gives rise to the return signal. Since the system is linear we can use the superposition principle.

The diagrams assume continuous-time signals described in the s-domain (Figure 12.2(a)) or discrete-time signals studied in the z-domain (Figure 12.2(b)). Even though both situations are common, the study that follows focuses on continuous-time schemes. Their sampled-data counterparts are analyzed in an equivalent manner. The study uses sampled-data operators instead of the corresponding continuous-time operators.

Like Figure 12.1, the diagram in Figure 12.2 uses the minus sign, because it is supposed to depict a negative feedback system. Moreover, the scheme uses the letter "A" in the forward block to indicate the need for a function typical of amplifiers. Another important point is that Figure 12.2 implicitly assumes that one block does not influence the other. For this, if the electrical quantity at the output of "A", whose output resistance is R_{out}, is a voltage, it is assumed that the block β does not load the forward block because its input resistance R_{in} is much higher than R_{out}. Similarly, if the output quantity is a current, then $R_{in} \ll R_{out}$. Other, similar, conditions must hold for the input side of the scheme concerning the load of the β block. Moreover, it is often assumed that the β block does not cause substantial delay or phase shift at the working frequency ($\beta(s) = \beta$). This condition holds for feedback blocks that are not intended to establish a frequency-dependent feedback.

Under the above assumptions, the scheme of Figure 12.2 is described by

$$x_{in} = x - x_f \tag{12.1}$$
$$x_f = \beta \cdot y \tag{12.2}$$
$$y = A(s)x_{in}, \tag{12.3}$$

which, combined together, yield

$$A_f(s) = \frac{y(s)}{x(s)} = \frac{A(s)}{1 + \beta \cdot A(s)}, \tag{12.4}$$

which defines A_f, the *feedback gain* of the scheme, also called *transfer gain*. The quantity βA is the *loop gain* because it corresponds to the gain around the feedback path. Its sign must be positive to secure negative feedback, or, more precisely, the phase shift must be low enough.

Notice that with negative feedback the modulus of $A_f(s)$ is lower than that of $A(s)$. This, generally speaking, makes sense, because having control means obtaining performance that is less than the maximum achievable. We can control something only if there is some margin above (but also below) the possible values of the controlled quantity. Moreover, we can expect that the larger the difference between maximum and obtained values, the better is the quality of control.

Consider equation (12.4) and observe that if $\beta A \gg 1$ the feedback gain is approximately

$$A_f \simeq \frac{1}{\beta}. \tag{12.5}$$

That, remember, disregards the dependance of β on s, because it supposes that the feedback block has a very high bandwidth. The approximation states that when the forward block has very high gain β must be small so as to give rise to large A_f gains. This remark is interesting for two reasons: first, with large forward gain the feedback block dominates the overall gain independently of the precise value of "A"; second, since we desire to secure gain, β must be more than 1, and a passive network can implement the feedback block.

A divider made, for example, by two resistors or two initially discharged capacitors, gives rise to an attenuated replica of the input. Since resistors and capacitors are much more linear than an active block, the gain is quite linear. In addition, passive networks do not need connections to the power supply, thus avoiding noise injection from the power lines.

A useful observation is that large forward gains make the input of the amplifier, $x_i = y/A(s)$, very small, since $A(s)$ is large in the band of the signal. Therefore

$$x - x_f \simeq 0 \rightarrow x_f \simeq x, \tag{12.6}$$

stating that the feedback block generates an almost identical replica of the input signal. The small difference between x and x_f is the signal used by the large gain of the forward amplifier to determine the output voltage.

The circuit with negative feedback subtracts the returning signal from the input because the forward gain $A(s)$ and the feedback β are both assumed to be positive. In reality what is required is to have an overall negative term. The use, for example, of a negative sign on the amplifier, which becomes inverting, together with a positive β but with an addition at input, is an equivalent solution. Other combinations of signs that preserve the overall operation are possible.

Be aware

Increasing frequency makes the feedback of any loop positive. That is unavoidable, but it is not a serious issue for stability if the high-frequency loop gain is low.

Notice that the study performed is not specific for negative feedback but is also valid for any type of feedback, including positive. In this case, it can be convenient to invert one of the signs around the loop to give rise, for the feedback gain with positive feedback, to

$$A_{f,p}(s) = \frac{y(s)}{x(s)} = \frac{A(s)}{1 - \beta \cdot A(s)}. \tag{12.7}$$

Equation (12.7) states that $A_{f,p}$ exceeds A if the feedback loop's gain βA is negative, with modulus less than 1. It goes to infinity when $|\beta A| = 1$ while above the $|\beta A| > 1$ limit the system is unstable: i.e., any temporary signal or spur affecting some point of the feedback loop is enough to make the output larger and larger (positive or negative), because feedback augments (or regenerates) the stimulus at input. However, obviously, for real electronic circuits the signal amplitude is limited to within the supply interval. When the signal approaches the rail voltages, the gain of any amplifier drops and the modulus of the loop gain may become equal to 1. This is a formal explanation of the fact that the output signal does not exceed the supply interval but sticks at one of the supply voltages.

12.3 PROPERTIES OF NEGATIVE FEEDBACK

Negative feedback, in some sense, trades a large gain of the forward amplifier against a number of valuable features. How the negative feedback reduces gain has been studied in a previous section. The feedback loop quantity βA, lower than the forward gain A, diminishes the forward gain A itself to determine a β-controlled amplification, namely almost $1/\beta$. Thus, for example, if the amplification needed is 50, the β factor must equal 0.02; in order to have a loop gain much larger than 1, say at least 40 times, it is necessary to use forward gains as high as 2000. That condition, well achievable at low frequencies, requires significant design care and consumes significant power if the circuit must operate at a high or very high frequency. The effort, obviously, is worth the equivalent benefits, as analyzed in the following subsections.

COMPUTER EXPERIMENT 12.1

Gain Sensitivity and Bandwidth Enhancement

With this computer experiment you can check out two relevant features of negative feedback: reduced gain sensitivity and increased bandwidth. As is well known, high gain amplifiers are not precise. On the other hand, the β determined by passive components is accurate. The ElvisLab board contains five mini-boards with two feedback networks, one continuous-time, the other discrete-time. The circuits on the mini boards are nominally equal but there are differences caused by fabrication inaccuracies. A clock generator at 200 MHz, not shown in the figure, determines the sampled-data rate of the discrete time scheme. The value of β can go down to zero to give you the ability to measure the amplifiers without feedback. In this case the full gain of the forward amplifier will obviously be required to use small input amplitudes. You can:

- set the same amplitude and frequency as the sampled data and continuous time sine wave (from 100 Hz to 100 MHz);
- change the value of the same feedback factor from 0 to 1 (the network used is passive);
- choose between the five available mini-boards;
- select the input displayed on the scope (continuous-time or discrete-time).

MEASUREMENT SET–UP

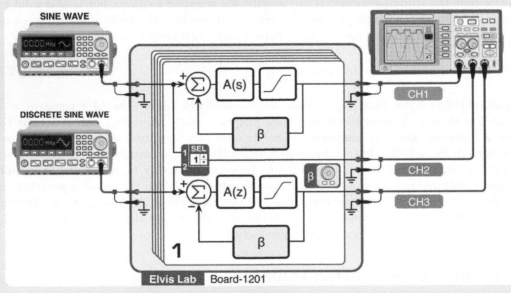

Elvis Lab Board-1201

TO DO

- Set the feedback factor to zero and use a very small input signal to stay far from saturation. Draw up Bode diagrams, using four points per decade. Go down with frequency until the gain remains constant. Measure the low frequency gain of the five mini-boards and relative errors.
- Use the following nominal values of feedback factor: β=1/100, β=1/10, β=1/5, β=1/2; measure the low-frequency gains. The matching accuracy of resistances used to realize the β block is 0.1%. What are the errors?
- Determine the frequency responses with a feedback factor equal to β=1/40 and compare the Bode diagrams with and without feedback.

12.3.1 Gain Sensitivity

A significant benefit of negative feedback is that the transfer gain is not very sensitive to the variations in forward gain. Equation (12.4) shows that the gain of the forward block appears in both the numerator and the denominator of a fraction; thus a change in the numerator is partially compensated for by a change in the same direction in the denominator. However, since equation (12.4) includes β, it is necessary to assess the effect of β errors on the final result. Actually, when an attenuator made of passive components is used to realize β the accuracy is remarkable, because the matching of concurrently fabricated passive components is better than that of active devices.

The result is made quantitative by differentiating equation (12.4) to yield

$$\delta A_f = \frac{\delta A}{(1 + \beta A)^2} - \frac{\delta \beta \cdot A^2}{(1 + \beta A)^2}, \tag{12.8}$$

leading to

$$\frac{\delta A_f}{A_f} = \frac{\delta A}{A} \cdot \frac{1}{1 + \beta A} - \frac{\delta \beta}{\beta} \cdot \frac{\beta A}{1 + \beta A}, \tag{12.9}$$

which provides the relative variation of transfer gain for changes in forward gain and feedback factor. One of the sensitivities is positive, and the other is negative.

For negative feedback the error due to variations in A is the relative error divided by the feedback loop gain, which is a large number. Therefore we have a desensitization of the transfer gain. This factor $(1 + \beta A)$ is often called the desensitivity factor. The sensitivity to β is, for large loop gain, almost -1. Therefore, the sensitivity from the one of the active block becomes that of the block typically realized with passive components.

12.3.2 Bandwidth Improvement

Any real amplifier has a limited speed, typically described in the s-domain by a dominant pole and non-dominant poles normally ahead of the frequency at which the Bode diagram crosses zero dB (the f_T). Since f_T occurs A_0 times the dominant pole, even with a large f_T the dominant pole is at low frequencies. Thus the gain is almost constant for a small frequency range. Feedback extends that interval.

For the sake of simplicity this study neglects the non-dominant pole contribution. Supposing that the dominant pole is at angular frequency $\omega_p = 1/\tau_p$, the forward transfer function is

$$A(s) = \frac{A_0}{1 + s\tau_p}. \tag{12.10}$$

Equation (12.4), using (12.10), gives rise to

$$A(s) = \frac{A_0}{1 + \beta A_0} \cdot \frac{1}{1 + s\tau_p/(1 + \beta A_0)}, \tag{12.11}$$

showing a pole at $\omega_{p,f} = (1 + \beta A_0)/\tau_p$. Thus the feedback network reduces the low-frequency gain by $(1 + \beta A_0)$ but increases the frequency of the dominant pole by the same amount. The combination of the two effects leaves f_T unchanged.

The amplitude Bode diagrams of the main amplifier and the feedback scheme look like the plots in Figure 12.3. The forward amplifier has very large gain, 97 dB, and unity gain frequency,

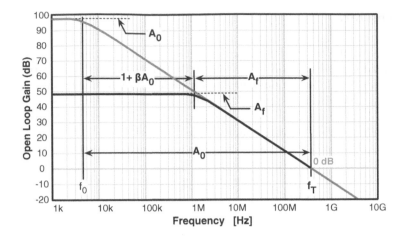

Figure 12.3   Bode diagram of a possible forward amplifier with dominant pole frequency limitation and Bode diagram of a resulting feedback scheme.

f_T, at 350 MHz. The dominant pole is at $f_T/A_0 = 5.55$ kHz. The transfer gain becomes 48 dB because of a loop gain (or the desensitizing factor) equal to 49 dB. The bandwidth increases by the same factor. Therefore, as expected, the unity gain frequency is unchanged.

An equivalent study with a two-pole forward amplifier gives rise to similar results. With the dominant pole at the angular frequency $\omega_d = 1/\tau_d$ and the non-dominant pole at $\omega_{nd} = 1/\tau_{nd}$, the forward gain is represented by

$$A = \frac{A_0}{(1 - s\tau_d) \cdot (1 - s\tau_{nd})}, \tag{12.12}$$

which, using equation (12.4), gives rise to the transfer gain. It is easy to verify that the dominant pole moves at higher frequency than the gain loop βA_0, while the non-dominant pole remains almost unchanged. In reality, as will be studied shortly when discussing stability, the non-dominant pole slightly shifts to lower frequencies.

Example 12.1

Determine the poles of a feedback network with $\beta = 1/100$. Use a gain forward amplifier with $A_0 = 10^4$ and two real poles at $f_d = -1$ Hz and $f_{nd} = -4 \cdot 10^4$ of the plane $s/2\pi$.

Solution

The time constants of the poles are

$$\tau_d = \frac{1}{2\pi f_d} = 0.1592 \text{ sec}$$

$$\tau_{nd} = \frac{1}{2\pi f_{nd}} = 3.979 \text{ μsec.}$$

The gain of a two-pole forward stage,

$$A(s) = \frac{A_0}{(1 - s\tau_d) \cdot (1 - s\tau_{nd})},$$

used in equation (12.4), gives rise to the feedback gain

$$A_f = \frac{A(s)}{1 + \beta A(s)} = \frac{A_0}{1 + \beta A_0} \frac{1 + \beta A_0}{1 + \beta A_0 - s(\tau_d + \tau_{nd}) + s^2 \tau_d \tau_{nd}}$$

$$= 99 \frac{1}{1 - s \cdot 1.6 \cdot 10^{-3} + s^2 \cdot 6.396 \cdot 10^{-9}}.$$

Solving this equation produces

$$f_{d,f} = 99.25 \text{ Hz}, \quad f_{nd,f} = 3.99 \cdot 10^4 \text{ Hz}.$$

Thus a loop gain equal to 100 makes the low-frequency gain a bit lower than the expected 100 ($10^4/101$). Moreover, the feedback increases the bandwidth by 99.25 and slightly moves the non-dominant pole (from 40 kHz to 39.9 kHz).

12.3.3 Reducing Distortion

The non-linear response of a processing block is easily calculated if the model of the transfer function is a polynomial expression and the input is a sine wave. For the input–output response of the forward amplifier assume that we use the function

$$V_{out}(V_{in}) = A_0(V_{in} + a_2 V_{in}^2 + a_3 V_{in}^3 + a_4 V_{in}^4 + \cdots) \tag{12.13}$$

and apply the sine wave $V_{in} = \overline{V} \sin(\omega_{in} t)$ at input.

Trigonometric functions determine the output as the superposition of sine waves at the input frequency and at multiples of the input frequency plus a DC term. Considering only the first four terms gives

$$V_{out}(t) = A_0 \left\{ \left[\frac{1}{2} a_2 \overline{V}^2 + \frac{3}{8} a_4 \overline{V}^4 \right] + \left[\overline{V} + \frac{3}{4} a_3 \overline{V}^3 \right] \sin(\omega_{in} t) \right.$$

$$\left. - \frac{1}{2} [a_2 \overline{V}^2 + a_4 \overline{V}^4] \cos(2\omega_{in} t) - \frac{1}{4} a_3 \overline{V}^3 \sin(3\omega_{in} t) + \frac{1}{8} a_4 \overline{V}^4 \cos(4\omega_{in} t) \right\}. \tag{12.14}$$

Let us consider just the output DC term, $V_{out,DC}$, that returns to input multiplied by $-\beta$. If it is small, we can neglect the harmonic components that it causes and state that it generates at output to $-A_0 \beta V_{out,DC}/(1 + \beta A_0)$. This contribution and $V_{out,DC}$ sums up to give to $V_{out,DC}/(1 + \beta A_0)$, much less than $V_{out,DC}$.

For the second harmonic term we can do a similar study and obtain the same attenuation. However a second harmonic spur feed back at input generates a DC term and harmonic at $4\omega_{in}, 6\omega_{in}, \ldots$. Since the amplitudes are small, they can be viewed as second order limits, or, possibly they can be accounted for, added to the main harmonic terms. The overall result, albeit approximated, shows that harmonic distortion diminishes by $(1 + \beta A_0)$.

Notice

The extent of any kind of negative benefit produced by feedback almost matches the gain reduction cost: the dimensionless quantity feedback loop gain.

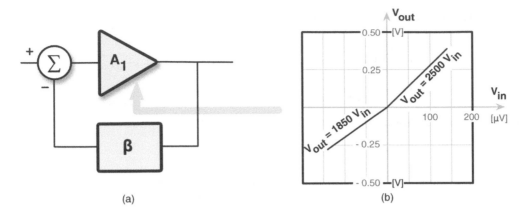

Figure 12.4 (a) Feedback loop using an op-amp whose (b) input–output characteristic has different slopes for positive and negative differential signals.

There are other sources of distortion; for example, a feedback loop like the one in Figure 12.4(a), which uses an op-amp whose input–output characteristic has different slopes for positive and negative differential signals, as illustrated in Figure 12.4(b). Feedback moderates the limit caused by slope discontinuity. To explain the benefit instead of deriving general equations we use a numerical example.

Suppose that the forward gain provided by the op-amp is 2500 with positive input and 1850 with negative input. The β of the feedback block is 0.1. Accordingly, since the loop gain is high the expected output is a little less than 10. If we estimate the output in more detail with an input signal equal to 10 mV and -10 mV we get at output, respectively,

$$V_{out}(10 \text{ mV}) = 10 \text{ m}\frac{2500}{1 + 250} = 99.6 \text{ mV} \tag{12.15}$$

$$V_{out}(-10 \text{ mV}) = -10 \text{ m}\frac{1850}{1 + 185} = -99.45 \text{ mV}. \tag{12.16}$$

Since the limit does not depend on amplitude of signal but on sign, the outcomes can be extended to any amplitude. Therefore, the result indicates that a ratio between gains with positive and negative inputs is equal to 1.35 for the forward block and becomes 1.0015 for the feedback scheme. The distortion caused by slope discontinuity is therefore greatly reduced.

The change of slope in the input–output characteristic can occur at inputs different from zero. In this case also, feedback reduces distortion. Again the improvement can be verified by observations based on numerical calculations. Suppose that a change in slope of the input–output response occurs at 1 mV:

$$V_{out} = \begin{cases} 1850 \ V_{in} & \text{for } V_{in} < 1 \text{ mV} \tag{12.17} \\ 2500(V_{in} - 0.001) + 1.85 & \text{for } V_{in} > 1 \text{ mV}, \tag{12.18} \end{cases}$$

then if the expected output is about ± 100 mV the input of the op-amp is ± 54 μV with a gain equal to 1850 and ± 40 μV for the 2500 gain region. Both values are much less than the point at which the slope changes and the output is not affected by the limit at all. The point at which the op-amp changes the slope of the input–output characteristic is at $V_{out} = 1.85$ V. It can be verified that, above that output amplitude, distortion is reduced by the loop gain.

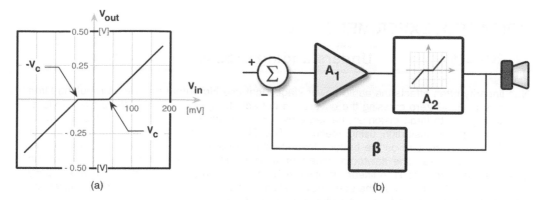

Figure 12.5   (a) Non-linear transfer characteristic of a possible forward amplifier; (b) feedback correcting crossover distortion of the power stage of an audio amplifier.

Another possible situation that takes advantage of feedback for improving linearity is when the forward amplifier is made up of a linear first stage followed by a second non-linear stage. A typical example is the power stage of audio amplifiers. The driver of loudspeakers must mainly provide power and, perhaps, some voltage gain. Very often the power stage has a non-linear feature like the ones in Figure 12.5(a). Outside a given region around zero the output voltage changes linearly in proportion to the input voltage. Around zero the stage does not respond properly because in that interval, called *crossover*, the bias of the power transistors is not enough to switch them on and deliver power properly. The output voltage is almost zero.

The feedback configuration of Figure 12.5(b) closes the loop with a β block, whose attenuation factor gives rise to the required voltage gain. The high gain of block A_1 and the feedback ensure linearity. Suppose that the crossover region is $\pm V_c$. At the zero crossing the output voltage of A_1 must jump from $-V_c$ to $+V_c$ (or vice versa) to switch off one of the power transistors likely to be used in the stage and to drive another on. Thanks to the high gain of A_1 the input change must be $2V_c/A_1$. For large gain, this can be very small. Therefore, since the input of A_1 is the difference between input and output, the required change in output voltage to compensate for the $\pm V_c$ crossover is the small output change $2V_c/A_1$.

12.3.4 Noise Behavior

The requirement for additional gain stages to provide high gain in the forward amplifier does not affect noise performance. Large gain requires a cascade of amplifiers with suitable frequency response, and, as is well known, noise generators at the input of each block describe the noise behavior. For a two-stage amplifier, for example, we have the configuration of Figure 12.6(a) with a single noise generator describing the noise performance of each block.

The noise generator of each block can be referred to the input of the previous stage and then combined with the noise source of that stage. Since the physical mechanisms that give rise to the noise of distinct devices are not related to one another, the combination of effects must be done on a statistical basis that involves adding power or performing a quadratic superposition. For the two-stage amplifier, transferring $v_{n,2}$ to the input of A_1 gives rise to $v_{n,2}/A_1$; the quadratic superposition, as shown in Figure 12.6(b) yields

$$v_{n,op} = \sqrt{v_{n,1}^2 + v_{n,2}^2/A_1^2}. \tag{12.19}$$

COMPUTER EXPERIMENT 12.2

Understanding Crossover Distortion

Non-linearity often affects the audio power stages that use big output transistors requiring large driving signals. At zero crossing the drivers are not able to operate properly the power devices, thus giving rise to large distortion, named crossover. This computer experiment studies how to limit that type of distortion using feedback. The ElvisLab board contains two equal power amplifiers. With the one on the bottom, loaded with the speaker, it is possible to determine its non-linear response. The sawtooth generator at input and the resulting signal displayed on Channel 3 give the result. A linear amplifier with variable gain (range 20 dB 60 dB) and unity gain frequency 60 kHz drives the equal power stage of the top diagram. A feedback block, whose β factor can change from 1/2 to 1/10, closes the loop. You can:

- set amplitude and frequency of the input sine wave (from 100 Hz to 40 kHz);
- set the sawtooth amplitude. The repetition rate is fixed at 100 Hz;
- select the gain of the linear amplifier and the β factor;
- switch the signal sent to Channels 1 and 2 of the scope to observe the signals from each scheme.

MEASUREMENT SET-UP

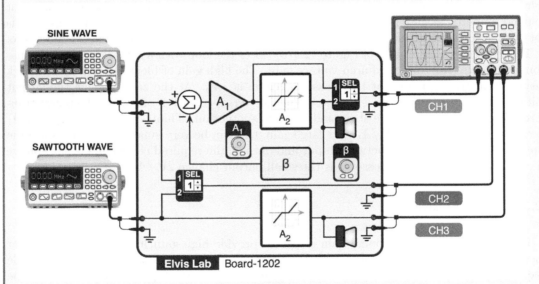

Elvis Lab   Board-1202

TO DO

- Set a suitable sawtooth amplitude and observe the non-linear response of the power amplifier. Measure the values and draft the input-output curve on paper.
- Use β = 0 and a 1 kHz sine wave. For various linear gains set the input amplitude that gives peak values at output equal to 0.3 V, 0.7 V, 2.4 V. Explain the results.
- Use β = 1/2 and observe the output voltage for different gains of the linear amplifier. Compare the input signal with the one at output of the linear stage.
- Determine the crossover intervals for different values of linear gain and β factor.
- Repeat the experiment using different feedback factors. Focus on performance at 20 kHz.

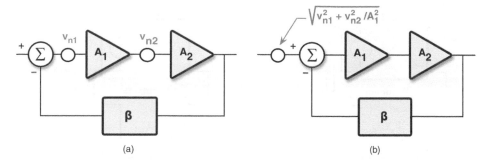

Figure 12.6 (a) Noise generators describing the noise behavior of the two-stage forward amplifier; (b) noise equivalent at input of the feedback block.

The equation can be generalized to a cascade of multiple stages. The procedure is the same, and the result is

$$v_{n,op} = \sqrt{v_{n,1}^2 + v_{n,2}^2/A_1^2 + v_{n,3}^2/(A_1 A_2)^2 + \cdots}. \qquad (12.20)$$

The contribution of the noise of the stages after the second one is further reduced, and in practice the input-referred noise mainly depends on the first stage with a minor increase caused by the second stage. The noise of the feedback block also contributes to noise performance. If the β block gives rise to $v_{n,\beta}$, the overall input-referred noise is

$$v_{n,in} = \sqrt{v_{n,op}^2 + v_{n,\beta}^2}, \qquad (12.21)$$

where the two input-referred terms are summed quadratically.

The result shows that if A_1 is large, the noise of the first stage and that of the feedback network determine noise performance. The contribution of extra gain stages, possibly required to give rise to a large forward gain, negligibly affects the noise.

12.4 TYPES OF FEEDBACK

The input and output quantities of a feedback network can be voltage or current. They determine the kind of amplifier and feedback block. If input is a voltage and output is also a voltage, the amplifier is a VCVS, which in a real scheme is an op-amp. Amplification and the β factor are dimensionless. If the input is a voltage and the output is current the amplifier is a transconductor (a CCVS), implemented at the circuit level by OTAs; the dimension of β in this case is resistance, to balance the dimensions around the loop, and so forth.

The four resulting topologies are shown in Figure 12.7. Circuit (a) applies feedback in series to input and measures the output by a parallel (or shunt) connection. Because of this, the scheme is normally referred to as a *series–shunt*. The configuration of Figure 12.7(b) is *series–series*, because at output there is the series of the output of the forward block and the input of the feedback block, while at input, again, the circuit applies feedback in series with input. Figure 12.7(c) is a *shunt–shunt* scheme and Figure 12.7(d) a *shunt–series* arrangement.

The four feedback architectures specify the input and output quantities but still use ideal elements. The schemes employ a resistance loading the output, just to indicate the subsequent stages. The value of the loading resistances is not relevant at this level, because the blocks

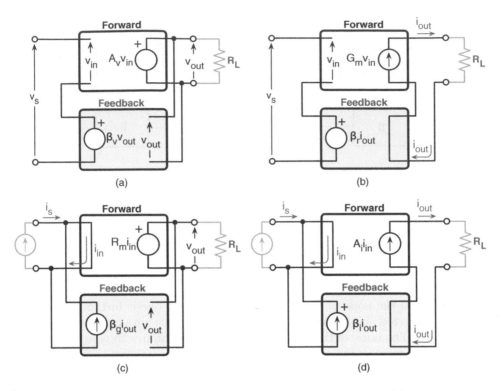

Figure 12.7   Feedback schemes realized with ideal building blocks and (a) voltages at input and output (*series–shunt*); (b) voltage at input and current at output (*series–series*); (c) current at input and voltage at output (*shunt–shunt*); (d) currents at input and output (*shunt–series*).

are ideal. The value is assumed to be non-zero when loading a voltage source and non-infinite when connected across a current source. The diagrams use small letters to emphasize that the circuit is presumed to be linear (or, better, that variables fluctuate in regions where the equations describing them can be approximated by linearization).

Let us consider in some detail the modules of the schemes. The forward block of Figure 12.7(a) uses an ideal voltage amplifier with infinite input impedance and zero output impedance. The input–output relationship is

$$v_{out} = A_v v_{in}. \tag{12.22}$$

The feedback block measures a voltage and generates a voltage, typically smaller than the input. The diagram represents that function by an infinite input impedance and an ideal voltage generator at output. This is formally correct, but having infinite resistance at the input of the forward block admits simplified β solutions with non-zero output resistance.

The forward block of Figure 12.7(b) measures a voltage and generates a current, the function of a transconductor. The infinite output impedance of the ideal current source makes it necessary to establish a path to ground by the non-infinite load R_L. The input–output relationship of the forward amplifier of Figure 12.7(b) is

$$i_{out} = G_m v_{in}. \tag{12.23}$$

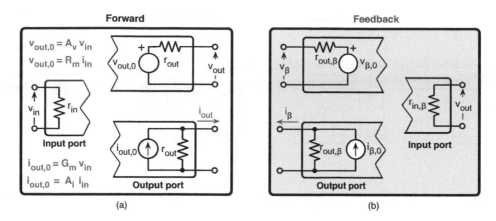

Figure 12.8   Equivalent circuit of input and output ports with Thévenin and Norton equivalents: (a) forward block; (b) feedback block.

The feedback block measures current and produces voltage, the function of a transresistance. Therefore, the dimension of β is Ω. The feedback block illustrated in the figure has zero input resistance and uses an ideal voltage generator for the output. Again, that is correct, but the infinite input resistance of the forward block permits the use of a simplified β network such as, for example, a resistance, in which a replica of the output current flows.

The forward block of Figure 12.7(c) is a transresistance, because it generates a voltage under the control of a current. The circuit implementation does not match any of the basic blocks already studied. However, designing a circuit that measures a current is not difficult. The use of a source or an emitter as input node grants low resistance. The input current possibly reaches a drain (or a collector) to be transformed into an output voltage. The required input–output relationship is

$$v_{out} = R_m i_{in}; \tag{12.24}$$

since the output of the β block is current and the input is a voltage, the dimension of β is $1/\Omega$.

The forward block of the scheme in Figure 12.7(d) has current at input and current at output. The required function is therefore current amplification. The relationship between input and output is

$$i_{out} = A_i i_{in}. \tag{12.25}$$

The feedback block measures current and generates current. The block β in the figure has zero input resistance and uses a current source at output. A good approximation of the function is given by a current mirror. The mirror factor gives rise to the β, a dimensionless quantity for this scheme.

12.4.1   Real Input and Output Ports

Real circuits used in feedback loops do not have the input and output resistance prescribed by the various configurations in Figure 12.7. Their real behavior, depending on the type of feedback, is conveniently modeled by Thévenin or the Norton equivalent. Figure 12.8 shows the result for input and output ports. Inputs are always simplified by an impedance or, for

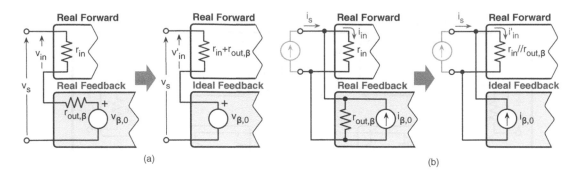

Figure 12.9   (a) Input section with serial input feedback; (b) input section with parallel input feedback.

low frequency, by a resistance. The value is close to zero for current sensing or very high for voltage sensing. The output ports use the Thévenin or the equivalent Norton counterpart, depending on whether the relevant output quantity is voltage or current. The voltage source gives rise to the nominal amplitude with the output terminals opened; the output current source generates the nominal current with the output terminals shorted. Depending on the use, the equivalent circuit of the block is the input section combined with one of the output sections. The relationships between input and output, for the four cases already indicated in equations (12.22), (12.23), (12.24) and (12.25), are referred to in Figure 12.8.

The use of the equivalent ports in Figure 12.8 gives rise to more realistic descriptions. The input section of Figure 12.7(a), for example, which foresees a voltage generator at input and receives a voltage as feedback signal, gives rise to Figure 12.9(a). In some situations it is useful to concentrate non-idealities in one of the ports. This is done in the scheme at the right-hand side of the arrow in Figure 12.9(a), which moves the series connection of r_{in} and $r_{out,\beta}$ inside the forward input port. The result is an ideal feedback port and a real forward port. The input voltages of the two schemes v_{in} and v'_{in} are related by

$$v_{in} = v'_{in} \frac{r_{in}}{r_{in} + r_{out,\beta}}, \qquad (12.26)$$

a relationship to be used when using the transformation that makes the feedback block an ideal one.

Figure 12.9(b) properly models the feedback input section when the input variable is current. The input resistance of the forward port is in parallel to the Norton equivalent of the feedback port. The feedback block is made ideal by moving $r_{out,\beta}$ in parallel to r_{in}, to give rise to the scheme at the right of the arrow in Figure 12.9(b). The relationship between i_{in} and the current of the transformed scheme i'_{in} is given by

$$i_{in} = i'_{in} \frac{r_{out,\beta}}{r_{in} + r_{out,\beta}}. \qquad (12.27)$$

The same procedure as modeling the input section of a feedback scheme is used for the output. We foresee two possible equivalent circuits: one with both ports real and the other with one of the ports incorporating the non-idealities of both.

When the output quantity is voltage, the scheme in Figure 12.10(a) results. The Thévenin transformation includes the non-ideal input resistance of the feedback block in the output

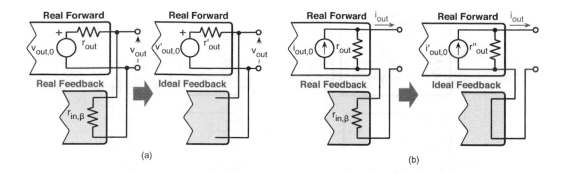

Figure 12.10   (a) Output section with serial input feedback; (b) output section with parallel input feedback.

forward port. The voltage generator and a resistor are given, respectively, by

$$v'_{out,0} = v_{out} \frac{r_{in,\beta}}{r_{out} + r_{in,\beta}}, \quad r'_{out} = \frac{r_{out} r_{in,\beta}}{r_{out} + r_{in,\beta}}. \tag{12.28}$$

If the quantity is current, the scheme of Figure 12.10(b) optimally models the section. The Norton equivalent incorporates $r_{in,\beta}$ in the output port of the forward block. The equivalent current and resistance are

$$i'_{out,0} = i_{out} \frac{r_{out}}{r_{out} + r_{in,\beta}}, \quad r''_{out} = r_{out} + r_{in,\beta}. \tag{12.29}$$

The above transformations make small-signal equivalent circuits available, with real forward and feedback blocks and, if they are more convenient, the corresponding schemes with a real forward and an ideal feedback block. The equivalent circuits serve for estimating the input and output impedance (or resistance) of feedback schemes with real blocks.

12.4.2   Input and Output Resistances

The equivalent resistance across two terminals of any electronic circuit is the ratio between a small signal voltage v_x applied across those terminals and the generated current i_x, provided that all the signal generators are sent to zero. That is the method used for determining the input and output resistance of the configurations given in Figure 12.11 when the blocks are real. This study employs the input and output equivalent circuits derived in the previous subsection.

The series–shunt configuration of Figure 12.7(a) leads to the equivalent circuits of Figure 12.11, two schemes useful for estimating the input and the output equivalent resistances. The load of Figure 12.11(a) is presupposed to be infinite (the scheme uses open output). The study with finite load R_L just requires the use of the Thévenin transformation, which substitutes the output resistance with the parallel connection of r'_{out} and R_L and attenuates v'_{out} by $R_L/(r'_{out} + R_L)$.

By inspection, the result is

$$v'_{in} = r'_{in} i_x = v_x - \beta_v v'_{out,0}, \quad r'_{in} = r_{in} + r_{out,\beta}. \tag{12.30}$$

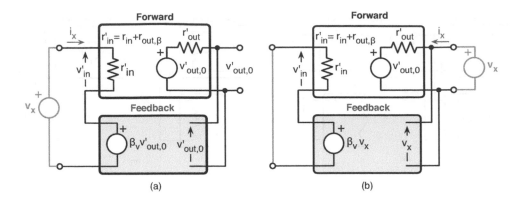

Figure 12.11   (a) Small-signal equivalent circuit used to estimate the resistance of a voltage amplifier in feedback configuration; (b) small-signal circuit for output resistance estimation of the same circuit.

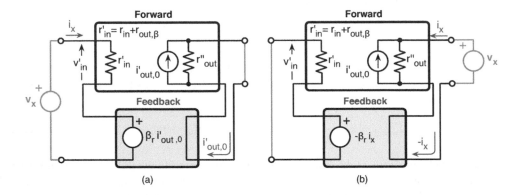

Figure 12.12   (a) Small-signal circuit used to estimate the input resistance of a transconductance amplifier in a feedback configuration; (b) small-signal circuit for output resistance estimation of the same circuit.

The use of equations (12.26) and (12.28) gives rise to

$$v'_{out,0} = v'_{in} A_v \frac{r_{in}}{r'_{in}} \frac{r_{in,\beta}}{r_{out} + r_{in,\beta}} = v'_{in} A_v \gamma_v, \quad \gamma_v = \frac{r_{in}}{r'_{in}} \frac{r_{in,\beta}}{r_{out} + r_{in,\beta}}. \tag{12.31}$$

Solving equations (12.30) and (12.31) gives rise to

$$v_x = i_x r'_{in} \{1 + \beta_v A_v \gamma_v\} = i_x r_{in,eq}. \tag{12.32}$$

That produces

$$r_{in,eq} = r'_{in} \{1 + \beta_v A_v \gamma_v\}, \tag{12.33}$$

showing an increase in input resistance by a factor equal to the loop gain $(1 + \beta_v A_v)$ if, as typically happens, $\gamma_v \simeq 1$ ($r_{in} > r_{out,\beta}$ and $r_{in,\beta} > r_{out}$). Thus, with large loop gains the input resistance significantly increases, as is required for measuring a voltage.

With the scheme of Figure 12.11(b) we estimate the output resistance of the series–shunt feedback scheme. The input terminals are shorted because the input generator is zero.

By inspection, we get the result

$$v_x = i_x r'_{out} + v'_{out}, \quad v'_{in} = -\beta_v v_x, \tag{12.34}$$

which, together with equation (12.31), gives rise to

$$v_x = i_x \frac{r'_{out}}{1 + \beta_v A_v \gamma_v}. \tag{12.35}$$

Therefore

$$r_{out,eq} = \frac{r'_{out}}{1 + \beta_v A_v \gamma_v}, \tag{12.36}$$

showing that the output resistance is r'_{out} divided by loop gain (under conditions that make $\gamma_v \simeq 1$). Thus, with large loop gains, feedback greatly diminishes output resistance, making the amplifier suitable for voltage generation as done by an ideal VCVS.

The series–shunt configuration of Figure 12.7(b) and the equivalent circuits of Figure 12.9(a) and Figure 12.10(b) give rise to the equivalent circuit of Figure 12.12. The estimation of input resistance (Figure 12.12(a)) uses zero resistive load. For the output resistance calculation the input terminals are shorted. If the load is finite, the Norton equivalence determines suitable attenuation factors for r''_{out} and i'_{out}.

The equations describing Figure 12.12(a) are

$$v'_{in} = r'_{in} i_x = v_x - \beta_r i'_{out,0}, \quad r'_{in} = r_{in} + r_{out,\beta}. \tag{12.37}$$

The use of equations (12.27) and (12.29) yields

$$i'_{out,0} = v'_{in} G_m \frac{r_{in}}{r'_{in}} \frac{r_{out}}{r_{out} + r_{in,\beta}} = v'_{in} G_m \gamma_i, \quad \gamma_i = \frac{r_{in}}{r'_{in}} \frac{r_{out}}{r_{out} + r_{in,\beta}}. \tag{12.38}$$

Solving (12.37) and (12.38) gives rise to

$$v_x = i_x r'_{in} \{1 + \beta_r G_m \gamma_i\} = i_x r_{in,eq}. \tag{12.39}$$

That yields

$$r_{in,eq} = r'_{in} \{1 + \beta_r G_m \gamma_i\}, \tag{12.40}$$

showing again an increase in input resistance by a factor approximately equal to the loop gain $(1 + \beta_r G_m)$, because $\gamma_i \simeq 1$ ($r_{in} > r_{out,\beta}$ and for a series–shunt scheme $r_{in,\beta} < r_{out}$). The increase in input resistance makes the input port suitable for measuring voltages.

The equivalent circuit of Figure 12.12(b) estimates the output resistance. The nodal equation of the output states:

$$i_x + i'_{out,0} - \frac{v_x}{r'_{out}} = 0. \tag{12.41}$$

Moreover,

$$i'_{out,0} = \beta_r G_m \gamma_i i_x, \tag{12.42}$$

which, combined with equation (12.41), gives rise to

$$v_x = i_x r'_{out} (1 + \beta_r G_m \gamma_i). \tag{12.43}$$

Therefore, the equivalent output resistance becomes

$$r_{eq,out} = r'_{out} (1 + \beta_r G_m \gamma_i), \tag{12.44}$$

COMPUTER EXPERIMENT 12.3

Measuring Input and Output Resistances

The input and output resistance are important parameters of electronic circuits. Ideally input resistance should be zero if the input variable is a current or infinite if it is voltage. By contrast, the output resistance should be zero or infinite, if the output variable is voltage or current respectively. These features are impossible to obtain; however, feedback improves performance. You can verify the benefits with this computer experiment that uses the four basic feedback schemes. It makes available a motherboard and four daughter boards. The motherboard provides the supply, input and output connections. The daughter board contains one of the feedback schemes, conceptually depicted in the figure. The parameters of the forward block (gain, input and output resistances and frequency of the dominant pole) are unknown. Instead you can change the parameters of the β block: $r_{1\beta}$, $r_{2\beta}$, and β. Moreover, you can:

- change the frequency of the input sine wave from 100 Hz to 200 MHz. Notice that the amplitude is 0.1 mV, a value that is small enough for ensuring linear responses under all operating conditions.

MEASUREMENT SET–UP

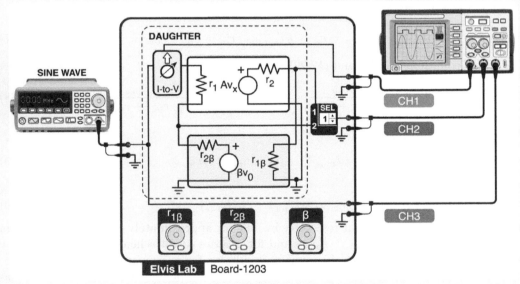

TO DO

- Pick one of the daughter boards (the first available) and identify the type of feedback. Anticipate the result you expect.
- Set the feedback factor to zero and measure input and output impedances at various frequencies, especially at very high ones.
- Use one of the following feedback factors: 1/100, 1/10, 1/5, and 1/2. Measure the input obtained and the output resistance at low frequency. Use the default resistance values.
- Set the resistances to a limit closer to ideal (zero or infinite) and observe the benefit.
- Change the input and output resistance of the feedback block and find out when its value affects the result.

a large increase compared with the output resistance of the current source with the input resistance of the feedback block in parallel.

A similar study for the shunt–shunt and the shunt–series feedback schemes gives rise to equivalent results. Input resistance for measuring a current, diminished by the loop gain and output resistance, increases or diminishes by a factor equal to the feedback loop, depending whether the output quantity is current or voltage.

Remark

Negative feedback increases or reduces equivalent resistances in a manner that improves the voltage or current port interfaces.

12.5 STABILITY

Negative feedback is good for controlling systems, but, if not properly designed, the circuit implementation can give rise to stability problems. It has already been mentioned that negative feedback becomes positive at high frequency. At a specific frequency it determines an overall phase shift equal to 2π (or 360°): the 180° prescribed for the negative loop plus an additional 180° caused by the phase lag. Stability is ensured if the loop gain at that frequency is lower than 1 (or 0 dB). The condition can be verified by looking at the Bode diagram of the loop gain. To be more specific, the points to observe are the crossing of the 0 dB axis on the amplitude diagram and the corresponding phase. The given result is not just information on stability status but, as we will see shortly, also a measure of stability robustness.

Control theory studies stability in great detail. There are specific courses and books that deal with the matter. What is done here is just to give you the first rudiments for understanding the issue. The elements that follow are restricted to linear approximations, also because more detailed studies often focus on linear time-invariant cases, i.e., systems for which any internal variable that decays to zero, $y_i(t)$, responds to any input $x_j(t)$ also decaying to zero

$$\lim_{t\to\infty} y_i(t) = 0, \quad \text{for any } \lim_{t\to\infty} x_j(t) = 0. \tag{12.45}$$

The stability of continuous-time systems is normally studied in the s-domain by an algebraic description of the behaviour and the use of various stability criteria. The most often used are the Hurwitz criterion, which looks at the position of the poles of the transfer function on the s-plane, the Routh criterion, which considers the coefficient of the transfer function, and the Nyquist criterion, which uses a graphic method of investigation.

12.5.1 Frequency Response of Feedback Circuits

The ratio of two polynomials represents any system transfer function in the s-domain. If the response of the forward block is described by

$$A(s) = A_0 \frac{P(s)}{Q(s)}, \tag{12.46}$$

with $P(0)/Q(0) = 1$, the transfer function of a feedback system is given by

$$A_f(s) = \frac{A(s)}{1 + \beta \cdot A(s)} = \frac{A_0 P(s)}{Q(s) + \beta A_0 P(s)}. \tag{12.47}$$

Thus, in a system with feedback (assumed to be implemented with a frequency-independent β), the zeros of the feedback transfer function are the zeros of the forward block, while the position of poles, which starts from the position of the poles of the forward block, depends on feedback.

Since, for any real system, the transfer function goes to zero at very high frequency

$$\lim_{\omega \to \infty} A(j\omega) = 0, \tag{12.48}$$

the number of poles is higher than the number of zeros. Therefore, the feedback scheme has the same number of zeros and poles as the forward transfer function has. Since stability requires finding possible singularities in loop gain, we focus on the denominator of equation (12.47). The expression depends on both poles and zeros of $A(s)$; however, the following analysis considers forward transfer functions that only have poles. The possible contribution of zeros can be analyzed as an extension of the more specific study.

The case of $A(s)$ with a single pole at $s_d = -1/\tau_d$ has already been discussed. The result showed that feedback moves that pole at high frequency by an extent equal to the loop gain. The feedback transfer function for $s = j\omega$ is

$$A_f(j\omega) = \frac{A_0}{1 + \beta A_0 + j\omega\tau_d} = \frac{A_0}{1 + \beta A_0}\left[\frac{1}{1 + j\omega\tau_d/(1 + \beta A_0)}\right]; \tag{12.49}$$

the denominator never goes to zero because the term controlled by β is able to change only the real (or the imaginary) part. Thus, any feedback factor does not cause a risk of instability.

Consider now a forward transfer function with two poles, at the negative angular frequencies $-\omega_1 = -1/\tau_1$ and $-\omega_2 = -1/\tau_2$:

$$A(s) = \frac{A_0}{(1 + s\tau_1)(1 + s\tau_2)}. \tag{12.50}$$

The feedback gain is

$$A_f(s) = \frac{A_0}{1 + \beta A_0 + s(\tau_1 + \tau_2) + s^2\tau_1\tau_2}, \tag{12.51}$$

giving rise to poles of $A_f(s)$

$$\omega_{1,2} = \frac{-(\tau_1 + \tau_2) \pm \sqrt{(\tau_1 + \tau_2)^2 - 4\tau_1\tau_2(1 + \beta A_0)}}{2\tau_1\tau_2}. \tag{12.52}$$

Using $\omega_1 = 1/\tau_1$ and $\omega_2 = 1/\tau_2$, that may also be written as

$$\omega_{1,2} = -\frac{1}{2}(\omega_1 + \omega_2) \pm \frac{1}{2}\sqrt{(\omega_1 + \omega_2)^2 - 4\omega_1\omega_2(1 + \beta A_0)}. \tag{12.53}$$

The above equation shows that the poles' position depends on the β factor. They are initially ($\beta = 0$) on the real axis of the s-plane at the angular frequencies $-\omega_1$ and $-\omega_2$; they move towards one another, with the pole at lower frequency "traveling" faster than the other. When the discriminant becomes zero, the poles are coincident at $-\frac{1}{2}(\omega_1 + \omega_2)$. Then the poles become complex conjugate, with the real part equal to $-\frac{1}{2}(\omega_1 + \omega_2)$.

The poles of the feedback transfer function, being the zero of $(1 + \beta A)$, provide a check of possible unstable conditions that can occur for $s = j\omega$. That condition happens if, for and

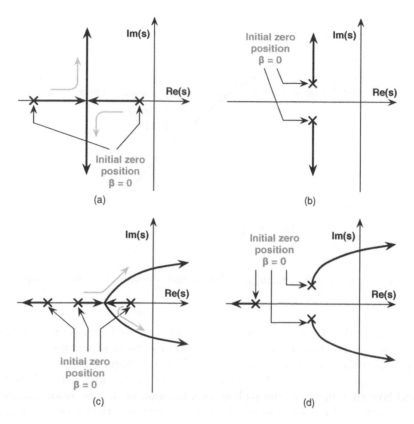

Figure 12.13   Root locus of a feedback scheme employing an $A(s)$: (a) with two real poles; (b) with two complex-conjugate poles; (c) with three real poles; (d) with one real and two complex-conjugate poles.

above a given β factor, one, or more likely a pair, of zeros move onto the right-hand side of the complex s-plane. From the above discussion it is evident that the poles always remain on the left-hand side of the s-plane and never cross the imaginary axis. Therefore, there is no angular frequency $j\omega$ for which the denominator goes to zero. This ensures that a feedback system whose forward amplifier has two poles in the left-hand s-plane is also stable.

Figures 12.13(a) and (b) show how the feedback factor moves the zeros of the denominator of $A_f(s)$ on the s-plane. The diagrams depict a trajectory called the *root locus*. The initial positions of zeros (with $\beta = 0$) are real or complex conjugate and are on the left-hand side of the s-plane. For both cases the poles remain on the left-hand side of the s-plane.

A forward transfer function with three poles has a third-order denominator:

$$A(s) = \frac{A_0}{1 + b_1 s + b_2 s^2 + b_3 s^3}. \tag{12.54}$$

The poles are all real, or one is real and two are complex conjugate; all of them are in the left-hand side of the s-plane. The denominator of the feedback transfer function is

$$D(s) = 1 + \beta A_0 + b_1 s + b_2 s^2 + b_3 s^3. \tag{12.55}$$

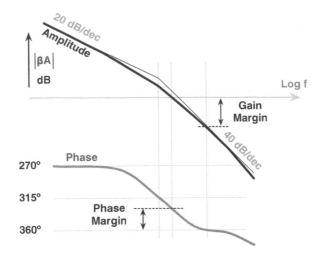

Figure 12.14   Phase and gain margin measured using modulus and phase Bode diagrams.

The task is to verify that the used value of β does not give rise to zeros of $D(s)$ on or beyond the imaginary axis $j\omega$. For this, it is necessary to find the roots of equation (12.55), a cubic equation with real coefficients. Real coefficients of $D(s)$ presuppose zeros that are all real or one real and two complex conjugate. The use of computer calculations determines where zeros are located and how they move on the s-plane as a function of β. The results show that with increasing β one of the real zeros moves to a high frequency while the two other, when complex conjugate, move to the right. If the zeros are initially real they first move towards one another; then they become coincident and turn into complex conjugate as shown in Figure 12.13(c). With initial complex-conjugate zeros the trajectories are like the ones in Figure 12.13(d). It is evident that a given value of β causes the complex zeros to cross the imaginary axis, entering the region of instability. Thus a forward transfer function $A(s)$ with three poles causes a risk of instability, if β is too large. The same obviously happens with more than three poles.

12.5.2   Gain and Phase Margins

In addition to verifying that feedback does not cause instability, it is useful to determine how robust the feedback control is or, in other terms, to measure how far away the operation is from unstable conditions. There are two parameters that provide a quantitative assessment: the gain and phase margins.

Consider the Bode diagrams of the feedback loop gain, $\beta A(j2\pi f)$, and focus on the frequency that gives rise to an overall phase shift of 2π (360°). This is the frequency at which the signal traveling around the loop shows up in phase with the initial signal. In order to avoid output starting to oscillate at that frequency with increasing amplitude, an attenuated returning signal is necessary. That attenuation, measured in dB, defines the *gain margin*.

The *phase margin* is the difference between 360° and the phase at which the Bode diagram of the module crosses the 0 dB axis. Since it indicates the phase difference between the operating conditions and the phase that would cause oscillations, it is, as indicated by its name, a safety limit. Figure 12.14 shows the gain and phase margins using the Bode diagrams.

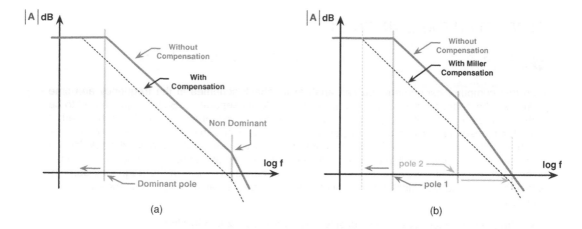

Figure 12.15   Bode diagram of amplifier without and with compensation: (a) with a non-dominant pole at high frequency; (b) in the case that requires pole-splitting because the non-dominant pole is not sufficiently spaced from the dominant one.

If the amplifier has just one pole the additional phase shift is 90°, giving rise to an excellent margin equal to 90°. For two poles the phase shift can be 180°, leading to a potential 0° phase margin. Even if a circuit is, in theory, stable up to a 360° phase shift, it is recommended to avoid situations too close to that limit. Typically there should be at least 60°. For three poles or more, suitable pole positions and β factors ensure the necessary phase margin.

12.5.3   Compensation of Operational Amplifiers

The requirement for operational amplifiers with very high gain often leads to schemes made from a cascade of two or more gain stages. The result gives rise to the required amplification, but the frequency response can be unsuitable for feedback because of bad location of poles.

The operation that improves the frequency response is called *compensation*. The goal is to have a dominant pole separated from the non-dominant poles by a frequency interval that is larger than the DC gain. The Bode diagram would roll off with a constant −20 dB slope until it crossed the 0 dB axis.

Therefore, the task requires the dominant pole to be moved to lower frequencies and, in some cases, the non-dominant poles to be pushed to high frequency. Let us assume we have an operational amplifier with non-dominant poles that are around the extrapolated 0 dB crossing, as shown in Figure 12.15(a). The phase shifts that they generate cause an insufficient or negative phase margin. The best compensation strategy is to shift the dominant pole to a lower frequency. Since a given time constant determines its position, probably controlled by the product of an equivalent resistance and a capacitance, compensation requires increasing the capacitance to give rise to a Bode plot like the dotted one in Figure 12.15(a).

A second possible situation is that of Figure 12.15(b). The separation between the first and second poles is relatively small and the Bode diagram quickly changes its slope from −20 dB to −40 dB. This situation occurs when cascading gain stages with similar frequency performances. The separation of the poles is too small for the pole just to be moved to a lower frequency, because the capacitance that might be required would be too large and the

COMPUTER EXPERIMENT 12.4

Studying the Feedback Loop

With this computer experiment you can analyze feedback schemes in the frequency and time domains. The board analyzes forward and feedback blocks separately. Both are presumed to be stable. The back panel of the spectrum analyzer provides a swept sine wave. It can be used at input for measuring the frequency responses. The second possible input is a sine wave. The feedback block can be controlled at the input terminal or the output of the forward block. With a sine wave at input it is possible to observe non-linearities, gain at defined frequencies and phase shifts. With the spectrum analyzer you have a performance overview in the frequency domain. Notice that the feedback block transfer function slightly changes at very high frequencies.
You can:

- change the frequency of the input sine wave from 100 Hz to 500 MHz;
- set the intervals of the frequency sweep of the spectrum analyzer;
- select the value of β.

MEASUREMENT SET–UP

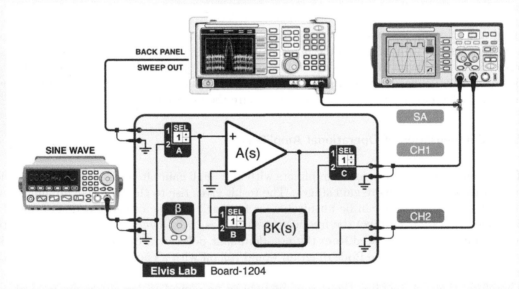

TO DO

- Use the swept signal at the input of the blocks. Observe the frequency responses and notice the differences in gain and bandwidth.
- Set a value of the feedback factor and determine the transfer function of the cascade of both blocks. Determine the frequency at which the loop gain crosses the 0 dB axis for various values of β.
- Use a sine wave at input. Measure the phase shift caused by each of the blocks and their cascade combination at various frequencies.
- Increase the frequency and estimate the value for which the loop gain becomes "1". Measure the phase shift at that frequency. Find out the limit of stability of the closed loop scheme. Use different β factors.

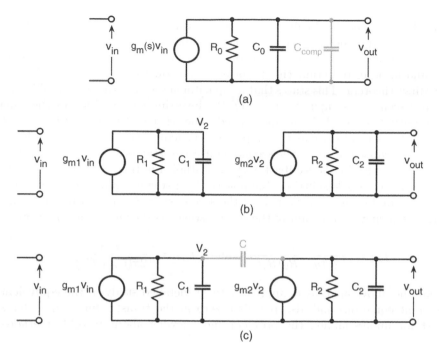

Figure 12.16   (a) Norton equivalent of the small-signal equivalent circuit of an amplifier with a dominant pole (the non-dominant poles are included in the transconductance gain); (b) Norton equivalent circuit of a two-pole op-amp; (c) circuit that gives rise to pole splitting in the Norton equivalent of scheme (b).

resulting frequency response too limited. In this case, it is more convenient to use the so-called pole splitting technique, which separates the poles. One goes to low frequency and the other is moved to high frequency. The dotted Bode diagram of Figure 12.15(b) illustrates the result. Pole splitting is a good compromise to give rise to proper pole location and a reasonable gain–bandwidth product.

The circuit implementation of the two possible solutions is conceptually illustrated in Figure 12.16. The scheme of Figure 12.16(a) depicts the equivalent circuit of an operational amplifier with a dominant pole. It is represented by a transconductance current generator, whose expression incorporates the non-dominant poles:

$$g_m(s) = \frac{g_{m,0}}{(1 + s\tau_{nd,1})(1 + s\tau_{nd,2})\cdots}. \qquad (12.56)$$

The dominant pole is explicitly represented by the RC output network.

The equivalent scheme, which is a good representation for some circuit op-amp configurations and OTAs, immediately shows that the use of a capacitive load comparable to or larger than C_{out} increases the time constant, thus achieving the goal. Notice that the DC gain of the amplifier is given by $g_{m,0}R_0$ and that the dominant pole is, before compensation, at $f_d = 1/(2\pi R_{out}C_{out})$.

Figure 12.16(b) illustrates the case of a cascade of two gain stages with similar frequency responses. The equivalent circuit of each gain stage is a transconductance generator that drives RC loads to model the dominant pole of the stage. Possible non-dominant poles are incorporated in the expression for the transconductance, when needed.

The capacitance that would be necessary for moving one of the poles towards low frequency is too large, and in order to permit the use of relatively low values the pole splitting method exploits the Miller theorem. This states that a capacitor across an inverting amplifier augments the capacitance seen at the input terminal by the inverting gain and leaves the capacitance seen from the output terminal almost unchanged. The scheme of Figure 12.16(c) exploits the benefit by using the second stage as the amplifier that determines the Miller amplification. In addition, since the capacitor across the second gain stage establishes feedback, the result is an increase in its bandwidth by the feedback factor, as studied in this chapter. The overall result is therefore the expected pole splitting: the pole of the first stage moves to lower frequency by the gain of the second stage, and the pole of the second stage (corrected by the action of C_c) moves to high frequency by the gain of the second stage, as shown by the equations

$$f_{p,1} = -\frac{1}{2\pi(g_{m2}R_2)R_1C_c} \quad \text{and} \quad f_{p,2} = -\frac{g_{m2}}{2\pi(C_c + C_2)}. \tag{12.57}$$

Notice that the above analysis is approximate. A detailed study of the equivalent circuit leads to different equations and also reveals a zero in the transfer function. This study is therefore just for understanding the problem and knowing about possible methods for its solution.

12.6   FEEDBACK NETWORKS

Many of the circuits studied in previous chapters use feedback. That was not clearly evident, but, actually, the output signal has been used as part of the input in many configurations. Here we recall some of them and consider other possible schemes that are able to produce the various types of feedback.

The four circuits of Figure 12.17 use an op-amp and various feedback networks. They realize the feedback schemes conceptually depicted in Figure 12.7. As is well known, it is required to measure the output voltage or the output current and to generate a voltage or a current for the input. The scheme of Figure 12.17(a) has, between the differential inputs of the op-amp, the signal voltage v_s minus a fraction of the output voltage, as established by the resistive divider $R_1 - R_2$. The feedback is negative because the feedback signal enters the negative terminal. Moreover, the β factor is

$$\beta = \frac{R_1}{R_1 + R_2}. \tag{12.58}$$

Since the signal generator v_s is at the non-inverting terminal of the op-amp, the input resistance is infinite, as required by a series–shunt configuration.

The scheme in Figure 12.17(b) shows the load resistance that drains the output current to ground. The voltage across the fraction R_{L2} of that load resistance gives rise to the voltage $i_{out}R_{L2}$. That voltage, in series with the input, is subtracted from v_s and determines the op-amp differential signal. The result is what is needed by a series–series feedback scheme. Even for this circuit the input resistance is infinite, being the signal source applied to the non-inverting input, while the feedback signal is at the inverting terminal. The transresistance of the feedback block, β_r, is R_{L2}.

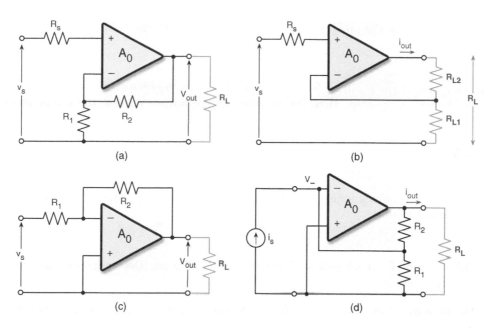

Figure 12.17   Feedback schemes made of an op-amp implementing: (a) series–shunt; (b) series–series; (c) shunt–shunt; (d) shunt–series.

Let us now consider the circuit of Figure 12.17(c). It is the inverting amplifier made with the op-amp already studied in Chapter 6. It is actually a feedback circuit, because the current entering the input node can be viewed as a current contribution coming from the output. Since the output feeds back to the input a current equal to the output voltage divided by R_2, the β_g factor is $1/R_2$. The input current is

$$i_s = \frac{v_s}{R_1}. \tag{12.59}$$

The input resistance is not the expected very low value but equals R_1. That relatively high input resistance is the price that we have to pay for using this type of voltage-to-current converter. Indeed, with a true current signal generator at input, the resistance R_1 is not necessary and the input resistance becomes equal to zero.

The scheme in Figure 12.17(d) uses the op-amp to implement the shunt–series scheme. The voltage of the inverting terminal is

$$v_- = i_s \frac{R_1 R_2}{R_1 + R_2} + v_{out} \frac{R_1}{R_1 + R_2}, \tag{12.60}$$

since the output current is

$$i_{out} = \frac{v_{out}}{R_L}. \tag{12.61}$$

The voltage at the inverting terminal of the op-amp is

$$v_- = i_s \frac{R_1 R_2}{R_1 + R_2} \left[i_s + i_{out} \frac{R_L}{R_2} \right]. \tag{12.62}$$

The loop is closed by the op-amp that gives rise at output to the voltage

$$i_{out} = -A_v v_-;$$ (12.63)

the feedback factor β_i, as resulting from equation (12.62), is R_L/R_2.

The circuit uses an op-amp. However, to generate the output current the use of an OTA would facilitate the task.

PROBLEMS

12.1   A negative feedback scheme has the following transfer functions of the forward and the feedback block:

$$A_0 = \frac{A}{1 + s\tau_1}, \quad \beta = \frac{1 + s\tau_2}{1 + s\tau_3}.$$

Calculate the feedback gain and plot the Bode diagram. Determine the phase at the zero dB crossing.

12.2   The forward block of a negative feedback scheme features one zero and two poles:

$$A(s) = \frac{A_0(1 + s\tau_1)}{(1 + s\tau_2)(1 + s\tau_3)}.$$

What is the feedback gain as function of frequency if the β is constant and equal to 0.032? Use $A_0 = 10^4$, $\tau_1 = 10^{-6}$, $\tau_2 = 10^{-4}$, and $\tau_3 = 10^{-9}$. What happens if τ_1 becomes 10^{-10} or $\tau_1 = -10^{-6}$? What are the τ_1 positive and negative limits that ensure stability?

12.3   Consider the schematic of Figure 12.7. The forward gain is 84 dB and $\beta_v = 0.01$. The output resistance of the forward block, r_{out}, is not zero and the input resistance of the feedback block $r_{in,\beta}$ is not infinite. We have $r_{in,\beta} = 10 \cdot r_{out}$. Calculate the feedback gain. Aging of the circuit degrades the forward gain by 12 dB and increases the r_{out} by 50%. What is the value of β_v that keeps unchanged the feedback gain?

12.4   The input resistance of the forward amplifier in the circuit of Problem 12.3 is

$$r_{in,\beta} = 10 \cdot r_{out}(1 + \gamma V_{out} + \delta V_{out}^2),$$

with $\gamma = 0.001$ and $\delta = 1/718$. Estimate the harmonic terms at the output with a 6.8 mV sine wave at input.

12.5   The input–output response of a power driver is

$$V_{out} = 8V_{in}^3$$

for $|V_{in}| < 0.5$ V, and outside that input voltage range has constant slope and non-discontinuity in the derivative for $|V_{in}| = 0.5$ V. The linear amplifier used in cascade with the power driver has 90 dB gain. The β of the feedback configuration, implementing the scheme of Figure 12.5, is 0.1. Determine the input–output transfer curve of the feedback scheme.

12.6   A forward amplifier is made by a cascade of five equal noisy stages with gain 12 dB. The β of the negative feedback is 0.1. Estimate the input-referred noise. Do not worry about possible problems due to stability associated with the use of this large number of stages. A special technique would avoid that problem.

12.7   Estimate the input and output resistance of a shunt–shunt and a shunt–series feedback scheme made from real blocks. Determine the ratio between the resistances without feedback and the resistances with feedback.

12.8   The forward block of the scheme in Figure 12.11(a) has gain

$$V'_{out,0} = \frac{A'_0}{1 + s\tau_1}$$

generated by a forward amplifier with low-frequency gain $A_0 = 8300$ and $\tau_1 = 10^{-9}$. Estimate the input resistance with $r'_{in} = 200$ kΩ and $\beta = 10^{-2}$.

12.9   The transfer functions of the forward and feedback blocks of a series–shunt feedback scheme are

$$A_0 = \frac{12000}{1 + s \cdot 10^{-8}}$$

$$\beta = \frac{1 + s \cdot 10^{-3}}{10 + s \cdot 10^{-4}}.$$

Plot the Bode diagram of the loop gain and determine the phase shift at the 0 dB crossing.

12.10   A series–shunt feedback loop uses the following blocks:

$$A(f) = \frac{2 \cdot 10^4}{(1 + jf/10^2)(1 + jf/2 \cdot 10^7)}$$

$$\beta = \frac{\beta_0}{1 + jf/10^8}.$$

Determine the phase margin as a function of β_0. What is the feedback factor β_0 that gives rise to 10 dB gain margin?

12.11   Consider the feedback scheme of Figure 12.17(a) and (b). What is the difference between the operations, assuming both are series–shunt schemes?

12.12   Replace R_2 with a capacitance C_2 in the diagram of Figure 12.7(a), and suppose that the forward amplifier has a dominant pole at $f = f_d$. What is the feedback gain? Draft the Bode diagram.

CHAPTER 13

POWER CONVERSION AND POWER MANAGEMENT

The supply of electrical power to any electronic system is, obviously, an essential function. The issue has not been discussed until this point, but, after studying systems and circuits down to the transistor level, it is necessary to study various techniques that, starting from different sources, generate the supply voltages needed by the system. This chapter discusses how to generate supply voltages and how to control circuits when the load changes, and it considers the general issue of optimal power use in electronic systems.

13.1 INTRODUCTION

One important task of electrical engineers is to provide the power needed by electronic circuits at the right voltage with the maximum efficiency. The power source can be an electrical plug, a battery, or a system that converts any form of energy into electricity. The latter kind of power acquisition is called power harvesting. Power sources include, but are not limited to: PhotoVoltaics (PV), wind, micro-turbines, fuel cells, and internal combustion engines. With these sources there are two possible situations: harvesting power for the electronic system itself, or transforming the power into electricity and transferring it to a remote location for generic use. When power is used directly by the system, the voltage must be well defined and stable. For remote use, the electronic circuits just control the harvesting devices to transform power into electricity with the amplitude and DC or AC format optimal for power transmission.

Understanding Microelectronics: A Top-Down Approach, First Edition. Franco Maloberti.
© 2012 John Wiley & Sons, Ltd. Published 2012 by John Wiley & Sons, Ltd.

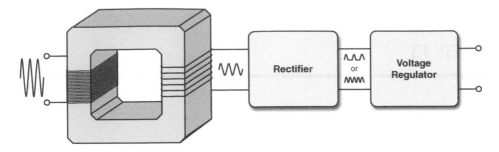

Figure 13.1   Block diagram of a supply generator based on voltage rectification.

The accuracy and quality (low noise, freedom from glitches, and temperature stability) are very important for supplying an electronic circuit, but for systems aimed at conversion and transmission, voltage quality and accuracy are less relevant.

There are four types of power conversion:

- **AC–AC**, normally obtained with a simple transformer;

- **AC–DC**, which uses a rectifier followed by proper electronic functions that reduce the voltage ripple and reject interfering terms;

- **DC–DC**, which directly transforms a DC voltage into another DC voltage (the output can be lower than, higher than, or with the opposite sign to, the input);

- **DC–AC**, which is a type of conversion typically used to facilitate power transmission and voltage amplitude transformation.

This chapter studies the methods for the second and third types of conversion. The first, which you probably already know about, is studied in basic courses. The last is a topic for a specialized course on power electronics and is not considered here.

Another task of electrical engineers is to ensure optimal use of power. Having low efficiency means that a significant fraction of the power becomes heat that, in electronic circuits, must be dissipated to avoid high temperatures damaging the parts. The requirement for high efficiency also comes from portable systems, which are becoming more widespread. Long operation time (or battery life) of portable systems requires, in addition to suitable power conversion, proper handling of power. That involves, for example, dynamic power reduction in some parts of systems, or the use of supply voltages that change dynamically with operational needs. These and other methods, to be discussed shortly, belong to the so-called power management discipline.

13.2   VOLTAGE RECTIFIERS

The distribution of medium–high power uses relatively high AC voltages at low frequency (50 or 60 Hz). The Root-Mean-Square (RMS) amplitude of the voltage available at home is 110 V or 220 V, a value that ensures relatively small currents for the power levels of commonly used electrical appliances. Copper cables that distribute electricity have a reasonable cross section, while having AC waveforms to facilitate amplitude transformation.

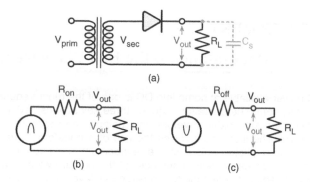

Figure 13.2   (a) Half-wave rectification of the voltage at output of a transformer; (b) simple equivalent circuit for forward bias; (c) simple equivalent circuit for reverse bias.

Electrical systems that use motors, heaters, or other simple components work well with AC voltages. Electronic circuits, on the other hand, need low voltages and DC supplies. Typically they use 3.3 V, 1.8 V, or 1.5 V. With advanced deep sub-micron circuits, the supply voltage is even less, below 1 V. The power consumed is low, rarely exceeding a few watts. It is therefore necessary to transform high AC amplitudes into small DC voltages with currents of a few amperes or less. The first step of the transformation is to reduce the AC line voltage to a level that is just a little more than what is required. An electrical transformer normally does that. It consists of a ferromagnetic core and two or more coils. The current in the primary coil creates an alternating magnetic field in the core. The magnetic field, coupled to secondary coils, induces in them alternating voltages. The ratio between the number of turns in the primary and secondary coils, also called windings, gives the ratios between the primary and secondary voltages. Since the power efficiency of a transformer is high, the current ratios are approximately the inverse of the turn ratios. After reducing the line voltage, a rectifier makes the alternating input unipolar. The result is then suitably regulated to the final value by other electronic blocks as shown in Figure 13.1.

13.2.1   Half-wave Rectifier

The key element used in voltage rectifiers is the diode. As we know, it is a non-linear device with low resistance when current flows in one direction and high resistance for reverse biasing. Rectification can be half-wave or full-wave. The scheme in Figure 13.2(a) depicts the half-wave rectifier and transformer. Full-wave rectifiers will be studied shortly.

The diode, biased in the forward direction as shown in Figure 13.2(b), is a resistance, R_{on}, much lower than the load, R_L. It becomes a high resistance, $R_{off} \gg R_L$, with reverse biasing (Figure 13.2(c)). The result is that an input sine wave gives at output the voltage established by two resistive dividers. For forward biasing we have

$$V_{out+} = V_{in} \frac{R_L}{R_L + R_{on}} \simeq V_{in} \left(1 - \frac{R_{on}}{R_L} \right), \tag{13.1}$$

and for reverse biasing

$$V_{out-} = V_{in} \frac{R_L}{R_L + R_{off}} \simeq V_{in} \frac{R_L}{R_{off}}. \tag{13.2}$$

COMPUTER EXPERIMENT 13.1

Understanding the Half-Wave Rectifiers

Transforming the AC power available at home into DC is one of the most frequent necessities for powering an electronic system. The operation requires three key actions: reducing the amplitude of the voltage, rectifying the sinusoidal waveform into unipolar and smooth, and adjusting the value to the desired level. The first step involves the use of a transformer, the second a diode rectifier, the third a voltage regulator. This computer experiment helps you in understanding the first two steps. There is a transformer with many coils at the primary winding and fewer coils at the secondary winding to reduce the voltage to a desired low level. A diode passes only positive swings and a simple filter limits the ripple. You can:

- change the load resistances;
- set the time constant of the RC filter;
- change the ratio of windings and choose between 110 V (60 Hz) and 220 V (50 Hz).

MEASUREMENT SET–UP

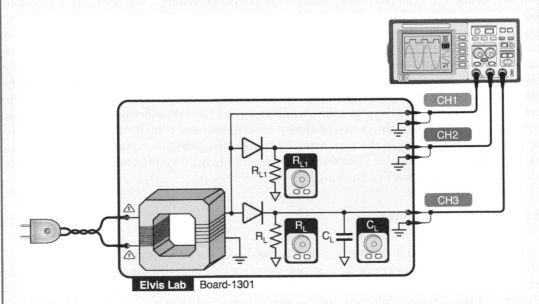

Elvis Lab | Board-1301

TO DO

- Set the resistances at the minimum allowed load. Notice that the maximum power which diode and resistances can dissipate is 1 W and 5 W respectively. If the power exceeds the limits, the increased temperature on the board activates a flashing red diode. Notice that the value of the voltage at the input of the transformer is RMS and that it can fluctuate.

- Use suitable resistance values and observe the waveforms with and without the filtering capacitor. Determine capacitance that gives rise to a 10% ripple at output. Change the resistance and explain why the output ripple changes.

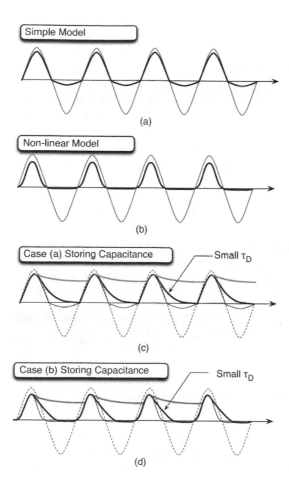

Figure 13.3 (a) Output waveform of a half-wave rectifier with diode described by on- and off-resistances; (b) waveform with non-linear model of diode; (c) and (d) waveforms (a) and (b) with two different storing capacitances.

The output voltage is a replica of the input with low attenuation for forward bias and high attenuation for reverse bias, as shown in Figure 13.3(a). Using a more accurate non-linear diode response produces a slightly different waveform, mainly because of the need of the diode to have a direct voltage of about 0.7 V for turning on. If the input is lower than the turn-on voltage, the on-resistance is high and the output voltage is near zero. For reverse biasing a more accurate model gives a different result. With a few kT/q the current is almost constant: the reverse current of the diode, I_S. Therefore, V_{out} is not voltage dependent but equal to $-I_S R_L$. The difference is, however, minor. Figure 13.3(b) displays the waveform. Remember that I_S is proportional to the junction area: using big diodes gives rise to higher reverse voltages.

Notice that the half-wave rectifier makes the output voltage unipolar but the amplitude changes in time, swinging from almost zero to a peak value and then back to almost zero. This is not the goal: the user does not just want unipolar voltage, but asks for, ideally, a DC constant signal. This requires a further step to transform the unipolar voltage into an almost constant waveform. A simple method that sustains output voltage is the use of a storing capacitor, C_S, connected across the output terminals, as shown in Figure 13.2(a). When the diode turns off, the storing element provides the current required by the load.

The capacitor discharges exponentially with time constant τ_D equal to $R_L C_S$ until the input voltage becomes higher than the output (a condition that biases the diode forward). Then the diode turns on and recharges the storing capacitance. The turn-on time does not occur at the maximum of the input voltage but a little way after the peak. This time is when the slope of the input equals the output voltage divided by the time constant. If τ_D is much less than the sine wave period, the diode remains on until the input voltage reaches zero. Thus the output voltage follows the input waveform. Instead, since we desire almost constant outputs, τ_D must be much larger than the sine wave period, which, remember, equals 20 or 16.66 ms for the 50 and 60 Hz line frequencies, respectively. Since load resistance can be a few ohms, the storing capacitor may be very large.

When the time constant is large, the diode turns on for small time intervals during which the current through the diode must provide all the energy needed for the entire period. Current in the secondary transformer winding therefore consists of large pulses. Their amplitude is approximately the average output current divided by the duty-cycle.

Example 13.1

Design a half-wave rectifier for a 1 kΩ load. The peak voltage of the secondary coil of the transformer is 5 V. Make sure there is a 0.4 V ripple. The diode has an on-voltage equal to 0.6 V. Moreover, input sine wave frequency is 50 Hz.

Solution

The low value of ripple enables us to solve the problem in an approximate manner. We assume that the diode switches on immediately after the peak and returns to on when the input voltage exceeds the output by 0.6 V. At this time V_{out} equals $(5 - 0.6 - 0.4)V = 4$ V and input must be 4.6.

The condition occurs ΔT before the peak:

$$\Delta T = \frac{T}{2\pi} \arcsin\left(\frac{4.6}{5}\right) = 0.186 \cdot T.$$

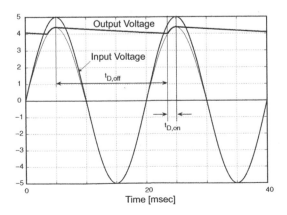

Since the interval between two peaks is 10 ms the discharge period is $\Delta T = 8.14$ ms. The exponential decrease is well approximated by the tangent. Therefore

$$V_{out}(t_{peak} + \Delta T) = V_{out}(t_{peak})\left(1 - \frac{\Delta T}{\tau}\right).$$

Since the drop voltage is 0.4 V, the time constant becomes

$$\tau = \Delta T \frac{V_{peak}}{\Delta V} = 8.14\frac{4.4}{0.4} = 101.75 \text{ ms.} \qquad (13.3)$$

Since the load resistance is 1 kΩ, that would need a filtering capacitance of about 100 µF, a value obtained with discrete electrolytic or tantalum components.

The figure on the previous page shows the output waveform. The output voltage is always 0.6 V below the input and drops to 4 V at the end of the discharge transient. The current in the diode flows only for a mere 1.86 ms out of an entire period of 20 ms.

13.2.2 Full-wave Rectifier

The previous subsection shows that the diode of a half-wave rectifier conducts current only for a small fraction of the half-period. During the other half there is no possible current because the input voltage reverse-biases the diode. Further, the low ripple requirement imposes the use of a storing capacitor that gives rise to a long time constant $R_L C_s$.

The limitation of large values of storing capacitance can be weakened by a factor of two while maintaining the same ripple, by using a full-wave rectifier. This scheme is shown in Figure 13.4(a). The secondary transformer coil uses a center-tapped connection to establish

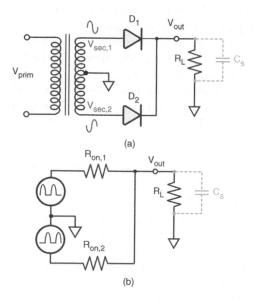

Figure 13.4   (a) Full-wave rectifier with center-tapped connection in the secondary transformer coil; (b) equivalent scheme using the on-resistances of diodes.

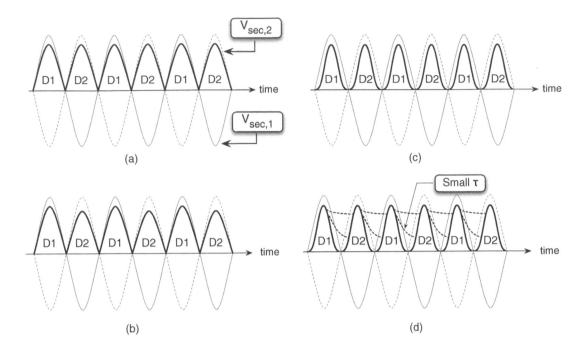

Figure 13.5   (a) Input and output waveforms of the equivalent circuit of Figure 13.4(b); (b) as (a) with different on-resistances of the two diodes; (c) output response with real diode characteristics; (d) responses with storing capacitance across output terminals.

the ground of the output voltage. The terminals of the secondary winding undergo sine wave variations with the same amplitude and opposite phases: $-V_{sec,1} = V_{sec,2}$. When one output is at the maximum positive, the other is at its minimum negative. The scheme operates like two half-wave rectifiers. The one on top furnishes current to the load during one half period and the other works during the other half period.

Figure 13.4(b) illustrates a simplified equivalent circuit that models the diode with an on-resistance in the forward bias region and predicts an open circuit in the reverse conditions. Actually, rather than making the off-resistance infinite, the scheme sends the input voltage to zero. This scheme allows for two resistive dividers, $R_{on,1} - R_L$ and $R_{on,2} - R_L$, that model the output voltage in the two phases of operation. If the two diodes have the same on-resistance, the output is the rectified and attenuated waveform of Figure 13.5(a). The gray curves are the voltages at the transformer terminals measured with respect to the center connection. If the diodes do not match but have different on-resistances, the attenuation factors of resistive dividers differ and the output looks like the waveform in Figure 13.5(b).

A more accurate diode model, able to account for the behavior at low forward biases, predicts small currents until the diode voltage reaches the on-level. A model with on-resistance that increases and goes to a very large value when the forward bias becomes zero describes this behavior. Figure 13.5(c) shows that across zero the waveform slope does not change sharply.

Notice that during one half-wave period only D_1 operates because the load current flows from the top of the secondary coil of the transformer. During the next half period D_1 goes off and only D_2 operates, with current exiting the bottom transformer section. The current in

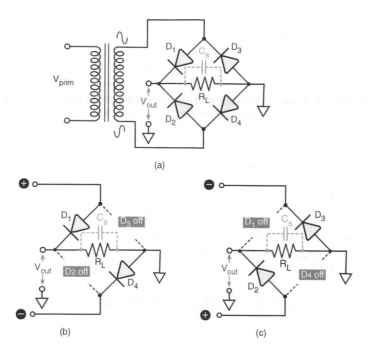

Figure 13.6 (a) Full-wave rectifier in bridge configuration; (b) path from positive terminal at top to negative terminal at bottom; (c) complementary path with positive voltage at bottom and negative at top.

the primary coil is almost a sine wave because it induces current in the top secondary winding during one half period and in the bottom secondary coil during the other half period. To be precise, there is some distortion caused by the non-linear response of the diode near zero crossings. Having a sinusoidal current in response to a sinusoidal voltage means that the power lines get a linear load. The distortion is not particularly relevant because power is typically low, but it implies the generation of harmonic terms at multiples of the line power frequency.

Even the full-wave rectifier uses capacitors across the output terminals to maintain the output voltage and limit ripple. The capacitor is left alone for a shorter time than with the half-wave because of the interleaved action of two diodes. Therefore, the amplitude of the output voltage, which depends on load, features a lower ripple. As already mentioned for the half-wave case, the output starts dropping after the positive peak of input voltage, but the exponential decrease ends before, because of the positive swing of the complementary voltage. The time during successive refuelings of the storing element is halved, thus making it possible to reduce its value by half while ensuring the same output ripple.

The key drawback of the scheme of Figure 13.4(a) is that it uses a transformer with a secondary center tap. Moreover, since the circuit interweaves two half secondary windings, the peak output voltage is half of the total induced voltage on the secondary windings.

The full-bridge rectifier of Figure 13.6(a) does not need the use of a center-tap connection in the transformer. The reason is that one secondary terminal establishes the current return path. If the top terminal is positive and the bottom is negative, current flows through diode D_1, biases the load and returns to the secondary coil through D_4. Diodes D_2 and D_3 are off,

COMPUTER EXPERIMENT 13.2

Full–Wave Rectifier

One limitation of the half–wave rectifier is that it uses only half of the input sinusoidal waveform. Since the current from transformer to load flows during half sine wave, the current in the diode is high and the filtering capacitor is left alone for relatively long periods. Using two diodes and a transformer with a center–tapped connection is also a problem. The full–wave rectifier studied in this computer experiment resolves the problems by using a bridge of diodes that reverses the path of current from the output of both transformer terminals. A current meter (I-to-V), in series with one terminal of the secondary winding, measures the current. The voltage across that current meter is almost zero while its maximum measured value is 1 A. The power dissipated by diodes and resistor must be lower than 1 W and 10 W respectively. In this virtual experiment you can:

- change the value of resistive load and filtering capacitor.
- change the winding ratio of the transformer
- select between 110 V (60 Hz) and 220 V (50 Hz).

MEASUREMENT SET–UP

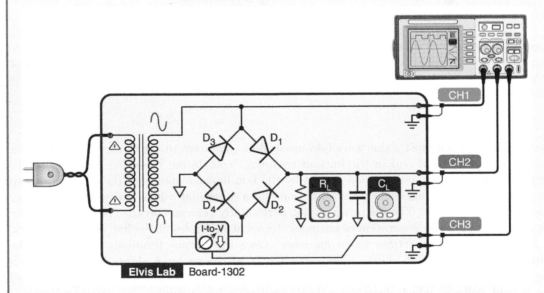

Elvis Lab   Board-1302

TO DO

- Estimate the minimum load resistance which does not damage any part of the circuit. Suppose that the voltage of the power line can differ from the nominal value by ±15%. Actually, this is what happens when you do the experiment. Use a winding ratio that gives rise to a nominal 5 V peak at output.
- Set the capacitance to zero and observe voltage and current.
- Choose three values of the load resistance and filtering capacitor and measure the output voltage ripple. Observe the current at the secondary winding and explain what happens.

as shown in Figure 13.6(b). The opposite polarity activates the path through D_2 and D_3, as shown in Figure 13.6(c), to establish a voltage with the same sign.

Notice that the ground connection is the reference for output. The terminal of the transformer that drains current away from ground with a forward-biased diode is one on-diode voltage below ground. The other terminal is one on-diode voltage above V_{out}. Every half clock period the DC level of each transformer terminal jumps from negative to positive. This is not a serious limitation because the secondary coil is electrically insulated from the primary winding. The only concern is that parasitic capacitances loading the secondary terminals consume power and are being continuously charged and discharged. However, the wasted power is low since the sine wave frequency is very low.

As noticed above, the output voltage of the full-wave rectifier is the voltage of the secondary coil minus the drop voltages of the two on-diodes:

$$V_{out} \leq V_{sec} - 2V_{on}. \tag{13.4}$$

Since V_{on} is about 0.65 V, the difference can be a large fraction of the output when generating low values. The voltage drop and power efficiency are lower than their half-wave counterparts because there are two diodes along the current path rather than one.

13.3 VOLTAGE REGULATORS

Voltage regulators are blocks used after rectifiers or, putting it more generally, after any generators whose output does not correspond to what is required. The voltage regulator generates a defined voltage amplitude starting from higher or lower levels. Ideally, the regulator gives rise to a drop or a boost exactly equal to the difference between input and desired output for any loading current. The ability to maintain constant output over a wide range of loads is a quality factor. It is often quantified by the load regulation parameter, defined as

$$LR = \left| \frac{V_{FL} - V_{ML}}{V_{NL}} \right| 100, \tag{13.5}$$

where V_{FL}, V_{ML}, and V_{NL} are the regulated voltages at full, minimum, and nominal load respectively. The load regulation is a dimensionless percentage factor that denotes good quality when its value is low.

For its proper operation any voltage regulator uses an input voltage somewhat different from the desired output. A second quality factor is the ability to operate with input voltages very close to output. Since the difference between input and output (often) denotes consumed power, it is necessary to allow minimal drops in voltage for optimal use of energy.

Voltage regulators are divided into two categories: linear regulators that generate voltages lower than input by controlling the drop between input and output with active devices and dissipative components, and switching regulators that use power switches and non-dissipative components under the control of a clock.

13.3.1 Zener Regulator

The simplest type of voltage regulator is the zener scheme of Figure 13.7(a). This circuit has been discussed previously to outline a possible use of the zener diode.

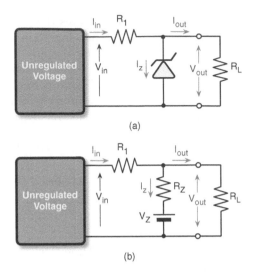

(a)

(b)

Figure 13.7   (a) Simple zener regulator; (b) equivalent circuit of the simple zener regulator.

The following equations describe the regulator, which operates under the obvious assumption that the zener current I_z is in the reverse direction and large enough to keep the diode in the zener region

$$I_{in} = \frac{V_{in} - V_{out}}{R_1}, \quad I_{in} = I_{out} + I_z. \tag{13.6}$$

The simple equivalent circuit of Figure 13.7(b) modeling the zener diode yields

$$V_{out} = V_z \frac{R_1}{R_1 + R_z} + V_{in} \frac{R_z}{R_1 + R_z} - \frac{R_1 R_z}{R_1 + R_z} I_{out} \tag{13.7}$$

$$V_{out} = R_L I_{out}. \tag{13.8}$$

The voltage regulator is simple but performance is limited. For good load regulation the second term should be low. Therefore, the resistance R_1 should be much higher than R_z. This is difficult to obtain because, even for good zener diodes, the value of R_z can become comparable to the load resistance. Another problem is that, since the zener current cannot be zero, the input current must always be larger than the full load current I_{FL}, giving rise to significant, constant power consumption.

The choice of R_1 depends on I_{FL}, I_z, and the minimum expected input voltage, $V_{in,min}$. R_1, that possibly includes the output resistance of the unregulated source, must satify the condition

$$R_1 < \frac{V_{in,min} - V_z}{I_z + I_{FL}}. \tag{13.9}$$

Therefore, the input current $I_{in} = (I_z + I_{FL})$ flowing through R_1 gives rise to the power that R_1 must dissipate,

$$P_{R_1} = \frac{(V_{in} - V_{out})^2}{R_1}. \tag{13.10}$$

The maximum power of the zener diode occurs when the output current is zero:

$$P_{z,max} \geq I_{in} V_z. \tag{13.11}$$

Even with full load current around 1 A and input voltages of a few volts the power levels that become heat are high and difficult to handle even for discrete implementations.

Example 13.2

Design a zener regulator that nominally generates 3.8 V and has full load current of 22 mA. The reverse current of the 3.8 V zener diode must be at least 3 mA. Determine the power dissipated by resistance and the maximum power consumed by the zener diode. The unregulated input voltage ranges from 4.6 V to 5.2 V; R_z is 2 Ω. What is the output voltage with minimum and maximum unregulated values? How does the voltage change when the current goes from zero to maximum?

Solution

The input current is 25 mA; 22 mA for the full load and 3 mA for the zener diode. The minimum of unregulated voltage and input current determines R_1:

$$R_1 = \frac{V_{in,min} - V_z}{I_{in}} = 32\ \Omega.$$

The input current with unregulated input at maximum, $(V_{in,max} - V_z)/R_1$, is 44.8 mA. Therefore, the maximum power in the zener diode and R_1 are 61.25 mW and 166.25 mW respectively. The use of equation (13.7) yields

$$V_{out} = V_z \cdot 0.941 + V_{in} \cdot 0.059 - 1.88 \cdot I_{out}.$$

The output voltage with zero output current ranges between

$$V_{out} = 3.847\ \text{V} \quad \text{and} \quad V_{out} = 3.882\ \text{V},$$

for minimum and maximum values of unregulated voltage respectively. The voltage drop by 41.4 mV caused by the full 22 mA current is comparable with the error caused by the unregulated input.

Self training

Do a search on the Web and download the specifications of three different zener diodes. Summarize in a short report their key features and recommended applications. Notice the power consumption limit and estimate the maximum current that the zener regulator can deliver while maintaining the nominal output voltage.

13.3.2 Series Linear Regulator

The needs of simple circuits that have fixed loads are well satisfied by the zener regulator studied in the previous subsection. We can set the current in the zener diode just at the right level to compensate for fluctuations of input voltage. When the load is not constant, the zener regulator becomes ineffective because total power dependent of load reduces the power

COMPUTER EXPERIMENT 13.3

Understanding Zener Regulation

The voltage at the output of half–wave or full–wave rectifiers or other equivalent power sources is not suitable for supplying electronic circuits, because of the value inaccuracy and/or short or long term fluctuations. For systems that need a well controlled supply voltage it is necessary to use electronic regulation of the generated voltage. This computer experiment enables you understand the operation of a simple voltage regulator: the zener type. It uses the special feature of controlled breakdown or the zener effect of a reverse–biased diode. To emulate an unregulated input signal the board uses a sine wave added to a DC component. The addition, made on the circuit board, enables a maximum current of 1 A. This ElvisLab board uses a simple method for measuring the current: a small test resistance R_T equal to 100 mΩ is placed in series with the zener and one of the scope channels measures the voltage across it. The zener voltage is 2.7 V. You can:

- set amplitude and frequency of the sine wave generator;
- set the DC level component with a voltage bias generator;
- change the value of the load resistance;
- short the test resistance with a mechanical switch (its on–resistance is negligible).

MEASUREMENT SET–UP

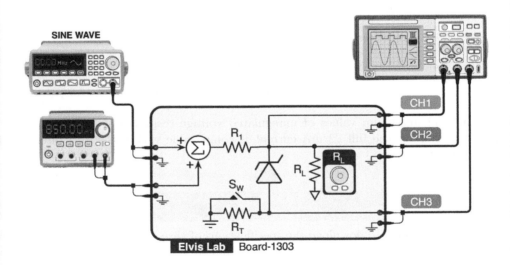

TO DO

- Set the DC voltage of the input generator to 2 V and the sine wave amplitude to 1 V. The frequency is 2 kHz. Observe the output and explain why you get that waveform.
- Set the input generators at 3.3 V + 0.2 V sine wave. Estimate the ripple of the output voltage and the current on the zener diode.
- Switch the test resistance used to measure current to zero and assess the error it caused in the output voltage.
- Repeat the experiment with various input waveforms.

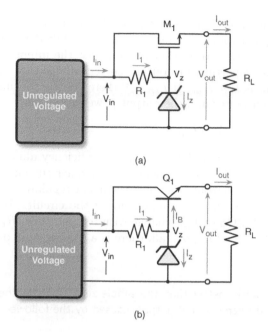

(a)

(b)

Figure 13.8   (a) Series linear regulator with MOS transistor; (b) series linear regulator with bipolar junction transistor.

effectiveness and gives rise to the need for power dissipation. As we have seen before, the power not transferred to output heats the zener diode, possibly damaging the part.

For more demanding situations the series linear regulator is more appropriate. Figure 13.8 shows two schemes, one with a bipolar transistor and the other with a MOS transistor. The bipolar transistor operates in the linear region and the MOS device works in saturation. The reason for the name "linear regulator" is that early schemes used bipolar transistors in the linear region of operation. Both circuits of Figure 13.8 employ followers; the one in Figure 13.8(a) uses a source follower and that in Figure 13.8(b) has an emitter follower. Moreover, both use a simple zener regulator to establish the follower control. The auxiliary zener regulators draw little current because what they need is just the current that keeps the zener diode in the breakdown region, plus, for the BJT scheme, the base current of the bipolar transistor. Therefore

$$I_1 = I_z \quad \text{or} \quad I_1 = I_z + \frac{I_{out}}{\beta}. \tag{13.12}$$

The β of a transistor is normally large enough to make the value of I_1 small. If necessary, a Darlington arrangement replaces the single bipolar transistor.

Observe that, since I_1 is low, the value of R_1 is large, even with relatively large drops in voltage across the auxiliary regulator. The voltage V_z controls the follower to give rise to its shifted-down replica. Using equation (13.7), we get

$$V_{out,MOS} = V_z \frac{R_1}{R_1 + R_z} + V_{in} \frac{R_z}{R_1 + R_z} - V_{GS1} \tag{13.13}$$

$$V_{out,BJT} = V_z \frac{R_1}{R_1 + R_z} + V_{in} \frac{R_z}{R_1 + R_z} - \frac{I_{out}}{\beta} \frac{R_1 R_z}{R_1 + R_z} - V_{BE1} \tag{13.14}$$

for the MOS and the BJT versions respectively.

Notice that a large resistance R_1 reduces the sensitivity to V_{in}. Moreover, the generated voltage changes with output current, because the follower shift depends on current. For the BJT scheme there is also an extra term that accounts for the minor change of the auxiliary zener output voltage. It is proportional to output current divided by the β of the transistor.

The unregulated current is $I_1 + I_{out}$. If $I_{out} = 0$ there is little consumed power, because I_1 is low. If $I_{out} > 0$, the input power and the output power go up to

$$P_{in} = V_{in}(I_{out} + I_z), \quad P_{out} = V_{out}I_{out}, \tag{13.15}$$

showing that if I_z is a small fraction of I_{out} the power efficiency almost equals $\eta = V_{out}/V_{in}$, the maximum a simple zener regulator can achieve (i.e., zener current almost zero).

In addition to the above, the output voltage of the linear regulator depends on temperature because of the dependencies of various components of the circuit. In particular, the zener voltage and the threshold of MOS or V_{BE} change with temperature. R_1 also varies because of its temperature coefficient. Therefore, overall, there is a temperature drift to account for.

Take note

The series linear regulator works better than the simple zener scheme does. The price to pay is the extra drop, in regulated reference and output, caused by the follower.

It is often necessary to know how the current changes the output voltage. As mentioned above, the output varies because V_{GS} and V_{BE} increase with source or emitter current. Moreover, what is normally done is to measure those changes with respect to a minimum current rather than zero. That makes sense because source or emitter opened do not give rise to any significant result. To ensure proper operation it is necessary to have a minimum current through the transistor.

For the MOS scheme, supposing it to be in saturation, we have

$$V_{GS} = V_{Th} + \sqrt{\mu C_{ox}\frac{W}{L}I_D}, \tag{13.16}$$

where, as we know, threshold voltage changes with V_{GB} and the overdrive increases as the square root of output current. If I_{ML} is the minimum load current, the follower shift is

$$V_{GS,ML} = V_{Th,ref} + \sqrt{\mu C_{ox}\frac{W}{L}I_{ML}}. \tag{13.17}$$

A larger current, $I_{out} > I_{ML}$, changes the output voltage by

$$\Delta V_{out} = -\Delta V_{Th} - \sqrt{\mu C_{ox}\frac{W}{L}}(\sqrt{I_{out}} - \sqrt{I_{ML}}), \tag{13.18}$$

where the first term is the change in threshold voltage caused by the body term.

For the bipolar scheme we use the relationship between V_{BE} and output current. For the minimum load current

$$I_{ML} \cong A_E I_{SS} e^{V_{BE,ML}/V_T}, \tag{13.19}$$

where A_E is the area of the base-to-emitter junction. The output voltage with $I_{out} > I_{ML}$ changes by

$$\Delta V_{out} = -V_T \ln \frac{I_{out}}{I_{ML}}, \tag{13.20}$$

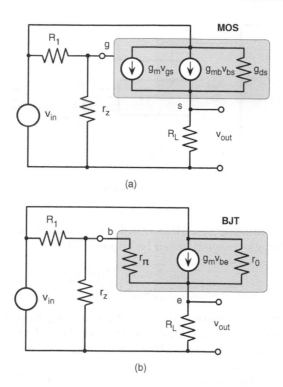

Figure 13.9   (a) Small-signal equivalent circuit of the series linear regulator with MOS transistor; (b) small-signal equivalent circuit of the series linear regulator with bipolar junction transistor.

showing that increasing the output current by 10 gives rise to a drop in voltage equal to $V_T \log_e 10 \cong 60$ mV, at room temperature.

One important quality feature is the rejection of disturbances to the unregulated supply. Equations (13.13) and (13.14) are for large signals. Assuming that spurs are small, it is more convenient to use the small-signal equivalent circuit. As usual, small letters denote small signals. Figure 13.9 shows the equivalent circuits of both linear regulators. The generator v_{in} represents small-signal disturbances to input.

For the circuit with the MOS transistor, the gate voltage is

$$v_g = v_{in} \frac{r_z}{R_1 + r_z}. \tag{13.21}$$

The nodal equation of output gives rise to

$$\frac{v_{out}}{R_L} = g_m(v_g - v_{out}) - g_{mb}v_{out} + g_{ds}(v_{in} - v_{out}), \tag{13.22}$$

which, together with equation (13.21), yields

$$v_{out} = v_{in} \frac{g_{ds} + g_m r_z/(R_1 + r_z)}{g_m + g_{mb} + g_{ds} + 1/R_L}. \tag{13.23}$$

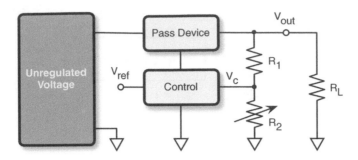

Figure 13.10   Conceptual block diagram of series linear regulator with adjustable voltage.

Since the transconductance gain of a MOS transistor in saturation is higher than g_{mb} and much higher than g_{ds}, the above equation can be approximated by

$$v_{out} \simeq v_{in} \frac{g_{ds} + g_m r_z / (R_1 + r_z)}{g_m + 1/R_L}, \tag{13.24}$$

showing that there are two disturbance paths from input to output, one from MOS drain and the other via the transistor's active operation. Fortunately large factors attenuate both. For a load higher than $1/g_m$ (or $g_m \gg 1/R_L$), they are g_{ds}/g_m and $r_z/(R_1 + r_z)$. Low output resistances make the attenuation factors large.

A similar study holds for the equivalent circuit in Figure 13.9(b). This gives rise to almost identical results, because in parallel to r_z we have the input resistance of the emitter follower. This resistance, as studied in Chapter 10, is large, being r_π plus the amplification of R_L:

$$r_{in} = r_\pi + (1 + g_m r_\pi) R_L. \tag{13.25}$$

Since r_{in} is typically much higher than r_z, the parallel connection almost equals r_z, leading to equations the same as (13.23) and (13.24).

13.3.3   Series Linear Regulator with Adjustable Voltage

Many electronic circuits require the use of a variable supply voltage for optimizing performance and ensuring flexibility. In those cases the voltage regulator must be able to adjust its output to a desired value within a given regulation range. A conceptual scheme for that function is that of Figure 13.10. The resistive divider $R_1 - R_2$, has an adjustable attenuation factor by the variable resistance R_2. The result is a variable fraction of the output voltage,

$$V_c = V_{out} \frac{R_2}{R_1 + R_2}. \tag{13.26}$$

Voltage V_c drives a control block together with another input: the reference voltage, V_{ref}. The block estimates the difference between V_{ref} and V_c,

$$\varepsilon_V = V_{ref} - V_c, \tag{13.27}$$

that drives a pass device whose output is V_{out}. A transistor that gives rise to the required voltage drop realizes the pass function. It can be a follower or an inverting configuration.

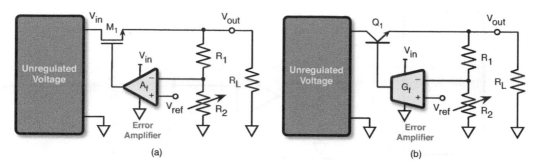

Figure 13.11 Block-level implementation of the series linear regulator with adjustable voltage: (a) with use of an NMOS pass transistor and (b) with an NPN-BJT pass transistor.

The sensing of output, its comparison with the reference, the control block, and the pass device make a feedback loop that tends to regulate the output voltage to the value

$$V_{out} = V_{ref} \frac{R_1 + R_2}{R_2}. \tag{13.28}$$

The accuracy of the regulated voltage depends, as already studied in Chapter 12, on the loop gain. Moreover, as is required in any feedback scheme, it is necessary to ensure loop stability by establishing a proper frequency response of the control block.

Equation (13.28) shows that a fraction of the output equals the reference voltage. This is a good feature because the designer can conveniently use reference generators whose value is lower than the output. For example, the on-voltage of a diode can be used; it is a cheap reference with reasonable performance. It does not change significantly with current and remains around 0.65 V. However, the on-voltage of the p–n junction changes with temperature (about -2.2 mV/°C). When temperature dependence is a concern, rather than V_D, other references are used. One of them is the band-gap voltage. It is generated by a relatively complex scheme and ensures very low temperature dependence.

The pass device can be a MOS transistor or a bipolar transistor in the follower configuration. The control is, respectively, by voltage or by current. Figure 13.11 shows two possible schemes establishing the control loop. The circuit with a MOS transistor employs an operational amplifier. The bipolar version utilizes an OTA. Since the follower is non-inverting, the feedback signal enters the negative terminal of the op-amp or the OTA so as to ensure negative feedback.

As is well known, the gain of a follower is less than 1. Moreover, if the load resistance is low, the gain is still lower than 1. Therefore, the op-amp gain or the transconductance gain of the OTA must be large, in order to guarantee a large forward gain, A_f. The β factor is

$$\beta = \frac{R_2}{R_1 + R_2}; \tag{13.29}$$

since the output of a feedback scheme is

$$V_{out} = V_{ref} \frac{A_f}{1 + \beta A_f}, \tag{13.30}$$

the regulated voltage is the expected result of equation (13.28) if $A_f \gg 1$.

Notice that possible errors or unwanted fluctuations of V_{ref} pass through the feedback loop as the reference voltage itself. They show up at output, amplified by $(R_1 + R_2)/R_1$. Because

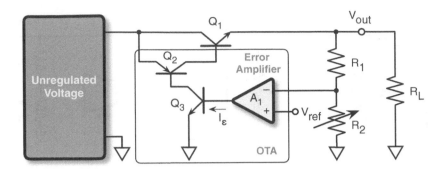

Figure 13.12   Possible implementation details of Figure 13.11(b) with a high transconductance gain OTA.

of the attenuation established by the resistive divider, it is good practice to use well-controlled and spur-free reference voltages.

13.3.4 Supply of Active Blocks and Drop-out Voltage

The schemes of Figure 13.11 use an active block that, obviously, needs its supply voltage. Since the gate of M_1 must be higher than the output by the drop voltage given by the follower, the op-amp supply must be higher than the regulated voltage. As a result, it is necessary to bias the op-amp or the OTA with the unregulated input, which is often affected by noise and spurs. It is therefore necessary to use a scheme with good power supply rejection, because interference and noise could reach the regulated terminal through the supply of the active block.

Another point to observe is that active blocks always generate signals below their supply voltage. Depending on the scheme, there is a minimum voltage overhead, ΔV_{ov}. On top of this the follower establishes another voltage shift, ΔV_{foll}. The maximum output voltage becomes

$$V_{out,max} = V_{in} - \Delta V_{ov} - \Delta V_{foll}. \tag{13.31}$$

Equation (13.31) shows that proper operation of the regulators demands a minimum drop between unregulated input and regulated output. This minimum voltage is called the *drop-out voltage*. It is also the voltage that determines the minimum power that can be consumed by the pass transistor for a given output current.

Suppose, for example, that ΔV_{ov} is 0.3 V and $\Delta V_{foll} = 0.7$ V. The drop-out voltage is as large as 1 V, a significant limit to power efficiency when handling low voltages.

Figure 13.12 shows a possible implementation that moderates the problem for the bipolar-based variable regulator with OTA. The output of the OTA is the collector of a PNP transistor Q_2 that gives rise to the base current of Q_1. In turn, transistor Q_3 controls the base current of Q_2 by the use of an error amplifier. The result is that the current at output of the error amplifier A_1, I_ε, is augmented by three βs, those of Q_3, Q_2, and Q_1, to give rise to the output current.

Notice that since the voltage at the output of error amplifiers is the V_{BE} of Q_3, the supply voltage of A_1 can be low, and therefore regulated, because even the input terminals are at low voltage: a fraction of the output. The good feature of the scheme is that it has only the emitter of Q_2 connected to an unregulated voltage.

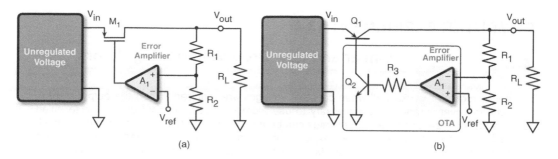

Figure 13.13   Low drop-out voltage regulator with (a) MOS transistor; (b) bipolar transistor.

The minimum drop between input and output in Figure 13.12 is the V_{BE} of Q_2 plus the V_{BE} of Q_1 (the V_{BC} of Q_2 is supposed zero). With large currents, rather than using transistor Q_1, it may be necessary to use a Darlington configuration. This requires an extra V_{BE} in the drop-out voltage. Therefore, the simple scheme of Figure 13.12 gives a drop-out voltage of $2\ V_D$; with a Darlington implementation V_{DROP} goes to $3\ V_D$. The value of V_{BE} is about $0.6 \div 0.7$ V but increases when temperature decreases. The result is that a large change of temperature gives rise to large variations in the drop-out voltage. With Darlington schemes it can go up to 2.5 V.

13.3.5   Low Drop-out (LDO) Voltage Regulator

A low drop-out voltage is required by many applications. A frequent strategy for low power is to regulate the supply voltage in two steps. The first uses a power effective regulator and the second a linear scheme that reduces ripple. In these cases the power consumed by the linear regulator must be minimal. For power efficiency, a low drop-out linear scheme (LDO) that regulates the voltage with minimum input–output drop is the appropriate solution.

Figure 13.13 illustrates a possible implementation of an LDO with MOS and bipolar transistors. The difference between the LDO schemes and the linear regulators of Figure 13.11 is that the pass device is an inverting amplifier instead of a follower. The error amplifier accounts for the minus sign of the inverter by reversing the input control, thus maintaining negative feedback.

Because of the change, the gate and base voltages are now lower than the input. The control does not require headroom and follower shift any longer. Therefore, the drop-out voltage can be to the lowest possible value of V_{DS} or V_{CE}, which are the saturation voltages for the MOS transistor and one diode-on voltage for the bipolar schemes. The result is that the drop-out voltage can be as low as 0.6–0.8 V for the bipolar scheme and even less for the MOS implementation.

As already mentioned, an LDO changes the function of a pass device from follower to inverting amplifier. The modification seems minimal, but in real situations it causes problems because of the added gain of the inverting amplifier. That is not a benefit, because the gain value depends on the load. The stability of feedback loops with gain that changes significantly is difficult to control. Details on how to study and resolve the problem are not given here, since they are the topic of more specific courses.

An alternative solution to the LDO schemes of Figure 13.13(a) exploits the fact that the pass device is a MOS transistor. It does not need an input current, as the resistance of the

COMPUTER EXPERIMENT 13.4

Linear and Low Drop Voltage Regulators

The zener regulator works well when constant voltages are required. When the output voltage must change over a range, it is necessary to use more sophisticated solutions. Even for fixed voltages having a zener at the desired value can be a problem. The use of op-amp and feedback enable flexible control. This computer experiment studies two types of voltage regulators: the linear and the low drop. They employ MOS transistors to establish a required voltage drop. Solutions with bipolar elements are also possible. The difference between the two schemes is that the linear regulator uses the source as output node. As you know, the LDO permits a lower voltage drop from input to output thanks to the use of the drain as output node. An operational amplifier establishes the control loop. An external voltage generator sets the voltage that is used as reference for both schemes. In this virtual experiment you can:

- set the input voltage as the superposition of a DC and an AC component;
- change amplitude and frequency of the sinusoidal component;
- set the reference voltage;
- set the resistive divider of output voltages and loads.

MEASUREMENT SET–UP

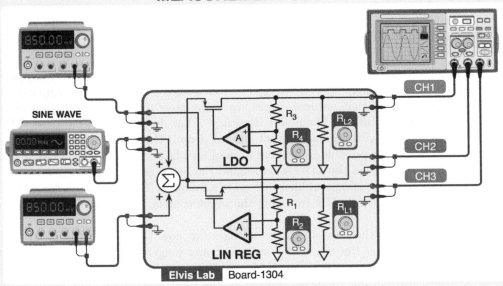

TO DO

- Set the input voltage at 4 V and zero the sine wave amplitude. Choose $R_1 = R_3 = 80$ kΩ, load resistance 10 kΩ and the reference voltage 1.2 V. Change the variable resistances from 640 kΩ to 20 kΩ. Plot the two output voltages as a function of the variable resistors. Notice the differences when the voltage drop between input and output becomes low.
- Add a 50 mV sinusoidal term at 1 kHz and observe the outputs.
- Change the value of the load resistances and measure the voltage drop at output. Estimate the output resistance of the regulators.

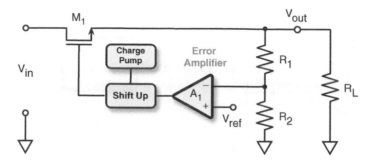

Figure 13.14   Low drop-out voltage regulator based on the linear regulator of Figure 13.11(a) and a boost of gate control.

gate is infinite. The scheme of Figure 13.14 illustrates the solution. Instead of an inverter the LDO of Figure 13.14 still uses a follower, but the voltage controlling the gate is boosted upwards. Having a gate voltage that possibly exceeds the supply does not create reliability problems unless the gate-to-channel voltage remains below the safety value.

The circuit may require the use of a charge pump, as indicated in the figure, to support the shifting-up block. Charge pumps are sampled-data circuits that give rise to voltages higher than the supply. They will be studied shortly.

13.3.6   Protection Circuits

Handling high power causes the risk of damage because of excessive voltage, large output current, or heating that causes high temperatures. In the case of a short circuit at output the regulator does not help because it controls currents poorly. The input of the error amplifier goes to zero and the control saturates. For the circuit of Figure 13.11(a) the gate of M_1 goes to the full V_{in} voltage. The output of the amplifier in the LDO scheme of Figure 13.13(a) drops to zero. The current even becomes very large, because the aspect ratio of M_1 is rather large. Similar situations occur for regulator schemes with BJTs. Since large currents damage metal connections or cause excessive heat that impairs the junctions, it is necessary to take actions that avoid dangerous situations. The protection schemes of many linear regulators provide:

- thermal shutdown;

- a current limiter.

The thermal shutdown operates when the temperature exceeds a safe limit (usually 160°C). The function is accomplished by a thermal sensor, which switches off the pass device when the junction temperature rises to values that can damage the part. A simple way to measure temperature is by using a monitoring diode-on voltage or the V_{BE} of a bipolar transistor. It has already been mentioned that the temperature coefficient of the diode-on voltage (and V_{BE}) is −2.2 mV/°C. At 160°C (130°C above the nominal room temperature) the temperature coefficient gives rise to a 286 mV reduction of voltage across a p–n junction with forward biasing. Such a large voltage change can easily be detected and monitored.

Consider, for example, the circuit of Figure 13.15. A zener regulator and a resistive divider $R_1 - R_2$ generate a voltage that is slightly temperature dependent. The value obtained (for example, 0.38 V in the circuit) is lower than the threshold voltage of the BJT inverter and the

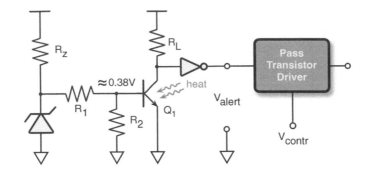

Figure 13.15   Temperature control circuit.

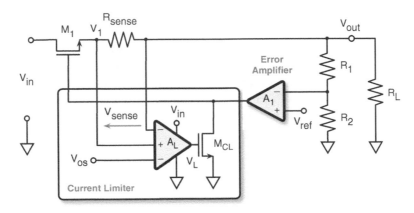

Figure 13.16   Current measure and current limiter circuit.

output is high. When the temperature rises to high values, the threshold V_{BE} becomes lower and switches on the alert voltage to indicate high temperature arriving at the control of the pass transistor. The driver reduces the output current or completely switches the device off.

The second built-in protection scheme is the current limiter. It prevents damage to an integrated circuit when the load impedence is too low or when there is a short circuit at output. This kind of protective circuit needs to measure the output current. This is normally done with a sense resistor of low value in series with the pass device. When the voltage drop across the sense resistor exceeds the limit, the pass device switches off.

A possible current limiter for a linear regulator with an N-channel MOS is shown in Figure 13.16. Since the voltage across the sense resistance is low, the offset of the amplifier is critical. Therefore, the circuit uses a suitable control that trims the offset V_{os}, used as threshold level. The output V_L, almost a logic signal used to drive the transistor M_{CL}, is the amplification of the difference between V_{sense} and V_{os}

$$V_L = A_L(V_{sense} - V_{os}). \tag{13.32}$$

When the voltage drop across R_{sense} is lower than V_{os}, V_L saturates to zero. When $V_{sense} > V_{os}$, the output of the amplifier, V_L, goes up, switching on the current limiter transistor M_{CL}. The output of the error amplifier is pulled down and the pass device M_1 lowers the output voltage as the current limiter takes control of the loop.

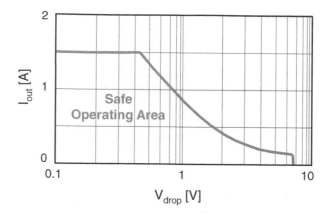

Figure 13.17 Safe operating area in the $V_{drop} - I_{out}$ plane.

Some integrated solutions use more sophisticated protection methods. They do not use fixed current limitations but take account for the voltage drop across M_1. Indeed, there are two things that matter: excessive power consumed by a power device and high current that can give rise to damage. Thus, if the voltage drop is low, higher currents are possible because the power is acceptable. However, when current exceeds a given limit it is necessary to limit it, because very high currents damage the circuit, even if the power is less than the maximum. Figure 13.17 indicates the safe operating area of the power transistor of a hypothetical integrated circuit. The maximum admitted power is 0.5 W and the highest usable voltage is 5 V. The curve, in addition to restraining thermal transfer capability, also indicates limitations from high voltages.

Self training

Search on the Web to learn more information about the protection circuits integrated with linear regulators. Find the schematic used for the thermal shutdown of the part and the current limiters. Describe in a short report the special features of one of the linear regulators, and indicate voltage and current limits and possible applications of the integrated circuit. Draft, if it is not provided by the data sheet, the safe area of operation.

13.4 SWITCHED CAPACITOR REGULATOR

A linear regulator or an LDO is, in practice, a variable "smart" resistance between input and output. The resistance value is exactly what is required to determine the voltage drop that makes the output equal to the desired regulated value. The scheme in Figure 13.18(a) illustrates the concept. Depending on load current the resistance must equal

$$R_{eq} = \frac{V_{in} - V_{out}}{I_{out}}. \tag{13.33}$$

Obviously, R_{eq} should change dynamically to follow the short- and long-term variations of input voltage and, possibly, the modified output voltage requirement.

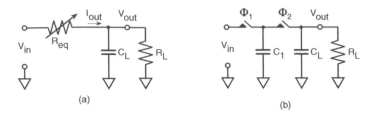

Figure 13.18   (a) Equivalent scheme of any linear voltage regulator made by a variable "smart" resistance between input and output; (b) switched-capacitor implementation of the linear regulator.

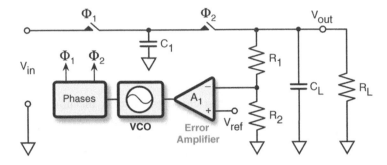

Figure 13.19   Switched-capacitor regulator. The feedback loop controls the VCO frequency.

Remember that the function of a resistance is to transfer charge from input to output as prescribed by Ohm's law. The conveyed charge is I_L coulomb per second, flowing in resistors continuously. However, almost the same operation can be attained in a sampled-data fashion by Switched-Capacitor (SC) schemes, like the ones studied in Chapter 5. Figure 13.18(b) illustrates the circuit. During phase Φ_1 the charge on capacitor C_1 is $Q_1 = C_1 V_{in}$. During Φ_2 it becomes $Q_2 = C_1 V_{out}$. Therefore, the charge conveyed every clock period is

$$\Delta Q = C_1(V_{in} - V_{out}), \tag{13.34}$$

which, on average, corresponds to a current equal to

$$\overline{I} = f_{ck}\Delta Q = f_{ck}C_1(V_{in} - V_{out}). \tag{13.35}$$

The result verifies what was already learned in Chapter 5: a switched-capacitor structure works as a resistance with value

$$R_{eq} = \frac{1}{f_{ck}C_1}. \tag{13.36}$$

The extra task here is to attain a control that makes the equivalent resistance "smart." This is done by changing the clock frequency. A higher frequency augments the transferred charge and, consequently, reduces the equivalent resistance. Thus the system must incorporate a controlled variable clock generator (or oscillator), necessary for the operation of the SC scheme.

Circuits capable of generating periodic waveforms with their frequency controlled by a voltage are called Voltage-Controlled Oscillators (VCOs). The next chapter studies them in some detail.

Figure 13.19 shows the block diagram of a switched-capacitor regulator. Again, a resistive divider $R_1 - R_2$ gives rise to a fraction of output voltage. The control established by the feedback makes that fraction equal to the reference; the amplified error changes the oscillating frequency of a VCO. The modified switching frequency changes the equivalent resistance and, in turn, regulates the output voltage.

The maximum frequency of the VCO, $f_{ck,M}$, gives the minimum equivalent resistance. Its value determines the drop-out voltage

$$V_{DROP} = \frac{I_{FL}}{f_{ck,M}C_1}. \tag{13.37}$$

To acquire familiarity with the numbers, suppose we have a clock frequency of 10 kHz and $C_1 = 5$ μF. The equivalent resistance is $R_{eq} = 1/(f_{ck}C_1) = 20$ Ω. With a full-load current of 2.5 mA the drop-out voltage is 50 mV.

13.4.1   Power Consumed by SC Regulators

A current through a resistance consumes power. That is an obvious assertion, and certainly the same must hold for the equivalent resistance of a switched-capacitor scheme. However, the statement is not immediately self-evident because capacitors are non-dissipative components.

To prove the point it is necessary to look carefully at the operation and observe that the switches connecting capacitor C_1 to V_{in} and V_{out} are dissipative components. Therefore their on-resistance is the element that consumes power.

When using a switched-capacitor circuit the charging and discharging of the capacitor gives rise to pulses of current, one positive entering one of the terminals and the other negative entering the other terminal (and, therefore, positive exiting). The current amplitude immediately after switching on is

$$I_0 = \pm \frac{V_{in} - V_{out}}{R_{on}}, \tag{13.38}$$

with a sign that depends on whether a charging or a discharging phase is being examined. The current goes to zero exponentially with time constant $t = R_{on}C_1$, making the energy consumed during each switching equal to

$$E_{sw} = \frac{(V_{in} - V_{out})^2}{R_{on}} \int_0^{T_{ck}/2} e^{-2t/\tau} dt = \frac{1}{2}C_1(V_{in} - V_{out})^2, \tag{13.39}$$

which assumes that Φ_1 or Φ_2 lasts for half of the clock period and that $T_{ck} \gg \tau$.

The power is the energy dissipated during charge and discharge switchings divided by the clock period (or multiplied by the clock frequency). This gives

$$P_{SC} = 2E_{sw}f_{ck} = \frac{C_1(V_{in} - V_{out})^2}{T_{ck}}, \tag{13.40}$$

which, using the expression for the switched-capacitor equivalent resistance $R_{sc,1} = T_{ck}/C_1$, yields

$$P_{SC} = \frac{(V_{in} - V_{out})^2}{R_{sc,1}}, \tag{13.41}$$

the power consumed by the equivalent SC resistance connected from input to output.

COMPUTER EXPERIMENT 13.5

Switched Capacitor (SC) Regulators

A linear regulator is equivalent to a resistance whose value is controlled by a feedback loop. The resistance gives rise to a controlled voltage drop equal to the difference between the unregulated input and the desired output. We can perform, on average, the same operation with a switched capacitor equivalent resistance. This computer experiment uses a DC term plus an AC component at input. The board comprises two schemes, one with a variable resistance whose value is changed in feedback to regulate the output voltage, the other with a switched capacitor circuit able of emulating the resistance operation. The frequency of an external square wave is properly multiplied by the feedback loop control (the display shows the multiplication factor). The new square wave signal serves to generate the phases for the switches. You can:

- set the input voltage by choosing the DC component and a low spur AC term; its sine wave frequency is fixed at 50 Hz;
- set the gain of the feedback loop controlling the series resistance and switched capacitor;
- change the value of the switched capacitor;
- regulate the equal loads.

MEASUREMENT SET–UP

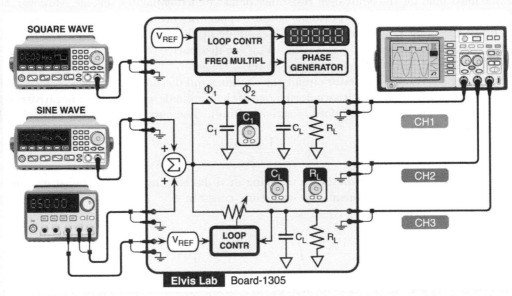

Elvis Lab Board-1305

TO DO

- Set the input voltage at 3 V plus 0.05 V sine wave. Use the default values of parameters and observe the output voltages. Change the gain feedback factor and measure the residual spur amplitude.
- Repeat the regulation with zero spur but change the switched capacitor value. Estimate the equivalent resistance. See what happens if you change the square wave frequency.
- Push the reset button, available on the ElvisLab console, and set the time span of the scope to observe the transients of the output voltages. Use different values of R_L and, in particular, observe what happens when load resistance is infinite.

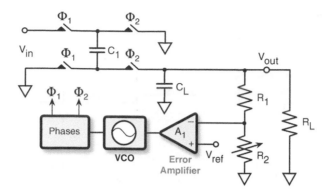

Figure 13.20   Generation of negative voltage with switched-capacitor regulator.

13.4.2   Generation of Negative Voltages

The use of a switched-capacitor circuit can give rise to negative voltages from positive inputs. It is worth observing that voltage is negative when current flows in the opposite direction to the one caused by positive voltage. For a sample-data scheme it is therefore necessary to invert the sign of the injected charge. The clock that controls the switches determines the transferred charge per unit time, albeit negative, thus establishing average current. Figure 13.20 illustrates the block diagram of a regulator that generates a negative voltage. During phase Φ_1 the switches on the input side charge capacitor C_1; during phase Φ_2 a switch connects the top plate to ground while the bottom plate feeds the load capacitance C_L.

> **Inverting capacitors**
>
> It is obvious but useful to observe that inverting a capacitor inverts the voltage across it and brings negative charges where positive charges were.

Observe that the action of switches virtually flips capacitor C_1 every half clock period. The top plate connects the capacitor to input; the bottom plate delivers charge to output. The charge transferred from input to output is

$$\Delta Q_1 = C_1(-V_{in} - V_{out}) = -C_1(V_{out} + V_{in}). \qquad (13.42)$$

The current in the load is, on average

$$\bar{I}_L = \frac{V_{out}}{R_L} = \frac{\Delta Q_1}{T_{ck}} = -C_1 f_{ck}(V_{out} + V_{in}); \qquad (13.43)$$

therefore

$$V_{out} = -V_{in}\frac{f_{ck}R_L C_1}{1 + f_{ck}R_L C_1}, \qquad (13.44)$$

which is an inverted fraction of the input voltage. If $R_L C_1/T_{ck} \gg 1$, the output voltage becomes an inverted replica of the input voltage.

The control circuit is the same as that of the voltage regulators already studied. The resistive divider with R_2 variable produces a fraction of the output voltage to be made equal to the reference. The error, amplified, controls the VCO that determines the clock frequency controlling the switches in the scheme.

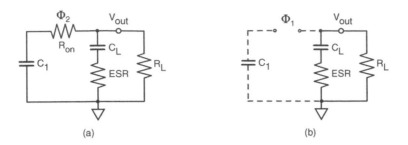

Figure 13.21   (a) Equivalent output section of SC voltage regulator during the charge transfer phase; (b) the same equivalent circuit during the complementary phase.

13.4.3   Voltage Ripple

The above study of switched-capacitor regulators supposes the output voltage to be constant. This assumption is evidently an approximation that holds for average quantities, because charging and discharging capacitors always involve transients. The load capacitor, C_L, receives charge from the input capacitor, C_1, (often called the *flying capacitor*) during one phase by a charging transient and delivers charge to the resistive load during the complementary phase by a discharging transient. Charge flows towards the resistive load with a flow that is constant on average but, in reality, changes in time causing fluctuations of output voltage, called *ripple*.

Two capacitors connected in parallel share their charges. If they are C_1 and C_L, charged at the voltages V_{in} and V_{out} respectively, the parallel connection gives rise to an instantaneous jump of V_{out} by

$$\Delta V_{out} = \frac{C_1(V_{in} - V_{out})}{C_1 + C_L} \qquad (13.45)$$

if the resistance of the connecting switch is small.

The use of a large storing capacitance, C_L, reduces that part of the ripple. For that action it is worth using discrete ceramic or tantalum capacitors that feature large capacitive values with relatively small volume. However, ripple also depends on the phase shift that the storing capacitor determines. It is accounted for by a resistance in series with the capacitor. Capacitor catalogs indicate that contribution as the *Equivalent Series Resistance* (ESR).

Figure 13.21 shows the equivalent circuits of the charge/discharge transients of C_L. R_{on} is the on-resistance of the switch and *ESR* is the ESR of C_1. During phase Φ_2 the flying capacitor C_1 feeds the resistive load and, at the same time, charges C_L. During phase Φ_1 a switch connects the flying capacitor to input and leaves the C_L alone, which starts a discharge transient. Current through C_L turns over and reverses the drop voltage across *ESR*. The change is $2I_L \cdot ESR$. If C_L is large, its discharge waveform is approximately linear with slope I_L/C_L. Therefore the ripple, neglecting the contribution given by equation (13.45), becomes

$$V_{ripple} = \frac{I_L}{C_L}\frac{T_{ck}}{2} + 2I_L R_{ESR}. \qquad (13.46)$$

The above equation is normally used for the design of SC regulators. It shows that, for a given equivalent series resistance, increasing the value of storing capacitance does not help in reducing ripple because the resistive drop becomes the dominant part.

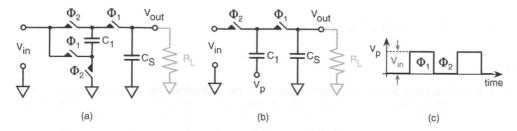

(a)   (b)   (c)

Figure 13.22   (a) One-stage charge pump with switched capacitor; (b) one-stage charge pump with a pushing-up square wave; (c) pumping signal.

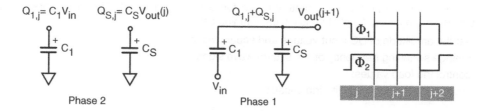

Figure 13.23   Capacitor configuration during phase Φ_2 and during the "pumping" phase Φ_1.

13.5   CHARGE PUMP

Switched-capacitor circuits are able to generate a voltage that is higher than the input supply. They are used when a high voltage, greater than that of the battery, is required. Among other circuits that need high voltages are those used to write data into non-volatile memories, and the boosters of MOS transistor voltage gates for low-drop performance.

Circuits that generate high voltages in a sampled-data manner are called *charge pumps*. They use capacitors, switches, and, in some cases, diodes. Charge pumps are a cascade of single stages, each of which increases the output voltage by a pushing-up action produced by capacitors. Each stage can, at most, augment the output voltage by the input supply.

Figure 13.22 shows the scheme of a one-stage charge pump. Switches controlled by phase Φ_2 charge C_1 at input rail voltage, V_{in}. Then, after Φ_2 goes down, the rising of Φ_1 pushes up the bottom plate of C_1 by V_{in}. For this action Figure 13.22(a) uses switches. Figure 13.22(b) employs a square wave that is at zero during Φ_2 and equal to the rail input, V_{in}, during Φ_1. The circuit transfers charge onto the storing capacitance, C_S, possibly increasing the output voltage.

Figure 13.23 shows the single-stage charge pump during phases Φ_2 and Φ_1 of two generic j and $(j+1)$ clock periods. At the end of phase Φ_2 (clock period j) the charges on C_1 and C_S are respectively

$$Q_{1,j} = C_1 V_{in} \quad \text{and} \quad Q_{S,j} = C_S V_{out}(j). \tag{13.47}$$

During the next Φ_1 (clock period $j+1$), C_1 and C_S, which are connected in parallel, share the total charge, while the voltage V_{in} applied to the bottom terminal of C_1 pushes up the output. The result, immediately after Φ_1 goes on, is

$$V_{out}(j) = 2\frac{C_1}{C_1 + C_S} V_{in} + \frac{C_S}{C_1 + C_S} V_{out}(j-1). \tag{13.48}$$

COMPUTER EXPERIMENT 13.6

Understanding Charge Pumps

Many situations need supply voltages higher than the available bias voltage. The current is normally negligible, because only high voltage is needed. For these kind of needs, charge pumps are a suitable solution. They generate a voltage higher than the input, thanks to the push–up action produced by switches or diodes and square waves. This computer experiment uses diodes to realize the cascade of two pumping cells. The circuit permits different amplitudes of input voltage. The phase generator gives rise, at output, to non-overlapping square waves with the same amplitude as the input square waves. Diodes are assumed to be an opened switch when reverse–biased and a 0.2 V battery in the forward bias condition. The capacitors of the pump cells are nominally 10 pF. In this experiment you can:

- set the amplitude of DC input voltage and square wave;
- change switching frequency of the square wave signal;
- control the load values;
- change the value of the filtering capacitor.

MEASUREMENT SET–UP

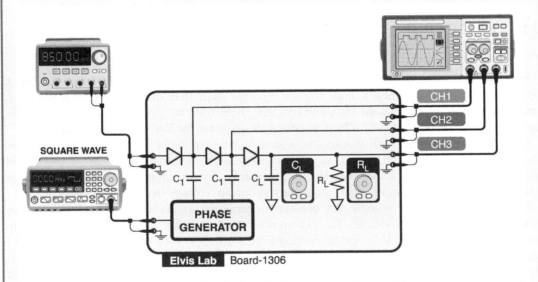

Elvis Lab   Board-1306

TO DO

- Choose amplitudes of the input voltage and the square wave signal. Use for example 1.2 V and 2 V, respectively. Measure the output transient with capacitive load equal to 1 nF and infinite load resistance.
- Reset the output and perform the experiment at different clock frequencies. Observe the transient that occurs after resetting.
- Change the values of load resistance and capacitance, and observe the output waveforms.
- Plot the output voltage as a function of the load resistance at various clock frequencies.

In steady-state conditions V_{out} is constant. Without the loss of charge that a possible resistive load would cause, the output voltage satisfies the condition

$$V_{out} = 2\frac{C_1}{C_1 + C_S}V_{in} + \frac{C_S}{C_1 + C_S}V_{out}, \qquad (13.49)$$

which yields

$$V_{out} = 2\,V_{in}. \qquad (13.50)$$

The above shows that a single-stage charge pump without load adds V_{in} to the input supply voltage, also equal to V_{in}, reaching $2\,V_{in}$. Another stage adds another V_{in}, and so on.

A load resistance, R_L, reduces the output voltage, because at every clock period it drains an amount of charge equal to

$$Q_S = \frac{V_{out}}{R_L}T_{ck}. \qquad (13.51)$$

The lost charge accounted for in equation (13.48) yields

$$V_{out}(j) = 2\frac{C_1}{C_1 + C_S}V_{in} + \frac{C_S}{C_1 + C_S}V_{out}(j-1) - \frac{Q_{out}}{C_1 + C_S}. \qquad (13.52)$$

In the steady states, this gives

$$V_{out} = 2V_{in}\frac{R_L C_1}{R_L C_1 + T_{ck}}. \qquad (13.53)$$

Using the expression for the equivalent SC resistance, $R_{eq} = T_{ck}/C_1$, the above becomes

$$V_{out} = 2V_{in}\frac{R_L}{R_L + R_{eq}}, \qquad (13.54)$$

which is twice the input supply voltage attenuated by the resistive divider $R_L - R_{eq}$.

Notice that we minimize the drop in output voltage with respect to the maximum with a low value of R_{eq}. This requires using large C_1 and/or high switching frequency.

The use of a cascade of equal stages, as shown in Figure 13.24(a), obtains higher and higher output voltages. The steady-state voltage of N stages can go up to

$$V_{out} = (N + 1)V_{in}, \qquad (13.55)$$

which is the value reached with infinite load resistance.

Since current on switches at the various stages always flows from a low voltage stage to the next with higher voltage, it can be convenient to replace the switches with diodes because they fulfill the same function. The resulting scheme is shown in Figure 13.24(b). The circuit is simpler because it does not need switches and drivers, but each stage needs a drop of about one on-diode voltage. This reduces the "push-up" capability of each cell and gives rise to extra power consumption when there is current flowing through the cells. As a result, the simplification costs a reduction in overall efficiency.

Be aware

Many electronic systems – for instance, color thin-film transistor panels and liquid crystal displays – use charge pumps. Typical currents are low but voltages can go up to 15 V, starting from a 3.5 V lithium-ion battery voltage.

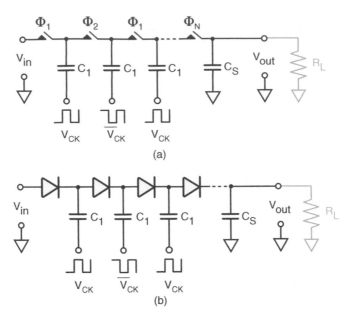

Figure 13.24 Cascade of charge pump stages: (a) with switches; (b) with diodes.

Remember that the use of capacitors enables the sign inversion of voltages, just by reversing connections. Therefore, the voltage boost of cells can be reversed, making possible the generation of high negative voltages. This option is not discussed here in detail. However, it is worth being aware of that possibility.

13.6 SWITCHING REGULATORS

One important category of power controller is the switching regulator, which, in addition to switches and capacitors, uses inductors. Switching regulators offer the advantage of good power efficiency with high voltage drop between input and output. They are able to give rise to DC voltages lower, higher, or with opposite polarity with respect to the input. We can have the following types of switching regulators:

- **Buck converter**, used to reduce the input voltage;

- **Boost converter**, able to generate a voltage higher than input;

- **Buck–boost converter**, suitable for generating voltages lower or higher than input with the same or inverted polarity.

In addition to the above types there are other categories of switching regulators. They are not discussed here because the scope of this book is not to study all possible solutions and architectures but to describe the most relevant of them, assuming that others are studied in specialized courses.

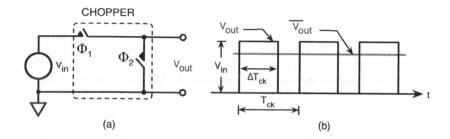

(a)   (b)

Figure 13.25   (a) Chopping of the supply voltage to obtain a lower regulated output; (b) output waveform and its average value.

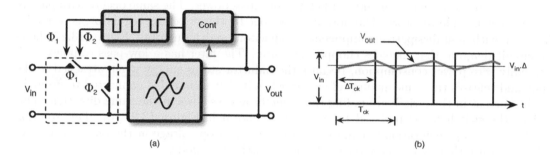

(a)   (b)

Figure 13.26   (a) Schematic diagram of a buck converter with control loop. (b) Waveforms at input and output of the filter.

13.6.1   Buck Converter

A simple way to obtain a voltage that, on average, is lower than input, is to chop the input using a square wave with a given duty-cycle, Δ. Figure 13.25(a) shows a simple scheme made from two switches. The output voltage equals V_{in} during phase Φ_1 and becomes zero during phase Φ_2. The clock frequency f_{ck} gives rise to a repeated waveform with period $T_{ck} = 1/f_{ck}$ and output equal to input for the time interval ΔT_{ck}. The average output voltage is

$$\overline{V}_{out} = V_{in}\frac{\Delta T_{ck}}{T_{ck}} = V_{in}\Delta, \tag{13.56}$$

a value that does not depend on clock frequency but only on the duty-cycle.

The converter's task is to regulate the duty-cycle and to extract the average value of the sequence of pulses at output. The use of a low-pass filter that removes high-frequency terms produces the result. Since the filter is never perfect the output voltage goes up and down around the average value, with some ripple. The architecture of the regulator includes a control feedback loop that maintains the output voltage at a required value independently of the load. The control measures output, compares it with the desired value and changes the duty-cycle accordingly, as sketched in the block diagram of Figure 13.26(a). Figure 13.26(b) depicts possible waveforms at input and output of the low-pass filter. The input is a sequence of pulses; the output amplitude is, as expected, $V_{in}\Delta T/T$ with some ripple around the average.

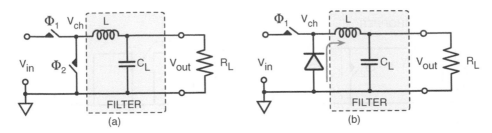

Figure 13.27   (a) Schematic diagram of a buck converter; (b) use of a diode to replace the switch controlled by phase Φ_2.

The design issue is the best way to make the filter with minimal power consumption. The use of active blocks, such as op-amps or OTAs, consumes power. The same occurs with passive circuits that include resistors. The optimum, as done by the switching regulator, is to realize the filter with non-dissipative components: capacitors and inductors. Figure 13.27 shows the circuit diagram that includes chopper and filter. The filter has a low-pass response with nominally zero power consumption, because the inductor stores energy while attached to the input and releases the same amount of energy when its left terminal is connected to ground.

The current in the switch grounding the inductor always flows in the same direction. This enables the switch to be replaced by a diode, as shown in Figure 13.27(b). This solution simplifies the implementation but must support a diode drop voltage in the on condition. In order to reduce that drop, many buck schemes use Schottky diodes.

Buck converters have two distinct modes of operation: Continuous Conduction Mode (CCM) and Discontinuous Conduction Mode (DCM). In the former case there is always a current flowing in through the inductor. In the latter case, current goes to zero (or becomes negative) for a fraction of the switching period.

The equations describing the circuits are

$$\frac{dI_L}{dt} = \frac{V_{ch} - V_{out}}{L} \tag{13.57}$$

$$\frac{dV_{out}}{dt} = \frac{I_L - V_{out}/R}{C}, \tag{13.58}$$

where the chopped voltage, V_{ch}, equals the input voltage during Φ_1 and becomes zero when the switch controlled by Φ_2 grounds the left-hand terminal of the inductor.

Equations (13.57) and (13.58) make a linear system of second-order differential equations. Even though it is not difficult, we do not solve the system here, but – more appropriately – we derive approximate solutions. Suppose that, as desired, the output voltage is almost constant. The derivative of inductance current, given as equation (13.57), is also constant and positive during Φ_1 and constant with negative value during Φ_2. That is, respectively,

$$\frac{dI_L}{dt} = \frac{V_{in} - V_{out}}{L} \quad \text{and} \quad \frac{dI_L}{dt} = -\frac{V_{out}}{L}. \tag{13.59}$$

The possible waveforms in the steady-state conditions are like the ones of Figure 13.28. The current has the same value at the beginning and end of the clock period; the slope is positive for the duty-cycle period and negative when the input switch goes off. Since the current never goes to zero, the waveform of Figure 13.28(a) refers to continuous-mode operation. With low required currents the converter goes into discontinuous mode. Its possible waveform is shown in Figure 13.28(b). The average value of current is what is delivered to the load.

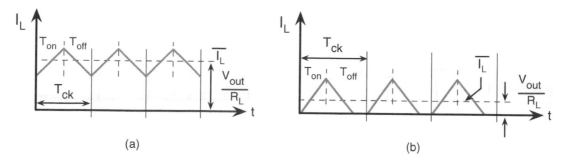

Figure 13.28 Steady-state waveforms of current in buck converter: (a) in continuous mode of operation; (b) in discontinuous mode.

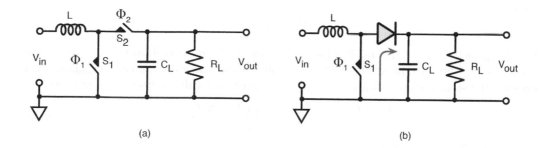

Figure 13.29 (a) Schematic diagram of a boost converter; (b) use of a diode to replace the switch controlled by phase Φ_2.

The duty-cycle, defined by equation (13.56), is the period of time during which the inductor current rises; the complementary part is that of falling current. In steady state, the rising part and the falling part must be equal. Therefore

$$\frac{(V_{in} - V_{out})\Delta T_{CK}}{L} = \frac{V_{out}(1 - \Delta)T_{CK}}{L}, \tag{13.60}$$

which satifies once more the condition $V_{out} = V_{in}\Delta$.

A feature of both modes of operation is the peak-to-peak difference of the inductor current. It depends on switching frequency and inductor value. These two design parameters are typically selected to maintain the current swing below 20% to 30% of the DC current value.

13.6.2 Boost Converter

The operation of buck converters relies on rising current in an inductor to induce a voltage drop, and on falling current to sustain the output voltage. The same behavior but with opposite roles can secure output voltages higher than input. For this, an inductor with falling current must determine a boosting rise from input to output. Then the current must climb without involving output. This is the basis of boost converters.

Figure 13.29 shows the scheme. During phase Φ_1 the inductor does not involve the output. Its connection to ground requires a voltage drop equal to input. The positive current derivative

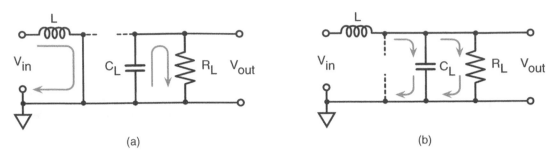

Figure 13.30   (a) Boost converter network during phase Φ_1: input charges the inductor and C_2 feeds the load; (b) network configuration during Φ_2: input charges capacitor and feeds the load.

is equal to

$$\frac{dI_L}{dt} = \frac{V_{in}}{L}. \tag{13.61}$$

When the switch controlled by Φ_2 closes, the current slope turns from positive to negative. In steady state the current at the end of Φ_2 must equal the value at the beginning of Φ_1. Therefore, the steady-state condition determines the output voltage from the equation

$$\frac{V_{in}\Delta T_{CK}}{L} = \frac{(V_{out} - V_{in})(1 - \Delta)T_{CK}}{L}, \tag{13.62}$$

which yields, after simple mathematics,

$$V_{out} = \frac{V_{in}}{1 - \Delta}. \tag{13.63}$$

Since current through the switch S_2 always flows from inductor to load, it can be replaced by a diode, as shown in Figure 13.29(b). This solution simplifies the design but, as happens for the buck converter, there is an undesired drop voltage across the diode.

Notice that during phase Φ_1 the storing capacitor C_L provides current to the load. Its function is not only necessary to reduce output ripple but essential to sustain output current while the inductor brings up its current.

Figure 13.30 depicts the configuration of the boost converter and shows the flow of currents during the two phases of operation. The presence of switches is such as to reconfigure the network, which is actually time variant. The study of the circuit requires the solution of two first-order differential equations during phase Φ_1,

$$\frac{dI_L}{dt} = \frac{V_{in}}{L} \quad \text{and} \quad \frac{dV_{out}}{dt} = \frac{V_{out}}{R_L C_L}, \tag{13.64}$$

and a second-order linear differential equation during phase Φ_2

$$\frac{d^2 V_{out}}{dt^2} L C_L + \frac{dV_{out}}{dt}\frac{L}{R_L} + V_{out} = V_{in}. \tag{13.65}$$

The output voltage and its derivative at the end of each time-slot are initial conditions for the other equation at the beginning of the successive time-slot.

The above is what is needed to calculate the transient waveform in the time domain. During phase Φ_1 the output voltage decreases exponentially. During the next Φ_2 phase it is necessary to solve a second-order differential equation. Here we do not go into further details. It is assumed that the elements given form a sufficient basis for a deeper study of boost converters.

COMPUTER EXPERIMENT 13.7

Working with Buck and Boost Converters

The main drawback of linear regulators or switched capacitor regulators is that power efficiency drops significantly when the difference between input and output voltage is large. Switching regulators employ inductors for securing higher efficiency in those conditions. This computer experiment enables you to study, using one single board, a buck and a boost converter with open loop control. The former generates a voltage lower than the input, while the second one gives rise to higher voltages. The two converters use the same values as inductor, loading a resistance and a capacitor. The switching frequency and duty-cycle are also the same. Two current meters (I-to-V converters), with zero voltage drop across the terminals, measure currents. Output voltages and currents can be observed on the scope via selectors. In this experiment you can:

- set the values of the inductance, capacitance and load;
- change the switching frequency of the square wave input signal;
- change the duty-cycle of complementary non-overlapped phases.

MEASUREMENT SET–UP

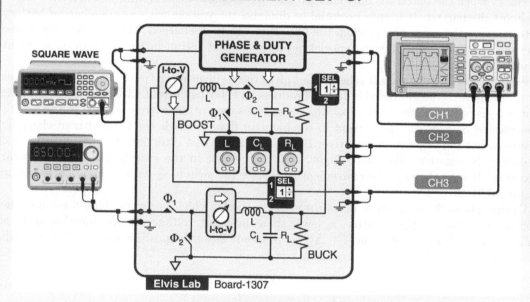

Elvis Lab   Board-1307

TO DO

- Set the supply voltage to 5 V and measure the output voltages and inductor currents with default values as parameters. Reset the output voltage (a short circuit for a very short period) and observe the voltage transient.
- Change the duty-cycle and observe the result. Compare the values of your measurement with those predicted by theory.
- Change loading conditions to better understand how things operate with various possible circuit responses.

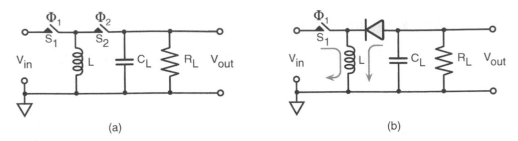

(a)   (b)

Figure 13.31 (a) Schematic diagram of an inverting buck–boost converter; (b) use of diode to replace switch controlled by phase Φ_2.

13.6.3   Buck–boost Converter

Buck–boost schemes satisfy the need for output voltages with amplitudes either higher or lower than input. The sign can be positive or negative. Having an output voltage that, depending on need, must be higher or lower than input is a frequent necessity. The voltage of a battery drops when the battery discharges. The generated voltage can be higher than necessary when the battery is fully charged and becomes lower than required when the battery is almost discharged. It is thus essential to diminish the voltage with a fully charged battery or to increase the output level when the battery is nearly discharged. Having inversion is also relevant and is desired for some special applications that use negative voltages on a circuit board.

Figure 13.31 shows a circuit scheme that gives rise to inverted output. The switch, S_1, that connects input to the grounded inductor produces an entering current with positive slope. The switch S_2 is open and the tank capacitor C_L supplies the output current. When the switch S_1 goes off and S_2 goes on, the current in the inductor flows towards the load without changing direction, but the slope turns negative. Because of the current direction the voltage generated at output is negative. For this scheme, too, the current in the switch flows with unipolar direction. It may therefore be convenient to replace the switch S_2 with a diode, as shown in Figure 13.31(b).

During phase Φ_1 no current flows through the diode, which is reverse-biased because the voltage at output, being negative, is lower than the input voltage. If the duration of Φ_1 is ΔT_{ck}, and the duration of Φ_2 is $(1 - \Delta)T_{ck}$, in steady state the positive current ripple equals the negative one. Therefore, the resulting equation is

$$\frac{V_{in}\Delta T_{ck}}{L} = \frac{-V_{out}(1 - \Delta)T_{ck}}{L}, \tag{13.66}$$

which yields

$$V_{out} = V_{in}\frac{-\Delta}{1 - \Delta}. \tag{13.67}$$

Thus the output voltage has opposite polarity to the input. The modulus of the generated voltage can be lower or higher than that of the input, depending on Δ. If the duty-cycle Δ is 1/2, the output is exactly an inverted replica of the input voltage.

Often it is required to have output voltages with the same sign as input. For this there are non-inverting buck–boost schemes that produce this result but need more switches for their operation. They are, in essence, the cascade of a buck converter and a boost converter. Figure 13.32 shows the conceptual steps that give rise to the circuit. A buck converter with infinite

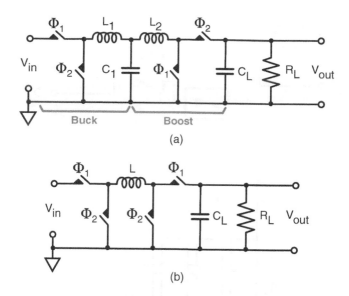

Figure 13.32   (a) Schematic diagram of the cascade of a buck and a boost converter; (b) scheme of non-inverter buck–boost obtained by removing C_1 from circuit (a).

load resistance feeds a boost scheme (Figure 13.32(a)). Then, as shown in Figure 13.32(b), the capacitor C_1 is removed and the two inductors merge together into a single inductor.

The voltage at output of a buck is Δ times the input. The voltage at output of a boost is $1/(1 - \Delta)$ times the input. Combining the two relationships, we obtain

$$V_{out} = V_{in} \frac{\Delta}{1 - \Delta}, \tag{13.68}$$

which is the same result as that of the inverting scheme but with positive sign.

The non-inverting buck–boost is less efficient than other converters studied above because the current flows through two switches. The finite resistance of switches causes a loss that affects the overall power efficiency.

13.6.4   Loop Control and Switches

The previous subsections show that the duty-cycle, Δ, establishes a link between input and output voltages. The relationships obtained are:

$$V_{out} = \begin{cases} V_{in}\Delta & \text{for buck converters;} & (13.69) \\[2mm] V_{in}\dfrac{1}{1 - \Delta} & \text{for boost converters;} & (13.70) \\[2mm] V_{in}\dfrac{\pm\Delta}{1 - \Delta} & \text{for buck–boost converters.} & (13.71) \end{cases}$$

As in the previous schemes studied, an open-loop circuit can determine the value of Δ, or, more properly, a suitable network establishes a feedback loop that produces the value of Δ that brings the output voltage to the desired level and maintains it there. The scheme used for

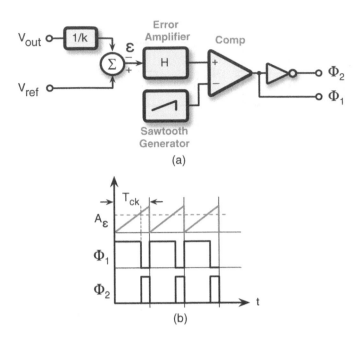

(a)

(b)

Figure 13.33 (a) Typical feedback control used in switching regulators; (b) output phases Φ_1 and Φ_2 have a suitable duty-cycle.

feedback control is like the one in Figure 13.33. The output voltage, possibly attenuated by a factor k to get a signal that is easier to handle, is subtracted from the reference to generate the error ϵ. The block "error amplifier" augments the error amplitude so that it can then be compared with a sawtooth waveform. When the amplifier error is higher than the sawtooth voltage, the output is high. At the crossing point the comparator output switches down to zero, ending the phase Φ_1. A logic inverter produces the other phase, Φ_2. Both control switch drivers.

As is necessary with any feedback scheme, it is necessary to ensure loop compensation. This is done by block H of Figure 13.33, which, in addition to amplification, provides filtering. Because of a time-variant operation of elements in the loop, stability control can be problematic. We do not go into detail on this issue because that topic is studied in advanced courses. It is enough to be aware that the loop control of switched regulators requires special attention.

For medium and high currents the signal generated by logic is not enough to drive the switches. With currents in the ampere range or higher, the bipolar or MOS transistors used as switches have large areas or large aspect ratio. It is therefore necessary to use specific drivers to generate the current or the voltage needed for operating the switch.

If a bipolar transistor makes the switch, the driving signal is the base current of the power transistor. Unfortunately, at high power the β of bipolar transistors is low. Since the base current becomes a significant fraction of the load, it may be necessary to use Darlington configurations to increase the effective β. This, however, augments the voltage drop between input and output and, consequently, reduces the maximum efficiency.

The use of a power MOS transistor requires signal generators capable of driving large parasitic gate capacitances. In order to roughly estimate the capacitance remember that the

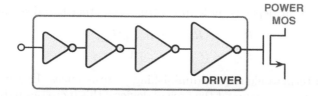

Figure 13.34   Driving a power transistor with a chain of tapered inverters.

MOS transistor operates in the triode region. Its on-conductance is approximated by

$$\frac{1}{R_{on}} = \mu C_{ox} \frac{W}{L} (V_{GS} - V_{th}).$$ (13.72)

This value depends on technology, overdrive voltage, and aspect ratio (W/L).

If the technology used has μC_{ox} products equal to 120 $\mu A/V^2$ and 50 $\mu A/V^2$ for the N-channel and P-channel transistors respectively, and the overdrive voltages $(V_{GS} - V_{th})$ equal 1 V, an on-resistance equal to 0.1 Ω requires

$$\left(\frac{W}{L}\right)_n = \frac{1}{(\mu C_{ox})_n R_{on}(V_{GS} - V_{th})} = 83\,350,$$ (13.73)

$$\left(\frac{W}{L}\right)_p = \frac{1}{(\mu C_{ox})_p R_{on}(V_{GS} - V_{th})} = 200\,000,$$ (13.74)

very large aspect ratios that give rise to large gate areas. The gate capacitance is proportional to the gate area by approximately $1fF/\mu^2$ (a parameter that can change according to the technology used). The result can be as high as several tens of pF.

For driving those large capacitances it is necessary to use a logic gate, typically an inverter, made up of very large transistors. The gate of that inverter, in turn, establishes a large capacitive load for the logic that controls it. Therefore it is necessary to use another logic gate with large transistors, of sizes typically lower than the driven inverter. This continues until the capacitive load is affordable for a normal logic circuit. The result is a chain of inverters with tapered sizes as pictured in the scheme of Figure 13.34.

Tapered inverters are also used for driving big logic ports such as the ones connected to the pads of integrated circuits. For all those circuits the design of the tapered chain aims at minimal power consumption while ensuring minimal delay and distortion of the driving signal.

13.6.5   Efficiency of Switching Regulator

The ratio between power delivered to the load and total power gives the efficiency of the switching regulator. Power is lost because of three terms: static power caused by the on-resistance of switches, dynamic power needed to charge and discharge parasitic capacitances, and power used by the control circuitry and auxiliary circuits, such as switch drivers. Therefore

$$\eta_{sw} = \frac{P_{out}}{P_{out} + P_{stat} + P_{dyn} + P_{aux}}.$$ (13.75)

Obviously, the output power is

$$P_{out} = V_{out} I_{out}.$$ (13.76)

P_{aux} is a fixed amount that depends on loop control. The other two dissipative terms are

$$P_{stat} = (R_{on,n} + R_{on,p})I_{out}^2 \qquad (13.77)$$

$$P_{dyn} = (C_{p,1} + C_{p,2})V_{DD}^2 f_{ck}. \qquad (13.78)$$

Reducing the static term would require diminishing on-resistances. However, for limiting the on-resistance it is necessary to augment the area of devices; this increases parasitic capacitance, and, in turn, the dynamic power consumption gets bigger. For a given switching frequency the dynamic power is constant, as shown by equation (13.78), while the static power increases quadratically with current. Therefore, power efficiency at low currents depends on dynamic and auxiliary power, while at high currents efficiency drops because of static losses.

Unlike that of switching regulators the efficiency of linear regulators or LDOs is almost independent of the supplied current. It mainly depends on the drop between input and output. It is approximately estimated by

$$\eta_{lin} = \frac{V_{out}I_{out}}{V_{in}I_{out} + P_{aux}}; \qquad (13.79)$$

the power needed by auxiliary functions limits the linear regulator efficiency at low currents.

Example 13.3

Estimate the efficiency of a buck converter used to generate 1.8 V with low–medium currents. The nominal input voltage is 2.5 V and the switching frequency is 5 MHz. Switches have a 0.15 Ω on-resistance and 20 pF and 60 pF parasitic capacitances for the N-channel and P-channel switches respectively. The power consumed by the auxiliary part is 1 mW. Compare the efficiency of the buck converter with that of a linear regulator that consumes the same auxiliary power. Consider a current in the range 1 mA – 10 A.

Solution

The on-resistance of the two switches gives rise to static power. The dynamic power depends on switching frequency, parasitic capacitances and voltage used to switch the transistors on and off. Assuming we drive switches with control phases with amplitude equal to the input voltage, 2.5 V, this gives

$$P_{dyn} = (C_{p,1} + C_{p,2})V_{in}^2 f_{ck} = 80 \cdot 10^{-12} \cdot 2.5^2 \cdot 5 \cdot 10^6 = 2.5 \text{ mW}.$$

The efficiency becomes

$$\eta_{sw} = \frac{1.8 \cdot I_{out}}{1.8 \cdot I_{out} + 0.3 \cdot I_{out}^2 + 3.5 \cdot 10^{-3}}.$$

The efficiency of a linear regulator counterpart is

$$\eta_{lin} = \frac{1.8 \cdot I_{out}}{2.5 \cdot I_{out} + 1 \cdot 10^{-3}}.$$

Figure 13.35 shows the plots of the two efficiencies. The linear regulator is superior to the buck converter at low and high currents. However, buck reaches much higher efficiencies, more than 90%, in an intermediate current range. Moreover, the efficiency of linear regulators drops at low output currents. Therefore, switching is the optimal solution for large input and output differences.

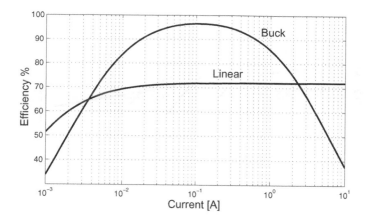

Figure 13.35   Efficiency of switching and linear regulator versus current.

13.7   POWER MANAGEMENT

Obtaining power conversion with high efficiency is important for portable applications. The need for long battery life and low power consumption is so important that analog and digital processors must be designed with power efficiency in mind. The multiple actions taken to ensure effective use of power are called *power management*.

The design of power-effective systems must start from the knowledge of the source. Portable systems use batteries, the most common type being lithium-based rechargeable ones. The second element to account for is the load. There are two possible situations: the load established by digital circuits, which mainly consume power because of charging and discharging of parasitic capacitances, and the load given by analog and RF circuits, which need power to support accuracy and bandwidth.

The third element is awareness of the effective power need. Supply voltages, currents used, and clock frequencies can exceed what is really necessary to achieve performances that, possibly, vary with time because of momentary changes in activity.

The electronic circuits used for power management account for the above elements and optimize the power in electronic systems.

13.7.1   Rechargeable Batteries

Rechargeable batteries are essential for portable devices and cordless power tools. The aim is to store in the smallest possible volume the highest amount of energy. This property is measured by the battery's *specific energy*, which is the stored energy per unity weight. Normally specific energy is measured in Wh/kg, i.e. how many watt-hours a 1 kg weight battery can furnish. The amount of energy stored in the battery is called its *capacity*. It is normally measured in Ah (or mAh) and corresponds to 20 times the current that the battery can constantly supply for 20 hours at room temperature. Faster discharge rates give rise to lower autonomy.

Rechargeable batteries are also called *secondary batteries* because they are not able to produce current immediately after fabrication. The other category is the disposable sort,

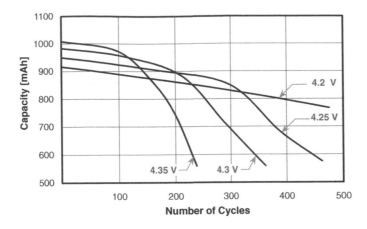

Figure 13.36 Battery cycle life at different charge voltages.

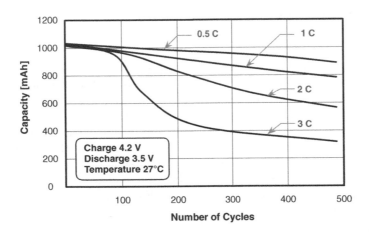

Figure 13.37 Aging of rechargeable battery caused by fast charge rate, C.

which are called *primary batteries* (here we have the zinc–carbon or alkaline type). They transform chemical energy furnished by a chemical reaction (like the one between zinc and manganese dioxide separated by an alkaline electrolyte) into electrical energy. Rechargeable batteries, on the other hand, furnish energy because of a reversible chemical reaction. They are electrochemical cells that transfer ions from a negative electrode to a positive one during discharge, and transfer ions the other way around during charge.

Be aware

Rechargeable batteries for electric vehicles are connected in series and in parallel to give output voltages that can be 300 V or even more. The electricity from the grid is transformed by the charger with about 90% efficiency. The efficiency of the Li-ion battery can be higher than 90%, yielding an overall result of up to 85% or more.

One important feature of a rechargeable battery as used in portable systems is its ability to be fabricated in any shape to fit volume needs. In addition, its specific energy must obviously be the highest possible with a very low self-discharge rate. Good batteries, like the Lithium-ion (Li-ion) type, have specific energy of hundreds of Wh/kg. However, developments in this field are quickly increasing that quantity. Another important property of batteries for portable devices is the number of discharge/charge cycles. Many batteries provide 300–500 discharge/charge cycles. These and other important operational parameters depend on the use of the battery and its management. Therefore it is important to know what the elements are that give rise to optimal performance to be controlled by the power management circuit. Some important issues are as follows.

- A well-controlled and defined voltage used in charging cycles preserves capacity and ensures long service life. Figure 13.36 shows the typical effect of charging voltage on capacity and battery life. High charging voltages boost the capacity at the beginning of the battery's life but diminish the number of charge/discharge cycles. The optimal charging voltage of a Li-ion battery is 4.2 V. Just 50 mV higher reduces battery life significantly.

- Overcharging with voltages higher than what is prescribed (for example, 4.3 V, which is just 0.1 V more than what is required by Li-ion cells) makes the battery unstable. Heat and permanent damage may occur.

- Too-low voltages caused by persistent discharge (less than 2.5 V for Li-ion cells) make the battery unusable. Its recharging is not possible any more.

- High temperatures are dangerous, especially for batteries with high specific energy. Li-ion cells start to be unstable at 130°C. A temperature of 150°C causes thermal runaway, which destroys a Li-ion battery within a few seconds.

- Charge and discharge rates influence the battery's longevity. The operation rate, called the *charge rate*, is a parameter often denoted by the symbol C. It indicates the inverse of the number of hours of charge used to reach battery capacity. For example, a battery charger that delivers 300 mA for two hours charges a 600 mAh battery at $C/2$. Figure 13.37 shows the effect on longevity of charge rate. As shown, a fast charging rate has a negative effect on battery life.

The above issues are important for operating battery chargers. They provide voltage and current in different manners depending on battery status. For example, they can have a pre-charge mode with constant current, used when the battery is considerably discharged. Then, at a given voltage, current increases to speed up the charging process. Finally, when the battery approaches its final voltage, the charger operates with constant voltage, starting to reduce the current until full charge is reached.

Self training

Search on the Web to understand better the features and limitations of rechargeable batteries. Understand the features, and the differences between, the following types:

- Nickel–Metal hydrate rechargeable battery;

- Nickel–Zinc rechargeable battery;

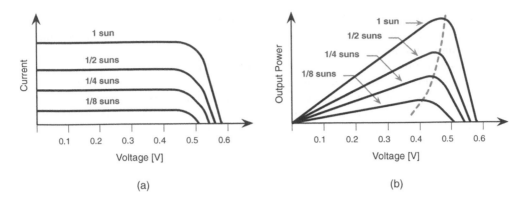

Figure 13.38   (a) Voltage–current response of solar cell at different levels of irradiation; (b) output power versus output voltage.

■ Lithium-ion rechargeable battery;

■ Conventional acid rechargeable battery.

What is, in your opinion, the best choice for a cell-phone or for a solar cell?

13.7.2   Power Harvesting

In addition to the power obtained from the household electricity supply or batteries, electronic systems secure power with techniques called energy harvesting (or energy scavenging). They collect energy from the environment, possibly to be stored in batteries or super capacitors. There are several power sources around. Perhaps the best known is the light transformed into electricity by solar cells. Other sources are the mechanical energy coming from vibrations and strain, thermal energy, and electromagnetic energy.

For all these sources it is necessary to know the best way in which energy is transformed into electricity. For example, the output characteristic of a solar cell depends on solar irradiation, as in Figure 13.38(a). This shows typical voltage–current plots in various illumination conditions, measured using the *sun* as the unit. Given that response, if the output voltage is zero (short circuit), the power at output is zero and all the received power becomes heat dissipated in the solar cell. The same is true for the open loop. The curves in Figure 13.38(b) show the power at output for different output voltages. It is evident that the output voltage for which the transferred power is maximal varies. Therefore, one of the goals of the power management of a solar cell is to track the maximum power point, extracting power at optimal output voltage.

Another possible energy source is electromagnetic radio or microwave signals. They are transformed into electrical signals by wideband antennas. Then a matching network, a rectifier, and a DC–DC converter change the power to a usable format. Figure 13.39 shows a schematic block diagram of a harvesting system. Power at output can be very low, and the power management circuit must be able to optimize power transfer in a similar way to what is done for solar cells and to control the load so as to avoid storage problems.

There are other methods of obtaining energy. One is to use piezoelectric devices. These are based on materials, such as quartz, that convert an applied strain into an electrical current.

COMPUTER EXPERIMENT 13.8

Analyzing Solar Cells

A solar cell is a device used for transforming solar energy into electrical energy. It is a p–n junction that, when illuminated, works as a current generator until the reverse voltage is less than a given value. Current and cut-off voltage depend on illumination (measured in "suns"). This computer experiment employs a light source that emulates the sun. When the power of the lamp is 300 W, the illumination received at the solar cell surface is equivalent to three suns. The bias of the lamp is made by a power supply generator controlled by a DC signal with a low frequency AC signal superposed. In this manner we modulate the illumination. The gain voltage of the power stage is 1 and the resistance of the lamp is 150 Ω. "Ideal" ammeters (I-to-V converters) measure the currents in the circuit. In this virtual experiment you can:

- set the DC level and the AC sine wave of the lamp control;
- change the value of load resistance;
- observe both solar cell voltage and its current, using the oscilloscope.

MEASUREMENT SET–UP

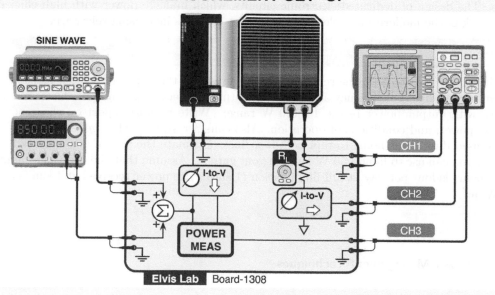

Elvis Lab Board-1308

TO DO

- Set the control of the lamp to obtain 1-sun illumination, and measure the solar cell voltage and current. Change the load resistance to drain lower or higher currents.
- Add modulation to the illumination, to estimate the effect of a cloud reducing the sun intensity by 50%.
- Plot efficiency versus illumination. The received power is 950 W/m² with 1 sun and the cell area is 100 cm².
- Set a value of load resistance that gives peak efficiency at illuminations ranging from 0.2 to 1.5 sun.

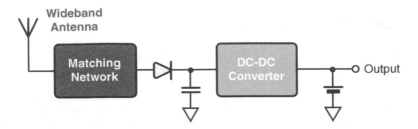

Figure 13.39   Schematic block diagram of a system harvesting power from electromagnetic waves.

Possibly the strain results from vibration of mechanical structures. Unfortunately, the amount of energy generated is small and the electronics they can power are only micro-power systems. Piezo materials can also be used in unconventional situations to capture energy. For example, a system built into a shoe can harvest the excess energy made available while walking.

The above just gives a glimpse of the possibilities offered by unconventional ways to power electronic systems. For all of them low power and power-conscious use are very important issues. The design of dedicated electronic circuits, which manage power with high efficiency, is one task of the modern microelectronics designer that has increasing relevance.

Self training

Search on the Web to learn more about the features and the types of solar cells. Look at key parameters like the efficiency and the temperature of operation. Find possible commercial devices with output power in the 10–100 W range. Write a short report on the generated voltage, power, and conditions of operation. Also consider systems that generate power that can be used directly at home to replace power lines. Estimate the size of a commercial solar cell that you can use to light a 50 W lamp in your garden. Assume that the equivalent number of sun hours is four per day at full illumination (1 sun). The power received at 1 sun is about 950 W/m^2.

13.7.3 Power Management Techniques

The methods used in power management circuits aim at controlling regulated voltages, temperature (for thermal management) and hardware (for changing operating conditions or reconfiguring architectures) with the goal of optimal performance/power trade-off. To illustrate various needs and the methods used, Figure 13.40 shows a block diagram of a data acquisition system with wireless communication. Signal conditioning and the A/D converter need a well-controlled supply voltage, free from noise. The power section uses an unregulated DC voltage at input and generates an intermediate voltage with a switching regulator. Then an LDO cascaded to the first regulator generates the quality supply voltage and a precise reference voltage needed for A/D operation. The accurate supply voltage serves the analog interface, the D/A converter and the RF section. If analog blocks need higher or lower supply voltages, the power section includes multiple LDOs that start from the same switched regulated voltage.

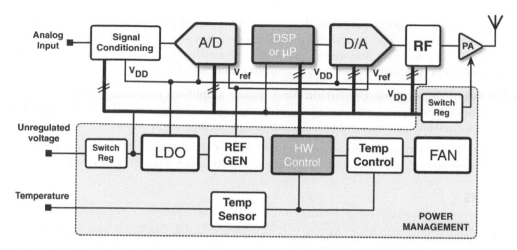

Figure 13.40   Block diagram of possible system and its power management section.

Requirements on the quality of the supply voltage of a DSP or a microprocessor are not so demanding as for analog sections. The voltage of a switching regulator can be enough. The system uses a single switching regulator for both analog and digital sections, or it may use two regulators, depending on voltage level needs. The supply voltage of the power amplifier driving the antenna can be the same as for the analog sections or, to optimize power consumption, can derive from a dedicated regulator that modulates the supply to follow the envelope of the transmitted signal. When amplitude transmission is low, the supply voltage can stay at a minimum level, thus saving power. A switching regulator, adapting output voltage to the level of transmitted power, carries out the task.

It may be important to keep temperature under control. One or more temperature sensors inform a temperature controller, which takes possible actions. These may be the activation of a fan or the generation of a warning signal for the hardware controller.

The hardware monitor and controller manages the system for optimal use of power. It controls the digital sections but may also regulate bias currents of analog circuits to adapt performance to needs.

Operations carried out by the hardware controller can involve the following functions.

- **Modules' sleep and wake-up** When a circuit stands idle, there is no need to power it. A controller reduces power to a level that depends on the time required to bring the circuit back into operation; for example, a power reduction by 20–40% gives rise to an almost immediate wake-up. A considerable power reduction, such as 85%, can require 1–2 milliseconds to have the circuit back in operation.

- **Variable supply voltage** The reduction of supply voltage obviously reduces power. Lower supply levels give rise to reduced performance but may preserve some of the circuit's operation if, for digital sections, the voltage reduction is within the limit allowed by the noise margin of the logic gates.

- **Variable clock frequency** This option relies on the different activity of separate sections in complex digital circuits. It can therefore be convenient to reduce the clock frequency of the system's sections where workload is low. This function can be combined with a supply voltage dynamic reduction.

- **Architecture reconfiguration** The demand for signal processing can depend on momentary activity. The hardware controller possibly reconfigures the hardware for use in a different manner and redistributes the processing load. The result is mainly a minimization of hardware but power consumption can also benefit.

- **Protocol-dependent operation** Since power amplifiers consume power, suspending their operation during idle periods saves power. This can be done if a protocol provides the information on when data transmission can occur.

The implementation of the above techniques requires the use of digital circuits that, obviously, consume power. It is therefore necessary to ensure a net gain in power reduction.

PROBLEMS

13.1   Design a half-wave rectifier that has at input a sine wave with peak amplitude 5 V. The diode is equivalent to an infinite resistance for bias lower than 0.6 V and for negative values; it is a resistance equal to 2 Ω if the forward bias is higher than 0.6 V. The load is a resistance equal to 10 Ω. Determine the capacitance in parallel to the load that gives rise to am output ripple lower than 10%.

13.2   Consider a full-wave rectifier made by a bridge of diodes. The I–V response of the diodes is

$$I_D = 0.2e^{(V_D - 0.5)/V_T}$$

for $V_D > 0.5$ and zero for $V_D < 0.5$. What is the drop voltage across the diodes connected to ground? The resistive load is 100 Ω and the peak sine wave at the secondary coil of the transformer is 24 V.

13.3   Design a zener regulator for a load of 5 Ω using a zener diode with $V_z = 3.3$ V. The minimum current in the zener is 1 mA. The unregulated voltage can range from 6 V to 4 V. The output voltage must always be higher than 3.3 V.

13.4   Consider the linear regulator of Figure 13.11(a). What is the maximum output voltage with the following design conditions?

$$V_{Th,n} = 0.7 \text{ V}, \quad R_1 = 10 \text{ k}\Omega, \quad R_L = 100 \text{ }\Omega, \quad V_{ov} = 0.2\sqrt{I_L} \quad \text{(where } I_L \text{ is in } A\text{).}$$

R_2 is variable from 1 kΩ to 100 kΩ and V_{ov} is the overdrive voltage of the MOS transistor. Determine the corresponding value of R_2.

13.5   Repeat Problem 13.4 but use the schematic of Figure 13.13(a). For this problem V_{ov} represents the overdrive voltage of the P-channel MOS transistor. Compare the results of the two problems.

13.6   Consider the circuit diagram of Figure 13.18(b) and assume the capacitance C_L is initially discharged. Estimate the output voltage at the end of the first ten clock periods, supposing $f_{CK} = 10$ MHz, $C_1 = 1$ μF, $C_L = 100$ μF, and $R_L = 1$ kΩ. Determine the asymptotic value of the output voltage.

13.7   A buck switching regulator operates at the switching frequency of 3 MHz. The on-resistance of the switches is 0.1 Ω and the parasitic capacitances are 40 pF. Determine the efficiency of the regulator as a function of the output current. The duty-cycle of the control waveform is 68% and the supply voltage is 5 V.

CHAPTER 14

SIGNAL GENERATION AND SIGNAL MEASUREMENT

Signals are the key ingredients of electronic circuits. They are not only what we process but also stimuli whose waveforms are needed to perform processing or verify the proper operation of a circuit. This chapter studies schemes and methods for generating different kinds of waveforms. We shall learn how to give rise to signals using pure analog or digitally controlled techniques. We shall study methods used to measure signals or, better, to determine the key features of signals. At the end, the chapter describes frequently used instruments and their principles of operation.

14.1 INTRODUCTION

After the design and fabrication of an electronic circuit or a complete system, it is necessary to verify that its operation corresponds to what was defined in advance. For that it is necessary to use experimental set-ups similar to those employed for study with this book by the electronic virtual student laboratory. The key goal of the virtual experiments is to understand the operation of circuits. Testing and verification, on the other hand, aim at validating the circuit correctness and design effectiveness. Nevertheless, the methods and instruments used are the same. Both strategies use signal waveforms to exercise input terminals and employ instruments to produce output for checking outcomes. The results of measurements are possibly transformed into digital form to be stored in a memory for online or offline post-processing.

Understanding Microelectronics: A Top-Down Approach, First Edition. Franco Maloberti.
© 2012 John Wiley & Sons, Ltd. Published 2012 by John Wiley & Sons, Ltd.

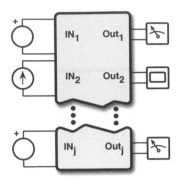

Figure 14.1   Generic set-up for performance verification of processing system with i inputs and j outputs.

Figure 14.1 illustrates the generic set-up used for the testing of a system with multiple inputs and multiple outputs. The scheme assumes that the signal generators and instruments do not disturb the circuit, because signal generators and measurement instruments are supposed to establish correct interfaces.

The above description highlights two necessities, as follows.

- Electrical signals that are able to drive the input ports of the device under test without signal alteration must be generated.

- The instruments must be capable of measuring output voltages or output currents with negligible disturbance (i.e., they must show an infinite impedance for voltages and zero impedance for currents).

Often, signals are used not just for testing but also for processing operations inside the circuit. For example, modulation requires a sine wave or a square wave at the input of a multiplier. A clock at a frequency locked with a bit stream is necessary to synchronize the data flow and its processing. For these needs, the waveform can be received from outside or we can generate the signal inside the circuit. Both solutions have advantages and disadvantages. Entering a signal into an integrated circuit means driving relatively large capacitances and facing problems caused by the inductance of wire bondings. Generating a signal inside a chip means encountering the problem of a limited protection from interfering signals.

14.2   GENERATION OF SIMPLE WAVEFORMS

Simple waveforms like a "single shot" pulse, a periodic sequence of a sawtooth signal, are routinely needed. The key ingredient for their generation is a time interval controlled by an electrical analog or digital quantity. With a clock available, it is easy to generate multiples of the clock period. However, the time discretization is given by the clock period itself, and having good resolution requires the use of very high clock rates. Analog circuits are more critical but give rise to better resolutions.

A generator of analog time interval is, for example, the scheme of Figure 14.2. It is called *monostable* because, after the circuit is switched on, it generates the pulse period only once.

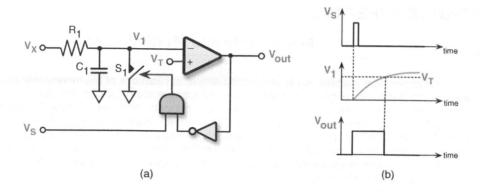

Figure 14.2   (a) Monostable pulse generator: simple pulse generator based on an RC network; (b) waveforms in relevant nodes of the circuit.

Then it requires a new trigger to give rise to another pulse. The signal V_s is the short pulse that starts the time interval. It closes the switch S_1, which suddenly discharges the capacitor C_1. The comparator generates a "1" at output. At the end of the V_s pulse the switch opens and the positive input voltage V_x starts charging the capacitor exponentially until it crosses V_T (another positive voltage smaller than V_x). The output of the comparator becomes "0". The simple logic made by the AND of the input pulse and the inverted output disables incoming trigger pulses until the end of the "single shot" output.

The crossing time of the threshold and the V_1 voltage occurs when

$$V_x(1 - e^{-T_p/R_1C_1}) = V_1 = V_T. \tag{14.1}$$

The equation shows that the concurrent action of two quantities controls the pulse duration: the ratio between the voltages, V_x and the threshold V_T, and the time constant, $\tau_1 = R_1C_1$. One of these variables adjusts the monostable pulse. The accuracy of the voltage ratio is normally much better than the accuracy of resistors and capacitors. Therefore, the precision of pulse duration is, in practice, determined by the accuracy of the time constant R_1C_1.

The monostable scheme is the basis of other signal generators. For example, it may give rise to a periodic sequence of pulses with controlled frequency and duty-cycle. The scheme in Figure 14.3 is a possible solution. The diagram uses two comparators to detect the crossing of two threshold voltages, $V_{T2} > V_{T1}$ ($V_x > V_{T2}$). When the voltage at the output of the first inverter, V_s, is "high" (reset of the flip-flop) the switch S_1 closes. Resistance R_2 discharges capacitor C with time constant R_2C. When V_1 crosses the V_{T1}, S_1 opens to begin a transient that charges C toward V_x through $R_1 + R_2$. When V_1 reaches V_{T2} the reset of the flip-flop makes V_s high, and the cycle starts again. The time constants of the charge and discharge phases are

$$\tau_{ch} = C(R_1 + R_2) \quad \text{and} \quad \tau_{dis} = CR_1. \tag{14.2}$$

Moreover, it is easy to verify that

$$T_1 = \tau_{ch} \ln\left(\frac{V_x}{V_x - V_{T2}}\right) \quad \text{and} \quad T_2 = \tau_{dis} \ln\left(\frac{V_{T2}}{V_{T2} - V_{T1}}\right), \tag{14.3}$$

leading to a pulse with duration $T = T_1 + T_2$ and duty cycle $\Delta = T_2/T$. Four quantities, two time constants and two voltage ratios, control the repetition frequency and duty cycle.

COMPUTER EXPERIMENT 14.1

Monostable and Multivibrator

This computer experiment enables you to understand simple applications of the monostable and multivibrator. The generated outputs are a single pulse or a sequence of pulses. The duration of pulses and their repetition rate depend on passive components and the defined threshold voltage(s). All the capacitors are fixed at 20 pF. Thus, three resistances control the time constants. The multivibrator (upper scheme) and the monostable (lower circuit), whose operations are described in this book, are built with discrete components. The measured time intervals are relatively low, because the board uses discrete elements. Much lower times can be obtained with integrated applications.

In this experiment you can:

- set the threshold voltages used by both circuits;
- control the resistors. Remember that these components together with the thresholds determine the timing of the monostable and the multivibrator;
- control the repetition rate and duration of the monostable trigger.

MEASUREMENT SET–UP

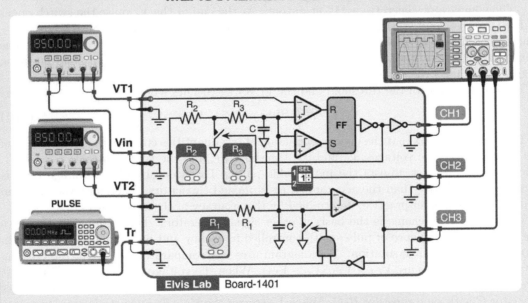

Elvis Lab Board-1401

TO DO

- Use the default values of resistors and observe output waveforms. Increase the repetition rate of the trigger until the period is lower than the monostable pulse. Explain the result.
- Restore the default values and change the thresholds.
- Focus on the multivibrator by changing resistances and thresholds. Understand why the output waveform has the resulting duty–cycle and frequency. Observing waveform across the capacitor helps understanding.
- Find the conditions for which the duty cycle is 50%. Change the input voltage and find at which threshold the duty cycle returns to 50%.

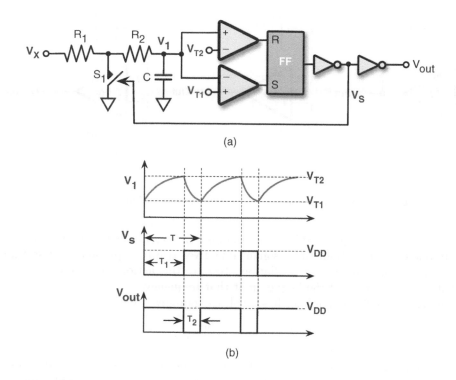

(a)

(b)

Figure 14.3 (a) Circuit diagram of a multivibrator; (b) waveform across the capacitor and logic signals at the output of the inverters.

Since the output voltage periodically swings (or vibrates) from zero to the supply voltage, the scheme is called a *multivibrator*. The circuit and its principle of operation have been known for many years. The popular "555" integrated circuit performs the function with few external components. This IC contains comparators, a flip-flop, an inverter, and a switch. Moreover, an internal resistive divider of the supply voltage, made from three equal resistors, gives rise to V_{T1} and V_{T2}. T_1 becomes $0.693\,(R_1 + R_2)C_1$, T_2 is $0.693\,R_1 C_1$ (0.693 is $\ln(2)$) and the period of the pulses gets to be $T = 0.693\,(2R_1 + R_2)C_1$.

14.3 OSCILLATORS

Electronic systems widely use periodic waveforms that oscillate with controlled periods. In this category we have sinusoidal, square, triangular, and sawtooth waveforms. The simple multivibrator just studied can be a possible starting element. It gives rise to periodic pulses that can generate the required waveform. Other solutions exploit positive feedback. Figure 14.4 illustrates the basic concept: a feedback loop made from an inverting gain amplifier and a frequency-dependent feedback block. The phase shift of $\beta(f)$ changes with frequency to give at a defined frequency the extra $180°$ that gives rise to instability. At this specific frequency, called the *resonant frequency*, the circuit oscillates. Notice that, as shown in Figure 14.4, a frequency lower than the resonant frequency gives rise to a longer period and, accordingly,

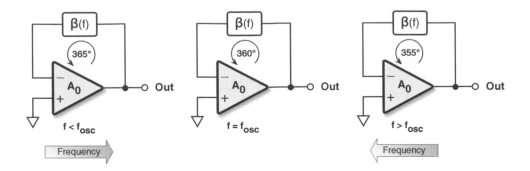

Figure 14.4   Feedback circuit around the oscillation frequency.

must give rise to an overall phase shift larger than 360°. This moves the frequency to a higher value. A frequency higher than the resonant one gives rise to a phase shift less than 360°, which brings back the frequency of oscillation. To have this action effective it is necessary to have a positive contribution of the loop gain at that frequency.

As we know, the feedback amplifier has a voltage gain given by

$$A_f(f) = \frac{A_0}{1 + \beta(f)A_0}. \tag{14.4}$$

When the denominator of equation (14.4) goes to zero the amplification becomes infinite and the scheme resonates at that frequency. It is enough to have a very small input signal at the resonant frequency to start oscillation, or, more precisely, to sustain the oscillations it is necessary that the loop gain be equal to or higher than -1. The condition $\beta(f)A_0 \geq -1$ is called the *Barkhausen criterion*. Indeed, if the modulus of the loop gain is much higher than 1, the loop is over-driven and the final result is a distorted output signal. A practical condition is to have a loop gain slightly higher than 1. A large amplitude at the output of real amplifiers results in saturation of the circuit; the gain diminishes, and, because of a reduced loop gain, the amplitude does not increase any more. Because of this non-linear amplitude control the waveform is not exactly a sine wave, but the distortion is negligible if the loop gain at small amplitudes is near 1. At power switch-on the output voltage is zero, but the circuit starts oscillating because the wide-band noise around the loop always has a sinusoidal component at the resonant frequency. The positive feedback enhances that frequency term, which quickly grows until it reaches a steady amplitude. Distortion and frequency "robustness" depend on the static input–output relationship of the feedback loop gain.

Observe

The Barkhausen condition, $\beta(f)A_0 \geq -1$, entails linear behavior. It would determine an infinite growth of the output voltage, which in real implementations stops when $\beta(f)A_0 = -1$.

Since operational amplifiers provide very large gains and wide bandwidth, the focus of the design is not on meeting the amplitude condition but rather on controlling the phase shift condition. The passive feedback block mainly determines the phase shift, but at very high frequency the op-amp starts lagging, giving rise to an additional phase component that influences the oscillation conditions. In those cases the best solution for the forward amplifier is a simple scheme with limited gain but large bandwidth.

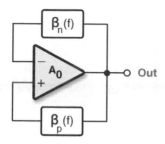

Figure 14.5   Double feedback circuit. The overall result is positive or negative depending on the βs.

Some circuits use a double feedback, with one on the positive path, β_p, and the other on the negative path, β_n. The degree of freedom gives rise to schemes with op-amp configurations like the one shown in Figure 14.5. The overall feedback is the superposition of the two β factors. The demand on loop gain, similar to the Barkhausen criterion, is

$$[\beta_n(f) - \beta_p(f)]A_0 = -1, \tag{14.5}$$

or, for a very large forward gain A_0, it is necessary to ensure that

$$\beta_n(f) - \beta_p(f) = 0 \tag{14.6}$$

or is slightly negative.

The control of two β factors realized with passive components is simpler and more accurate than controlling a feedback loop gain. Its use facilitate the generation of waveforms that are "almost" sine waves.

14.3.1 Wien-bridge Oscillator

The scheme of Figure 14.6 gives an example of double feedback. It is the popular Wien-bridge oscillator. The feedback towards the inverting input is a plain resistive divider, R_1 and R_2, that gives rise to

$$\beta_n(f) = \frac{R_1}{R_1 + R_2}. \tag{14.7}$$

The other feedback, as detailed in Figure 14.6(b), is the cascade of a high-pass passive and a low-pass filter made by equal resistors and capacitors. The feedback network is a bandpass because, at low frequency, the series capacitance opens the input–output connection while at high frequency the across capacitor shorts the feedback signal.

A simple circuit analysis of the positive loop yields

$$\beta_p(f) = \frac{V_f}{V_{out}} = \frac{j2\pi fRC}{1 + 3j(2\pi fRC) - (2\pi fRC)^2}. \tag{14.8}$$

The phase of $\beta_p(f)$ goes to zero at

$$f_0 = \frac{1}{\sqrt{2}\pi RC}; \tag{14.9}$$

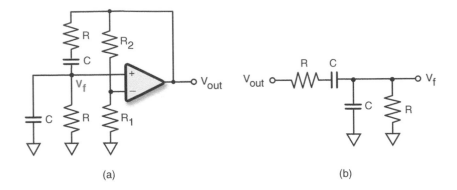

Figure 14.6   (a) Wien-bridge oscillator; (b) feedback network towards the non-inverting input.

moreover, at f_0, the modulus of $\beta_p(f)$ is $1/3$. To sustain oscillation it is necessary that $\beta_n(f)$ equals or is slightly less than $1/3$. The equality condition leads to

$$R_2 = 2R_1. \tag{14.10}$$

To determine the amplitude of the output it is necessary to use equation (14.5), which accounts for a limited finite gain of the op-amp or exploiting the non-linearity of passive elements. Suppose that the resistance R_1 of the negative feedback path is non-linear with a value that augments with the output voltage. The negative β increases with amplitude until it becomes equal to the value of the positive β at the resonant frequency.

14.3.2   Phase-shift Oscillator

The feedback block of oscillators with a single negative feedback must give rise to $180°$ phase shift. The circuit of Figure 14.7 achieves the result with the cascade of three RC low-pass filters. As we know, the phase shift of a passive RC filter may go from zero to $90°$. Two cells are not enough, but three can build up the necessary $180°$ phase shift. Notice that the three cells of Figure 14.7 are directly cascaded onto each other and use the same values of resistors and capacitors.

Assuming the effect of the input resistance of the inverting amplifier, R_1, is negligible, the feedback voltage is calculated by

$$V_2 = V_f SC \left(R + \frac{1}{SC} \right) = V_f(1 + SRC) \tag{14.11}$$

$$V_1 = (V_2 + V_f)SRC + V_2 = V_f[(1 + SRC)^2 + SRC] \tag{14.12}$$

$$V_{out} = (V_1 + V_2 + V_f)SRC + V_1 = V_f[(1 + SRC)^3 + 2(SRC)^2 + 3SRC], \tag{14.13}$$

which yields

$$\beta(f) = \frac{1}{(1 + j2\pi fRC)^3 2(j2\pi fRC)^2 + 3j2\pi fRC}. \tag{14.14}$$

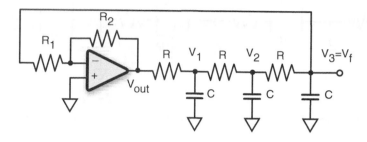

Figure 14.7   Phase-shift oscillator with three passive low-pass filters making the feedback block.

The 180° phase shift occurs at $f_0 = 1/(2\pi RC\sqrt{6}) = 0.065/RC$. The magnitude of β at this frequency is

$$|\beta(f_0)| \simeq \frac{1}{29}. \tag{14.15}$$

Accordingly, in order to secure oscillations, the modulus of the inverting amplification, established by R_2/R_1, must be slightly higher than 29.

The assumption that the input of the amplifier does not load the feedback block is not exactly satisfied, even when using values of R_1 much higher than the filter resistor, R. The load established by R_1 reduces the phase shift of the last cell and increases the attenuation. The 180° shift occurs at lower frequency and the circuit needs a larger gain. The use of a unity gain buffer separating the feedback network from the amplifier eliminates the problem.

The solution of Figure 14.7 uses three cells. That is the minimum; other oscillator schemes of the same type use more than three RC cells and may separate the sections with unity gain buffers.

14.3.3   Ring Oscillator

The use of a cascade of multiple cells and the distribution of the gain around the loop gives rise to a ring oscillator. This uses an odd number of simple gain stages, like an inverter, to form a feedback loop. The overall gain matches the required gain, and the odd number provides inversion. For the phase shift the connection between the simple amplifiers is by RC filters, as shown in Figure 14.8. Also for this configuration the non-immediate response of the inverters, described by the propagation delay T_d, changes the condition of oscillation.

A 180° phase shift corresponds to half the oscillation period. This is the propagation delays plus the delays determined by the RC filter at the oscillation frequency:

$$\frac{1}{2\omega_{osc}} = 3T_d + 3\arctan(\omega_{osc}RC). \tag{14.16}$$

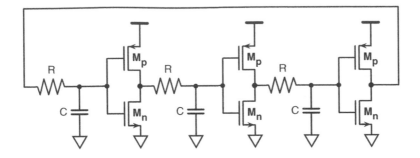

Figure 14.8   Ring oscillator with three inverters and distributed delay.

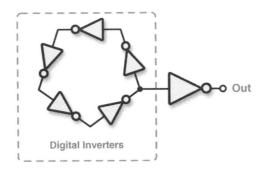

Figure 14.9   Ring oscillator with five digital inverters along the loop and an output inverter.

Thus, reducing the time constant RC increases the oscillation frequency. The maximum occurs when RC goes to zero, or the the phase shift cells are not used at all. The period of oscillation is controlled by the number of inverters multiplied by the propagation delay with another inverter as load. Three inverters is the minimum number of cells. Often, for generating relatively low frequencies, it is preferable to use a large number of cells instead of the hybrid scheme of Figure 14.8. Figure 14.9 shows the block diagram of the ring oscillator with five inverters. In addition, the scheme uses an extra inverter inserted in a generic point of the ring for minimizing the output load and generating a signal with the required strength.

The frequency of oscillation depends only on the propagation delay of each inverter given by the charging or the discharging time of the parasitic capacitive load. Since C_p is typically very low, in the fF range, the inverter's delay is very short. The frequencies obtained can be very high but frequency control, accuracy, and stability are often problematic. Some schemes adjust the oscillation frequency by changing the supply voltage or controlling the bias current of one or more inverters in the loop.

A precise estimation of the propagation delay can only be done with computer simulations, because the non-linear nature of the charging and the discharging transients is difficult to model. However, approximately, we can suppose that the transistors of the inverter are in saturation for about 50% of the time and stay in the linear region for the second half of the switching transients. When the transistor is in saturation, the current depends on the overdrive voltage and transistor sizes. It obeys

$$I_{D,sat} = \frac{\mu C_{ox}}{2} \frac{W}{L} (V_{DD} - V_{Th})^2, \tag{14.17}$$

COMPUTER EXPERIMENT 14.2

Understanding the Ring Oscillator

A ring oscillator is a cascade of inverters, each of them determining a delay (or a phase shift). The frequency at which the overall phase shift is 360° gives rise to the oscillation frequency, provided that the loop gain is higher than 1. The minimum number of cells that determine a 360° shift is three. This computer experiment uses that minimum number and, in addition, employs RC cells to control the phase shifts (the inverter delay is very small). One method used to control the ring oscillator is to change the supply voltage of the inverter. This ElvisLab board permits you to do that for the first inverter. A two-output bias generator regulates the supply voltage of all the inverters. The board includes two circuits; the one on the top is a normal ring oscillator, the scheme on the bottom has the loop opened and uses a square wave at the input. You can:

- set supply voltage and bias of the first inverter;
- set the equal value of resistance and capacitance used in all the cells of the two schemes;
- set the trigger of the oscilloscope on Channel 1 or Channel 3.

MEASUREMENT SET–UP

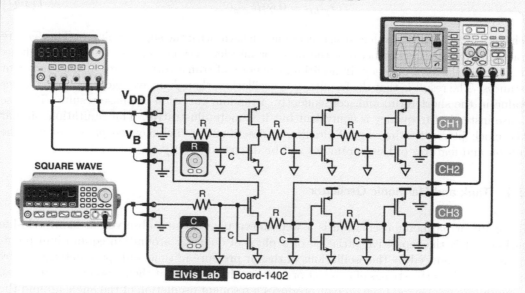

Elvis Lab   Board-1402

TO DO

- Use the resistor default value and set both supply voltages at 1.2 V. Observe the pseudo sine wave at the ring oscillator output. Change the resistor values and notice how the waveform changes. Generate a signal close to a square wave.
- For the square wave generator, use a frequency half that of the ring oscillator. Measure the delay between the waveforms on Channels 2 and 3. Change the supply of the first inverter, and observe waveforms and delay.
- Change the supply of the first inverter and determine the frequency sensitivity. Change the other supply and repeat the sensitivity measure.
- Change the resistance and observe the waveforms generated. Notice what effect input capacitance has on the inverter.

and gives rise to a linear charging or discharging transient whose slope is

$$\frac{\Delta V}{\Delta t} = \frac{I_{D,sat}}{C_p} = \frac{\mu}{2L^2}(V_{DD} - V_{Th})^2, \qquad (14.18)$$

which is large if the transistor length is at the minimum and the supply voltage is large.

When the transistor is in the triode region it behaves like a resistor whose value is

$$R_{on} = \frac{V_{DD} - V_{Th}}{\mu C_{ox}(W/L)}. \qquad (14.19)$$

R_{on}, together with C_p, determines the time constant of an exponential transient.

Often, simplified calculations use overall on-resistances, $R_{eq,n}$ and $R_{eq,p}$, to depict charging and discharging waveforms. The propagation time is conventionally assumed as the time required to charge or discharge the node at half of the supply voltage. Since $e^{-t/\tau}$ describes the discharge and $(1 - e^{-t/\tau})$ characterizes the charge transient, the propagation times are

$$t_{p,HL} = 0.69 R_{eq,n} C_p \qquad (14.20)$$
$$t_{p,LH} = 0.69 R_{eq,p} C_p, \qquad (14.21)$$

which change with the equivalent resistances and, in turn, with the supply voltage. This feature has a negative consequence because the noise or interferences corrupting the supply voltage affects the oscillation frequency. In addition, the noise of transistors changes in the equivalent resistances. The result is that the frequency of oscillation, though on average constant, changes its value in the short term, and, consequently, a random variation of phase results. The use of phase instead of frequency is equivalent but it is more illustrative. This limitation, similar to the clock jitter affecting digital circuits, is described by the so-called *phase noise*. The definition and features of this parameter will be discussed shortly.

14.3.4   Tank and Harmonic Oscillator

Resonant or harmonic oscillation is a well-known phenomenon observed in many real-life situations. It is the repetitive variation of a physical quantity around an equilibrium point. The sound is produced by the oscillations of the air pressure in time and space determined by generators, such as the strings of a guitar or a violin, that vibrate at their resonant frequencies. The pendulum suspended from a pivot produces a resonant oscillation of the angle around the equilibrium position, and so forth.

Oscillations occur when any real system exhibits a resonant frequency. That is the special frequency at which some form of resistance is at minimum, ideally zero. Oscillations always follow a stimulus but are not persistent. The amplitudes fade away because the system loses energy due to friction or dispersions. Therefore, in order to sustain oscillations it is necessary to provide the system with the lost energy. Since the necessary energy comes from a tank of energy, this category of oscillator, rather than being called *harmonic*, is a *tank oscillator*.

To sustain oscillation it is necessary to periodically transfer energy from one type to another type (as in the pendulum) or from a storing element to another storing element with minimal (or zero) losses in the transfer. Electrical components that store energy are the capacitor and the inductor, and, as we all know, a series or a parallel connection of capacitor and inductor creates a resonant circuit because the impedance or admittance of the series or the parallel

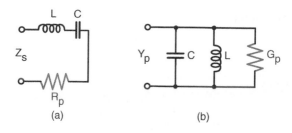

Figure 14.10   (a) Series resonant circuit; (b) parallel resonant circuit.

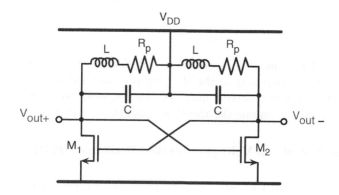

Figure 14.11   Balanced parallel resonant oscillator with loss compensation realized by a negative active resistance.

(see Figure 14.10) are

$$Z_s = R_p + j\left(WL - \frac{1}{WC} \right) \qquad (14.22)$$

$$Y_p = G_p + j\left(WC - \frac{1}{WL} \right), \qquad (14.23)$$

which become zero at the resonant frequency, if the series parasitic resistance R_p or the parallel parasitic conductance are zero, respectively.

Often, an electronic tank oscillator uses the parallel resonator scheme rather than the series configuration. Moreover, in order to sustain oscillations, it is necessary to use a suitable method that compensates for the loss caused by the parallel conductance. The method used to counterbalance parasitic conductance is to connect in parallel with the resonant circuit a negative conductance with equal but opposite value.

Creating a negative conductance (or a negative resistance) seems difficult, but a circuit that uses feedback makes it possible. A typical topology is the one shown in Figure 14.11(a). The two transistors connected in positive feedback give rise to the required negative conductance. They are placed across the parallel resonant circuit that, in the schematic of the figure, is divided into two equal parts to determine balanced oscillations of the two outputs while biasing the circuit with V_{DD}. The scheme assumes that the only losses come from the inductances. This is represented by the series resistances, R_p.

The quality factor of the inductor is

$$Q = \frac{W_0 L}{R_p},$$

(14.24)

and the oscillation frequency is

$$\omega_0 = \frac{1}{\sqrt{LC}} \sqrt{1 - \frac{R_p^2 C}{L}}.$$

(14.25)

The extent of negative conductance depends on the transconductance of transistors M_1 and M_2. If it is too low, oscillations do not take place. To sustain oscillation it is necessary to satify the condition

$$g_m = \frac{R_p C}{L}.$$

(14.26)

The above equation determines the average current that must flow through the two equal transistors. The current also depends on the W/L ratio, whose value is defined by the design. The current gives rise to the overdrive voltage of the transistor and, in turn, the quiescent value of the output voltage. It equals the threshold plus the overdrive voltage.

14.3.5   Digitally Controlled and Voltage-controlled Oscillator (VCO)

Equation (14.25) shows that the oscillation frequency of a tank oscillator depends on the inductor and the capacitance. It is often desirable to change the oscillation frequency within a given range. Circuits that realize this function are called Voltage Controlled Oscillators (VCOs). An effective way to fulfill that requirement is to change the capacitance by using, for example, the scheme of Figure 14.12. The circuit uses a capacitor with fixed value C and in parallel has an array of small elements whose connections depend on the logic signals that control the switches. The capacitance that can change is the interval

$$C_{MIN} = C, \quad C_{MAX} = C + \sum_{0}^{n-1} C_i.$$

(14.27)

The variable part can be made of an array of binary weighted capacitors: $C_u, 2C_u, 4C_4, \ldots,$ $2^{(n-1)} C_u$. The maximum total extra capacitance is $(2^n - 1)C_u$; its actual value depends on the bit controlling the switches. The solution gives rise to 2^n possible steps of regulation with increments determined by the minimum capacitor of the array, C_u.

The switches used to establish the connections of capacitors are not ideal components. In the on condition they give rise to an on-resistance in series with the capacitor. Since resistances are non-conservative elements, the loss that they produce requires additional compensation determined by a higher negative conductance of the positive feedback loop.

Changing the oscillation frequency in a digital manner complies with the need to control the operation of circuits by microprocessors or DSPs. The method is simple, but obtaining a small step of frequency change and relatively large tuning ranges requires the use of very large capacitive arrays. For this reason some systems prefer to use a continuous control of frequency, made possible by capacitors whose value changes with the dc voltage across terminals. These components, called *varicaps* (or *varactors*), are special p–n reverse-biased junctions.

It is known that the reverse voltage applied across a p–n junction gives rise to depletion, essentially a region at the p–n interface depleted by carriers. The depleted separation behaves

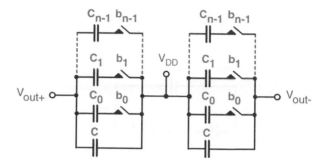

Figure 14.12 Variable capacitance with digitally controlled steps.

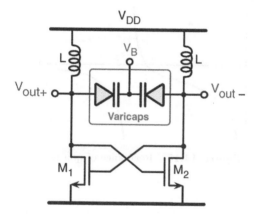

Figure 14.13 Use of varicap to permit tuning of the oscillation frequency.

like the dielectric of a capacitor. Since the width of the depletion region increases with the reverse biasing the structure is like a parallel-plate capacitor whose plate separation is voltage controlled. The voltage–charge relationship depends on the transition between p-doping and n-doping. Special doping profiles of varicaps give rise to good voltage sensitivities.

The scheme in Figure 14.13 shows the balanced oscillator of Figure 14.11 made with varicaps. The figure uses a special symbol to depict a varicap: a diode combined with a capacitor. The voltage V_B, higher than the maximum output, reverse biases the varicaps. The difference between average output and V_B determines the varicap capacitance and, consequently, the output frequency. If necessary, fixed capacitors in parallel to the varicaps linearize the operation at the cost of a lower tuning range.

Other oscillator schemes studied above can also become VCOs. As already mentioned, changing the bias voltage or the supply current of a ring oscillator controls the propagation delay of each inverter. The range in variation of the loop delay around the ring determines the tuning range of the oscillator.

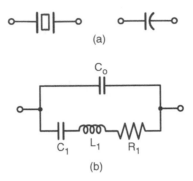

(a)

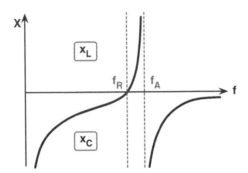

(b)

Figure 14.14   (a) Symbols for quartz; (b) equivalent circuit of quartz.

Figure 14.15   Reactance of quartz.

14.3.6   Quartz Oscillator

A resonant circuit is made from an LC pair connected in series or in parallel. A special device featuring resonance in some ways equivalent to an LC scheme is the quartz. This is the two-terminal device mentioned at the beginning of this book for its use in electronic watches.

Quartz has piezoelectric properties when cut in given crystallographic directions. It can feature very low temperature coefficients with very high quality factors. Figure 14.14(a) shows the symbols for quartz. Figure 14.14(b) provides the equivalent circuit. It is the parallel connection of a capacitor and a series resonant circuit. The impedance across the two terminals is

$$Z(s) = \left(R_1 + sL_1 + \frac{1}{sC_1}\right) \left\|\left(\frac{1}{sC_0}\right),\right.$$
(14.28)

which can be written as

$$Z(s) = \frac{s^2 + s(R_1/L_1) + 1/(L_1C_1)^2}{sC_0\{s^2 + s(R_1/L_1) + [(C_1 + C_0)/(L_1C_1C_0)]^2\}}.$$
(14.29)

Equation (14.29) shows two resonances, one given by the numerator (series resonant), and the other by the denominator (parallel resonant). The reactance, X (imaginary part of Z), is zero at the series resonant frequency and becomes infinite at the parallel resonant frequency.

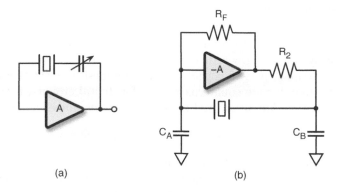

(a)   (b)

Figure 14.16   (a) Resonant feedback loop with quartz; (b) Pierce oscillator made with quartz.

Figure 14.15 plots the reactance of quartz. The frequencies f_R and f_A are called resonant and anti-resonant. Outside the interval between those two frequencies, quartz has capacitive behavior; within a small interval it operates as an inductor.

Since quartz at the resonant frequencies has no phase shift, it can be used as an element of a feedback loop to give rise to oscillation. This can occur at the series resonant frequency

$$f_r = \frac{1}{2\pi\sqrt{L_1 C_1}},\tag{14.30}$$

as shown in the scheme of Figure 14.16(a). The feedback network includes a variable capacitance in series with the quartz to change the resonant frequency slightly. This may be needed to compensate for the temperature dependence (albeit very low) of the resonant frequency of the quartz.

The inductive behavior of quartz in the small frequency region f_R–f_A can give rise to a 180° phase shift. This is what is obtained by the so-called Pierce oscillator, which uses a feedback π network made of two capacitors C_A and C_B together with quartz. There are two paths from output to input, one through the resistance R_F and the other via the resistance R_2 and the π network. The resistance R_F serves to establish a dc bias of the input of the amplifier; the other feedback path dominates at the resonant frequency.

The resistance R_2 and capacitance C_B determine some phase shift that becomes the required 180° because of the inductive behavior of the quartz and the capacitance C_A. The oscillation takes place at a frequency between the resonant frequency and the anti-resonant frequency

$$f_A = \frac{1}{2\pi\sqrt{L_1 C_1 C_0/(C_1 + C_0)}}.\tag{14.31}$$

It can be shown that the oscillation frequency is approximately given by

$$f_{osc} = \frac{1}{2\pi\sqrt{L_1 C_1 C_T/(C_1 + C_T)}},\tag{14.32}$$

where the capacitance C_T is

$$C_T = C_0 + \frac{C_A C_B}{(C_A + C_B)}.\tag{14.33}$$

The above equations, which hold in the $f_R \div f_A$ interval, indicate it is possible to trim the oscillation frequency. This is often useful for adjusting circuit operation.

14.3.7   Phase Noise and Jitter

The operation of many pieces of electronic apparatus depends on an internal or external time base. For such systems it is worth using a parameter that quantifies the quality of oscillators. A quantity frequently used is *phase noise*, an undesirable entity represented in the frequency domain. Its time-domain counterpart, used for digital circuits, is the *jitter*.

Consider a real oscillator. Amplitude and phase fluctuations make the output voltage

$$V_{out}(t) = (V_0 + \delta V_0(t)) \sin(2\pi f_{out} t + \Phi(t)), \qquad (14.34)$$

where V_0 is the amplitude and f_{out} is the frequency.

Both amplitude and phase errors affect timing accuracy. However, often the fluctuations on amplitude are less important than the phase variations. The error on phase $\Phi(t)$ is therefore the one that we consider. It typically changes randomly in time and for this reason is called phase noise. Neglecting $\delta V_0(t)$, equation (14.34) becomes

$$\begin{aligned} V_{out}(t) &\simeq V_0 \sin(2\pi f_{out} t + \Phi(t)) \\ &\simeq V_0 [\sin(2\pi f_{out} t) \cos(\Phi(t)) + \cos(2\pi f_{out} t) \sin(\Phi(t))], \qquad (14.35) \end{aligned}$$

for small values of phase noise $\cos(\Phi(t)) \simeq 1$ and $\sin(\Phi(t)) \simeq \Phi(t)$. Therefore, equation (14.35) is simplified as

$$V_{out}(t) \simeq V_0 \sin(2\pi f_{out} t) + V_0 \Phi(t) \cos(2\pi f_{out} t). \qquad (14.36)$$

That is an ideal sinusoidal wave plus the phase noise $\Phi(t)$ modulated by a phase-shifted version of the same sine wave (it is a cosine rather than a sine).

The spectral representation of phase noise is, as is well known, the Laplace transform $\mathcal{L}[\Phi(t)] = \Phi(f)$. This is the quantity normally used to denote how the output spectrum skirts the two sides of the oscillation frequency. Figure 14.17 shows a typical spectrum of a real oscillator and its phase noise. The real oscillator does not show a sharp line (a delta in the ideal case) but exhibits a frequency interval around the oscillation frequency with non-negligible spectral components. This behavior is better shown by the phase noise diagram, which details the behavior of the phase noise spectrum before its co-sinusoidal modulation.

Several factors cause phase noise, among others the flicker noise of the transistors used in the circuit and a limited Q factor of the resonant circuits. For a detailed study of the issue and description of the circuit solutions that the designer uses to moderate the limit you can refer to specific books or attend advanced courses.

The parameter that describes the time quality of clock signals is jitter. The period of a clock, $\bar{T}_{ck}$, is ideally constant. For real clock generators the cycle duration of a generic period, j, differs from the ideal value, as depicted in Figure 14.18. Clock jitter is

$$\delta T_j = \bar{T}_{ck} - T_{ck}(j), \qquad (14.37)$$

where δT_j has zero mean and looks like a random variable. The root mean square of δT_j is the time jitter

$$T_{ji} = \sqrt{\langle \delta T_j \rangle^2}. \qquad (14.38)$$

We can define clock jitter not just for the real square waves used for clocks but also for any periodic waveform. The crossings of the periodic waveform through a given constant level establish the periods. If we do that for a sinusoidal signal it is possible to link the phase noise and the jitter.

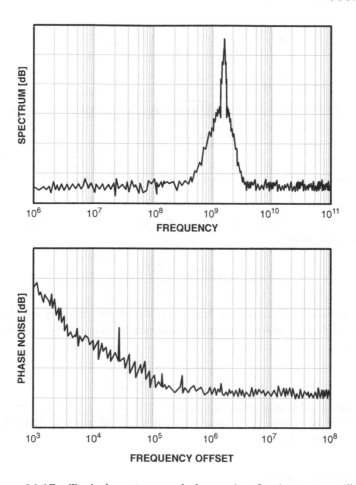

Figure 14.17   Typical spectrum and phase noise of a sine wave oscillator.

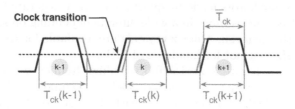

Figure 14.18   Clock signal affected by jitter.

If the threshold is the mid-amplitude of the sine wave, the ideal crossings occur at the times for which

$$2\pi f_{out}t_j = 2n\pi. \tag{14.39}$$

For a sine wave affected by phase noise the crossings occur when

$$2\pi f_{out}t_j + \Phi(t_j) = 2n\pi. \tag{14.40}$$

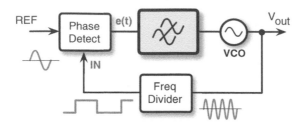

Figure 14.19   Block diagram of a Phase-Locked Loop (PLL).

The period of the jth waves, $T_{ck}(j)$, is the difference $(t_j - t_{j-1})$. Therefore, phase noise and clock jitter are related by

$$\delta T(j) = \frac{\Phi(t_j) - \Phi(t_{j-1})}{2\pi f_{out}}, \tag{14.41}$$

which uses two samples of the time-domain description of the phase noise. Since both jitter and phase error are like noise, the two successive samples of phase noise are uncorrelated. The time jitter becomes

$$T_{ji} = \sqrt{\langle (\delta T_j)^2 \rangle} = \frac{\sqrt{\langle \Phi_j \rangle^2}}{\sqrt{2}\pi f_{out}}, \tag{14.42}$$

an equation often used to obtain one parameter after measuring another. The result also shows that a clock generated by squaring the sine wave of an oscillator has jitter that depends on both phase noise and clock frequency.

14.3.8   Phase-locked Oscillator

As already mentioned, oscillators made by quartzes are very accurate because the Q factor is very high. Resonant frequencies range in a wide interval depending on size, shape, assembly of the part, and cut of the material. Quartz crystals are available for frequencies from a few tens of kHz to ten of MHz, but not for hundreds of MHz.

When it is necessary to generate a precise and very high frequency it is possible to exploit the accuracy of the quartz by using the so-called "locking technique." This method keeps in phase a sub-multiple of the high-frequency output with the phase of the quartz oscillator. The high-frequency generator must be a VCO, tunable by analog or digital control. After a division by a suitable number, it gives rise to the signal used to close a loop so that the feedback locks the divided frequency to that of the quartz. In this manner the much higher frequency takes advantage of the accuracy and stability of the lower-frequency quartz.

The circuit performing the operation is called a Phase-Locked Loop (PLL). Figure 14.19 shows its block diagram. The core is a VCO with at output a high-frequency sine wave (or a pseudo-sine wave). A frequency divider reduces the VCO frequency to a lower value, giving rise to a square wave. The divided signal and the quartz reference (possibly transformed into a square wave) are the inputs of a phase detector (or a Phase-Frequency Detector, PFD). The output is proportional to the phase difference between the two inputs. A filter removes the high-frequency components and generates an analog or digital control for the VCO.

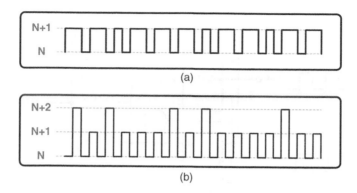

Figure 14.20   Fractional control of frequency divider of a PLL: (a) two-level division; (b) three-level division.

The study of PLLs involves complicated analysis, out of the scope of this book. Indeed, the aim here is to give you a wide but not too deep description of some special topics in microelectronics. It is assumed that a qualitative description and a few additional details about key parts of a special function are enough of an introduction to more advances studies.

An important distinction between PLLs concerns integer and fractional division of frequency. Integer, obviously, means that the ratio between the high frequency and that of the quartz is an integral number. Fractional is when the ratio is not an integer. Dividing a frequency by an integer factor, N, is not difficult, especially when the signal is a square wave. A digital divider or a counter by N performs the task. The only concern is the power consumed, especially at very high frequency. Often two steps make up the division: a pre-scaler by N_1 reduces the high frequency to an affordable level and a second division by N_2 gives rise to the overall division by $N = N_1 N_2$. Power diminishes because the high-frequency section is simple.

The main limit of integer division is the frequency granularity. The difference between divisions by N and $N + 1$ is $f_{in}/[N(N + 1)]$. Applications that require resolution in the ppm range lead to very high division factors, such as almost to undo the locking benefit.

The so-called fractional division, instead of using a fixed integer, N, dynamically alternates divisions by N and $N + 1$. The result gives rise, on average, to a fractional number in the interval $(N, (N + 1))$. For example, we obtain the fractional value 67.312 by dividing the VCO frequency by 67 and 68. The division by 67 must occur for 68.8% of the time and the division by 68 takes place for 31.2% of the time. The result is, on average, $N_{frac} = 67(68.8/100) + 68(31.2/100) = 67.312$, as needed. What is important is to properly distribute over time the divisions by 67 and 68. The goal is to give rise to minimal spur or phase-noise degradation.

The use of a $\Sigma\Delta$ modulator favors a noise-like choice of the division coefficients. However, a constant value at input does not always ensure good outcomes, especially with critical fractional values. It may be more profitable to extend the division factors to more than two. Figure 14.20 shows two examples of fractional divisions. Figure 14.20(a) uses two division levels, N and $(N + 1)$. Since there are 11 divisions by N and 21 divisions by $(N + 1)$, the fraction is $(N + 0.583)$. Note that the periods after divisions seem to be equal. In reality the time span of one pulse is N clock periods of the VCO and of the other is $(N + 1)$ periods. The difference is not noticeable because N is assumed to be large.

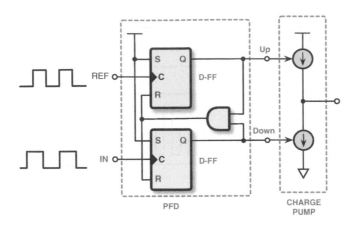

Figure 14.21   PFD and the charge pump of a PLL.

Figure 14.20(b) uses three division factors: N, $(N + 1)$, and $(N + 2)$. The result is a more busy sequence: five divisions by $(N + 2)$ that, added to 11 divisions by $(N + 1)$ and 16 divisions by N, give rise to an average fraction equal to $(N + 0.583)$ – the same as the previous case. However, since the division factor changes more frequently, there are fewer low-frequency spectral components. The result still generates spur terms but they are pushed to a high frequency, far from the signal band. The optimum would be a shaped noise-like spectral distribution.

Knowing how to implement a phase detector (or a phase–frequency detector) is instructive and is briefly discussed here. Consider the scheme of Figure 14.21 with square waves at inputs. Two edge-triggered flip-flops and a Boolean AND construct the logic part of the PFD. There is a second part, which is analog, called the charge pump. The outputs of the PFD named *up* and *down* switch on a positive or a negative current generator in the charge pump. The reference and the input signal make the clock of two D-type flip-flops (labeled *D-FF*). The first of the two clocks that comes in raises its logic output, activating the *up* or the *down* control. The condition holds until the rising edge of the other input occurs. This sets its *D-FF* "on" while the AND immediately generates a reset command for both *D-FF*s. The reset of the Q output switches off the current source.

Figure 14.22 shows output logic waveforms for two different situations. The first has an input frequency lower than the reference. The outputs are only *up* pulses with different durations, depending on the phase relationship between the two inputs. The case of Figure 14.22(b) assumes that the input frequency is lower than the reference at the beginning, becomes higher in the middle, and returns to lower at the end of the time interval. The output logic is made by both *up* and *down* commands. *Up* occurs until the input frequency is lower than the reference. Then there are some *down* pulses aimed at diminishing the frequency. An *up* pulse near the end of the time interval indicates a diminished input frequency. The *up* and *down* signals operate current sources to produce an analog signal. A low-pass filter smooths the result before the VCO control. The response of that filter is important for performance. A tight filter prevents quick frequency hopping, often required in communications; a relaxed filter can give rise to a weak frequency lock. In addition, it is necessary to consider practical problems possibly caused by the requirement for large resistive or capacitive values that lead to an excessive use of silicon area or require the use of out-of-chip components.

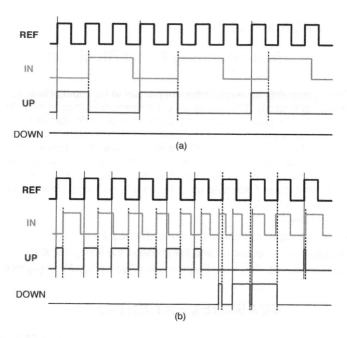

Figure 14.22   Input and output logic signals of the phase–frequency detector of Figure 14.21: (a) case with input frequency lower than reference; (b) case with input frequency changing from lower to higher and again to lower than reference.

The PLL scheme of Figure 14.19 is almost analog. However, it can be partially or totally implemented in a digital manner. The VCO that generates the output frequency controls a varactor. It becomes "digital" by using an array of digitally selected capacitors to realize the tunable part of the tank capacitor. A digital filter generates the tuning code of the digital VCO as shown in Figure 14.23. The output frequency, after a division with a pre-scaler, becomes digital thanks to a time-to-digital converter that measures the scaled output period using the period of the quartz oscillator as a precise time ruler. The processing of the measured output period together with the fractional request gives rise to a signal that, after digital filtering, controls the digital VCO.

The digital scheme of Figure 14.23 is attractive because of the advantages of digital processing. It allows complex filtering; moreover, since the filter function is software defined, it can be easily reconfigured. Furthermore, the critical operation of phase/frequency comparison is better implemented in the digital domain. The result is robust and flexible; however, the limit is the "noise" caused by the discretization in time and amplitude of the processed quantities.

Self training

Use the Web to learn more about analog and digital PLLs. They can have an internal or external loop filter. They can be integer or fractional. Account for the following features:

- number of bits of the integer and the fractional parts;

COMPUTER EXPERIMENT 14.3

Understanding Basic Blocks of the PLL

A PLL uses a VCO for generating an output frequency equal to the input reference multiplied by a desired number (integer or fraction). This computer experiment examines key blocks of a PLL, one cascading to the other. The experiment does not include frequency divider and does not close the loop. The inputs are two square waves used to mimic the reference and divided outputs. The two inputs control the phase frequency detector, whose outputs are logic signals. One is called "up" and the other "down". The up signal occurs when one of the inputs is high before the other and returns to zero when the other also becomes high. The down pulse does the opposite. The up and down signals drive the two current sources used in the charge pump. Finally, the output of the charge pump, after filtering, drives the VCO.

In this virtual experiment you can:

- set the value of up and down currents used in the charge pump;
- set the input square wave frequencies and the phase of the first generator;
- control the VCO parameters. The VCO frequency is given by $f_{FR}+kV_{in}$ (f_{FR}, free running).

MEASUREMENT SET–UP

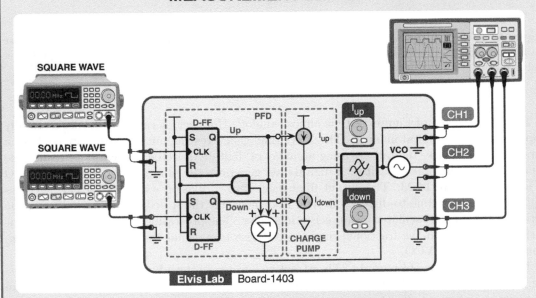

Elvis Lab   Board-1403

TO DO

- Use the same frequency for two square waves and set their phase difference to zero. Observe that output is an 800 MHz sine wave.
- Change the phase difference and observe the outputs with positive or negative phase shifts.
- Change one of the two frequencies and observe Channel 3. Explain the result, remembering the phase relationship between square waves at different frequencies.
- Relate the filter output with the VCO frequency. Remember that the output result is not locked, because the loop is opened.

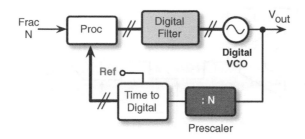

Figure 14.23   Possible block diagram of digital PLL.

- jitter of the output voltage in ps;

- power dissipation for various frequencies of operation.

Describe the special features of one of the identified parts, indicate the possible options, and draft the block diagram with the main flow of the architecture indicated.

14.4   DAC-BASED SIGNAL GENERATOR

A digital processor, a memory, and a DAC can determine periodic waveforms and, in particular, sine waves. A reconstruction filter makes the sampled-data analog waveform generated by the DAC into a continuous-time one. The technique, called Direct Digital Synthesis (DDS), employs a generic architecture like that of Figure 14.24. A counter determines a digital sawtooth signal used as the address of a look-up table with m bits at input and n bits at output. The LUT output drives a DAC that generates an analog signal. The output is assumed to be periodic with a periodicity given by the clock period multiplied by the counter sequence length. The DAC accuracy determines analog accuracy and, consequently, the SNR at output. The digital staircase lasts for 2^m clock periods and the content of the LUT is the digital portrait of the desired output voltage.

The method is flexible because changing memory content allows us to generate any kind of waveform, but the memory size (equal to the number of points, 2^m, multiplied by the number of bits of the DAC, n) can be problematic. Symmetries in the output waveforms can reduce the LUT size. For example, a sine wave reduces the memory size by four because a quarter of it is enough. The other quarters result from reversing the sign or complementing the address.

The clock of Figure 14.24(a) determines the time interval between output samples. If the waveform is a sine wave the clock also provides the phase increments. In order to adjust the frequency, since the full scale is 2^m, we can accommodate an integer number, k, of sinusoidal periods in the 2^m series. The period and frequency of the sine wave are

$$kT_{out} = 2^m T_{ck}, \quad f_{out} = \frac{k}{2^m} f_{ck}. \tag{14.43}$$

The above equations show that there are not very many possible frequencies. We increase their number by making the full scale of the counter programmable, as shown in Figure 14.24(b). The circuit uses a reset signal that sends to zero the counter content when it

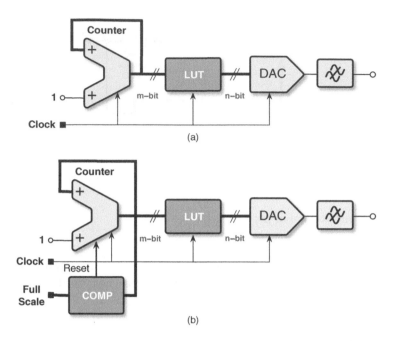

Figure 14.24   (a) Block diagram of a simple DDS signal generator; (b) the same diagram with the addition of the counter full-scale control.

is equal to the programmed full scale. The method works but is not practical, because it is necessary to reprogram the LUT at every change of sine wave frequency. Tuning the clock frequency is another solution, but the use of variable oscillators is expensive when the generated frequency must be very precise (the scheme must use a fractional PLL).

A more effective method is the one shown in Figure 14.25, which defines the phase increment, $\Delta\Phi$, during each clock period with a multi-bit digital quantity. The role of the accumulator changes; from being a meter of time it becomes a gauge of phase. The number of bits, j, used to measure the phase and its increment i, is much higher than m, the number of bits utilized in Figure 14.24 for measuring time. A simple block truncates some of the bits j at the output of the phase accumulator to give rise to the LUT control. It transforms the phase into an n-bit code driving the DAC. The output low-pass filter smoothes the analog waveform.

Since i is the number of bits representing $\Delta\Phi$ and the phase accumulator uses j bits, the maximum output frequency is

$$f_{out,max} = \frac{2^j}{2^i} f_{ck} = 2^{(j-i)} f_{ck}. \tag{14.44}$$

Moreover, since 2^j corresponds to 2π, $2^{(j-i)}$ indicates how many samples are used by the DAC to represent a maximum-frequency sine wave.

The i-bit word of the phase increment $\Delta\Phi$ determines the output frequency. With $\Delta\Phi = N$, the generated frequency is

$$f_{out} = \frac{2^j}{N} f_{ck}. \tag{14.45}$$

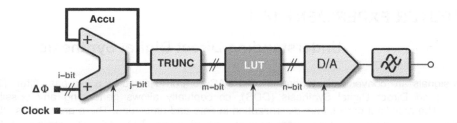

Figure 14.25   DDS based on phase increment.

Changing $\Delta\Phi$ by one yields

$$\Delta f_{out} = -\frac{f_{out}}{N+1};$$ (14.46)

therefore, in order to determine small frequency steps it is necessary to use a large number of bits for the phase increment and, consequently, for the phase accumulator.

Truncation is just done by discharging the LSBs from j to m. The operation reduces the accuracy of the phase or, in other words, reduces the phase quantization to the level established by the LUT. In radians, it is given by

$$\Delta\Phi_{out} = \frac{2\pi}{(2^m - 1)}.$$ (14.47)

The number of bits n at the output of the LUT gives rise to the amplitude accuracy of the analog waveform. If the full-scale voltage is V_{FS}, the amplitude quantization is $V_{FS}/(2^n - 1)$.

14.5   SIGNAL MEASUREMENT

The verification and validation of circuits need instruments that determine a static, dynamic, or frequency-dependent quantification of input and output electrical quantities. This book proposes several virtual experiments done with virtual instruments. For those it was only necessary to indicate essential features. Now, at the end of the book, it is worthwhile to comprehend instruments better, because for real testing we use real and not virtual instruments.

The following section considers three categories of instruments:

■ those that perform DC measures and show the result on a numerical display with a given number of digits;

■ instruments making time-domain measurements with results presented on a display that plots signal versus time or signal versus signal (the X–Y mode);

■ instruments performing measurements in the frequency domain with the display of spectra.

We shall discuss how the important ones work and what is the (initial) knowledge required for their proper use.

COMPUTER EXPERIMENT 14.4

Understanding Direct Digital Synthesis

Periodic signals are conveniently generated by a DAC control followed by a low pass filter. This method, called Direct Digital Synthesis (DDS), conceptually allows generating any possible waveform. The only limit comes from discretization in time and amplitude. There are various DDS architectures. This computer experiment studies two of them. The first (up diagram in the measurement setup) uses a digital sawtooth generated by an accumulator with "1" at the input. The result is the address of a LUT that contains the digital portrait of the output. The accumulator of the second method (bottom diagram) employs the phase increment $\Delta\Phi$. The high resolution phase, after truncation, provides the LUT address. This experiment makes three different types of waveform available, one of which is a sine wave. You can load one of them on the LUT. The output of the DAC passes through a low-pass filter, whose specifications depend on the spur requirement. The filter used by this board gives rise to almost linear interpolation between successive output samples. You can:

- define clock frequency, phase increment (for the second method) and number of bits of the accumulator (for the second method that is typically a large number);
- choose the number of bits used by various digital parts. The value of n is equal to 10;
- select the type of output waveform.

MEASUREMENT SET–UP

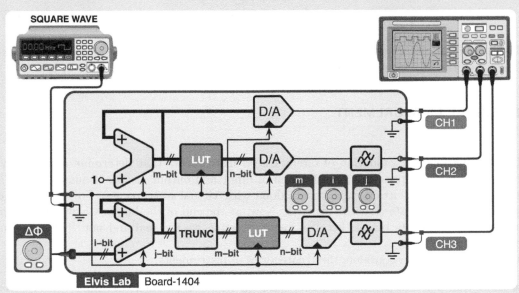

TO DO

- Set the clock frequency to 400 MHz and, using the sawtooth scheme, generate all the frequencies admitted in the 10–100 MHz range. Do the same with the phase increment method and a 16-bit accumulator. Estimate system accuracy when generating 32 MHz.
- Choose an output frequency in the 10–30 MHz range. Estimate the minimum number of bits of the phase accumulator for a given frequency accuracy. Set the scope in the free-running mode and take various snapshots of the scope screen. Calculate the actual periods, measuring the zero crossings. Draw a statistical distribution of the periods with data from 20 traces (or more).

14.5.1 Multimeter

The multimeter is an instrument that allows multiple measurements: current, voltage, resistance, diode polarity, continuity, and others. There are two basic types of multimeter: digital and analog. The difference is obviously evident from the display. Analog multimeters have a needle indicating the value of the measure on a scale. Digital multimeters have an LCD or a LED display with its number of digits giving the resolution.

An important feature of multimeters is the input impedance. Typically the input impedance is in the 10 MΩ range, meaning that with 1 V signal the drained current is 0.1 μA, a negligible value in most practical cases. However, it is important to know the value of the input impedance and to be aware of possible changes with different full scale settings.

Many multimeters have an "auto-range" mode that selects the instrument's full scale automatically. The decimal point and the units (V, mV, µV) change automatically.

A simple use of a multimeter is the continuity check. Continuity means that two nodes are electrically connected. Many multimeters have a piezo buzzer, which beeps if resistance between the two terminals is lower than a defined value. The continuity check serves in debugging test boards or PCBs. Components soldered on a board can be badly connected because of cold solder connections, a trace can be broken, or a piece of solder can short two connections. The continuity check permits technicians to verify such cases. It is only necessary to remember to power off the circuit before checking (to avoid any possible consequences). The continuity check also serves to verify the polarity of a diode. The multimeter provides a voltage with current limitation and polarity given by colors on the terminals. When the current is larger than a limit the piezo buzzer beeps. A forward-biased diode gives continuity. Reverse biasing indicates an open circuit.

The continuity check is one of the possible functions, which are normally selected with a function switch. Others typically include AC voltage, DC voltage, resistance, AC current, and DC current. The use of the meter for most functions is straightforward. The only point to clarify concerns AC operation.

The parameters describing an AC signal are the peak voltage, V_{pk}, the average voltage, V_{av}, and the Root-Mean-Square (RMS) voltage, V_{rms}. An AC signal swings around zero and reaches alternately a peak positive and a peak negative, both in modulus equal to V_{pk}. The average voltage is defined by

$$V_{av} = \frac{1}{T} \int_{t_1}^{t_1+T} |V(t)| \, dt, \tag{14.48}$$

where T is the period of the alternate waveform and t_1 is a generic time.

The RMS defines an equivalent heating power: i.e., the DC voltage that delivers the same power to a resistive load. The following equation defines it:

$$V_{rms} = \frac{1}{T} \sqrt{\int_{t_1}^{t_1+T} [V(t)]^2 \, dt}. \tag{14.49}$$

Measuring the peak or average value is not difficult. A peak detector or a full-wave rectifier followed by a low-pass filter procures the result. The estimation of the RMS is more difficult. In the past, multimeters used a transformation of voltage into heat, thus changing the temperature of a heated element. The increase in temperature, proportional to power, becomes an electrical quantity by using thermocouples and analog processing. That method is no longer used because, thanks to its modern digital capability, the multimeter now estimates the RMS after the A/D conversion of the input voltage.

The display of a multimeter can show the peak, the average, and the RMS of the measured signal. Remember that a sine whose peak amplitude is equal to 1 has $\pi/2 = 0.6366$ as its average value. The average power is $1/2$, leading to an RMS of $\sqrt{2}/2$. For a triangular wave with peak amplitude equal to 1, the average is $1/2$ and the power integral over a period is $T/3$, leading to an RMS of $\sqrt{3}/3 = 0.5774$. The square wave has equal peak, average, and RMS values.

14.5.2   Oscilloscope

The oscilloscope is a very common instrument used to observe signal wave shapes. It has a display similar to that of a computer and a number of control knobs and switches. At the beginning the core of the oscilloscope (also called a *scope*) was a Cathode Ray Tube (CRT) with an electron beam lighting a dot by hitting the screen, which was covered with phosphorous. Two plates, one horizontal and the other vertical, were used to deflect the beam. Now that kind of instrument is archeology. Oscilloscopes use a liquid crystal or LED display and signal processing to reproduce waveforms on the screen after an A/D conversion of their inputs.

The main function of an oscilloscope is to capture and accurately reproduce an input waveform. The accuracy of signal representation is normally referred to as *signal integrity*, a general concept applied to all situations that presuppose some risk of signal degradation. For an oscilloscope two parts are key for ensuring signal integrity: the electronics of the oscilloscope itself and the probe. Accuracy of an electronic circuit mainly depends on the precision of analog blocks, namely the voltage references, the definition of time, and the data converters. However, when the signal contains high frequencies, accurate electronics is not enough. It is also relevant in which manner the signal enters the oscilloscope. Typically, using simple wires to connect the scope and the circuit under measure (normally called the DUT, or Device Under Test) is inappropriate because without adequate precautions the input gets distorted. Probes avoid this problem. A coaxial cable connection and impedance matching, maintained with periodic probe calibrations, preserve the signal waveform while it travels from the tip of the probe to the input terminal of the oscilloscope.

We can use active or passive probes. Active means that near the tip, biased through a cable connected to the scope, an active electronic circuit facilitates the matching of the network performing the measurement. There are current probes, voltage probes, and logic probes, all of which, as their names imply, are specialized to ensure the integrity of various types of signal. Some probes improve frequency performance at the cost of signal attenuation. For example, probes 2× or 10× give rise to attenuations by 2 or 10 respectively but ensure better signal integrity.

As mentioned above, signal integrity results from a proper combination of oscilloscope and probe. The right pair depends on various elements: signal (voltage, current, optical, etc.), bandwidth, rise and fall time, source impedance (resistive, capacitive, mixed), signal amplitude, and its range. An obvious condition is that the frequency limitation and load of the probe are negligible with respect to the signal band and equivalent impedance. The simple circuit of Figure 14.26(a) approximates the source. The passive probe, also described by a simple equivalent circuit, matches impedance at high frequency; therefore, not just components but also board design and execution are relevant. The active probe reduces the load capacitance because its buffer sustains the capacitive load of the connecting cable.

Figure 14.26(b) shows possible scope traces obtained with the same oscilloscope but using different probes. The traces are shifted on the vertical axis to make the differences more

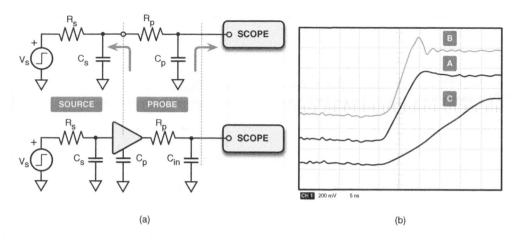

Figure 14.26   (a) Equivalent circuit of source and probe for a measurement with an oscilloscope; (b) three different traces of the same signal with suitable and unsuitable probes.

visible. The trace A is optimal. Traces B and C refer to overcompensated and low-speed probes, respectively.

Let us go back to the oscilloscope, a graphical-display device that typically uses time on the X-axis and voltage on the vertical axis (Y). However, there is the option to disregard the time and to display on both axes, the X-axis and the Y-axis, voltage signals.

The instrument has many uses, as you have seen several times when doing virtual experiments with the *ElvisLab* system. The oscilloscope allows you to measure and display:

- the time and voltage value of a signal;

- the period or frequency of an oscillating waveform;

- the DC or AC components of a signal;

- the level of noise and, consequently, SNR.

For the use of oscilloscopes it is important to distinguish between periodic and non-periodic signals. Periodic waveforms repeat themselves over and over again. Non-periodic signals are single shot, because they are normally caused by asynchronous excitations of a circuit. In addition it is necessary to distinguish, if displaying multiple inputs, between synchronous and asynchronous inputs. Defined time correlations link multiple synchronous signals. For asynchronous inputs the various waveforms change in time without any interdependence.

These two features are important for the display on the monitor. Repetitive waveforms can be viewed as the recurrence of signal portions. If the starting point of a portion is suitably chosen, the result on the display looks like a static waveform. For example, if the input is a voltage sine wave and its display starts with a given amplitude, say 0.2 V, with given slope, say positive, the result is the repetition of equal traces on the display. In the case of non-periodic inputs, such as a pulse arriving at undefined times, it is important to start the acquisition at the right time, immediately after a relevant feature occurs (it may be started by an excitation of the system), and possibly to store the waveform on a memory and show it as a snapshot of the event.

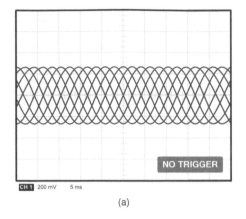

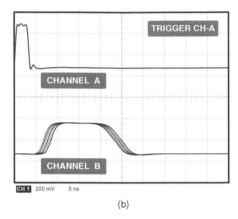

(a)

(b)

Figure 14.27 (a) Plot of the oscilloscope display with a free-running input and no trigger control; (b) display of two traces, a pulse that defines the trigger, and another signal not fully synchronous with the first input.

The above possibilities are, in practice, realized by the trigger control. This is the function that defines the time at which traces begin, based on the crossing with the given slope of a defined threshold, provided that the displayed trace has reached the end of the time interval shown on the screen. When there are multiple inputs to be displayed, one of them establishes the trigger.

Figure 14.27 shows two situations that can be viewed on the oscilloscope screen. Figure 14.27(a) depicts a situation for which the trigger is not able to fix the waveform. The lack of trigger control produces a free-running result. The second display (Figure 14.27(b)) refers to a two-input case. Channel A defines the trigger. The signal sent to Channel B is almost synchronous but some time uncertainties give rise to multiple pictures of the second waveform. This is what typically happens when there is some time jitter blurring the timing scale of one signal.

Self training

Use the Web or go to a real laboratory to understand more about oscilloscopes. Perhaps you can find an old oscilloscope with a CRT tube. Look at the front panel and try to understand the various functions of the knobs. Find a detailed description of the following features:

- the trigger of a multichannel oscilloscope;

- AC and DC inputs, and control of the Y position;

- the X–Y mode of operation and its simple use with two sine waves at input.

Write a short report on your findings, including the features of three possible oscilloscopes, one for the 20 MHz signal band, one able to display signals up to 200 MHz, and a top oscilloscope with extremely high-frequency capabilities.

Digital storage oscilloscopes satisfy the need to display non-periodic or asynchronous inputs. They can capture the waveform and store it in digital format. The detected waveform can

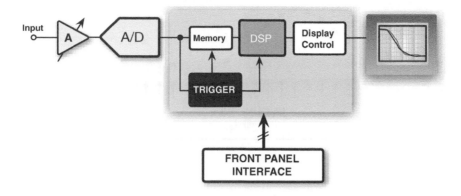

Figure 14.28   Block diagram of a digital oscilloscope.

be displayed, processed, or archived in the memory of the oscilloscope or within a connected computer. Figure 14.28 shows a block diagram of a digital oscilloscope. The analog input, after variable amplifications, is transformed into digital form. The digital result enters a memory when the trigger control establishes that the input signal is suitable for storing. The digital processing section, under the control of the front panel, performs the required signal manipulation and generates the X and Y signals for the display.

Another important class of oscilloscope is the sampling type. The functioning of this category of oscilloscopes exploits downsampling to display repetitive waveforms at very high frequencies. Figure 14.29 illustrates the principle of operation. The input signal is a sine wave at extremely high frequency, say 5.032 GHz. A sampler running at 334 MHz collects samples, as shown in Figure 14.29(a). Notice that the input frequency is a little more than 15 times the sampling rate. Since it is in the 16th Nyquist interval, the sampling gives rise to a folded tone at $f' = (f_{in} - 15F_s)$, 22.266 MHz. The display visualizes the set of samples whose plot is the shifted replica of the high-frequency input signal.

14.5.3   Logic Analyzer

A logic analyzer is an oscilloscope specialized for digital signals. The task of the digital designer in debugging circuits is the analysis and verification of numerous signals with a focus on time transitions. While an oscilloscope displays details on signal amplitude, rise and fall time, ringing and other characteristics for a limited number of signals (the number of channels is up to four), a logic analyzer does not reveal details that are important for analog applications, but, instead, detects threshold logic levels and their transition times for a large number of inputs. Logic analyzers typically have more than 32 channels and may have over 128 channels.

The use of a logic analyzer is therefore mainly for verifying and debugging digital designs. With the logic analyzer it is possible to verify that the digital circuit works properly. For this it is necessary to capture and observe several signals at once, to check the logic correctness, and to examine the timing relationships between critical signals.

The large number of inputs requires special attention to the acquisition probes. They can be general-purpose probes intended for single-point accesses or multi-channel probes to be connected to the system under test with dedicated connectors. Obviously, the impedance of

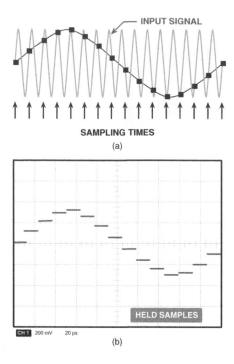

Figure 14.29   (a) High-frequency sine wave and its sampling times; (b) display of the trace on the sampling oscilloscope.

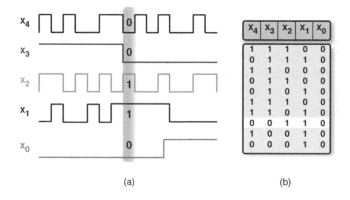

Figure 14.30   (a) Waveform display of five-bit logic data; (b) its binary list representation.

the probes gives rise to some load on the circuit. It is therefore necessary to use methods that introduce minimal loading conditions or to account for the load in performance estimations.

The input data, possibly stored in a memory, is used in a variety of formats. One is the waveform display to show the multi-channel time relationship of the signals, similar to what is done on the display of an oscilloscope. Another possibility is a list provided in binary or hexadecimal form. Figure 14.30 shows an example of two displays of logic data.

14.6   SPECTRUM ANALYZER

The spectrum analyzer is an important instrument for the testing and verification of circuits and systems in the frequency domain. The instrument is complex, and for our purposes it is just enough to know the fundamental elements. We refer here to instruments that provide the spectrum almost in real time, because it is obviously possible to perform signal processing using the information about a signal after it is digitized and stored in a memory. The real-time measurement is indispensable when verifying and debugging the operation of circuits by observing snapshots of the signal in the frequency domain.

There are two types of spectrum analyzer: the swept Spectrum Analyzer (SSA) and the Vector Signal Analyzer (VSA). The swept spectrum, or superheterodyne, analyzer is a well-known traditional "analog" architecture, even though modern instruments include digital parts such as data converters, microprocessors, or DSPs. A VSA moves to the digital domain after a little analog processing, and determines the spectrum with digital processing techniques. Figure 14.31 shows the possible block diagrams of the two types. Both have the same input sections before the down converter, a mixer used to move the spectrum of the signal to a lower frequency. The difference is in the local oscillator. For the SSA the frequency is swept by the control of a sweep generator. The band of the input signal moves down by a variable quantity. The portion that falls in the band of a bandpass (BP) filter (it may also be a low-pass one) is amplified and passed through a detector that extracts the power. Then a video filter constructs the displayed signal. The width of the BP filter determines the resolution of the SSA. A wide filter gives rise to large but low-resolution signals (the power in the frequency interval is higher than that in a narrow band). When a good resolution is required the BP filter is more selective, but the signal is lower. Moreover, the slope of the sweep generator must be reduced.

The measurement assumes that the analyzer can complete the frequency sweep without significant changes to the input signal. That means that it depends on relatively stable input signals. In the case of rapid changes in the signal, the result is not valid.

The BP filter used in the VSA has a different role. It defines the bandwidth of interest and just removes the spectral components that the user does not want to measure. This operation improves the performance because the following ADC receives a signal whose amplitude is just determined by the band of interest. The down converter also operates in a different way because the local oscillator is at a fixed frequency. The ADC digitizes all of the RF power within the passband of the instrument and temporarily stores the digitized waveform in memory. The subsequent processing – further downconversion, filtering, detection, and FFT – are done digitally by powerful DSPs. The transformation into the digital domain benefits the user because much additional information can be provided.

The architectures described above are, strictly speaking, not "real time" because there is some delay for the output generation. There are more sophisticated instruments that are able to address the difficulty associated with the spectral representation of transient and dynamic RF signals better.

Self training

Use the Web to understand the spectrum analyzer better. You can find good tutorials that describe the theory and use of this kind of complicated instrument. Find details on the use of the instrument for:

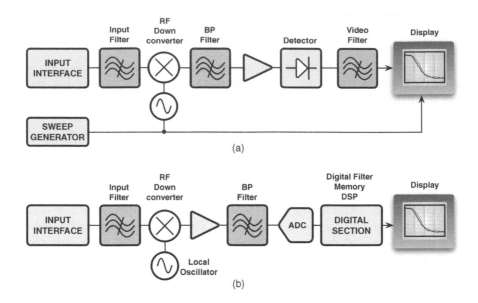

Figure 14.31   (a) Block diagram of a Swept Spectrum Analyzer (SSA); (b) Vector Signal Analyzer (VSA).

- monitoring a wanted signal plus adjacent spurs;

- measuring the Bode diagram of an amplifier;

- estimating the signal-to-noise ratio of an analog processor.

Write a short report on your study or, if possible, acquire practical experience or watch a demonstration.

PROBLEMS

14.1   The monostable circuit of Figure 14.2 uses $R_1 = 100$ kΩ and $C_1 = 10$ pF. The on-resistance of the switch is $R_{on} = 1$ kΩ and $V_T = V_x/2 = 0.9$ V. What is the pulse duration? Estimate the error of the pulse duration caused by an input trigger arriving just 15 ns after the end of the previous pulse.

14.2   The multivibrator of Figure 14.3 uses the following component values: $R_1 = R_2 = 1$ kΩ, $C = 20$ pF. The on-resistance of the switch is $R_{on} = 1$ kΩ and $V_{T1} = 0.7$ V, $V_{T2} = V_x/2 = 0.9$ V. Determine the period of the generated waveform. (Hint: when the switch is on, estimate and use the Thévenin equivalent.)

14.3   In the Wien bridge oscillator of Figure 14.6 a parasitic capacitance C_p loads the inverting node of the op-amp. Estimate the new $\beta_n(f)$ factor and apply the Barkhausen criterion to determine the oscillation frequency.

14.4 Design a phase-shift oscillator with three equal low-pass RC filters decoupled from one another by unity gain buffers. The input impedance of the buffers is infinite, and the output impedance is zero. That gives rise to an ideal separation between the phase shift cells. Estimate the oscillation conditions and the required forward gain.

14.5 Suppose you are to use a varicap in the design of a VCO. The varicap junction area is 100 times smaller than the one whose non-linear capacitive behavior is shown in Figure 9.9. What is the inductance that gives rise to a 1 GHz oscillation frequency? The nominal reverse bias voltage of the varicap is 2 V. Estimate the tuning range, supposing that the reverse biasing can change in the range ± 0.5 V.

14.6 Do a search on the Web and find the features of a commercial quartz with 1 MHz oscillation frequency. The aim of the search is to finding a part suitable for a VCO that must run at 2.2 GHz. What is the temperature accuracy of the VCO measured in kHz/$^\circ$C?

14.7 Find on the Web an online catalogue of instruments that describes a probe for an oscilloscope suitable for measuring a 5 GHz sine wave signal. Take note of the recommendations for its optimal use and write a short summary.

Index

Understanding Microelectronics: A Top-Down Approach, First Edition. Franco Maloberti.
© 2012 John Wiley & Sons, Ltd. Published 2012 by John Wiley & Sons, Ltd.

Printed and bound by CPI Group (UK) Ltd, Croydon, CR0 4YY

27/10/2024

14580374-0005